Basic Mathematics for Engineers and Technologists

Basic Mathematics for Engineers and Technologists

Alan Jeffrey

University of Newcastle upon Tyne

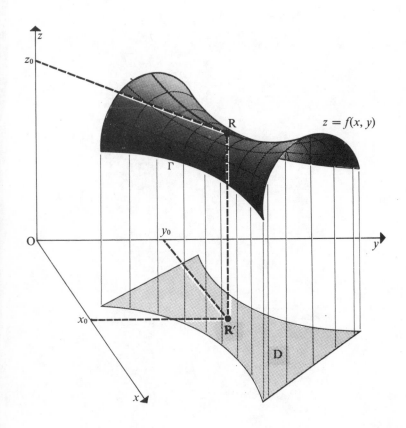

Nelson

Thomas Nelson and Sons Ltd
36 Park Street London W1Y 4DE

Nelson (Africa) Ltd
PO Box 18123 Nairobi Kenya

Thomas Nelson (Australia) Ltd
171–175 Bank Street South Melbourne Victoria 3205

Thomas Nelson and Sons (Canada) Ltd
81 Curlew Drive Don Mills Ontario

Thomas Nelson (Nigeria) Ltd
PO Box 336 Apapa Lagos

First published in Great Britain by Thomas Nelson and Sons Ltd, 1974

ISBN 017 771007 1

Printed in Great Britain by The Whitefriars Press Ltd, London and Tonbridge

Preface

This book has been designed to provide a reasonably concise account of the basic mathematical needs of all first-year engineering students. It should also fulfil the same role for the many technology students whose need is for a sound coverage of those mathematical ideas that are absolutely essential to any first course. In preparing this new book, full use has been made of the author's more comprehensive account *Mathematics for Engineers and Scientists* also published by Thomas Nelson. However, the chapters of that book have been extensively edited to retain only the essentials, and in the process some material has been rearranged and some rewritten. A new chapter has been added on probability and statistics.

The new chapter reflects the growing need to include both probability and some statistics in a first-year course, even though these topics might be covered again more fully later. This chapter develops and expands the account of discrete probability given in the earlier book. Thus the Poisson and normal distributions are discussed, together with the approximation to the binomial distribution by means of the normal distribution. The rudimentary account of statistics is centred around the notion of a confidence interval and its determination, particularly in the important practical case of small samples. Linear regression is discussed in this same context with the emphasis being placed on the confidence interval for the gradient of the regression line, since in experimental work this usually has considerable physical significance.

Many features of the earlier book have been retained in the belief that they form a necessary part of a course on basic engineering mathematics. Thus a proof is still given wherever possible, the introduction of each new result is invariably accompanied by a worked example and a good supply of problems still follows each chapter. The order of presentation which seemed logical and satisfactory in the earlier book has been retained here, and answers are still provided to odd numbered problems.

A. J.

Contents

1 Introduction to sets and numbers

1·1 Sets and algebra

In applications of mathematics to engineering and science, we often use the properties of real numbers. Many of these properties are intuitively obvious, but others are more subtle and depend for their proper use on a simple understanding of the mathematical basis of the so-called real number system. This chapter describes the elements of the real number system in a straightforward manner for subsequent use throughout the book.

We must first define the term *set* for which the alternative terms *aggregate*, *class*, and *collection* are also often used. Our approach will be direct and pragmatic and we shall agree that a set comprises a collection of objects or elements, each of which is chosen for membership of the set because it possesses some required property. Membership of the set is determined entirely by this property; an object only belongs to the set if it possesses the required property, otherwise it does not belong to the set. The properties of membership and non-membership of a set are mutually exclusive.

An important numerical-set which we shall often have occasion to use is the set N of natural numbers 1, 2, 3, . . ., used in counting. In future the symbol N will always be used to signify this natural set of positive integers. Notice that there can be no greatest member m of this set, since however large m may be, $m + 1$ is larger and yet is also a member of the set N. Accordingly, when we use a number m that is allowed to increase without restriction, it will be convenient to imply this by saying that 'm tends to infinity', and to write the statement in the form $m \to \infty$. Notice that infinity is not a number in the usual sense, but just the outcome of the mathematical process of allowing m to increase without bound. It is always necessary to relate the symbol ∞ to some mathematical expression, since by itself it has little or no meaning.

N is only one type of set however, and from the wording of our definition it is apparent that the elements of a set need not be numerical. Thus in statistics one is concerned with sets of events which may or may not be numerical, whereas in the analysis of logical operations one is concerned with sets of decisions.

To simplify the manipulation of these ideas we must introduce a notation for elements of a set, for sets themselves, and for the membership of an element to a set. It is customary to denote general elements of sets by lower case letters $a, b, . . ., x, . . .$, and sets themselves by capital letters $A, B, . . ., S, . . .$. If a is a member of set A we shall write

$$a \in A.$$

This is usually read 'a is an element of A'. Conversely, if a is not an element of A we shall write

$a \notin A.$

In this notation we have $3 \in \mathbf{N}$, but $\pi \notin \mathbf{N}$, where $\pi = 3 \cdot 1415\ldots$, and $\mathbf{N}$ is the set of natural numbers.

If a set only contains a small number of elements it is often simplest to define it by enumerating the elements. Hence, for a set S comprising the four integer elements 3, 4, 5, and 6 we would write $S = \{3, 4, 5, 6\}$. This set is a *finite* set in the sense that it comprises a finite number of elements. Conversely, the set $\mathbf{N}$ of natural numbers is an *infinite* set since it contains an infinite number of elements.

Often it is useful to have a notation which indicates the membership criterion that is to be used for the set. Thus, if we were interested in the set B of positive integers n whose squares lie between the positive numbers m and $2m$, we would write

$B = \{n \,|\, n \in \mathbf{N}, m < n^2 < 2m\}.$

Here we have used the convention that the symbol n to the left of the vertical rule signifies a general element of the set in question, whilst the expressions to the right of the rule express the membership criteria for the set. There, of course, the symbol $<$ when used in conjunction with numbers a and b in the form $a < b$ is to be read 'a less than b'.

An important set that is frequently used is the set of *ordered pairs*. An element of this set will be written (m, n), where m and n are not necessarily numbers and the element (m, n) is different from the element (n, m) unless m and n are identical. An important use of this set is in the construction of tables, when the ordered pair becomes an ordered number pair, the first member of which is usually the argument and the second member the functional value.

Ordered number pairs are also encountered when constructing graphs of functions where the convention is usually that (a, b) signifies the point with x-coordinate a and y-coordinate b. Thus the graph of the function $y = f(x)$ for which x is between a and b could be written in set notation

$S = \{(x, f(x)) \,|\, a < x < b\}.$

The notation of an ordered pair as an element of a set readily extends to an ordered triple (m, n, r), which again need not necessarily involve numerical quantities, nor need it be determinate. Again, two ordered triples will only be identical if their corresponding entries are identical. Ordered number triples of a determinate kind occur when considering the graph of a function of two independent variables as, for example, the equilibrium temperature at a given point of a cross-section of a very long metal bar. We will see later that ordered pairs and triples of an indeterminate kind occur in statistics.

It is often necessary to study relationships between sets and for this purpose an algebra of sets must be constructed. The simplest situation that can occur is that from a set A, a new set B is formed, such that all elements of B are also elements of A. Such a set B will be called a *subset* of A. This result will be written

$B \subseteq A$,

which is to be read 'B is a subset of A'.

If x is an element of A, so that we may write $x \in A$, then either $x \in B$, or $x \notin B$. When there are some elements $x' \in A$ which are not to be found in B, so that $x' \notin B$, then B is called a *proper subset* of A, the result being written

$B \subset A$.

The definition of a subset B of A does not preclude the possibility that for every element $x \in A$ it is also true that $x \in B$. When this occurs sets A and B have the same elements and are said to be *equal*, the result being written

$A = B$.

It is clear from the definition of equality that when $A = B$ both the statements $A \subseteq B$ and $B \subseteq A$ must be true. These last two statements are often useful as an alternative definition of equality between sets.

With the above definitions it is clear that if $A = N$ and $B = \{1, 2, 3, 4, 5\}$, then $B \subset A$; whereas if $A = \{4, 7, 3, 5, 9\}$ and $B = \{7, 4, 5, 9, 3\}$, then $A \subseteq B$ and $B \subseteq A$ so that $A = B$.

A more general situation arises when two sets A and B are involved, each of which possesses elements which are not common to the other so that neither statement $A \subset B$, nor $B \subset A$ is true. The set of elements C that is common to these two sets A and B will be called the *intersection* of the sets A and B and is written

$C = A \cap B$.

Sometimes this is read 'A cap B' with the understanding just defined.

In the event that there are no elements common to the sets A and B we shall write

$A \cap B = \phi$,

with the understanding that ϕ is the *null set*, which we define to be the set containing no elements. Under these circumstances the sets A and B are said to be *disjoint*.

By way of example, if $A_1 = \{a, b, 1, 3, 5, 7\}$ and $B_1 = \{a, c, d, e, 3, 7, 9\}$, then $A_1 \cap B_1 = \{a, 3, 7\}$; whereas if $A_2 = \{1, 3, 7\}$ and $B_2 = \{0, 4, 9, 11\}$, $A_2 \cap B_2 = \phi$.

Another important set related to sets A and B is the set C containing all the elements belonging to A, to B or to both A and B. This is called the *union* of sets A and B and is written

$$C = A \cup B;$$

which reads 'A cup B'. With the sets defined above we obviously have $A_1 \cup B_1 = \{a, b, c, d, e, 1, 3, 5, 7, 9\}$ and $A_2 \cup B_2 = \{0, 1, 3, 4, 7, 9, 11\}$. Clearly, for any set A we have $\phi \subseteq A$, $A \cup \phi = A$, and $A \cap \phi = \phi$.

We now leave the algebra of sets to introduce a notation that will clarify our mathematical reasoning. The need is for a notation that will stress the logical consequences of an argument. Accordingly, it is necessary to appreciate clearly the implication of any statement that may be made in the derivation of a result. These statements may either be 'one way' implications or 'two way' implications in the following sense. An implication will be said to be one way if it is a simple statement of the form 'result A implies result B'. This statement is usually written symbolically in the concise form

$$A \Rightarrow B.$$

A two way implication arises if from the above statement it also follows that 'result B implies result A', so that in addition to the previous statement it is also permissible to write

$$B \Rightarrow A.$$

Rather than write for a two way implication the two results $A \Rightarrow B$ and $B \Rightarrow A$, the notation is contracted so that the two way implication may be written concisely in the form

$$A \Leftrightarrow B.$$

The symbol $\Leftrightarrow$ is usually read 'implies and is implied by'.

1·2 Integers, rationals and arithmetic laws

The reader will already be familiar with the fact that if the arithmetic operation of addition is performed on the natural numbers, or the positive integers as they are often called, the result will also be a positive integer. Written symbolically this statement becomes $a, b \in \mathbf{N} \Rightarrow (a + b) \in \mathbf{N}$. However the arithmetic operation of subtraction is less simple, since we know from direct experience that even when $a, b \in \mathbf{N}$, this does not necessarily imply that $a - b$ is a positive integer. Indeed, in general $a - b$ may be equal to some positive or negative integer or to zero.

Thus an attempt always to express the result of subtraction of natural numbers in terms of the natural numbers themselves must fail. This is usually expressed by saying that the system of natural numbers $\mathbf{N}$ is not *closed* with respect to subtraction. The difficulty is of course resolved by supplementing the set of natural numbers $\mathbf{N}$ by the set $\mathbf{N}^* = \{\ldots, -3, -2, -1, 0\}$ of negative integers and zero. If now in place of **N** we use the complete set of integers $I = N^* \cup N$ already encountered in Problem 1·1, the assertions $a, b \in I \Rightarrow (a + b) \in I$ and $a, b \in I \Rightarrow (a - b) \in I$ become unconditionally true.

The need to generalize the notion of the natural numbers **N** to the complete set of integers **I** is thus seen to arise as a natural result of seeking a number system in which the usual arithmetic operation inverse to addition is always true; namely the operation of subtraction. However, the set of numbers **I** is still far from adequate to enable everyday practical arithmetic to be performed. To see this it is only necessary to comment that although the product of two integers belonging to **I** itself lies in **I**, the quotient of two integers belonging to **I** does not necessarily lie in **I**. Thus the complete set of integers **I** is not closed with respect to division. Symbolically we can write this as $a, b \in I \Rightarrow ab \in I$, but $a, b \in I \Rightarrow a/b \in I$ only if $b \neq 0$ and $a = kb$ with $k \in I$. The symbol $\neq$ used here is to be read 'not equal to' and the condition involving k simply ensures that the quotient a/b is integral.

Here again the operations of multiplication and division are inverse arithmetic operations. To remove the artificial restriction placed on division, so that the quotient of any two non-zero integers becomes a number in some number system, we must still further extend the system **I** of integers. This is achieved by introducing the familiar system **R*** of *rational numbers*, which is defined as the set of all numbers of the form a/b, where $b \neq 0$ and $a, b \in I$. Obviously, since integers are just a special case of rational numbers and, for example, 2 is represented by any of the rationals 2/1, 4/2, 10/5, . . ., the set **R*** also contains all the integers and so we may write $I \subset R^*$.

It might, at first sight, seem that the rationals **R*** must contain all possible numbers. In fact this is far from the truth since it is possible to show that numbers exist which are not expressible as a rational fraction and yet which lie *between* two rationals, however close they may be. For obvious reasons they are called *irrational* numbers, and to substantiate our assertion we now prove the existence of one such number.

We will show that $\sqrt{2}$ is irrational or, to phrase the statement more precisely, that there is no fraction of which the square is 2. The argument starts from a given assumption and then produces a contradiction, thereby showing that the original assumption must be false. It is called an argument by contradiction and is a device frequently used in higher mathematics.

Suppose that m/n is such that m and n are integers having no common factor and $(m/n)^2 = 2$. Then $m^2 = 2n^2$ so that m^2 must be even and hence m itself is even. Because m is even we may set $m = 2r$, where r is some integer. (Why?) Then $4r^2 = 2n^2$, or $2r^2 = n^2$, which now shows that n^2 and hence n must be even. The fact that n is even now allows us to set $n = 2s$ and thus the numbers m and n have a common factor 2, contradicting the initial assumption. Hence the original assumption that $\sqrt{2}$ is capable of representation in the rational form m/n is false. We have thus proved that $\sqrt{2}$ is an irrational number.

If the set **R*** of rational numbers is supplemented by the inclusion of the irrational numbers, the resulting set **R** is called the *real number system* or, the *field* of real numbers. The fact that **R** contains all possible types of real numbers is expressed by saying that the set of real numbers **R** is *complete*. Consequently, until we have occasion to consider entities such as $\sqrt{-1}$ there will be no need for us to work outside the real number system **R**. The *transcendental* numbers form an important subset of the irrational numbers. These are numbers like e and π which are not defined as the root of a polynomial with rational coefficients (cf. § 2·3).

For future reference it will be useful to summarize the basic properties of the field of real numbers already known to the reader. We now do this making full use of the mathematical shorthand so far introduced.

Additive properties

A·1 $a, b \in \mathbf{R} \Rightarrow (a + b) \in \mathbf{R}$; **R** is closed with respect to addition.

A·2 $a, b \in \mathbf{R} \Rightarrow a + b = b + a$; addition is *commutative*.

A·3 $a, b, c \in \mathbf{R} \Rightarrow (a + b) + c = a + (b + c)$; addition is *associative*.

A·4   For every $a \in \mathbf{R}$ there exists a number $0 \in \mathbf{R}$ such that $0 + a = a$; there is a *zero* element in **R**.

A·5   If $a \in \mathbf{R}$ then there exists a number $-a \in \mathbf{R}$ such that $-a + a = 0$; each number has a negative.

Multiplicative properties

M·1 $a, b \in \mathbf{R} \Rightarrow ab \in \mathbf{R}$; **R** is closed with respect to multiplication.

M·2 $a, b \in \mathbf{R} \Rightarrow ab = ba$; multiplication is *commutative*.

M·3 $a, b, c \in \mathbf{R} \Rightarrow (ab)c = a(bc)$; multiplication is *associative*.

M·4   There exists a number $1 \in \mathbf{R}$ such that $1 . a = a$ for all $a \in \mathbf{R}$; there is a *unit* element in **R**.

M·5   Let a be a non-zero number in **R**, then there exists a number $a^{-1} \in \mathbf{R}$ such that $a^{-1}a = 1$; each non-zero number has an *inverse*. Usually we shall write $1/a$ in place of a^{-1}, so that the two expressions are to be taken as being synonymous.

Distributive property

D·1 $a, b, c \in \mathbf{R} \Rightarrow a(b + c) = ab + ac$; multiplication is *distributive*.

The above results are self-evident for real numbers and are usually called the *real number axioms*. They are used by mathematicians as the logical basis for our number system.

It is an immediate consequence of these axioms that commonplace arithmetic operations may be performed without question. For example, it is fundamental to arguments that $a - b = 0 \Leftrightarrow a = b$, and $a\xi = a\eta \Rightarrow \xi = \eta$

if $a \neq 0$. These, and other elementary results of similar form, follow directly as a result of simple applications of the axioms.

So far our list of properties of real numbers has been concerned only with equalities. The valuable property of real numbers that they can be arranged according to size, or *ordered*, has so far been overlooked. It is of course this property that allows us to represent real numbers by points on a line and thereby to construct graphs and other valuable geometrical representations. Ordering is achieved by utilizing the concept 'greater than' which when used in the form 'a greater than b', is denoted by $a > b$. Hence to the other real number axioms must be added:

Order properties

O·1   If $a \in \mathbf{R}$ then exactly one of the following is true; either $a > 0$ or $a = 0$ or $-a > 0$.

O·2 $a, b \in \mathbf{R}, a > 0, b > 0 \Rightarrow a + b > 0$, and $ab > 0$.

We now define $a > b$ and $a < b$, the latter being read 'a less than b', by $a > b \Rightarrow a - b > 0$ and $a < b \Rightarrow b - a > 0$. The following results are obvious consequences of the real number system and are called *inequalities*. In places they also involve the symbol $\geq$ which is to be read 'greater than or equal to'.

Elementary inequalities in *R*

I·1 $a > b$ and $c \geq d \Rightarrow a + c > b + d$.

I·2 $a > b \geq 0$ and $c \geq d > 0 \Rightarrow ac > bd$.

I·3 $k > 0$ and $a > b \Rightarrow ka > kb$.

I·4 $a > b \Rightarrow -a < -b$.

I·5 $a < 0, b > 0 \Rightarrow ab < 0; a < 0, b < 0 \Rightarrow ab > 0$.

I·6 $a > 0 \Rightarrow a^{-1} > 0; a < 0 \Rightarrow a^{-1} < 0$.

I·7 $a > b > 0 \Rightarrow b^{-1} > a^{-1} > 0; a < b < 0 \Rightarrow b^{-1} < a^{-1} < 0$.

An important use of inequalities is in defining intervals on a line and regions in a plane. Using the order property of numbers to associate numbers with points on a line, an *interval* on a line may be considered to be a segment of the line between two given points or numbers, a and b, say. Three cases arise according as to whether (a) both end points are included in the interval,

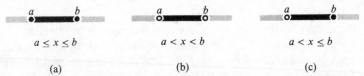

$$a \leq x \leq b \qquad\qquad a < x < b \qquad\qquad a < x \leq b$$

(a)   (b)   (c)

Fig. 1·1 Intervals on a line: (a) closed interval $a \leq x \leq b$; (b) open interval $a < x < b$; (c) semi-open interval $a < x \leq b$.

(b) both end points are excluded from the interval, or (c) one is included and one is excluded. These are called, respectively, (a) a *closed* interval, (b) an *open* interval, (c) a *semi-open* interval. Namely, an interval is closed at an end which contains the end point, otherwise it is open at that end. In terms of the points a and b and the variable x representing an arbitrary point on the line these are written:

(a) $a \leq x \leq b$; closed interval;

(b) $a < x < b$; open interval;

(c) $a < x \leq b$ or $a \leq x < b$; semi-open interval.

Thus $1 \leq x < 2$ defines the semi-open interval containing the point $x = 1$ and the points up to, but not including, $x = 2$. These are represented in Fig. 1·1 in which a solid line represents points in the interval, a circle represents an excluded point, and a dot an included point.

Special cases occur when one or both of the end points of the interval are at infinity. The intervals $-\infty < x < a$ and $b < x < \infty$ are called *semi-infinite* intervals and $-\infty < x < \infty$ is an *unbounded* interval or, more simply, the complete real line.

We illustrate the corresponding definition of a region in the (x, y)-plane by considering the three inequalities $x^2 + y^2 \leq a^2$, $y < x$, $x \geq 0$. The first defines the interior of a circle of radius a centred on the origin, the second defines points below, but not on, the straight line $y = x$, and the third defines points in the right half of the (x, y)-plane including the points on the y-axis

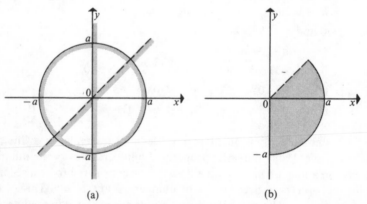

(a)   (b)

Fig. 1·2 Regions in plane: (a) region boundaries $x^2 + y^2 = a^2$, $y = x$, and $x = 0$; (b) region $x^2 + y^2 \leq a^2$, $y < x$, $x \geq 0$.

itself. These curves represent boundaries of the regions in question and the boundary points are only to be included in the region when possible equality is indicated by use of the signs $\geq$ or $\leq$. The three regions are indicated in Fig. 1·2 (a) in which a full line indicates that points on it are to be included, a dotted line indicates that points on it are to be excluded, and shading indicates the side of the line on which the region in question must lie. Fig. 1·2 (b) indicates the region in which all the inequalities are satisfied.

Simple inequalities of the form $(x + 1)(x + 3) > (x - 1)(x - 2)$ also define intervals. For, clearing the brackets, we 'see that $x^2 + 4x + 3 > x^2 - 3x + 2$ which, by simple application of the elementary inequalities just listed, reduces to $x > -1/7$ defining a semi-infinite interval, open at the end $x = -1/7$.

The elementary inequalities may often be used to advantage to simplify complicated algebraic expressions by yielding helpful qualitative information as the following example indicates.

Example 1·1 Prove that if $a_1, a_2, \ldots, a_n$ and $b_1, b_2, \ldots, b_n$ are positive real numbers, then

$$\min_{1 \leq r \leq n} \left(\frac{a_r}{b_r}\right) \leq \frac{a_1 + a_2 + \cdots + a_n}{b_1 + b_2 + \cdots + b_n} \leq \max_{1 \leq r \leq n} \left(\frac{a_r}{b_r}\right).$$

Here the left-hand side of the inequality is to be interpreted as meaning the minimum value of the expression (a_r/b_r), with r assuming any of the integral values between 1 and n and the right-hand side is to be similarly interpreted reading maximum in place of minimum. The result follows by noticing that

$$\frac{a_1 + a_2 + \cdots + a_n}{b_1 + b_2 + \cdots + b_n} = \frac{1}{\sum\limits_{r=1}^{n} b_r} \left[b_1 \cdot \left(\frac{a_1}{b_1}\right) + b_2 \left(\frac{a_2}{b_2}\right) + \cdots + b_n \left(\frac{a_n}{b_n}\right) \right],$$

where $\sum\limits_{r=1}^{n} b_r = b_1 + b_2 + \cdots + b_n$. For if each of the expressions (a_1/b_1), $(a_2/b_2), \ldots, (a_n/b_n)$ is replaced by the smallest of these ratios, which could be the value taken by all the expressions if $a_1 = a_2 = \cdots = a_n > 0$ and $b_1 = b_2 = \cdots = b_n > 0$, then

$$\frac{a_1 + a_2 + \cdots + a_n}{b_1 + b_2 + \cdots + b_n} \geq \min_{0 \leq r \leq n} \left(\frac{a_r}{b_r}\right) \left[\frac{(b_1 + b_2 + \cdots + b_n)}{\sum\limits_{r=1}^{n} b_r} \right]$$

$$= \min_{0 \leq r \leq n} \left(\frac{a_r}{b_r}\right),$$

which is the left half of the inequality. The right half follows by identical reasoning if maximum is written in place of minimum.

1·3   Absolute value of a real number

DEFINITION 1·1 The *absolute value* $|a|$ of the real number a provides a measure of its size without regard to sign, and is defined as follows:

$$|a| = \begin{cases} a \text{ when } a \geq 0 \\ -a \text{ when } a < 0. \end{cases}$$

Thus if $a = 3$, then $|a| = 3$ and if $a = -5 \cdot 6$ then $|a| = 5 \cdot 6$.

There are three immediate consequences of this definition which we now enumerate as

THEOREM 1·1 If $a, b \in \mathbf{R}$ then

(a) $|ab| = |a|\,|b|$,

(b) $|a + b| \leq |a| + |b|$,

(c) $|a - b| \geq ||a| - |b||$.

The proof is simply a matter of enumerating the possible combinations of positive and negative a and b, and then making a direct application of the definition of the absolute value. We shall only illustrate the proof of (a).

There are three cases to be considered; firstly $a \geq 0$, $b \geq 0$, secondly $a \geq 0$, $b < 0$, and thirdly $a < 0$, $b \geq 0$. If $a \geq 0$, $b \geq 0$ then $ab \geq 0$ and so $|ab| = ab = |a|\,|b|$. The second and third situations are essentially similar so we shall discuss only the second. As $a \geq 0$, $b < 0$ we have $ab \leq 0$, whence $|ab| = -ab = a(-b) = |a|\,|b|$, establishing (a). For reasons we give later, result (b) is usually called the triangle inequality.

The absolute value may also be used to define intervals since an expression of the form $|a - x| \geq 2$ implies two inequalities according as $a - x$ is positive or negative. If $a - x > 0$ then $|a - x| = a - x$ and we have $a - x \geq 2$ or $x \leq a - 2$. However if $a - x < 0$, then by the definition of the absolute value of $a - x$ we must have $|a - x| = -(a - x)$ showing that $-(a - x) \geq 2$, or, $x \geq a + 2$. Taken together the results require that x may be equal to or greater than $2 + a$ or equal to or less than $a - 2$. x may not lie in the intervening interval of length 4 between $x = a - 2$ and $x = a + 2$. This is illustrated in Fig. 1·3 (a) where a solid line is again used to indicate points in the interval satisfied by $|a - x| \geq 2$ and the dots are to be included in the appropriate intervals.

By exactly similar reasoning we see that if we consider the inequality $1 < |x + 1| \leq 2$, then if $x + 1 > 0$, $|x + 1| = x + 1$ and the inequality becomes $1 < x + 1 \leq 2$. Hence the interval is $0 < x \leq 1$. However, if $x + 1 < 0$, then $|x + 1| = -x - 1$ and so the inequality becomes $1 < -x - 1 \leq 2$ giving rise to the interval $-3 \leq x < -2$. These intervals are shown in Fig. 1·3 (b) with circles indicating points excluded from the end of the solid line intervals and dots indicating points to be included.

$a+2$ $a-2$ -3 -2 0 1

$|a-x| \geq 2$ $1 < |x+1| \leq 2$

(a)   (b)

Fig. 1·3 Intervals on a line: (a) $|a - x| \geq 2$; (b) $1 < |x + 1| \leq 2$.

1·4   Mathematical induction

Mathematical propositions often involve some fixed integer n, say, in a special role and it is desirable to infer the form taken by the proposition for arbitrary integral n from the form taken by it for the specific value $n = n_1$. The logical method by which the proof of the general proposition, if true, may be established, is based on the properties of natural numbers and is called *mathematical induction*.

In brief, it depends for its success on the obvious fact that if A is some set of natural numbers and $1 \in A$, then the statement that whenever integer $n \in A$, so also does its successor, implies that $A = \mathbf{N}$, the set of natural numbers.

The formal statement of the process of mathematical induction is expressed by the following theorem where, for simplicity, the mathematical proposition corresponding to integer n is denoted by $S(n)$.

THEOREM 1·2 (mathematical induction)   If it can be shown that,

(a) when $n = n_1$, the proposition $S(n_1)$ is true,

and

(b) if for $n \geqslant n_1$, when $S(n)$ is true then so also is $S(n + 1)$,

then the proposition $S(n)$ is true for all natural numbers $n \geqslant n_1$.

A simple illustrative example will help here and we now prove inductively that the sum $\sum_{r=1}^{n} r$ of the first n natural numbers is given by $n(1 + n)/2$. In other words, in this example the proposition denoted by $S(n)$ is that the following result is true:

$$1 + 2 + \cdots + n = n(1 + n)/2.$$

Proof, step (a)   First the proposition must be shown to be true for some specific value $n = n_1$. Any integral value n_1 will suffice but if we set $n_1 = 1$ the proposition corresponding to $S(1)$ is immediately obvious. If, instead, we had chosen $n_1 = 3$, then it is easily verified that proposition $S(3)$ is true, namely that $1 + 2 + 3 = 3(1 + 3)/2$.

Proof, step (b)   We must now assume that proposition $S(n)$ is true and attempt to show that this implies that the proposition $S(n + 1)$ is true. If $S(n)$ is true then

$$1 + 2 + \cdots + n = n(1 + n)/2$$

and, adding $(n + 1)$ to both sides, we obtain

$$1 + 2 + \cdots + n + (n + 1) = n(1 + n)/2 + (n + 1)$$
$$= (n + 1)(2 + n)/2.$$

However, this is simply a statement of proposition $S(n + 1)$ obtained by replacing n by $n + 1$ in proposition $S(n)$. Hence $S(1)$ is true and $S(n) \Rightarrow S(n + 1)$ so, by the conditions of Theorem 1·2, we have established that $S(n)$ is valid for all n.

Later we shall use this form of proof in cases less trivial than the above example which simply involved establishing the sum of an arithmetic progression.

An important and useful result that can be established by induction is the *binomial expansion theorem* for integral n. We leave this as an exercise for the reader and simply quote the result. In formulating this we use the number called *factorial n* which is written $n!$. This is defined as follows: $n! = n(n - 1)(n - 2)\ldots3.2.1$, so that, for example, $3! = 3.2.1 = 6$ and $5! = 5.4.3.2.1 = 120$. Here we adopt the usual convention that $0! = 1$.

THEOREM 1·3   (Binomial theorem for positive integral n)

If a, b are real numbers and n is a positive integer, then

$$(a + b)^n = a^n \left[1 + n\left(\frac{b}{a}\right) + \frac{n(n - 1)}{2!}\left(\frac{b}{a}\right)^2 \right.$$
$$\left. + \frac{n(n - 1)(n - 2)}{3!}\left(\frac{b}{a}\right)^3 + \cdots \right]$$

This expression, which contains only a finite number of terms when n is a positive integer, can be shown to be true for any real n, positive or negative. However, if n is not a positive integer then it contains an infinite number of terms, and for the sum of the right-hand side to be finite the numbers a, b must be such that $|b/a| < 1$. When n is real and positive the result is also true for $|b/a| = 1$.

When $|b/a|$ is very much less than unity the right-hand side of this expansion is often approximated by retaining only the first two terms. So that if $|b/a| \ll 1$ we have

$$(a + b)^n \approx a^n \left(1 + n\frac{b}{a} \right).$$

The coefficients in the expansion for positive integral n are called *binomial coefficients* and the coefficient of $(b/a)^r$ is often written $\binom{n}{r}$, where

$$\binom{n}{r} = \frac{n!}{(n - r)!\, r!}.$$

In this notation the binomial expansion for positive integral n takes the form

$$(a + b)^n = a^n \left[1 + \binom{n}{1}\left(\frac{b}{a}\right) + \binom{n}{2}\left(\frac{b}{a}\right)^2 + \cdots + \binom{n}{r}\left(\frac{b}{a}\right)^r + \cdots \right].$$

PROBLEMS

Section 1·1

1·1 Enumerate the elements in the following sets in which **I** signifies the set of natural positive and negative integers including zero:

(a) $S = \{n \mid n \in \mathbf{I}, \quad 5 < n^2 < 47\}$;

(b) $S = \{n^3 \mid n \in \mathbf{N}, \quad 15 < n^2 < 40\}$;

(c) $S = \{(m, n) \mid m, n \in \mathbf{I}, \quad 12 < m^2 + n^2 < 18\}$;

(d) $S = \{(m, n, m + n) \mid m, n \in \mathbf{N}, \quad 45 < m^2 + n^2, \quad 3 < m + n < 9\}$;

(e) $S = \{\dot{x} \mid x \in \mathbf{N}, \quad x^2 + 0 \cdot 1x - 1 \cdot 1 = 0\}$.

1·2 Express the following sets in the notation of the previous question:

(a) the set of positive integers whose cubes lie between 7 and 126;

(b) the set of integers which are the squares of the integers lying between M and N $(0 < N < M)$;

(c) the points in the plane that lie between circles of radii 1 and 3 drawn about the origin and which have x-coordinates greater than 0·5.

1·3 State the relationships between the sets A and B if:

(a) $A = \mathbf{N}, \quad B = \{2n \mid n \in \mathbf{N}\}$;

(b) $A = \{\sin x \mid x = (1 + 12n)\frac{1}{6}\pi, \quad n \in \mathbf{N}\}, \quad B = \{\frac{1}{2}\}$;

(c) $A = \{1, 2, 3, 4\}, \quad B = \{5, 7, 9, 11\}$.

1·4 Form the union and intersection of the sets A and B if:

(a) $A = \mathbf{N}, \quad B = \{2n \mid n \in \mathbf{N}\}$;

(b) $A = \{a, b, c, 0, 2, 4\}, \quad B = \{d, e, f, 1, 3, 6, 7\}$;

(c) $A = \{1, \sqrt{2}, 2, 3, \sqrt{5}, 6\}, \quad B = \{0, \sqrt{2}, \sqrt{5}\}$.

Sections 1·2 and 1·3

1·5 Use the fact that $\sqrt{2}$ is irrational to prove that if α is a rational number, then $\alpha + \sqrt{2}$, $\alpha\sqrt{2}$ and $\sqrt{2}/\alpha$ are also irrational. Would the results still be true if $\sqrt{2}$ were replaced by any other irrational number, and would your proof still suffice?

1·6 Indicate by means of a diagram the intervals defined by the following expressions, using a dot to signify an end point belonging to an interval and a circle to indicate an end point excluded from the interval:

(a) $(x + 2)(x + 3) \leq (x - 1)(x - 2)$;

(b) $0 < |x - 3| \leq 1$;

(c) $|x| \leq 2$;

(d) $0 < |2x + 1| < 1$;

(e) $|3x + 1| \geq 2$;

(f) $\dfrac{x + 1}{2x + 2} < \dfrac{x}{2(x - 1)}$.

1·7 Prove that if $a > b > 0$ and $k > 0$ then

$$\frac{b}{a} < \frac{b + k}{a + k} < 1 < \frac{a + k}{b + k} < \frac{a}{b}.$$

1·8 Prove Theorem 1·4 (b) by considering separately the cases $a \geq 0, b > 0$; $a < 0, b < 0; a \geq 0, b < 0; a < 0, b \geq 0$.

1·9 Identify the regions in the (x, y)-plane determined by the following inequalities. Mark a boundary that belongs to the region by a full line; a boundary that does not by a dotted line; an end point that is included in an interval by a dot; an end point excluded from an interval by a circle:

(a) $x^2 + y^2 < 1$; $x < 0$; $y < -x$;
(b) $y \leq \sin x$; $x^2 + y^2 \geq \pi^2$; $y \leq \frac{1}{2}$;
(c) $\frac{1}{4}x^2 + y^2 > 1$; $|y| \leq \frac{1}{2}$;
(d) $y \geq x^2$; $|x - 1| \leq 1$; $y \leq 4$.

Section 1·4

1.10 Give an inductive proof that

$$(a) \quad \sum_{r=0}^{n-1} (a + rd) = \frac{n}{2} [2a + (n - 1)d];$$

(Arithmetic Progression)

$$(b) \quad \sum_{r=1}^{n} r^2 = \frac{n(n + 1)(2n + 1)}{6}.$$

(Sum of Squares)

1·11 Give an inductive proof of the results

$$(a) \quad \sum_{s=0}^{n-1} r^s = \frac{1 - r^n}{1 - r},$$

(Geometric Progression)

$$(b) \quad \sum_{s=1}^{n} s^3 = \left[\frac{n(n+1)}{2} \right]^2.$$

(Sum of Cubes)

2 Variables, functions, and mappings

2·1 Variables and functions

In the physical world the idea of one quantity depending on another is very familiar, a typical example being provided by the observed fact that the pressure of a fixed volume of gas depends on its temperature. This situation is reflected in mathematics by the notion of a *function*, which we shall now discuss in some detail.

The modern definition of a function in the context of real numbers is that it is a relationship, usually a formula, by which a correspondence is established between two sets A and B of real numbers in such a manner that to each number in set A there corresponds only one number in set B. The set A of numbers is the *domain* of the function and the set B of numbers is the *range* of the function.

If the function or rule by which the correspondence between numbers in sets A and B is established is denoted by f, and x denotes a typical number in the domain A of f, then the number in the range B to be associated with x by the function f is written $f(x)$ and is read 'f of x'. The numbers x and $f(x)$ are *variables* with x being given the specific name *independent variable* and $f(x)$ the name *dependent variable*. The independent variable is also often called the *argument* of the function f.

It is often helpful to construct the graph of f which mathematically is the set of ordered number pairs $(x, f(x))$, where x belongs to the domain of f. Geometrically the graph of f is usually represented by a plane curve, drawn relative to an origin defined by the intersection of two perpendicular straight lines called axes. The process of construction is as follows. A distance proportional to x is measured along one axis and a distance proportional to $f(x)$ along the other axis. Through each resulting point on an axis is then drawn a line parallel to the other axis and these two perpendicular lines intersect at a unique point in the plane of the axes. This point of intersection is the point $(x, f(x))$ and the graph of f is defined to be the locus or curve formed by joining up all such points corresponding to the domain of f, as illustrated by Fig. 2·1.

However, it is not necessary to use axes of this type, called rectangular Cartesian axes, and any other geometrical representation which gives unique representation of the points $(x, f(x))$ would serve equally well. Thus the axes could be inclined at an angle $\alpha \neq \frac{1}{2}\pi$ and the scale of measurement along them need not be uniform. For example, it is often useful to plot the logarithm of x along the x-axis, rather than x itself. This compresses the x scale so that large values of x may be conveniently displayed on the graph together with small values. Another possible representation involves the use of curved reference axes and leads to *curvilinear coordinates*. This will be taken up again later in connection with conformal mapping.

Not every function can be represented in the form of an unbroken curve, and the function

$$f(x) = \begin{cases} 0 \text{ when } x \text{ is rational,} \\ 1 \text{ when } x \text{ is irrational,} \end{cases}$$

provides an extreme example of this situation. Here, although the graph would look like a line parallel to the x-axis on which all points have the value unity, in reality the infinity of points with rational x-coordinates would be missing since they lie on the x-axis itself. The domain is all the real numbers **R** and the range is just the two numbers zero and unity.

Because f transforms one set of real numbers into another set of real numbers a function is sometimes spoken of as a *transformation* between sets of real numbers. On account of the restriction to real numbers or, more explicitly, to real variables, the function $f(x)$ is called a *function of one real variable*. Another name that is often used for a function is a *mapping* of some set of real numbers into some other set of real numbers. This name is of course suggested by the geometrical illustration of the graph of a function and we shall return more than once to the notion of a mapping. In this terminology, $f(x)$ is referred to as the *image* of x under the mapping f.

Since the domain and range of f occur as intervals on the x- and y-axes, it is convenient to use a simplified notation to identify the form of the interval that is involved. We now adopt the almost standard notation summarized below in which a round bracket indicates an *open* end of an interval, and a square bracket indicates a *closed* end of an interval:

$$(a, b) \Leftrightarrow a < x < b,$$
$$[a, b] \Leftrightarrow a \leq x \leq b,$$
$$(a, b] \Leftrightarrow a < x \leq b,$$
$$[a, b) \Leftrightarrow a \leq x < b,$$
$$(-\infty, a] \Leftrightarrow x \leq a,$$
$$[a, \infty) \Leftrightarrow a \leq x,$$
$$(-\infty, \infty) \Leftrightarrow \text{all } x \in \mathbf{R}.$$

As the definition of open and closed intervals is only a matter of considering the behaviour of the end points, we shall define the length of all the intervals (a, b), $[a, b)$, $(a, b]$, and $[a, b]$ to be the number $b - a$. This is consistent with the obvious result that the length of an 'interval' comprising only one point is zero.

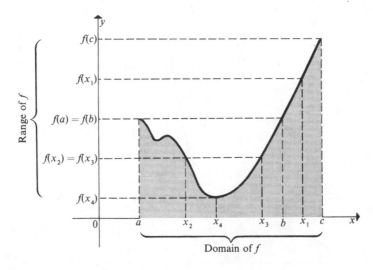

Fig. 2·1   Domain, range, and graph of $f(x)$.

It may happen that when x lies within some interval, as for example the interval $(b, c]$ in Fig. 2·1, each point x is associated with a unique image point $f(x)$ and, conversely, each image point $f(x)$ is associated with a unique point x. Such a mapping or function f is then said to be *one–one* in the domain in question.

However, there is another possibility that can arise and that is that in some interval of the x-axis, more than one point x may correspond to the same image point $f(x)$. This is again well illustrated by Fig. 2·1 if now we consider the interval $[a, b]$ and the points x_2 and x_3, both of which have the same image point since $f(x_2) = f(x_3)$. In situations such as these the mapping or function f if said to be *many–one* in the domain in question.

A specific example might help here and we choose for f the function $f(x) = x^2$ and the two different domains $[0, 3]$ and $[-1, 3]$. A glance at Fig. 2·2 shows that f maps the domain $[0, 3]$ onto the range $[0, 9]$ one–one, but that it maps the domain $[-1, 3]$ onto the same range $[0, 9]$ many–one. Expressed another way, the range $[0, 1]$ shown as a solid line in the figure is mapped twice by points in the domain $[-1, 3]$; once by points in the sub-domain $-1 \leq x < 0$ and once by points in the sub-domain $0 < x \leq 1$. Again considering the domain $[-1, 3]$, the function $f(x) = x^2$ maps the sub-domain $1 < x \leq 3$ onto the range $(1, 9]$ one–one.

In many older books the term function is used ambiguously in that it is sometimes applied to relationships which do not comply with our definition

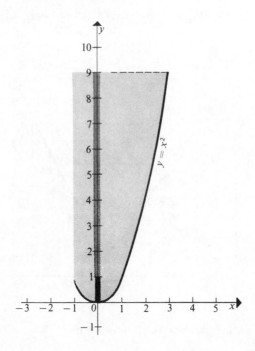

Fig. 2·2 Example of many–one mapping in shaded range and a one–one mapping in the hatched range.

of a function. The most familiar example of this is the 'function' $y = \sqrt{x}$, which fails to comply with our definition because to every positive x there correspond *two* values for y, namely the positive and negative square roots of x which are equal in magnitude but opposite in sign. A mapping of this kind is one–many in the sense that to one value of x there correspond more than one image point $f(x)$, and although it is permissible to describe this relationship as a mapping, it is incorrect to term it a function.

Nevertheless, the square root operation is fundamental to mathematics and we must find some way to make it and similar ones legitimate. The difficulty is easily resolved if we consider how the square root is used in applications. In point of fact two different relationships are always considered which together are equivalent to $y = \sqrt{x}$. These are $y_1 = +\sqrt{x}$ and $y_2 = -\sqrt{x}$, where the square root is always to be understood to denote the positive square root and the sign identifies the relationship being considered. Each of the mappings $y_1(x)$ and $y_2(x)$ of the domain $(0, \infty)$ are one–one as Fig. 2·3 shows, so that they may each be correctly termed a function, the particular one to be used in any application being determined by other considerations, such as that the result must be positive or negative. These ideas will arise again later in connection with inverse functions.

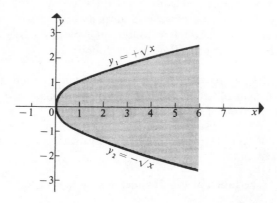

Fig. 2·3   The square root function.

In general, if the domain of function f is not specified then it is understood to be the largest interval on the x-axis for which the function is defined. So if $f(x) = x^2 + 4$, then as this is defined for all x, the largest possible domain must be $(-\infty, \infty)$. Alternatively the function $f(x) = +\sqrt{(4 - x^2)}$ is only defined in terms of real numbers when $-2 \leq x \leq 2$ showing that the largest possible domain is $[-2, 2]$. Similarly, the function $f(x) = 1/(1 - x)$ is defined for all x with the sole exception of $x = 1$ so that the largest possible domain is the entire x-axis with the single point $x = 1$ deleted from it.

A function need not necessarily be defined for all real numbers on some interval and, as in probability theory, it is quite possible for the dependent and independent variables to assume only discrete values. Thus the rule which assigns to any positive integer n the number of positive integers whose squares are less than n, defines a perfectly good function. Denoting this function by f we have for its first few values $f(1) = 0$, $f(2) = 1$, $f(3) = 1$, $f(4) = 1$, $f(5) = 2$, $f(6) = 2$, $f(7) = 2$, $f(8) = 2$, $f(9) = 2$, $f(10) = 3$, ... Clearly, both its domain and its range are the set $\mathbf{N}$ of natural numbers and the mapping is obviously many–one.

Before examining some special functions let us formulate our definition of a function in rather more general terms. This will be useful later since although in the above context the relationships discussed have always been between numbers, in future we shall establish relationships between quantities that are not simply real numbers. When we do so, it will be valuable if we can still utilize the notion of a function. This will occur, for example, when we establish correspondence between quantities called vectors which although obeying algebraic laws are not themselves real numbers.

The idea of a relationship between arbitrary quantities is one which we have already started to examine in the previous chapter in connection with sets. As might be expected, set theory provides the natural language for the formulation and expression of general ideas associated with functions, and indeed we have already used the word 'set' quite naturally when thinking of a set of numbers. A more general definition follows.

DEFINITION 2·1   A *function* f is a correspondence, often a formula, by which each element of set A which is called the domain of f, is associated with only one element of set B called the range of f.

To close this section we now provide a few examples illustrating some of the ideas just mentioned.

Example 2·1   The function $y = f(x)$ defined by the rule

$$f(x) = \frac{1}{(x-1)(x-2)}$$

is defined for all real x with the exception of the two points $x = 1$ and $x = 2$. The domain of f is thus the set of real numbers **R** with the two numbers 1 and 2 deleted. The range of f is **R** itself.

Example 2·2   A discrete valued function may be defined by a table which is simply an arrangement of ordered number pairs in a sequence.

Table 2·1

x	0	1	3	7
$f(x)$	2·1	4·2	1·0	6·3

Example 2·3   This example is a final illustration of our more general definition of a function. Take as the domain of the function f the set A of all people, and as the range B of the function f the set of all towns in the world. Then for the function f we propose the rule that assigns to every person his place of birth.

Clearly this function defines a many–one mapping of set A onto set B, since although a person can only be born in one place, many other people may have the same place of birth. This example also serves to distinguish clearly between the concept of a 'function' which is the rule of assignment, and the concept of the 'variables' associated with the function which here are people and places.

2·2   Inverse functions

In the previous section we remarked that a typical example of a correspondence between physical quantities was the observed fact that the pressure of a fixed volume of gas depends on its temperature. Expressed in this form we are implying that the dependent variable is the pressure p and the independent variable is the temperature T, so that the law relating pressure to temperature has the general form

$$p = \phi(T),\tag{A}$$

where ϕ is some function that is determined by experiment.

However, we know from experience that in thermodynamics it is often necessary to interchange these roles of dependent and independent variables and sometimes to regard the temperature T as the dependent variable and the pressure p as the independent variable, when the temperature–pressure law then has the form

$$T = \psi(p),\tag{B}$$

where, naturally, the function ψ is dependent on the form of the function ϕ. Indeed, formally, ϕ and ψ must obviously satisfy the identity $\phi[\psi(p)] \equiv p$ for all pressures p in the domain of ψ.

The relationships (A) and (B) are particular cases of the notion of a function and its inverse and the idea is successful in this context because the correspondence between temperature and pressure is known to be one–one.

Consider a general case of a function

$$y = f(x)\tag{2·1}$$

that is one–one and defined on the domain $[a, b]$, together with its inverse

$$x = g(y)\tag{2·2}$$

which has for its domain the interval $[c, d]$ on the y-axis.

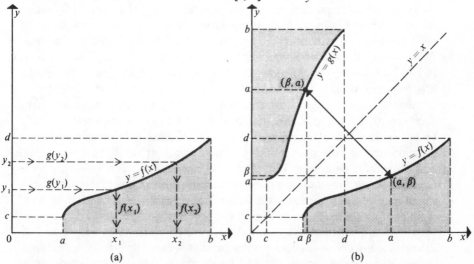

Fig. 2·4   (a) Inversion through the graph of $f(x)$; (b) inversion by reflection in $y = x$.

Graphically the process of inversion may be accomplished point by point as indicated in Fig. 2·4 (a). This amounts to selecting a point y in $[c, d]$ and then finding the corresponding point x in $[a, b]$ by projecting horizontally from y until the graph of f is intercepted, after which a projection is made vertically downwards from this intercept to identify the required point on the x-axis.

The relationship between a function and its inverse is represented in Fig. 2·4 (b). In this diagram we have used the fact that when a function is represented as an ordered number pair, interchange of dependent and independent variables corresponds to interchange of numbers in the ordered number pair. The lower curve represents the function $y = f(x)$ and the upper curve represents the function $y = g(x)$, with the function g inverse to f; both graphs being plotted using the same axes. The line $y = x$ is also shown on the graph to emphasize that geometrically the relationship between a one–one function and its inverse is obtained by reflecting the graph of either function in a mirror held along the line $y = x$. Henceforth such a process will simply be termed *reflection in a line*. Notice that when using this reflection property to construct the graph of an inverse function from the graph of the function itself, *both* functions are represented with y plotted vertically and x plotted horizontally. This follows because the range of f is the domain of g, and vice versa.

No difficulty can arise in connection with a function and its inverse because of the one–one nature of the mapping. Expressed more precisely, we have used the obvious property illustrated by Fig. 2·4 (a) that a one–one function f with domain $[a, b]$ is such that $f(x_1) = f(x_2) \Rightarrow x_1 = x_2$ for all x_1 and x_2 in $[a, b]$.

In graphical terms this result can only be true if the graph of f either increases or decreases steadily as x increases from a to b. When either of these properties is true of a function then it is said to be *strictly monotonic*. In particular, if a function f increases steadily as x increases from a to b, as in Fig. 2·4 (a), then it is said to be *strictly monotonic increasing* and, conversely, if it decreases steadily then it is said to be *strictly monotonic decreasing*.

Slightly less stringent than the condition of strict monotonicity is the condition that a function f be just *monotonic*. This is the requirement that f be either non-decreasing or non-increasing, so that it is permissible for a function that is only monotonic to remain constant throughout some part of its domain of definition. The adjectives increasing and decreasing are again used to qualify the noun monotonic in the obvious manner. Representative examples of monotonic and strictly monotonic functions, all with domain of definition $[a, b]$ are shown in Fig. 2·5.

The example of a strictly monotonic decreasing function shown in Fig. 2·5 (b) has also been used to emphasize that a function need not be represented by an unbroken curve. The curve has a break at the single point $x = \alpha$ where it is defined to have the value $y = \beta$. However, as the value β

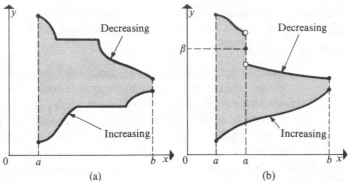

(a)   (b)

Fig. 2·5   Monotonic and strictly monotonic functions: (a) monotonic; (b) strictly monotonic.

lies between the functional values on adjacent sides of $x = \alpha$ the function is still strictly monotonic decreasing. Had we set $\beta = 0$, say, then the function would be neither strictly monotonic nor even monotonic on account of this one point!

It is sometimes useful to relate a function and its inverse by essentially the same symbol and this is usually accomplished by adding the superscript minus one to the function. Thus the function inverse to f is often denoted by f^{-1} which is not, of course, to be misinterpreted to mean $1/f$. Before examining some important special cases of inverse functions when many–one mappings are involved, let us formalize our previous arguments.

DEFINITION 2·2   Let the set onto which the one–one function f with domain $[a, b]$ maps the set S of points be denoted by $f(S)$. Then we define the *inverse mapping* f^{-1} of $f(S)$ onto S by the requirement that $f^{-1}(y) = x$ if and only if $y = f(x)$ for all x in $[a, b]$.

It now only remains for us to consider how some important special functions such as $y = x^2$, $y = \sin x$, and $y = \cos x$, together with other simple trigonometric functions which are all many–one mappings, may have unambiguous inverses defined.

Firstly, as we have already seen, the function $y = x^2$ gives a many–one mapping of $[-a, a]$ onto $[0, a^2]$. Here the difficulty of defining an inverse is resolved by always taking the *positive* square root and defining two *different* inverse functions

$$x = +\sqrt{y} \quad \text{and} \quad x = -\sqrt{y},$$

which are then both one–one mappings of $(0, a^2]$. The inversion must thus be regarded as having given rise to two different functions; the one to be selected depending on other factors as mentioned in connection with Fig. 2·3. If we recall that the domain of definition of a function forms an intrinsic part of the definition of that function, then $y = x^2$ may be regarded as two one–one mappings in accordance with the two inverses just introduced.

This is achieved by defining the many–one function $y = x^2$ on the domain $[-a, a]$ as the result of the two different one–one mappings

$$y = x^2 \text{ on } -a \le x < 0 \qquad \text{and} \qquad y = x^2 \text{ on } 0 < x \le a,$$

the difference here being only in the domains of definition. The point 0 is excluded from both domains since that single point maps one–one. By means of this device we may, in general, reduce many–one mappings to a set of one–one mappings so that the inversion problem is always straightforward.

It will suffice to discuss in detail only the inversion of the sine function, after which a summary of the results for the other elementary trigonometric functions will be presented in the form of a table. In general, as shown in Fig. 2·6 (a), the function $y = \sin x$ maps an argument x in the set **R** of real numbers onto $[-1, 1]$ many–one, but it maps any of the restricted domains $[(2n - 1)\tfrac{1}{2}\pi, (2n + 1)\tfrac{1}{2}\pi]$ corresponding to integral n onto $[-1, 1]$ one–one.

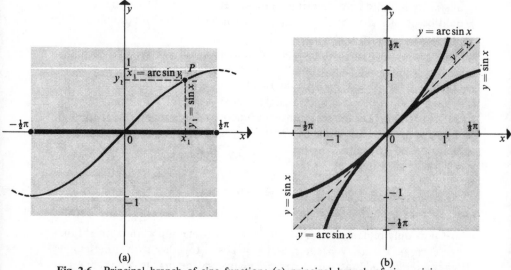

(a)   (b)

Fig. 2·6 Principal branch of sine function: (a) principal branch of $\sin x$ giving one–one mapping in $[-\tfrac{1}{2}\pi, \tfrac{1}{2}\pi]$; (b) inversion of $\sin x$ by reflection in $y = x$.

Now in line with our approach to the inverse of the square root function, the ambiguity as regards the function inverse to sine may be completely resolved if we consider the many–one function $y = \sin x$ with $x \in \mathbf{R}$ as being replaced by an infinity of one–one functions $y = \sin x$, with domains $[(2n - 1)\tfrac{1}{2}\pi, (2n + 1)\tfrac{1}{2}\pi]$. For then in each domain corresponding to some integral value of n, because the mapping there is one–one, an appropriate inverse function may be defined without difficulty.

The intervals are all of length π and are often said to define different *branches* of the inverse sine function. In general, when no specific interval is named we shall write $x = \text{Arcsin } y$, whenever $y = \sin x$. The function Arcsine thus denotes an arbitrary branch of the inverse sine function.

Because of the periodicity of the sine function, when considering the inverse function it is only necessary to study the behaviour of one branch of Arcsine. As is customary, we arbitrarily choose to work with the branch of the inverse sine function associated with the domain $[-\tfrac{1}{2}\pi, \tfrac{1}{2}\pi]$, calling this the *principal* branch and denoting the inverse function associated with this branch by arcsine. Hence for the inverse we shall always write $x = \arcsin y$ when $y = \sin x$ and $-\tfrac{1}{2}\pi \le x \le \tfrac{1}{2}\pi$.

In Fig. 2·6 (b) is shown in relation to the line $y = x$ the function $y = \sin x$ with domain of definition $[-\tfrac{1}{2}\pi, \tfrac{1}{2}\pi]$ and the associated function $y = \arcsin x$ with domain of definition $[-1, 1]$. The reflection property of inverse functions utilized in connection with Fig. 2·4 (b) is again apparent here. It should perhaps again be emphasized that when an inverse function is obtained by reflection in the line $y = x$, then in both the curves representing the function

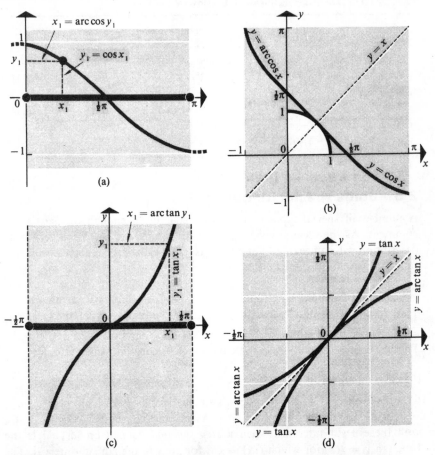

Fig. 2·7 Principal branches of inverse cosine and tangent functions: (a) principal branch of $\cos x$; (b) inversion of $\cos x$ by reflection in $y = x$; (c) principal branch of $\tan x$; (d) inversion of $\tan x$ by reflection in $y = x$.

and its inverse, the variable y is plotted as ordinate (i.e. vertically) and the variable x as abscissa (i.e. horizontally).

Table 2·2 summarizes information concerning the most important inverse trigonometric functions and should be studied in conjunction with Fig. 2·7. In general the notation for a function inverse to a named trigonometric function is obtained by adding the prefix *arc* when referring to the principal branch and *Arc* otherwise. In other books the convention is often to add the superscript minus one after the named function, distinguishing the principal branch by use of an initial capital letter when writing the function. Thus, for example, some authors will write Sin^{-1} in place of arcsine and $\sin^{-1}$ in place of Arcsine. Unfortunately notations are not uniform here and so when using other books the reader would be well advised to check the notation in use.

Table 2·2   Trigonometric functions and their inverse functions

Function	Domain	Inverse function	Branch	Domain
$y = \sin x$	$[-\frac{1}{2}\pi, \frac{1}{2}\pi]$	$y = \arcsin x$	Principal	$[-1, 1]$
$y = \sin x$	$[(2n - 1)\frac{1}{2}\pi, (2n + 1)\frac{1}{2}\pi]$	$y = \text{Arcsin } x$	Any	$[-1, 1]$
$y = \cos x$	$[0, \pi]$	$y = \arccos x$	Principal	$[-1, 1]$
$y = \cos x$	$[n\pi, (n + 1)\pi]$	$y = \text{Arccos } x$	Any	$[-1, 1]$
$y = \tan x$	$(-\frac{1}{2}\pi, \frac{1}{2}\pi)$	$y = \arctan x$	Principal	$(-\infty, \infty)$
$y = \tan x$	$((2n - 1)\frac{1}{2}\pi, (2n + 1)\frac{1}{2}\pi)$	$y = \text{Arctan } x$	Any	$(-\infty, \infty)$

2·3   Some special functions

A number of special types of function occur often enough to merit some comment. As the ideas involved in their definition are simple, a very brief description will suffice in all but a few cases. To clarify these descriptions, the functions are illustrated in Fig. 2·8.

(a) Constant function

The *constant* function is a function $y = f(x)$ for which $f(x)$ is identically equal to some constant value for all x in the domain of definition $[a, b]$. Thus a constant function has the equation $y \equiv$ constant, for $x \in [a, b]$.

(b) Step function

Consider some set of n sub-intervals or partitions $[a_0, a_1)$, $[a_1, a_2)$, $[a_2, a_3)$, . . ., $[a_{n-1}, a_n]$ of the interval $[a_0, a_n]$. Associate n constants $C_1, C_2, . . ., C_n$ with these n sub-intervals. Then a *step function* defined on $[a_0, a_n]$ is the function $y = f(x)$ for which $f(x) \equiv C_r$, for all x in the rth sub-interval. The

Fig. 2·8 (opposite) Some special functions: (a) constant function; (b) step function; (c) $y = |x|$; (d) even function; (e) odd function; (f) bounded function on $[a, b]$.

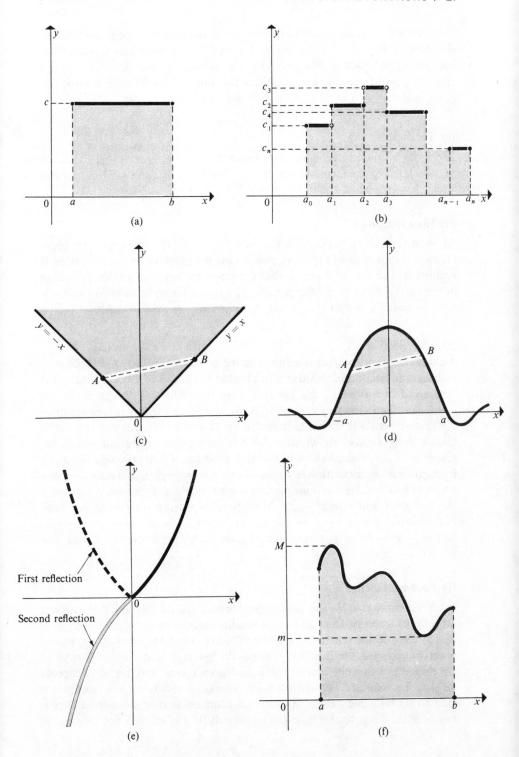

(a)

(b)

(c)

(d)

First reflection

Second reflection

(e)

(f)

function will be properly defined provided a functional value is assigned to all points x in $[a_0, a_n]$ including end points of the intervals. Usually it is immaterial to which of two adjacent sub-intervals an end point is assigned and one possible assignment is indicated in Fig. 2·8 (b), where a deleted end point is shown as a circle and an included end point as a dot.

(c) The function $|x|$

From the definition of the absolute value of x it is easily seen that the graph of $y = |x|$ has the form shown in Fig. 2·8 (c). It is composed of the line $y = x$ for $x \geq 0$ and the line $y = -x$ for $x < 0$.

(d) Even function

An *even* function $y = f(x)$ is a function for which $f(-x) = f(x)$. The geometrical implication of this definition is that the graph of an even function is symmetrical about the y-axis so that the graph for negative x is the reflection in the y-axis of the graph for positive x. Typical examples of even functions are $y = \cos x$, $y = 1/(1 + x^2)$ and the function $y = |x|$ just defined.

(e) Odd function

An *odd* function $y = f(x)$ is a function for which $f(-x) = -f(x)$. The geometrical implication of this definition is that the graph of an odd function is obtained from its graph for positive x by first reflecting the graph in the y-axis and then reflecting the result in the x-axis. In Fig. 2·8 (e) the result of the first reflection is shown as a dotted curve and its reflection in the x-axis gives a second curve shown as a full line in the third quadrant which, together with the original curve in the first quadrant, defines the odd function. By virtue of the definition we must have $f(0) = -f(0)$, showing that the graph of an odd function must pass through the origin. Typical odd functions are $y = \sin x$ and $y = x^3 - 3x$. Most functions are neither even nor odd. For example, $y = x^3 - 3x + 1$ is not even, since $y(-x) = (-x)^3 - 3(-x) + 1 = -x^3 + 3x + 1 \neq y(x)$, nor, by the same argument, is it odd, for $y(-x) \neq -y(x)$.

(f) Bounded function

A function $y = f(x)$ is said to be *bounded* on an interval if it is never larger than some value M and never smaller than some value m for all values of x in the interval. The numbers M and m are called, respectively, *upper* and *lower* bounds for the function $f(x)$ on the interval in question. It may of course happen that only one of these conditions is true, and if it never exceeds M then it is said to be bounded above, whereas if it is never less than m it is said to be bounded below. A bounded function is thus a function that is bounded both above and below. The bounds M and m need not be strict in the sense that the function ever actually attains them. Sometimes when the bounds are strict they are only attained at an end point of the domain of definition of the function.

Of all the possible upper bounds M that may be assigned to a function that is bounded above on some interval, there will be a smallest one M', say. Such a number M' is called the least upper bound or the *supremum* of the function on the interval and the name is usually abbreviated to i.u.b. or to sup. Similarly, of all the possible lower bounds m that may be assigned to a function that is bounded below on some interval, there will be a largest one m', say. Such a number m' is called the greatest lower bound or the *infimum* of the function on the interval and the name is usually abbreviated to g.l.b. or to inf.

Not all functions are bounded either above or below, as evidenced by the function $y = \tan x$ on $(-\frac{1}{2}\pi, \frac{1}{2}\pi)$, though it is bounded on any closed subinterval not containing either end point. Typical examples of bounded functions on the interval $(-\infty, \infty)$ are $y = \sin x$ and $y = \cos x/(1 + x^2)$. The function $y = 1/(x - 1)$ is bounded below by zero on the interval $(1, \infty)$ but is unbounded above, whereas the function $y = 2 - x^2$ is strictly bounded above by 2 but is unbounded below on the interval $(-\infty, \infty)$.

(g) Convex and concave functions

A *convex* function is one which has the property that a chord joining any two points A and B on its graph always lies above the graph of the function contained between those two points. Similarly, a *concave* function is one which has the property that a chord joining any two points A and B on its graph always lies below the graph of the function contained between those two points. Thus the function $y = |x|$ shown in Fig. 2·8 (c) is convex on the interval $(-\infty, \infty)$ whereas the function shown in Fig. 2·8 (d) is only concave on the closed interval $[-a, a]$.

(h) Polynomial and rational functions

A *polynomial* of *degree n* is an algebraic expression of the form

$$y = a_n x^n + a_{n-1} x^{n-1} + \cdots + a_1 x + a_0,$$

where n is a positive integer and it is defined for all x.

A *rational* function is a function which is capable of expression as the quotient of two polynomials and so has the form

$$y = \frac{b_m x^m + b_{m-1} x^{m-1} + \cdots + b_1 x + b_0}{a_n x^n + a_{n-1} x^{n-1} + \cdots + a_1 x + a_0},$$

and is defined for all values of x for which the denominator does not vanish.

An example of a polynomial of degree 2 is the quadratic function $y = x^2 - 3x + 4$; a typical rational function is

$$y = \frac{3x^2 - 2x - 1}{4x^3 + 11x^2 + 5x - 2},$$

which is defined for all values of x apart from $x = -2$, $x = -1$, and $x = \frac{1}{4}$, at which points the denominator vanishes. For this reason these values are called the *zeros* of the polynomial forming the denominator and they arise directly from its factorization into the form

$$4x^3 + 11x^2 + 5x - 2 \equiv (4x - 1)(x + 2)(x + 1).$$

(i) Algebraic function

An *algebraic* function arises when attempting to form the inverse of a rational function. The function $y = +\sqrt{x}$ for $x \geq 0$ provides a typical example here. More complicated examples are the functions:

$$y = x^{2/3} \qquad y = x^2 + 2\sqrt{x} - 1 \qquad y = x\sqrt{x}/(2 - x).$$

More precisely, we shall call the function $y = f(x)$ algebraic if it may be transformed into a polynomial involving the two variables x and y, the highest powers of x and y both being greater than unity. This criterion may easily be applied to any of the above examples. In the case of the last example, a simple calculation soon shows that it is equivalent to the polynomial $2y^2 - 2xy^2 - x^3 = 0$, which is of degree 2 in y and 3 in x.

(j) Transcendental function

A function is said to be *transcendental* if it is not algebraic. A simple example is $y = x + \sin x$, which is defined for all x but is obviously not algebraic.

(k) The function $[x]$

On occasions when working with quantities that may only assume integral values it is useful to write $y = [x]$ with the meaning that we assign to every real number x the greatest integer y that is less than or equal to it. Thus, for example, we have $[-3] = -3$, $[-1.3] = -2$, $[0] = 0$, $[0.92] = 0$, $[\pi] = 3$, and $[17] = 17$.

2·4 Parameterization of a curve

We have seen that when a curve is represented by an explicit equation of the form $y = f(x)$, then for inversion reasons the mapping must be one–one. In other words, either f must be strictly monotonic in its domain of definition or, if not, it must be expressible piecewise as a set of new functions which are strictly monotonic on suitably chosen domains.

A more general representation of a curve that overcomes the necessity for sub-division of the domain, and even allows curves with loops, may be achieved by the introduction of the notion of *parametric representation of a curve*. The idea here is simple and is that instead of considering x and y to be directly related by some function f, we instead consider x and y separately to be functions of the variable parameter α. Thus we arrive at the pair of equations

$$x = s(\alpha) \qquad y = t(\alpha), \tag{2·3}$$

with $a \leq \alpha \leq b$, say, which together define a curve. For any value of α in $[a, b]$ we can use these equations to determine unique values of x and y, and hence to plot a single point on the curve represented parametrically by Eqn (2·3). The set of all points described by Eqn (2·3) then defines a curve.

As a simple example of a curve without loops we may consider the parametric equations

$$y = \alpha^2 \qquad x = \alpha \qquad \text{for } -\infty < \alpha < \infty.$$

These obviously define a parabola that lies in the upper half plane and is symmetrical about the y-axis with its vertex passing through the origin. Elimination of α is easy here and results in the explicit representation $y = x^2$. In more complicated cases the parameter cannot usually be eliminated and, indeed, this should not be expected since parametric representation is more general than explicit representation.

An important consequence of the parametric representation of a curve is that increasing the value of the parameter defines a sense of direction along the curve which is often very useful in more advanced applications of these ideas. An example of a curve containing a loop is provided by the parametric equations

$$x = \alpha^3 - \alpha \qquad y = 4 - \alpha^2 \qquad \text{for } -2 \leq \alpha \leq 2,$$

which is shown in Fig. 2·9 together with the sense of direction defined by increasing α.

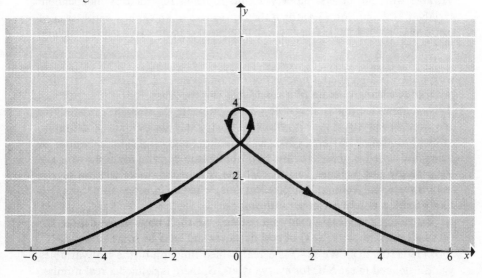

Fig. 2·9   Parameterization of a curve defining sense of direction.

It is implicit in the concept of the parametric representation of a curve that a given curve may be parameterized in more than one way. Hence changing the variable in a parameterization will give a different parametric representation of the same curve. Thus if in the example above we replace the parameter α by the parameter β using the relationship $\alpha = \beta + 1$, then it is readily seen that

$$x = \beta^3 + 3\beta^2 + 2\beta \qquad y = 3 - 2\beta - \beta^2 \qquad \text{for } -3 \le \beta \le 1.$$

This is an alternative parameterization of the same curve shown in Fig. 2·9.

2·5 Functions of several real variables

In physical situations, to say that a quantity depends only on one other quantity is usually a gross oversimplification. Indeed, this was so in the thermodynamic illustration used to introduce the notion of a function of one real variable, because we insisted on maintaining a constant volume of gas. In general the pressure p of a given gas will depend on both its temperature T and its volume v. Here we would say that there was a functional relationship between p, T, and v which, in an implicit form, may be expressed by the equation

$$f(p, T, v) = 0. \tag{2·4}$$

The function f occurring here is a *function of three real variables* and obviously depends for its form on the particular gas involved.

Usually one of the three quantities, say p, is regarded as a dependent variable with the others, namely T and v, being regarded as independent variables. Solving Eqn (2·4) for p then gives rise to an explicit expression of the form

$$p = g(T, v), \tag{2·5}$$

with g then being called a *function of two real variables*.

Just as with a function of a single real variable, in addition to specifying the functional form it is also necessary to stipulate the domain of definition of the function. Thus Eqn (2·5), which in thermodynamic terms would be called the *equation of state* of the gas, would only be valid for some range of temperature and volume. In this case the reason for the restriction on the temperature and volume is a physical one, whereas in other situations it is likely to be a purely mathematical one.

Extending the ideas already introduced we shall now let $\mathbf{R}^2$ denote the set of all ordered pairs (x, y) of real numbers and let S be some subset of $\mathbf{R}^2$.

DEFINITION 2·3 We say f is a real valued function of the real variables x and y defined in set S if, for every $(x, y) \in S$, there is defined a real number denoted by $f(x, y)$.

As is the case with a function of one variable, when the domain of definition of a real valued function of two or more real variables is not specified it is to be understood to be the largest possible domain of definition that can be defined. Thus, for example, the largest subset $S \subset \mathbf{R}^2$ in which the function $f(x, y) = \sqrt{(1 - x^2 - y^2)}$ is defined is given by

$$S = \{(x, y) \in \mathbf{R}^2 \mid x^2 + y^2 \leq 1\}.$$

This concept of a function immediately extends to include functions of more than two variables. Using $\mathbf{R}^n$ to denote the set of all ordered n-tuples $(x_1, x_2, \ldots, x_n)$ of real numbers of which S is some subset, this definition can be formulated.

DEFINITION 2·4   We shall say that f is a real valued function of the real variables $x_1, x_2, \ldots, x_n$ defined in set S if, for every $(x_1, x_2, \ldots, x_n) \in S$, there is defined a real number denoted by $f(x_1, x_2, \ldots, x_n)$.

A typical example of a function of the three variables x, y, z is provided by $f(x, y, z) = \sqrt{(2 - x)} + \sqrt{(9 - y^2)} + \sqrt{(16 - z^4)}$. The largest subset $S \subset \mathbf{R}^3$ for which this function may be defined is obviously

$$S = \{(x, y, z) \in \mathbf{R}^3 \mid x \leq 2; -3 \leq y \leq 3; -2 \leq z \leq 2\}.$$

The geometrical idea underlying the graph of a function of a single variable also extends to real functions f of two real variables x, y. Denote the value of the function f at (x, y) by z, so that we may write $z = f(x, y)$. Then with each point of the (x, y)-plane at which f is defined we have associated a third number $z = f(x, y)$. Taking three mutually perpendicular straight lines with a common origin 0 as axes, we may then identify two of the axes with the independent variables x and y and the third with the dependent variable z. The ordered number triples $(x, y, z) \equiv (x, y, f(x, y))$ may then be plotted as points in a three-dimensional geometrical space. The set of points (x, y, z) corresponding to the domain of definition of the function $f(x, y, z)$ then define a surface which, in practice, usually turns out to be smooth. It is conventional to plot z vertically.

On account of the geometrical representation just described, even in $\mathbf{R}^n$ it is customary to speak of the ordered n-tuple of numbers $(x_1, x_2, \ldots, x_n)$ as defining a 'point' in the 'space' $\mathbf{R}^n$.

By way of illustration of a graph of a function of two variables we now consider

$$f(x, y) = \frac{x^2}{4} + \frac{y^2}{9} \quad \text{with} \quad \frac{x^2}{4} + \frac{y^2}{9} \leq 2,$$

where the inequality serves to define a domain of definition for the function. The surface described by this function has the equation $z = x^2/4 + y^2/9$ and the domain of definition is the interior and boundary of the curve

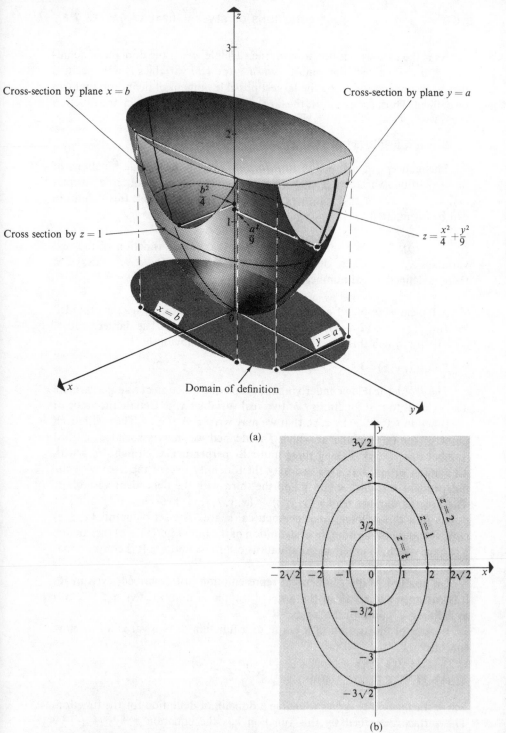

Cross-section by plane $x = b$

Cross-section by plane $y = a$

Cross section by $z = 1$

$z = \dfrac{x^2}{4} + \dfrac{y^2}{9}$

$\dfrac{b^2}{4}$

$\dfrac{a^2}{9}$

$x = b$

$y = a$

Domain of definition

(a)

$z = 2$

$z = 1$

$z = \frac{1}{2}$

(b)

Fig. 2·10 Surfaces and level curves: (a) representation of surface; (b) level curves.

$x^2/4 + y^2/9 = 2$. If this latter expression is rewritten in the form $x^2/8 + y^2/18 = 1$ then it can be seen that the domain of definition of f is in fact the interior of an ellipse in the (x, y)-plane having semi-minor axis $2\sqrt{2}$ and semi-major axis $3\sqrt{2}$, and being centred on the origin. As $f(x, y)$ is an essentially positive quantity it follows directly that $0 \leq z \leq 2$ in the domain of f.

To deduce the form of the surface, two further geometrical concepts are helpful. The first is the notion of the curve defined by taking a cross-section of the surface parallel to the z-axis. The second is the notion of a *contour line* or *level curve*, defined by taking a cross-section of the surface perpendicular to the z-axis.

To examine a cross-section of the surface by the plane $y = a$, say, we need only set $y \equiv a$ in $f(x, y)$ to obtain $z = x^2/4 + a^2/9$, showing that the curve so defined is a parabola with vertex at a height $z = a^2/9$ above the y-axis. A similar cross-section by the plane $x = b$ shows that the curve so defined is $z = b^2/4 + y^2/9$, which is also a parabola, but this time with its vertex at a height $z = b^2/4$ above the x-axis. (See Fig. 2·10 (a).) If desired, sections by other planes parallel to the z-axis may also be used to assist visualization of the surface.

The curve defined by a section of the surface resulting from a cross-section taken perpendicular to the z-axis is called a *contour line* or *level curve* by direct analogy with cartography, where such lines are drawn on a map to show contours of constant altitude. Level curves are obtained by determining the curves in the (x, y)-plane for which $z = $ constant, and it is customary to draw them all on one graph in the (x, y)-plane with the appropriate value of z shown against each curve. (See Fig. 2·10 (b).)

Let us determine the level curve in our example corresponding to $z = \frac{1}{2}$ which is representative of z in the range $0 \leq z \leq 2$. We must thus find the curve with the equation $x^2/4 + y^2/9 = \frac{1}{2}$, which we choose to rewrite in the standard form $x^2/2 + y^2/(9/2) = 1$. This shows that it describes an ellipse centred on the origin with semi-minor axis $\sqrt{2}$ and semi-major axis $3\sqrt{2}$. It is not difficult to see that all the level curves are ellipses; the one corresponding to $z = 2$ being the boundary of the domain of f and the one corresponding to $z = 0$ degenerating to the single point at the origin.

PROBLEMS

Section 2·1

2·1 Sketch the graphs of these functions:

(a) $f(x) = x^2 - 3x + 2$ $(-1 \leq x \leq 3)$;

(b) $f(x) = x + \sin x$ $(-\pi/2 \leq x \leq \pi/2)$;

(c) $f(x) = x^3$ $(-2 \leq x \leq 2)$;

(d) $f(x) = x^2 + 1/x$ $(0{\cdot}2 \leq x \leq 2)$;

(e) $f(x) = x + 1/x^2$ $(0{\cdot}5 \leq x \leq 5)$.

2·2 Determine the domain and the range of each of functions (a) to (e) defined above.

2·3 Determine the range of the function $f(x) = x^2 + 1$ corresponding to each of the following domains and state when the mapping is one–one and when it is many–one:

(a) $[-1, 1]$; (b) $(2, 4)$; (c) $[-2, 4]$; (d) $[-3, 1]$.

2·4 Find the largest domain of definition for each of the following functions:

(a) $f(x) = x^3 + 3$; (b) $f(x) = x^2 + \sqrt{(1 - x^2)}$;
(c) $f(x) = x^2 + \sqrt{(1 - x^3)}$; (d) $f(x) = 1/(x^2 - 1)$;
(e) $f(x) = x + 1/x$; (f) $f(x) = x^2/(1 + x^2)$.

2·5 Let $f(n)$ denote the function that assigns to any positive integer n the number of positive integers whose square is less than or equal to $n + 2$. By enumerating the first few values of $f(n)$ deduce the values of n for which $f(n) = 3$.

2·6 An integer m is said to be a *prime number* if its only factors are 1 and m. Given that $f(n)$ is the function that associates with n the number of primes less than or equal to $2n + 1$, enumerate the first ten values of $f(n)$.

Section 2·2

2·7 Sketch the graphs of the following functions in their stated domains of definition and in each case use the process of reflection in the line $y = x$ to construct the graph of the inverse function:

(a) $f(x) = x^3$ with $x \in [-2, 2]$;
(b) $f(x) = x + \sin x$ with $x \in [0, \pi/2]$;
(c) $f(x) = x/(1 + x^2)$ with $x \in [-1, 2]$.

2·8 Where appropriate, classify the following functions as either monotonic or strictly monotonic increasing or decreasing on the stated domains of definition:

(a) $f(x) = x^2$ for $x \in [-1, 2]$;
(b) $f(x) = x^2$ for $x \in [-1, 0)$;
(c) $f(x) = \sin x$ for $x \in [-3\pi/4, \pi/4]$;
(d) $f(x) = \cos x$ for $x \in [0, \pi]$;
(e) $f(x) = \tan x$ for $x \in [-\pi/4, \pi/4]$;

(f) $f(x) = \begin{cases} x \text{ for } x \in [0, 1] \\ 1 \text{ for } x \in (1, 2] \\ x^2/4 \text{ for } x \in (2, 6]; \end{cases}$

(g) $f(x) = \begin{cases} x \text{ for } x \in [1, 2) \\ x^2 \text{ for } x \in [2, 4]. \end{cases}$

2·9 Complete the entries in this table:

f	S	$f(S)$	Is mapping one–one	f^{-1} when it exists
x	$[-3, 1]$			
x^3		$(2, 4)$		
$1/(1 + x)$	$[1, 3]$			
$\sin x$	$[-\tfrac{1}{2}\pi, \tfrac{1}{4}\pi]$			
$\cos (x + \tfrac{1}{4}\pi)$	$[0, \pi]$			
$\tan [x - \tfrac{1}{4}\pi]$	$[0, \tfrac{1}{2}\pi]$			

Section 2·3

2·10 Sketch these functions in their associated domains of definition:

(a) $f(x) = |2x|$ for $x \in [-2, 2]$;

(b) $f(x) = x + |x|$ for $x \in [-2, 2]$;

(c) the step function assuming the values 1, 2, -3, 2, 4 on the x intervals $[0, 1)$, $[1, 2]$, $(2, 3\cdot5)$, $[3\cdot5, 4]$, and $(4, 5]$, respectively. Identify end points belonging to a line by a dot and end points deleted from a line by a circle.

(d) $f(x) = \begin{cases} |x| & \text{for } x \in [0, 1) \\ |x - 1| & \text{for } x \in [1, 2) \\ |x - 2| & \text{for } x \in [2, 3]. \end{cases}$

2·11 Where appropriate, classify the following functions as even or odd:

(a) $f(x) = x + |x|$;

(b) $f(x) = x + \sin 2x$;

(c) $f(x) = x^2 + \sin x$;

(d) $f(x) = 1/x$;

(e) $f(x) = x^2/(1 + x^2)^2$;

(f) $f(x) = x^5 - x^3 + x$;

(g) $f(x) = 2 \cos x + \sin x$.

It is obvious that any arbitrary function $f(x)$ which is defined in an interval $\mathscr{I}$ containing the origin may be written in the form

$$f(x) = \tfrac{1}{2}(f(x) + f(-x)) + \tfrac{1}{2}(f(x) - f(-x)),$$

in any interval $\mathscr{J} \subset \mathscr{I}$ that is symmetric about the origin. Such an interval $\mathscr{J}$ is said to be *interior* to $\mathscr{I}$. This shows that any such $f(x)$ is expressible as the sum of an even function $\tfrac{1}{2}(f(x) + f(-x))$, and an odd function $\tfrac{1}{2}(f(x) - f(-x))$ within $\mathscr{J}$. Apply this result to display the following functions as the sum of even and odd parts, in each case stating the largest interval $\mathscr{J}$ for which the result is true:

(h) $f(x) = 1 + x^3 + x \sin x$ for $-2\pi \le x \le 3\pi$;

(i) $f(x) = 1 + x + |x| \sin x$ for $-3\pi \le x \le 3\pi$;

(j) $f(x) = 1 - x + 2x^2 + 4x^3$ for $-4 \le x \le 3$.

2·12 Determine if upper and lower bounds exist for the following functions and, when appropriate, state their values and where they occur on the respective domains of definition:

(a) $f(x) = 1/x$ for $x \in [1, 4]$;

(b) $f(x) = 1/x$ for $x \in (0, 3]$;

(c) $f(x) = 1 + x^2$ for $x \in [-2, 1]$;

(d) $f(x) = \sin x$ for $x \in [0, 3\pi/2]$;

(e) $f(x) = \tan x$ for $x \in (-\pi/2, \pi/2)$.

Section 2·4

2·13 Sketch the curve represented by the parametric equations $x = 2 \cos \alpha$, $y = \sin \alpha$ for $-\pi/2 \le \alpha \le \pi/2$.

2·14 Sketch the curve represented by the parametric equations $x = \alpha^3 + \alpha^2 - 2\alpha$, $y = 5 - \alpha^2$ for $-3 \le \alpha \le 2$. Indicate by arrows on the curve the sense of direction corresponding to increasing α.

2·15 Sketch the curve represented by the parametric equations $x = \cos \alpha + 4 \cos (\alpha/3)$, $y = \sin \alpha + 4 \sin (\alpha/3)$ for $0 \le \alpha \le 3\pi/2$. Use arguments involving even and odd functions to deduce the form taken by the curve for $0 \leq \alpha \leq 6\pi$.

Section 2·5

2·16 What are the largest domains of definition for the following functions of several variables:

(a) $f(x, y) = 1 + x^2 + y^2$; .
(b) $f(x, y) = (x^2 + y^2)/\sqrt{(1 - x^2 - y^2)}$;
(c) $f(x, y) = \sin xy/(x^2 + y^2 + 1)$;
(d) $f(x, y) = 3x^2 + y^2 + \sqrt{(2 - y)} + \sqrt{(4 - x^2)}$;
(e) $f(x, y, z) = \sqrt{(3 - x)} + x\sqrt{(9 - y)} + y\sqrt{(1 - z^2)}$;
(f) $f(x, y, z) = \sqrt{(x^2 + y^2 - 1)} + \sqrt{(4 - x^2 - y^2 - z^2)}$.

2·17 The function $f(x, y) = x^2 y$ has for its domain of definition the rectangle in the (x, y)-plane defined by $|x| \leq 3, |y| \leq 2$. Deduce the shape of the curves defined by cross-sections of the surface $z = f(x, y)$ taken by the three planes $x = -2, x = 0,$ and $x = 2$ that are parallel to the (y, z)-axes and by the three planes $y = -2, y = 0,$ and $y = 2$ that are parallel to the (x, z)-axes, using your results to sketch the surface. Sketch on one diagram the level curves corresponding to $z = -4, z = -2, z = 0,$ and $z = 6$.

2·18 Sketch the surface $z = f(x, y)$ defined by the function $f(x, y) = 1/(1 + x^2 + y^2)$ in the domain $|x| \leq 4, |y| \leq 4$. Draw the level curves corresponding to $z = 1/9, z = 1/3, z = 2/3,$ and $z = 1$.

3 Sequences, limits, and continuity

3·1 Sequences

The notion of a 'sequence' is a constantly recurring one in everyday life, where it usually implies the ordering of some set of events with respect to time. The sets of events that are so ordered, or arranged, are very varied and may be either numerical or non-numerical in nature. Typical examples of commonplace sequences in these categories are these:

(a) the sequence of months in a year;
(b) the sequence of digits identifying a telephone subscriber;
(c) the sequence of machining operations required to make a certain component.

However, sequences are not necessarily decided by the chronological order of events and they are often determined instead by some attribute possessed by the members of the set to be ordered. Thus, for example, two commonly occurring sequences to be found in any library are the entries in the alphabetic catalogues of authors and titles, neither of which are in the chronological order of acquisition of the books. Although these general ideas could be discussed at greater length, such an examination is inappropriate here, and it must suffice that these few examples show that sequences are commonplace in the world around us, and that they need not necessarily involve numbers.

These ideas find an immediate parallel in mathematics, where the natural order existing in **R** combined with the arithmetic properties discussed in Chapter 1 enables us to deal very successfully and in great detail with questions relating to mathematical sequences. Our main pre-occupation in this book will be with sequences of numbers and sequences of functions so we must first make the mathematical notion of a sequence more precise. Before doing this however we must first issue a word of warning concerning the colloquial usage of the words *sequence* and *series*, and on their mathematical usage which is quite different. Colloquially the words sequence and series are often used interchangeably, but in mathematics they have two quite different meanings which must never be confused. In brief, in mathematical terms a sequence is a set of quantities that is enumerated in a definite order, whereas a series involves the sum of a set of quantities. Thus $1, 3, 5, 7, 9, \ldots$ is a sequence but $1 + \frac{1}{2} + \frac{1}{4} + \frac{1}{8} + \frac{1}{16} + \cdots$ is a series.

If a sequence is composed of elements or *terms u* belonging to some set S, then it is conventional to indicate their order by adding a numerical suffix to each term. Consecutive terms in the sequence are usually numbered sequentially, starting from unity, so that the first few terms of a sequence involving u would be denoted by u_1, u_2, u_3, ... Rather than write out a number of terms in this manner this sequence is often represented by $\{u_n\}$, where u_n is the nth term of the sequence. The sequence depends on the set chosen for S and the way suffixes are allocated to elements of S. A sequence will be said to be *infinite* or *finite* according as the number of terms it contains is infinite or finite and, unless explicitly stated, all sequences will be assumed to be infinite. The notation for a sequence is often modified to $\{u_n\}_{n=1}^{N}$ when only a finite number N of terms is involved.

As an example of an infinite numerical sequence, let S be the set of real numbers and the rule by which suffixes are allocated be that to each integer suffix n we allocate the number $1/2^n$ which belongs to $\mathbf{R}$. We thus arrive at the infinite sequence $u_1 = 1/2$, $u_2 = 1/2^2$, $u_3 = 1/2^3$, . . ., which could either be written in the form

$$\frac{1}{2}, \frac{1}{2^2}, \frac{1}{2^3}, \cdots, \frac{1}{2^n}, \frac{1}{2^{n+1}}, \cdots,$$

or, more concisely, in the form

$$\{1/2^n\}.$$

Had the set S still been the set $\mathbf{R}$ of real numbers, but the rule of allocation of suffixes been changed, so that to each integer suffix n chosen from the first N natural numbers we allocated the number $1/(2n + 1)$, then the finite sequence

$$\frac{1}{3}, \frac{1}{5}, \frac{1}{7}, \cdots, \frac{1}{(2N + 1)}$$

would have resulted.

If we use the notion of a function $f(x)$ which is defined only for integral values of the argument x, the following concise definition can be formulated.

DEFINITION 3·1   In mathematical terms a *sequence* is a function f defined only for integer values of its argument and having for its range an arbitrary set S.

Hence the first sequence that was displayed could be regarded as resulting from the function $f(x) = 1/2^x$ with $u_n = f(n)$, where n is always a positive integer. By exactly similar reasoning, the second sequence can be derived from the function $f(x) = 1/(2x + 1)$ by setting $u_n = f(n)$.

The connection between functions and sequences that is established in this definition makes it appropriate to describe numerical sequences in the same terms as would be used to describe the function giving rise to them.

Thus if the terms of a sequence $\{u_n\}$ are such that $m < u_n < M$ for all values of n then the sequence is said to be *bounded*, whilst if $u_{n+1} > u_n$ for all n then the sequence is said to be *strictly monotonic increasing*. The terms bounded above, bounded below, unbounded, strictly monotonic decreasing, monotonic, and oscillating, etc., can also be used in the obvious manner as shown below.

Example 3·1

(a) $\{1/n\}_1^\infty$ is a bounded, strictly monotonic decreasing sequence. The upper bound 1 is strict but the lower bound 0 is never actually attained.

(b) $\left\{\dfrac{1}{\sin(1/n)}\right\}_1^\infty$ is a strictly monotonic increasing sequence, strictly bounded below by $(\sin 1)^{-1}$ but unbounded above.

(c) $\left\{\dfrac{(-1)^n}{n}\right\}_1^\infty$ is a bounded sequence with strict upper bound $\tfrac{1}{2}$ and strict lower bound -1.

(d) $\{u_n\}_1^\infty$ where $u_{2m-1} = m/(m+1)$ and $u_{2m} = u_{2m-1}$. The first six terms of this sequence are $\tfrac{1}{2}, \tfrac{1}{2}, \tfrac{2}{3}, \tfrac{2}{3}, \tfrac{3}{4}, \tfrac{3}{4}$ corresponding pairwise, respectively, to $m = 1$, 2, and 3. The sequence is thus both bounded and monotonic increasing. It is not strictly monotonic increasing because pairs of terms are equal. The lower bound $\tfrac{1}{2}$ is strict, but the upper bound 1 is never actually attained.

(e) $\{(-1)^n\}$ is an oscillating but bounded sequence with strict upper bound 1 and strict lower bound -1.

(f) $\{(-2)^n\}$ is an oscillating but unbounded sequence.

Just as a graph proved to be useful when representing functions, so also may it be used to represent sequences. Exactly the same method of representation can be adopted, but this time, since the domain of the function defining the sequence is the set of natural numbers, the graph of a sequence will be a set of isolated points. A typical example is the graph of the first few terms of the sequence $\{u_n\}$ with $u_n = [n + (-1)^n]/n$ which are shown as dots in Fig. 3·1.

An even simpler graphical representation than this is the one often used in which the values of successive terms in the sequence are plotted one-dimensionally as points on a straight line relative to some fixed origin. Because of the identification of the numerical value of a term of the sequence with a point on a line, the behaviour of a sequence is often spoken of in terms of the behaviour of the points in this representation (that is; there is a one–one mapping of $\{u_n\}$ onto the straight line). In terms of this representation, the same sequence that gave rise to Fig. 3·1 will appear as in Fig. 3·2. This could also have been obtained from Fig. 3·1 by projecting the points of the graph horizontally across to meet the vertical axis.

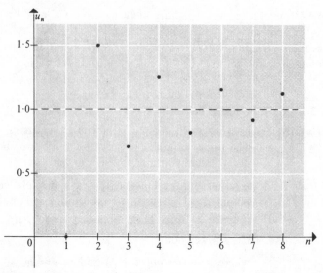

Fig. 3·1 Graph of sequence $\left\{1 + \dfrac{(-1)^n}{n}\right\}$.

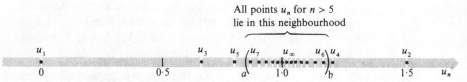

Fig. 3·2 Sequence $\left\{1 + \dfrac{(-1)^n}{n}\right\}$ plotted on line.

In each of these two representations, the tendency for the points of the sequence $\{1 + (-1)^n/n\}$ to cluster around the value unity as n increases is obvious and clearly expresses an important property possessed by the sequence. We shall now explore this more fully.

In the sequence just discussed it is obvious that as n increases, so the points of the sequence cluster ever closer to the unit point in Fig. 3·2. If we adopt the convention of calling an open interval (a, b) containing some fixed point a *neighbourhood* of that point, then it is not difficult to see that any neighbourhood of the point unity will contain an infinite number of points of the sequence $\{u_n\}$. In fact in this case we can assert that no matter how small the length $b - a$ of the neighbourhood, there will always be an infinite number of points in (a, b) and there will always be a finite number of points outside (a, b). This is even true when $b - a$ shrinks virtually to zero!

The fact that any neighbourhood of the value unity has the property that an infinite number of points of the sequence are contained within it, whereas only a finite number of points lie without it, is recognized by saying that the *limit* of the sequence is unity. On account of this name the point corresponding to the value unity in Fig. 3·2 is called a *limit point* of the sequence. We shall examine the idea of a limit in the next section, and so for the moment will confine discussion to limit points. Later on we shall require the notion of a sub-sequence. Henceforth, by a *sub-sequence* we shall mean a sequence $u_{n_1}, u_{n_2}, \ldots, u_{n_m}, \ldots$, of terms belonging to the sequence $\{u_n\}$, where $n_1, n_2, \ldots, n_m, \ldots$ is some numerically ordered set of integers selected from the complete set of natural numbers. Thus $u_2, u_9, u_{27}, u_{31}, \ldots$ is a sub-sequence of $u_1, u_2, u_3, \ldots$ and obviously $\{u_2, u_9, u_{27}, u_{31}, \ldots\} \subset \{u_n\}$.

In sequences involving only one limit point the sequence will be said to *converge* to the value associated with the limit point. This value will be called the *limit* of the sequence.

Not all sequences have limit points and the following examples exhibit sequences having three, one, and no limit points, respectively.

Example 3·2

(a) $\left\{ \sin \left(\dfrac{n^2 + 1}{2n} \right) \pi \right\}$ has the three limit points -1, 0, and 1, of which 0 is a member of the sequence and the other two are not. The sequence does not converge.

(b) $\left\{ \dfrac{1}{n} \sin \left(\dfrac{n\pi}{2} \right) \right\}$ has only one limit point at zero which is a member of the sequence. The sequence converges to zero.

(c) $\{n^2\}$ has no limit point and so the sequence does not converge.

One of the most important applications of the notion of a sequence is to the study of series. The difficulty here is to give a meaning to the sum of an infinite number of terms. What, for example, is the meaning of

$$\sum_{n=1}^{\infty} \frac{1}{n!}. \tag{A}$$

The solution is to be found in the behaviour of the sequence $\{s_m\}$ defined by

$$s_m = \sum_{1}^{m} \frac{1}{n!}.$$

The first few terms of the sequence $\{s_m\}$ are

$$s_1 = 1, \quad s_2 = 1 + \frac{1}{2!}, \quad s_3 = 1 + \frac{1}{2!} + \frac{1}{3!}, \quad s_4 = 1 + \frac{1}{2!} + \frac{1}{3!} + \frac{1}{4!}$$

and obviously all such terms s_m will only involve the sum of a finite number of numbers. For obvious reasons s_m is called the *m*th *partial sum* of the series (A). The interpretation of the infinite sum (A) is to be found in the behaviour of the Nth term of $\{s_m\}$, namely the Nth partial sum s_N, as N tends to infinity. If $\{s_m\}$ has only one limit point at which s_m tends to some number S, then this will be called the *sum* of the series. If S is infinite the series will be said to *diverge*. A moment's reflection will show the reader that this is the practical approach to the problem, since the term s_N is the sum of the first N terms of the infinite series (A), and it seems reasonable to assume that when the value of (A) is finite, it must be close to the value s_N, when N is suitably large.

These preliminary ideas on series must suffice for now, but we shall take them up again later and devise tests to determine whether series are convergent or divergent.

3·2   Limits of sequences

The term limit was first introduced intuitively in the previous section in connection with a sequence $\{u_n\}$ which had only one limit point. As n increases so the points representing the terms u_n cluster ever closer to the limit point whose value L, say, is the limit of the sequence. This idea of a limit is correct in spirit but it is not very satisfactory from the mathematical manipulative point of view since the phrase 'cluster ever closer to' is far too vague. The difficulty of making the expression 'limit' precise is connected with the exact meaning we give to this phrase.

Our difficulty can be resolved if we recall that any neighbourhood of a limit point will contain an infinite number of points of the sequence and, if there is only one limit point, will exclude only a finite number of points. Thinking in terms of numbers rather than points, a neighbourhood of a limit point is simply an open interval of the line on which the numbers u_n are plotted and we already have a notation for representing such an interval. Suppose, for convenience, that the neighbourhood is symmetrical about the number L and of width 2ε, where ε is some arbitrarily small positive number. Then a variable u will be inside this neighbourhood if $L - \varepsilon < u < L + \varepsilon$. Recalling the definition of 'absolute value', this inequality can be rewritten concisely as $|u - L| < \varepsilon$. Different values of $\varepsilon > 0$ determine different neighbourhoods, and if u is identified with the term u_n of the sequence, then L is the limit of the sequence if, no matter how small ε may become, only a finite number of terms u_n lie outside the neighbourhood and an infinite number lie within it.

We can now give a proper definition of a limit.

DEFINITION 3·2   The sequence $\{u_n\}$ will be said to tend to the *limit* L if, and only if, for any arbitrarily small positive number ε, there exists an integer N such that

$$n > N \Rightarrow |u_n - L| < \varepsilon.$$

Let us test our definition on the sequence $\{u_n\}$ with $u_n = 1 + (-1)^n/n$.

We already know that this sequence has only one limit point at the value unity, and consequently our definition should show that the limit is unity. Suppose, for the sake of argument, that we check to see that the definition is satisfied if $\varepsilon = 1/100$. To do this we must find a number N such that when $n > N$ we have

$$\left| \left(1 + \frac{(-1)^n}{n} \right) - 1 \right| < \frac{1}{100}.$$

This result is obviously equivalent to the requirement that $(1/n) < 1/100$ which will be true for any value of n greater than 100. Hence if we take $N = 100$ the conditions of the definition are satisfied. There are thus 100 terms outside the neighbourhood and an infinite number within it.

Had we demanded a much smaller value of ε, say $\varepsilon = 10^{-6}$, the identical argument would have shown that the definition is satisfied if $N = 10^6$. There would now be a very large number of terms outside the neighbourhood $0.999999 < u_n < 1.000001$, in fact 10^6 in all, but this is still a finite number whereas the number of terms within the neighbourhood is still infinite. Clearly, however small the value of ε, the conditions of the definition will still apply showing that it is in accord with our earlier intuitive ideas.

In general, when the sequence $\{u_n\}$ has a limit L, so that we say it converges to L, we shall write

$$\lim_{n \to \infty} u_n = L.$$

Whenever using this notation for a limit the reader must always keep in mind the underlying formal definition just given.

The definition and the illustrative example just given show that when a sequence has only one limit point, then it must converge to the value associated with that limit point. Any sequence such as $\{u_n\}$ with $u_n = \sin\{\pi(n^2 + 1)/2n\}$ cannot have a limit, for it has three limit points at -1, 0, and 1 and any small neighbourhood taken about any one must, of necessity, exclude the infinitely many terms associated with the other two. Such a sequence does not converge.

Frequently the limit of a sequence is of more importance than its individual terms, and in such circumstances the notation $\lim_{n \to \infty} u_n$ is advantageous in that it focusses attention on the general term u_n of the sequence. The result of the limiting operation is often readily deduced from the general term as these examples indicate.

Example 3·3 Determine the limits in each of the following:

(a) $\displaystyle \lim_{n \to \infty} \left[\frac{(2n - 1)(n + 4)(n - 2)}{n^3} \right]$;

(b) $\displaystyle \lim_{n \to \infty} \left[\frac{1}{n^2} + \frac{2}{n^2} + \cdots + \frac{n - 1}{n^2} \right]$;

(c) $\lim\limits_{n \to \infty} \left[\dfrac{5^{n+1} + 7^{n+1}}{5^n - 7^n} \right]$;

(d) $\lim\limits_{n \to \infty} \left[\dfrac{1 + 2^2 + 3^2 + \cdots + n^2}{n^2} \right]$.

Solution (a)   The general term is $u_n = [(2n - 1)(n + 4)(n - 2)]/n^3$, so that expanding the numerator and dividing by n^3 gives

$$u_n = 2 + \frac{3}{n} - \frac{18}{n^2} + \frac{8}{n^3}.$$

Obviously, as n increases, the last three terms comprising u_n approach zero, and in the limit we have

$$\lim_{n \to \infty} \left[\frac{(2n - 1)(n + 4)(n - 2)}{n^3} \right] = 2.$$

Solution (b)   The general term is $u_n = [1 + 2 + \cdots + (n - 1)]/n^2$, in which the numerator is the sum of an arithmetic progression. Now it is readily verified that $1 + 2 + \cdots + (n - 1) = n(n - 1)/2$ so that

$$u_n = \left(\frac{n - 1}{2n} \right).$$

Using the same argument as in (a) above we see at once that as n increases so u_n approaches the value $\frac{1}{2}$, whence

$$\lim_{n \to \infty} \left[\frac{1}{n^2} + \frac{2}{n^2} + \cdots + \frac{n - 1}{n^2} \right] = \frac{1}{2}.$$

Solution (c)   The general term here is $u_n = (5^{n+1} + 7^{n+1})/(5^n - 7^n)$ and by dividing numerator and denominator by 7^n it may be written:

$$u_n = \frac{5(5/7)^n + 7}{(5/7)^n - 1}.$$

Now $5/7 < 1$ so that $(5/7)^n$ will tend to zero as n increases. Thus u_n will approach the value -7. In this case we may write

$$\lim_{n \to \infty} \left[\frac{5^{n+1} + 7^{n+1}}{5^n - 7^n} \right] = -7.$$

Solution (d)   The general term is $u_n = [1^2 + 2^2 + \cdots + n^2]/n^2$, in which the numerator is the sum of the squares of the first n natural numbers. Using the familiar result

$$1^2 + 2^2 + \cdots + n^2 = \frac{n(n + 1)(2n + 1)}{6}$$

enables us to write

$$u_n = \frac{(n + 1)(2n + 1)}{6n}.$$

It is obvious that the numerator is quadratic in n whereas the denominator is first degree or linear in n. Hence as n increases without bound, so will u_n. This sequence diverges and we write

$$\lim_{n \to \infty} \left[\frac{1^2 + 2^2 + \cdots + n^2}{n^2} \right] \to \infty.$$

Notice that we do not use the equality sign in connection with the symbol ∞, in accordance with the idea that infinity is not an actual number but essentially a limiting process.

Before continuing our discussion of limits, let us introduce a useful notation. In the examples above it is apparent that the value of the limit of a sequence involving the ratio of two expressions as n increases, is entirely determined by the ratio of the most significant terms in the numerator and denominator. In the case of a polynomial involving n, the most significant term as n increases is obviously the highest degree term in which it appears. Thus in (a), an inspection of the brackets in the numerator shows the most significant term to be $2n^3$, and as the denominator only involves n^3, it is at once obvious that for large n the ratio will approach $(2n^3/n^3) = 2$.

To streamline limiting arguments of this type, and yet to preserve something of the effect of the less significant terms, we now introduce the so-called 'big oh' notation appropriate to functions.

DEFINITION 3·3 We say that function $f(x)$ is *of the order of* the function $g(x)$, written $f(x) = O(g(x))$ if, for some set of values of x, usually $x \to \infty$

(a) $g(x) > 0$

and

(b) $|f(x)| < Mg(x)$,

where M is some constant.

The value of the constant M is usually unimportant as for most arguments it suffices that such an M should exist. We have these obvious results:

$$\left. \begin{aligned} 2x^3 + 2x + 1 &= O(x^3) \\ 3x + \sin x &= O(x) \\ \sin x &= O(1) \end{aligned} \right\} \text{for large } x,$$

where the symbol $O(1)$ has been used to denote a constant.

In terms of this notation we may write the general term u_n in Example 3.3 (a) in the simplified form

$$u_n = \frac{2n^3 + O(n^2)}{n^3} \qquad \text{whence} \qquad u_n = 2 + \frac{O(n^2)}{n^3}. \qquad (A)$$

By virtue of the definition of the symbol 'big oh', $O(n^2)$ implies an expression that is bounded above by Mn^2, so that $O(n^2)/n^3 \Rightarrow (Mn^2)/n^3$. However, $M/n \to 0$ as n increases without bound, so that

$$\lim_{n \to \infty} u_n = 2. \qquad (B)$$

Normally the argument just outlined would be omitted, so that result (B) would be written down immediately after (A).

Implicit in the examples just examined are results which we now combine.

THEOREM 3·1   If it can be shown that $u_1, u_2, u_3, \ldots$ and $v_1, v_2, v_3, \ldots$ are two sequences such that $\lim_{n \to \infty} u_n = L$ and $\lim_{n \to \infty} v_n = M$, then

(a) $u_1 + v_1, \quad u_2 + v_2, \quad u_3 + v_3, \ldots$ is a sequence such that
$\lim_{n \to \infty} (u_n + v_n) = L + M$;

(b) $u_1 v_1, u_2 v_2, u_3 v_3, \ldots$ is a sequence such that $\lim_{n \to \infty} u_n v_n = LM$;

(c) provided $M \neq 0$, $u_1/v_1, u_2/v_2, u_3/v_3, \ldots$ is a sequence such that
$\lim_{n \to \infty} (u_n/v_n) = L/M$.

These assertions are virtually self-evident and so we prove only the first result, making full use of our definition of a limit and of the triangle inequality of Theorem 1·1.

Suppose ε is given. Then because $\{u_n\}$ converges to the limit L, there exists a number N_1 such that $n > N_1 \Rightarrow |u_n - L| < \frac{1}{2}\varepsilon$. By the same argument there exists another number N_2 such that $n > N_2 \Rightarrow |v_n - M| < \frac{1}{2}\varepsilon$. Now $|(u_n + v_n) - (L + M)| = |(u_n - L) + (v_n - M)| \leq |u_n - L| + |v_n - M|$, and so $n > \max(N_1, N_2) \Rightarrow |(u_n + v_n) - (L + M)| < \frac{1}{2}\varepsilon + \frac{1}{2}\varepsilon$. Thus, taking $N = \max(N_1, N_2)$, and given an arbitrarily small positive number ε, we have

$$n > N \Rightarrow |(u_n + v_n) - (L + M)| < \varepsilon$$

or

$$\lim_{n \to \infty} (u_n + v_n) = L + M.$$

In effect, this theorem justifies any argument in which it is asserted that, if a is close to A and b is close to B, then $a + b$ is close to $A + B$, ab is close to AB, and, provided b and $B \neq 0$, a/b is close to A/B.

At this stage in our discussion of sequences the following result should be self evident and we state it in the form of a postulate, rather than prove it.

POSTULATE   Every increasing sequence which is bounded above tends to a limit.

The proof of this postulate is outlined in Problem 3·9 at the end of the chapter. The details are left to the reader, together with the task of showing the consequence that every decreasing sequence which is bounded below must also tend to a limit.

It is this postulate that validates the usual arithmetic procedure for finding a square root. In the procedure an additional digit is added to the approximation at each stage, thereby giving rise to an increasing sequence that is bounded above. With a number such as $\sqrt{2}$ which we know to be irrational, this same postulate also justifies its successive approximation by the increasing sequence $\{u_n\}$ of rational numbers 1, 1·4, 1·41, 1·414, 1·4142, . . ., u_n, ... In this case an irrational number $\sqrt{2}$ is determined as the limit of a sequence of rationals. The implications are important, since although irrational numbers are of frequent occurrence, in our world in which we live we can only undertake practical calculations using rationals!

3·3   The number e

Later we shall use an important mathematical constant that is always denoted by the symbol e. This number is both irrational and transcendental, and for reference purposes its value to ten decimal places is

e = 2·7182818284.

There are numerous different ways of defining this constant, but although these are interesting, our real concern later in this book will be with the mathematical use of the constant e. We shall, for example, see how it is of fundamental importance in the study of differential equations and in the definition of important mathematical functions like the natural logarithm and the hyperbolic functions sinh x, cosh x, and tanh x.

However, the real purpose of this section will not be to study these applications, but to examine one interesting definition of e as the limit of a particular sequence. This problem provides both a first encounter with e, and also a useful illustration of how approximate information may be extracted from the properties of a difficult sequence. We shall prove that if

$$e = \lim_{n \to \infty}\left[\left(1 + \frac{1}{n}\right)^n\right],\tag{3·1}$$

then $2 < e < 3$. The problem of determining e correctly to any given number of figures will be deferred until we are better equipped for the task.

Consider the sequence $\{u_n\}$ with the general term

$$u_n = \left(1 + \frac{1}{n}\right)^n.$$

We will first establish that u_n is a strictly increasing sequence, so that $u_{n+1} > u_n$, and then show that the sequence $\{u_n\}$ is bounded above by the number 3. The postulate of the previous section then establishes that the limit e exists and is such that e < 3. Finally, the lower bound 2 will be added as a trivial consequence of the proof used to establish the upper bound.

First let us expand u_n by the binomial theorem:

$$\left(1 + \frac{1}{n}\right)^n = 1 + n\left(\frac{1}{n}\right) + \frac{n(n-1)}{2!}\left(\frac{1}{n}\right)^2 + \cdots$$
$$+ \frac{n(n-1)\ldots[n-(n-1)]}{n!}\left(\frac{1}{n}\right)^n.$$

Now rewrite this:

$$u_n = 1 + 1 + \frac{1}{2!}\left(1 - \frac{1}{n}\right) + \frac{1}{3!}\left(1 - \frac{1}{n}\right)\left(1 - \frac{2}{n}\right) + \cdots$$
$$+ \frac{1}{n!}\left(1 - \frac{1}{n}\right)\left(1 - \frac{2}{n}\right)\cdots\left(1 - \frac{n-1}{n}\right). \quad (3\cdot2)$$

An exactly similar argument applied to u_{n+1} then gives

$$u_{n+1} = 1 + 1 + \frac{1}{2!}\left(1 - \frac{1}{n+1}\right) + \frac{1}{3!}\left(1 - \frac{1}{n+1}\right)\left(1 - \frac{2}{n+1}\right) + \cdots$$
$$+ \frac{1}{n!}\left(1 - \frac{1}{n+1}\right)\left(1 - \frac{2}{n+1}\right)\cdots\left(1 - \frac{n-1}{n+1}\right)$$
$$+ \frac{1}{(n+1)!}\left(1 - \frac{1}{n+1}\right)\left(1 - \frac{2}{n+1}\right)\cdots\left(1 - \frac{n}{n+1}\right).$$

Now all the terms in u_n and u_{n+1} are positive and u_{n+1} has one more term than u_n. In addition, terms in u_{n+1} that are associated with factorials are larger than the corresponding terms in u_n because of the obvious inequalities

$$\left(1 - \frac{1}{n+1}\right) > \left(1 - \frac{1}{n}\right);$$
$$\left(1 - \frac{1}{n+1}\right)\left(1 - \frac{2}{n+1}\right) > \left(1 - \frac{1}{n}\right)\left(1 - \frac{2}{n}\right); \cdots$$

Hence $u_{n+1} > u_n$, showing that $\{u_n\}$ is a strictly increasing sequence.

To show that $\{u_n\}$ is bounded above we must try to sum the finite series for u_n and then examine the behaviour of the sum as n increases. As the finite series (3·2) stands we can make no progress, but an overestimate of this sum can easily be obtained if the terms of the series are simplified. This approach will suffice for our purposes, since to prove that the limit e exists, we only need to prove that $\{u_n\}$ is strictly increasing and bounded above; a strict upper bound is *not* necessary here. It is only needed when the exact value of the limit is to be determined.

If we use the obvious inequalities

$$1 > \left(1 - \frac{1}{n}\right) > \left(1 - \frac{1}{n}\right)\left(1 - \frac{2}{n}\right) > \cdots$$

$$> \left(1 - \frac{1}{n}\right)\left(1 - \frac{2}{n}\right) \cdots \left(1 - \frac{n-1}{n}\right),$$

it follows at once from Eqn (3·2) that

$$u_n < 1 + 1 + \frac{1}{2!} + \frac{1}{3!} + \cdots + \frac{1}{n!}. \tag{3·3}$$

This is still too difficult to sum explicitly, so using the observation:

$$\frac{1}{3!} < \frac{1}{2^2}; \frac{1}{4!} < \frac{1}{2^3}; \cdots; \frac{1}{n!} < \frac{1}{2^{n-1}},$$

we further simplify Eqn (3·3) to the form

$$u_n < 1 + 1 + \frac{1}{2} + \frac{1}{2^2} + \frac{1}{2^3} + \cdots + \frac{1}{2^{n-1}}. \tag{3·4}$$

This can now be summed, since after the first term the remaining terms form a geometric progression. We arrive at the result

$$u_n < 1 + \frac{1 - (\frac{1}{2})^n}{1 - \frac{1}{2}},$$

whence $\lim_{n \to \infty} u_n < 3$.

The conditions of our postulate are satisfied, so we may conclude that $\{u_n\}$ has a finite limit e and, furthermore, that e < 3. Examination of Eqn (3·2) shows that $u_n > 2$ for all n so that finally we have established our claim that

2 < e < 3.

The form of argument used to overestimate series (3·2) is often useful and the final inequality (3·4) is usually called a *majorizing* series.

Closely related to limit (3·1) is the sequence $\{v_n(x)\}$ with general term

$$v_n(x) = \left(1 + \frac{x}{n}\right)^n. \tag{3·5}$$

To establish the relationship that exists between e and the limit of $\{v_n(x)\}$ let us first denote the limit by $E(x)$, so that

$$E(x) = \lim_{n \to \infty} \left[\left(1 + \frac{x}{n} \right)^n \right]. \tag{3.6}$$

Suppose $x > 0$ to be any rational number and define an increasing sequence $\{n_k\}$ of natural numbers by the requirement that the numbers n_k/x are integral. Henceforth we shall set $N_k = n_k/x$. Then by restricting n to be a member of $\{n_k\}$ we may define a sub-sequence $\{v_{n_k}(x)\}$ of $\{v_n(x)\}$ for which Eqn (3.5) may be written in the form

$$v_{n_k}(x) = \left(1 + \frac{1}{N_k} \right)^{N_k x} = \left[\left(1 + \frac{1}{N_k} \right)^{N_k} \right]^x. \tag{3.7}$$

Using the definition of u_n we see that

$$v_{n_k}(x) = (u_{N_k})^x,$$

so that taking the limit as $n_k \to \infty$ we have

$$\begin{aligned}
E(x) &= \lim_{n_k \to \infty} v_{n_k}(x) \\
&= \lim_{N_k \to \infty} (u_{N_k})^x \\
&= [\lim_{N_k \to \infty} u_{N_k}]^x = e^x.
\end{aligned}$$

Whence the important result

$$E(x) = e^x. \tag{3.8}$$

With a more subtle argument it can be established that Eqn (3.8) is generally true without the restriction of n to the sequence $\{n_k\}$. This implies that the result is true for *all* real x.

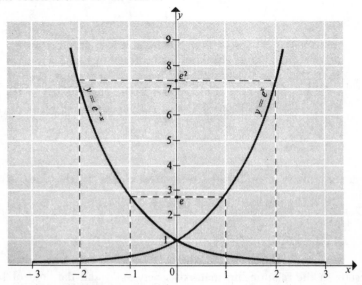

Fig. 3.3 Graph of the functions e^x and e^{-x}.

The function e^x is one of the most important functions in mathematics and it is called the *exponential function*. Fig. 3·3 shows its behaviour with x. Notice that it is an essentially positive function which is strictly monotonic increasing with x. Also shown on the figure is the associated function e^{-x}.

3·4　Limits of functions—continuity

The notion of the limit of a function $f(x)$ as x tends towards some value a

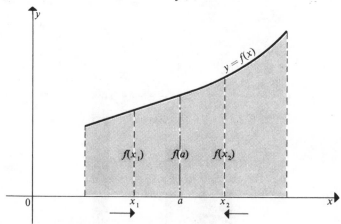

Fig. 3·4　Function $f(x)$ with unbroken graph.

is intuitively obvious in the case of functions whose graph is an unbroken curve. A typical function of this kind is illustrated in Fig. 3·4 from which it is easily seen that if x is considered to be a moving point, then $f(x)$ will approach the value $f(a)$ as x approaches a from either the left or the right. In this case $f(x)$ actually attains the value $f(a)$, and we shall speak of $f(a)$ as the 'limit of $f(x)$ as x tends to a' and write

$$\lim_{x \to a} f(x) = f(a).$$

Thus, if $f(x) = x^3 - 2x^2 + x + 3$, then clearly in this case $\lim_{x \to 2} f(x)$ $= 5 = f(2)$. A slightly less obvious example involves finding $\lim_{x \to 1} f(x)$ when

$$f(x) = \frac{\sqrt{x} - 1}{x - 1},$$

since the formal substitution of $x = 1$ in $f(x)$ seems to yield $0/0$ which is meaningless as it stands. The difficulty here is easily resolved by cancelling a factor $(\sqrt{x} - 1)$ in the numerator and denominator to give

$$f(x) = \frac{1}{\sqrt{x} + 1},$$

from which it is apparent that $\lim_{x \to 1} f(x) = \frac{1}{2}$.

In effect, the intuitive notion involved in the limit of a function is essentially the same as that for the limit of a sequence. Namely, we say that L is the limit of $f(x)$ as x tends to a if, for all x sufficiently close to a, $f(x)$ is close to L. In fact, the determination of the value of the limit L involves the behaviour of $f(x)$ *near* to $x = a$, but does not consider the actual value of $f(x)$ *at* $x = a$.

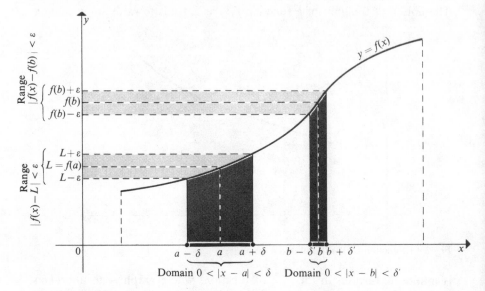

Fig. 3·5 Function $f(x)$ has a smooth graph and attains the limit L at $x = a$.

Whether or not $f(a)$ is actually equal to L, as was the case above, is immaterial. By only slightly modifying our definition of the limit of a sequence, we arrive at the following definition of the limit of a function, which is illustrated in Fig. 3·5, and will be used for our subsequent discussion of continuity.

DEFINITION 3·4 The function $f(x)$ will be said to tend to the limit L as x tends to a if, and only if, for any arbitrarily small positive number ε, there exists a small positive number δ such that

$$0 < |x - a| < \delta \Rightarrow |f(x) - L| < \varepsilon.$$

The significance of the condition $0 < |x - a| < \delta$ is that the value $f(a)$ is specifically excluded from consideration as being irrelevant to the determination of the limit. Thus, if

$$f(x) = \begin{cases} 1 + x^2 & \text{for } x \neq 1, \\ 5 & \text{for } x = 1, \end{cases}$$

then $\lim_{x \to 1} f(x) = 2$, despite the fact that $f(1) = 5$.

If the graph of a function $f(x)$ is not unbroken then more care must be exercised when discussing the notion of a limit. The reason can be seen after examination of Fig. 3·6 in which the graph has a break at $x = c$, at which point the functional value $f(c)$ has been allocated arbitrarily. This graph defines a perfectly satisfactory function, but as x approaches c from either the left or the right, so $f(x)$ approaches either the value L_- or L_+ which are

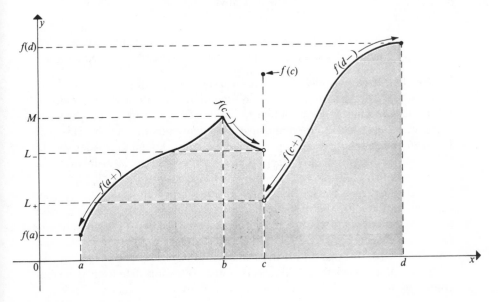

Fig. 3·6 Function $f(x)$ has broken graph.

obviously limits in some sense. Furthermore $L_- \neq L_+$ and neither is equal to $f(c)$. To take account of this, we introduce the concepts of a limit from the left and a limit from the right.

To simplify the explanation we shall write $x \to a-$ in place of 'x tends to a from the left' and $x \to a+$ in place of 'x tends to a from the right'. In terms of this notation the function $f(x)$ in Fig. 3·6 has the property that $\lim_{x \to c-} = L_-$ and $\lim_{x \to c+} = L_+$ which is indicated in the diagram by means of arrows. Once again, in arriving at the limits from the left and right of a point, the functional value itself at that point is not involved. It may or may not equal one of the two limits so defined. These ideas may be expressed formally as a definition.

DEFINITION 3·5 The function $f(x)$ will be said to have the *left-hand limit*, or limit from the left, L_- as $x \to a-$ if, and only if, for any arbitrarily small positive number ε, there exists a small positive number δ such that

$$0 < a - x < \delta \Rightarrow |f(x) - L_-| < \varepsilon.$$

A corresponding definition exists for the *right-hand limit*, or limit from the right, as $x \to a+$ in which L_- is replaced by L_+.

Notice that the function $f(x)$ in Fig. 3·6 only has one-sided limits at $x = a$ and $x = d$ and, even though $f(x)$ has a cusp at $x = b$, and so is not smooth there, it nevertheless still has a limit in the ordinary sense at that point. This is because of the following obvious result.

THEOREM 3·2 If $f(x)$ has identical left- and right-hand limits at a point $x = a$ so that $L_- = L_+ = L$, say, then $\lim_{x \to a} f(x)$ exists and is also equal to L.

We shall usually resolve limit problems of the type just discussed either intuitively or by appeal to a graph. An exception is the discussion of *indeterminate forms* which involve finding the limit of a quotient as x approaches some value at which both numerator and denominator vanish. This will be taken up again later as an application of calculus though the reader should notice that we have already resolved one such simple problem involving a limit of the form $0/0$.

In the physical world functional relationships are usually smoothly changing in the sense that a small change in the independent variable usually produces only a small change in the dependent variable. This smoothness-of-change property is given the mathematical name *continuity* and plays an important part throughout all mathematical analysis. If the reader pauses to think for a moment he will see that the following definition describes continuity in terms of the left- and right-hand limits.

DEFINITION 3·6 The function $f(x)$ is said to be *continuous* at $x = x_0$ if:

(a) $\lim_{x \to x_0-} f(x) = \lim_{x \to x_0+} f(x) = L$

and

(b) $f(x_0) = L$.

In this definition, (a) demands the equality of the left- and right-hand limits and (b) ensures that there is no 'gap' in the graph of $f(x)$ at $x = x_0$. That is to say that the point $(x_0, f(x_0))$ lies on an unbroken curve and so coincides with the limits (a). An alternative, but equivalent, definition of continuity that is often used replaces (a) by the requirement that $\lim_{x \to x_0} f(x)$ $= L$ but still retains (b). Either form of definition is equally good but we have chosen to emphasize the ideas of left- and right-hand limits since they find important applications in engineering and physics.

Continuity essentially describes a property of a function in the neighbourhood of a point of interest and not just at the point itself. Accordingly, a function will be said to be continuous in the interval (a, b) if it is continuous at all points x within (a, b).

Notice that the effect of condition (b) of our definition on a function such as

$$f(x) = \begin{cases} x^3 + 1 & \text{for } x \neq 1 \\ 6 & \text{for } x = 1 \end{cases}$$

is to show that $f(x)$ is continuous everywhere except at $x = 1$.

Let us paraphrase the notion of continuity. In effect, by requiring that a function $f(x)$ be continuous at $x = a$, we are insisting that if the variation of the function about the value $L = f(a)$ does not exceed $\pm \varepsilon$, where $\varepsilon > 0$ is arbitrary, then we can find an x-interval of width 2δ centred on $x = a$ within which this property is always true. This is illustrated by Fig. 3·5, which also indicates that in general the number δ depends on both ε and the value of x at which $f(x)$ is continuous. Thus for the same value of ε, the interval about $x = a$ is of width 2δ, whereas the interval about $x = b$ is of width $2\delta'$, with $\delta' \neq \delta$.

There are a number of immediate consequences of the definition of a limit of a function and of the definition of continuity which we now state as two important theorems.

THEOREM 3·3 (limits)   Suppose that $\lim\limits_{x \to x_0} f(x) = L$ and $\lim\limits_{x \to x_0} g(x) = M$, then

(a) $\lim\limits_{x \to x_0} [f(x) + g(x)] = L + M$;

(b) $\lim\limits_{x \to x_0} f(x)g(x) = LM$;

(c)   provided $M \neq 0$, $\lim\limits_{x \to x_0} [f(x)/g(x)] = L/M$.

The proof of these results is similar in all respects to the proof of Theorem 3·1 and since a representative example was presented there we shall not repeat the argument again.

THEOREM 3·4 (continuity)   If $f(x)$ and $g(x)$ are continuous at $x = x_0$, then so also are the functions

(a) $f(x) + g(x)$;

(b) $f(x)g(x)$;

(c) $f(x)/g(x)$, provided $g(x_0) \neq 0$.

If, furthermore, $f(x)$ is continuous at $x = x_0$ and $g(u)$ is continuous at $u = f(x_0)$, then the continuous function of a continuous function $g[f(x)]$

is continuous at $x = x_0$. Once again the proof of this theorem is similar in all respects to the proof of Theorem 3·1.

Arguments involving continuity usually rely for their success on the knowledge that certain familiar functions are continuous. Once a small list of such functions has been established it can then be considerably enlarged by repeated applications of Theorem 3·4. Accordingly, we present below a table of functions, in each case stating the intervals in which they are continuous. No proof will be given for most entries since the results are obvious from the graphs but for the sake of completeness we shall formally prove the first three entries.

Example 3·4

(a) *Given that $C = $ constant, the function $f(x) = C$ is continuous everywhere.*

The proof is trivial, since for any $x = x_0, f(x_0) \equiv C$ showing that the definition is always satisfied.

(b) *The function $f(x) = x$ is continuous everywhere.*

The proof is again trivial, but let us indicate how the alternative definition of continuity may be used. We must prove that for all x_0, $\lim_{x \to x_0} f(x)$ exists and is equal to $f(x_0)$. Now it is obvious from the definition of $f(x)$ that $f(x_0) = x_0$. Also, for any $x = x_0$ and given $\varepsilon > 0$, $|f(x) - f(x_0)| \equiv |x - x_0| < \varepsilon$ $\Rightarrow |x - x_0| < \varepsilon$ so that in this case the quantity $\delta = \varepsilon$. The function is thus continuous at $x = x_0$ and, as x_0 was arbitrary, it finally follows that $f(x) = x$ is continuous everywhere.

(c) *The function $f(x) = x^n$ with n a positive integer is continuous everywhere.*

We give a proof by induction. Suppose the result is true for some n so that x^n is continuous at $x = x_0$ for all x_0. Now $x^{n+1} = x \cdot x^n$, and we have just proved that x is continuous at x_0. Hence, using Theorem 3·3 (b), x^{n+1} is continuous. The result is true for $n = 1$ and so by the principle of induction it is true for all n. With a little more care this result can be shown to be true for *any* real positive n and not just for n a natural number.

The information contained in this table is likely to be useful on many occasions and so should be memorized. Its application, together with Theorem 3·4, to questions of continuity is usually immediate. Thus, for example, the function $f(x) = 1/x + \sin x$ is continuous everywhere except at the point $x = 0$, and $f(x) = (x^m + a_1 x^{m-1} + \cdots + a_m)/\sin x$, with $m > 0$, is continuous everywhere except at the points $x = n\pi$ for which n is an integer.

Table 3·1   Short list of continuous functions

Function $f(x)$	*Interval over which $f(x)$ is continuous*
C (constant)	$(-\infty, \infty)$
x	$(-\infty, \infty)$
$x^n \ (n > 0)$	$(-\infty, \infty)$
$x^{-n} \ (n > 0)$	$(-\infty, \infty)$ excluding point $x = 0$
$\lvert x \rvert$	$(-\infty, \infty)$
$x^n + a_1 x^{n-1} + \cdots + a_n \ (n > 0)$	$(-\infty, \infty)$
$\dfrac{x^n + a_1 x^{n-1} + \cdots + a_n}{x^m + b_1 x^{m-1} + \cdots + b_m}$	$(-\infty, \infty)$ excluding the zeros of the denominator
$\sin x$	$(-\infty, \infty)$
$\cos x$	$(-\infty, \infty)$
$\tan x$	$(2n - 1)\dfrac{\pi}{2} < x < (2n + 1)\dfrac{\pi}{2}$, integral n
$\sec x$	$(2n - 1)\dfrac{\pi}{2} < x < (2n + 1)\dfrac{\pi}{2}$, integral n
$\operatorname{cosec} x$	$n\pi < x < (n + 1)\pi$, integral n
$\cot x$	$n\pi < x < (n + 1)\pi$, integral n

3·5   Functions of several variables—limits, continuity

The related concepts of a limit and the continuity of a function extend without difficulty to functions of more than one independent variable, provided only that the notion of the proximity of two points is suitably extended. The ideas involved here can best be appreciated if we confine attention to functions $f(x, y)$ of the two independent variables x and y.

Let us suppose that $f(x, y)$ has for its domain of definition some region D in the (x, y)-plane and that (x_0, y_0) is some point interior to D. Then, before considering $f(x, y)$, we must first make clear what is to be meant by $x \to x_0$, $y \to y_0$ in D.

An inspection of Fig. 3·7 shows that starting from the points P and Q in D, both the full curve and the dotted curve describe possible paths by which x and y may tend to x_0 and y_0. In general, we shall write $x \to x_0$, $y \to y_0$, or, say that the point (x, y) tends to the point (x_0, y_0), if $\rho \to 0$, where $\rho = \sqrt{[(x - x_0)^2 + (y - y_0)^2]}$ is the distance between the moving point (x, y) and the fixed point (x_0, y_0). This simple device then allows us to interpret a statement about the two variables x and y in terms of a statement about the single variable ρ. By confining attention to a circular region of radius δ centred on (x_0, y_0) we may conveniently define a neighbourhood of the point (x_0, y_0). Any rectangle or other simple closed geometrical curve

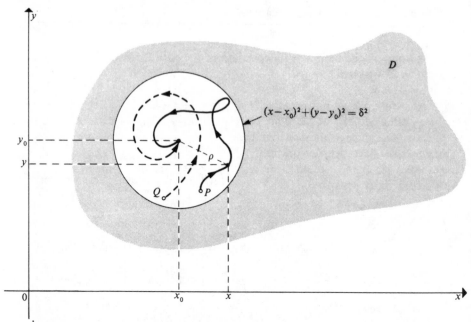

Fig. 3·7 Paths for which the point $(x, y) \to (x_0, y_0)$.

containing (x_0, y_0) would, of course, serve equally well to define a neighbourhood of (x_0, y_0). When using such a neighbourhood it may or may not be necessary to exclude the boundary and the point (x_0, y_0) itself from the definition of the neighbourhood.

Thus, for example, the square $x = 0$, $y = 0$, $x = 1$, and $y = 1$ defines a neighbourhood of the point $(\frac{1}{2}, \frac{1}{2})$. The function

$$f(x, y) = 1/\{xy(x - 1)(y - 1)(x - \tfrac{1}{2})(y - \tfrac{1}{2})\}$$

is defined *in* this neighbourhood, but not *at* $(\frac{1}{2}, \frac{1}{2})$, *on* the boundary or on $x = \frac{1}{2}$, $y = \frac{1}{2}$.

Definition 3·7 is now proposed, with this interpretation of $x \to x_0$, $y \to y_0$ firmly in mind.

DEFINITION 3·7 The function $f(x, y)$ will be said to tend to the limit L as $x \to x_0$ and $y \to y_0$, and we shall write

$$\lim_{\substack{x \to x_0 \\ y \to y_0}} f(x, y) = L,$$

if, and only if, the limit L is independent of the path followed by the point (x, y) as $x \to x_0$ and $y \to y_0$.

As before, we do not necessarily require that $f(x_0, y_0) = L$, as the functional value actually at the limit point (x_0, y_0) is not involved in the limit process. If it can be established that the result of the limiting operation depends on the path taken then, demonstrably, the function has no limit. The following examples make these ideas clear and, on account of their simplicity, are offered without proof.

Example 3·5

(a) If $f(x, y) = \dfrac{2x}{x^2 + y^2 + 1}$, then $\displaystyle\lim_{\substack{x \to 1 \\ y \to 3}} \dfrac{2x}{x^2 + y^2 + 1} = \dfrac{2}{11}$;

(b) if $f(x, y) = \dfrac{xy + 1}{x^2 + y^2}$, then $\displaystyle\lim_{\substack{x \to \infty \\ y \to 1}} \dfrac{xy + 1}{x^2 + y^2} = 0$;

(c) if $f(x, y) = \dfrac{\sin xy}{x^2 + y^2 + 1}$, then $\displaystyle\lim_{\substack{x \to \frac{1}{2}\pi \\ y \to 1}} \dfrac{\sin xy}{x^2 + y^2 + 1} = \dfrac{4}{8 + \pi^2}$;

(d) if $f(x, y) = \dfrac{x(y - 1)}{y(x - 1)}$, then $\displaystyle\lim_{\substack{x \to 1 \\ y \to 1}} f(x, y)$ does not exist since

$\displaystyle\lim_{\substack{x \to 1 \\ y \to 1}} f(x, y) = 1$ if taken along the line $y = x$, but $\displaystyle\lim_{\substack{x \to 1 \\ y \to 1}} f(x, y) = -1$

if taken along the line $y = 2 - x$.

As might be expected, the concept of continuity of a function $f(x, y)$ of two variables then follows as a direct extension of the definition of a limit.

DEFINITION 3·8   The function $f(x, y)$ will be said to be continuous at the point (x_0, y_0) if:

(a) $\displaystyle\lim_{\substack{x \to x_0 \\ y \to y_0}} f(x, y) = L$ exists

and

(b) $f(x_0, y_0) = L$.

We shall say that $f(x, y)$ is continuous in a region if it is continuous at all points (x, y) belonging to that region. Notice that condition (a) demands that $f(x, y)$ has a unique limit as $x \to x_0$ and $y \to y_0$, and condition (b) then ensures that there is no 'hole' in the surface $z = f(x, y)$ at the point (x_0, y_0). The continuity of a function $f(x, y)$ is illustrated in Fig. 3·8 where a circular neighbourhood of the point (x_0, y_0) is shown in relation to the surface. In effect, continuity of $f(x, y)$ is simply requiring that a small change in location of the point (x, y) will cause only a small change in $z = f(x, y)$.

In Fig. 3·8 the point (a, b) has been deliberately detached from the otherwise unbroken surface $z = f(x, y)$, so that the function $f(x, y)$ does not

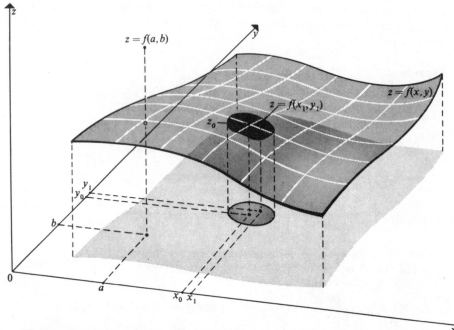

Fig. 3·8　Continuity of $f(x, y)$ at (x_0, y_0) and discontinuity at (a, b).

satisfy the definition there and hence is not continuous at that single point. In general, a function of one or more variables which is not continuous at a point will be said to have a *discontinuity* at that point or, alternatively, to be discontinuous there. Thus the function of one variable shown in Fig. 3·6 has a discontinuity at $x = c$ and the function of two variables shown in Fig. 3·8 is discontinuous at $x = a$, $y = b$.

These ideas also extend to functions of several real variables in an obvious manner once the 'distance' between two points has been defined satisfactorily. For functions $f(x, y, z)$ of the three independent variables x, y, z a suitable *distance function* between points (x_1, y_1, z_1) and (x_0, y_0, z_0) is the linear distance between them when plotted as points relative to three mutually perpendicular Cartesian axes. The distance ρ is then given by the Pythagoras rule as $\rho = \{(x_1 - x_0)^2 + (y_1 - y_0)^2 + (z_1 - z_0)^2\}^{1/2}$.

Again the determination of the regions in which any given function is continuous will usually be done either on an intuitive or on a graphical basis. Thus, in Example 3·5 it is easily seen that:

(a) $f(x, y) = \dfrac{2x}{x^2 + y^2 + 1}$ is continuous everywhere;

(b) $f(x, y) = \dfrac{xy + 1}{x^2 + y^2}$ is continuous everywhere except at $x = 0$, $y = 0$;

(c) $f(x, y) = \dfrac{\sin xy}{x^2 + y^2 + 1}$ is continuous everywhere;

(d) $f(x, y) = \dfrac{x(y - 1)}{y(x - 1)}$ is continuous everywhere except at $(0, 0)$ and $(1, 1)$
and along $x = 1$ and $y = 0$.

3·6   A useful connecting theorem

By now it will have become apparent that there is a strong connection between theorems concerning limits of sequences and the corresponding theorems concerning limits of functions. In fact, with only trivial modification, most limit theorems that are true for sequences are also true for functions. Naturally this is no coincidence and the reason is explained by this connecting theorem.

THEOREM 3·5   Let $f(x)$ be a function defined for all x in some interval $a \leq x \leq b$. Further, let $\{x_n\}$ be a sequence defined in the same interval which converges to a limit α that is not a member of the sequence. Then if, and only if, $\lim\limits_{n \to \infty} f(x_n) = L$ for each such sequence $\{x_n\}$, it follows that $\lim\limits_{x \to \alpha} f(x) = L$.

The proof of this connecting theorem comprises two distinct parts. First it must be established that if $\lim\limits_{x \to \alpha} f(x) = L$, then sequences $\{x_n\}$ exist having the required property. Second, the converse result must be proved; that if the required sequences $\{x_n\}$ exist, then $\lim\limits_{x \to \alpha} f(x) = L$. Together, these two results will ensure that the theorem works in both directions, so that corresponding function and sequence limit theorems satisfying the necessary conditions may be freely interchanged without further question. We leave the details of this to any interested reader as an exercise.

To close this chapter, we shall use this theorem together with geometrical arguments to establish the three useful limits:

$$\lim_{\theta \to 0} \left(\frac{\sin \alpha\theta}{\theta} \right) = \alpha; \tag{3·9}$$

$$\lim_{\theta \to 0} \left(\frac{1 - \cos \alpha\theta}{\theta} \right) = 0; \tag{3·10}$$

$$\lim_{\theta \to 0} \left(\frac{1 - \cos \alpha\theta}{\theta^2} \right) = \frac{\alpha^2}{2}. \tag{3·11}$$

We shall establish that they are all related to the single limit

$$\lim_{\theta \to 0} \left(\frac{\sin \theta}{\theta} \right) = 1,$$

which we prove first.

Consider Fig. 3·9 which represents a circular arc of unit radius with its centre at O, inscribed in the right-angled triangle OAB.

Then it is obvious that

Area of triangle OAC < Area of sector OAC < Area of triangle OAB.

Expressed in terms of the angle θ measured in radians this becomes

$$\tfrac{1}{2}\sin\theta < \tfrac{1}{2}\theta < \tfrac{1}{2}\tan\theta,$$

from which we see that

$$\cos\theta < \frac{\sin\theta}{\theta} < 1. \tag{A}$$

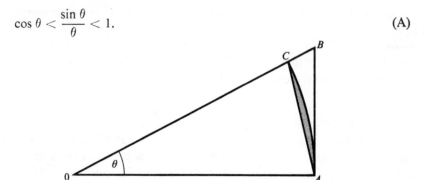

Fig. 3·9 Area inequalities.

This result must be true for all acute angles θ and, in particular, for the values of the sequence $\{\theta_n\}$ defined by $\theta_n = 1/n$. Thus (A) takes the form

$$\cos\theta_n < \frac{\sin\theta_n}{\theta_n} < 1 \tag{B}$$

and, since $\lim\limits_{n\to\infty}\theta_n = 0$ where the limit is not a member of the sequence, we may combine **(B)** with Theorem 3·5 to deduce that

$$\lim_{\theta\to0}\left(\frac{\sin\theta}{\theta}\right) = 1. \tag{3·12}$$

To establish limit (3·9) it is only necessary to replace θ in Eqn (3·12) by $\alpha\theta$, giving rise to

$$\lim_{\alpha\theta\to0}\left(\frac{\sin\alpha\theta}{\alpha\theta}\right) = 1$$

or, equivalently,

$$\lim_{\theta\to0}\left(\frac{\sin\alpha\theta}{\theta}\right) = \alpha.$$

The limits (3·10) and (3·11) then follow by using the identity $1 - \cos\alpha\theta$ $= 2\sin^2\tfrac{1}{2}\alpha\theta$ to form the expressions

$$\frac{1 - \cos \alpha\theta}{\theta} = 2 \sin \tfrac{1}{2}\alpha\theta \left(\frac{\sin \tfrac{1}{2}\alpha\theta}{\theta}\right),$$

and

$$\frac{1 - \cos \alpha\theta}{\theta^2} = 2 \left(\frac{\sin \tfrac{1}{2}\alpha\theta}{\theta}\right)^2.$$

Applying result (3·9) to these we finally arrive at the required results

$$\lim_{\theta \to 0} \left(\frac{1 - \cos \alpha\theta}{\theta}\right) \to 0 \,.\, \alpha = 0$$

and

$$\lim_{\theta \to 0} \left(\frac{1 - \cos \alpha\theta}{\theta^2}\right) \to 2 \,.\, \left(\frac{\alpha}{2}\right)^2 = \frac{\alpha^2}{2}.$$

PROBLEMS

Section 3·1

3·1 Give an example of a numerical sequence and of a non-numerical sequence.

3·2 Use the terms bounded, unbounded, strictly monotonic increasing, and strictly monotonic decreasing to classify the sequences $\{u_n\}$ which have the following general terms:

(a) $u_n = (-n)^{n+1}$;

(b) $u_n = \left(n - \dfrac{1}{n}\right)^2$;

(c) $u_n = \sin (1/n)$;

(d) $u_n = 2 + (-1)^n$;

(e) $u_n = \dfrac{n + 1}{2n + 3}$;

(f) $u_n = \dfrac{2n + 3}{n + 1}$.

Section 3·2

3·3 Name the limit points of the sequence $\{u_n\}$ with the general term $u_n = \sin [(n^2 + n + 1)/2n]\pi$. Identify the sub-sequences that converge to these limit points.

3·4 Give examples of sequences having (a) no limit point, (b) one limit point, (c) two limit points.

3·5 Name the limit points of the sequence $\{u_n\}$ which has the general term

$$u_n = \begin{cases} 1 - \dfrac{1}{3^{2n}} & \text{for } n \text{ even} \\[2mm] \dfrac{1}{3^{2n+1}} & \text{for } n \text{ odd}. \end{cases}$$

State whether or not the limit points belong to the sequence.

3·6 Determine the following limits:

(a) $\lim\limits_{n \to \infty} \dfrac{(3n + 1)(2n - 1)(n - 1)}{n^3}$;

(b) $\lim\limits_{n \to \infty} \dfrac{(2n^2 + n - 1)(n + 2)}{(3n^2 + 7n + 11)}$;

(c) $\lim\limits_{n \to \infty} \dfrac{n + (-1)^n}{n - (-1)^n}$;

(d) $\lim\limits_{n \to \infty} \dfrac{n + (-2)^n}{n - (-2)^n}$;

(e) $\lim\limits_{n \to \infty} \left(\dfrac{1^2 + 2^2 + 3^2 + \cdots + n^2}{2n^3} \right)$.

3·7 Give an expression for the nth term of the sequence $\sqrt{2}$, $\sqrt{(2\sqrt{2})}$, $\sqrt{[2\sqrt{(2\sqrt{2})}]}$, ... Use your result to deduce the limit of the sequence.

3·8 Determine the limits:

(a) $\lim\limits_{n \to \infty} (\sqrt{(n + a)} - \sqrt{n})$, where $a > 0$ is any real number;

(b) $\lim\limits_{n \to \infty} \dfrac{n(2 \sin n - 3 \cos 2n)}{n^2 + 2n + 1}$;

(c) $\lim\limits_{n \to \infty} \left(\dfrac{3^{n+2} + 5^{n+2}}{3^n - 5^n} \right)$;

(d) $\lim\limits_{n \to \infty} {}^n\sqrt{(1 + a^n)}\,(a \geq 0)$.

3·9 Let $\{u_n\}$ be an increasing sequence bounded above by m. Let this bound m, together with the members u_n of the sequence, be represented by points on a line. Then, either the mid-point $m_1 = \frac{1}{2}(u_1 + m)$ of the line segment between u_1 and m is an upper bound of $\{u_n\}$, or it is not. According as m_1 is, or is not, an upper bound of $\{u_n\}$, take for the next point m_2 the mid-point of the half line segment to the left or right of m_1, respectively. Next, according as m_2 is or is not, an upper bound of $\{u_n\}$, take for the next point m_3 the mid-point of the quarter line segment to the left or right of m_2, respectively. Repeat this process indefinitely to generate an infinite sequence of points $\{m_r\}$ as indicated in the diagram.

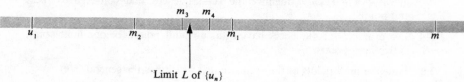

Give reasons why

(a) $\{m_r\}$ has a single limit point L;

(b) the fact that $\{u_n\}$ is an increasing sequence implies that $\lim\limits_{n \to \infty} u_n = L$.

Section 3·4

3·10 Determine the following limits of functions:

(a) $\lim\limits_{x \to a} x^3 - x^2 + x + 1$;   (b) $\lim\limits_{x \to 3} \dfrac{x^2 + x + 1}{x^3 - 1}$;

(c) $\lim_{x \to 3} \dfrac{\sqrt{(x^2 - 6)}}{x^2 + 1}$;

(d) $\lim_{x \to -2} \dfrac{x^3 + x^2 - x - 2}{(x + 1)(x + 2)}$;

(e) $\lim_{x \to 0} \dfrac{(x + h)^3 - x^3}{h}$;

(f) $\lim_{x \to \infty} \{\sqrt{(x^2 + 1000)} - \sqrt{(x^2 - 1000)}\}$;

(g) $\lim_{x \to \infty} x[\sqrt{(x^2 + 3)} - x]$.

3·11 Determine these limits when they exist:

(a) $\lim_{x \to 1} f(x)$ where $f(x) = \begin{cases} x^3 + x - 1 \text{ for } x \le 1 \\ 1 + \sin (x - 1) \text{ for } x > 1; \end{cases}$

(b) $\lim_{x \to 1} \dfrac{x - 1}{x^2 - 1}$;

(c) $\lim_{x \to \frac{1}{2}\pi} f(x)$ where $f(x) = \begin{cases} x^2 + \sin \frac{1}{2}\pi x \text{ for } x \le 3 \\ 4 + x^2 \text{ for } x > 3; \end{cases}$

(d) $\lim_{x \to 1} |x^2 - 1|$;

(e) $\lim_{x \to \frac{1}{2}\pi} \dfrac{1 + \cos x}{1 - \sin x}$.

3·12 Determine the left- and right-hand limits of these functions at the stated points:

(a) $\lim_{x \to 2\pm} \dfrac{3^{x+1} + 5^{x+1}}{3^x + 5^x}$;

(b) $\lim_{x \to \frac{1}{2}\pi \pm} f(x)$ where $f(x) = \begin{cases} 1 + 2 \sin x \text{ for } x \le \frac{1}{2}\pi \\ \text{cosec } x \text{ for } x > \frac{1}{2}\pi; \end{cases}$

(c) $\lim_{x \to 2\pm} |x^2 + x - 1|$;

(d) $\lim_{x \to 0\pm} f(x)$ where $f(x) = \begin{cases} -2 \text{ for } x \le 0 \\ x + |x| \text{ for } x > 0; \end{cases}$

(e) $\lim_{x \to 3\pm} \dfrac{x}{3 - x}$.

3·13 Determine the domains of definition for which these functions are continuous:

(a) $f(x) = x + |x|$;

(b) $f(x) = 1/(x^2 - 1)$;

(c) $f(x) = \dfrac{x^5 + x^2 - 1}{4 + \sin x - 2 \cos x}$;

(d) $f(x) = \dfrac{x^3 + 4x^2 + x - 6}{(x - 1)(x + 4)}$;

(e) $f(x) = \begin{cases} 2x + \sin x \text{ for } x \ne n\pi/2 \\ \dfrac{n^2 + 1}{2n^2 + 3} \quad \text{for } x = n\pi/2. \end{cases}$

Section 3·5

3·14 Determine these limits when they exist:

(a) $\lim_{\substack{x \to 1 \\ y \to 2}} \dfrac{3x^2y}{2x^2 + 2y^2 + 1}$;

(b) $\lim_{\substack{x \to \infty \\ y \to 2}} \dfrac{2x^2 + xy + 1}{x^2 + 2xy + y^2}$;

(c) $\lim\limits_{\substack{x \to 2 \\ y \to 1}} \dfrac{(y - 1) \sin x}{x^2 - 4}$;

(d) $\lim\limits_{\substack{x \to 0 \\ y \to \frac{1}{2}\pi}} \left[\dfrac{1 + 2 \cos xy + \sin xy}{2 + xy} \right]$.

3·15 State the largest neighbourhood about the stated points P in which the following functions are defined. Also state if they are defined at P and on the boundary of the neighbourhood:

(a) $f(x, y) = 1/\{xy(2x - 1)(y + 2)(x + 1)(y - 2)\}$ taking point P as $(-1, 2)$;

(b) $f(x, y) = \dfrac{1}{1 - x^2 - y^2}$ taking point P as $(0, 0)$;

(c) $f(x, y) = \dfrac{1}{1 + x^2 + y^2}$ taking point P as $(2, 3)$.

3·16 Find the points or lines of discontinuity of these functions:

(a) $f(x, y) = \begin{cases} 0 \text{ for } x^2 + y^2 = 1 \\ \dfrac{x \sin xy}{1 - x^2 - y^2} \text{ elsewhere}; \end{cases}$

(b) $f(x, y) = \begin{cases} 3 \text{ for } x = 1, y = 2 \\ \dfrac{xy}{1 + 2x^2 + y^2} \text{ elsewhere}; \end{cases}$

(c) $f(x, y) = \dfrac{x^3 + 2xy + 1}{y - 1}$;

(d) $f(x, y) = \dfrac{x^2 \sin y + y^2 \sin x + 2}{x^4 + 2x^2y^2 + y^4 + 1}$.

Section 3·6

3·17 Apply the results of Section 3·6 to determine these limits:

(a) $\lim\limits_{x \to 0} \dfrac{x}{\sqrt{(1 - \cos x)}}$;

(b) $\lim\limits_{x \to \frac{1}{4}\pi} \dfrac{1 - \sqrt{2} \cos x}{\sqrt{[2 \sin (x - \frac{1}{4}\pi)]}}$;

(c) $\lim\limits_{h \to 0} \dfrac{\sin (x + h) - \sin x}{h}$;

(d) $\lim\limits_{x \to 0} \dfrac{2 \sin^3 (x/4)}{x^3}$;

(e) $\lim\limits_{x \to 0 \pm} \dfrac{|\sin x|}{x}$.

3·18 Apply the results of Section 3·6 to determine these limits:

(a) $\lim\limits_{h \to 0} (x^2 + hx + 1) \left(\dfrac{\cos (x + h) - \cos x}{h} \right)$;

(b) $\lim\limits_{x \to a} \dfrac{\sin x - \sin a}{x - a}$;

(c) $\lim\limits_{x \to 1} \left(\dfrac{\sin \pi x}{\sin \alpha \pi x} \right)$;

(d) $\lim\limits_{x \to \infty} \left(x \sin \dfrac{1}{x} \right)$.

4 Complex numbers and vectors

4·1 Introductory ideas

A number of important properties of the real number system have already been considered, and we shall now examine to what extent quantities representable as displacements in space may be incorporated into a number system. The name *vector quantity* is reserved for all quantities that are representable as a displacement in space or, more exactly, as a directed line element. Familiar vector quantities are force, magnetic field and velocity, which are all representable by a line whose length is proportional to their magnitude and whose direction is parallel to the direction of the original quantity. In addition, the line of action of a vector has a *sense* associated with it, which means that we must specify a direction along the line to indicate the way in which the vector acts.

Thus to represent a velocity of 3 ft/s in an easterly direction we would first adopt a convenient length scale, say 1 in to represent 1 ft/s and then, after marking the points of the compass on our paper, we would draw a line 3 in long in an east–west direction. Finally we would add an arrow to the line pointing eastwards to indicate the sense of the velocity. This line could be located anywhere on our paper since it does not represent a velocity that is associated with any particular point. Reversal of the arrow would correspond to a reversal of the direction of the velocity, so that the line would then represent a velocity of 3 ft/s in a westerly direction.

Not all quantities are vectors, and another important group are called scalars. The word *scalar* describes any quantity that has magnitude but no direction. Typical scalar quantities which have units are temperature, mass and pressure. The real numbers are themselves scalars, and are used to describe the numerical magnitudes of both scalar and vector quantities, irrespective of whether units may be involved. The terms scalar and vector describe collectively two important groups of quantities in the real world. It should, however, be added that they do not jointly give a complete description of all possible physical quantities. Others exist that are neither scalar nor vector, though this need not be elaborated here.

In giving meaning to the square root operation when applied to negative numbers, we shall see that a special kind of two-dimensional vector arises. Its value in mathematics has proved to be so great that although such vectors are restricted to describing vector quantities in a plane, they have been given a special name, *complex numbers*. Because of this restriction, in addition to studying complex numbers, we shall need a more general theory of vectors so that we can describe the cited examples of vector quantities, and any others that may arise, in all possible situations and not just in a plane.

Despite this limitation of complex numbers, their vector properties are still important enough in special situations for them to be in this chapter. Their value elsewhere in mathematics however is even greater, and makes them a discipline in their own right. The main reason for this is to be found in their relationship to real numbers and in the consequences of their intro- duction into functional relationships in the roles of independent and dependent variables. This latter aspect will be pursued later when we discuss another valuable geometrical idea, a conformal transformation. In the meantime we shall develop the vector properties and algebra of complex numbers to the point of general usefulness in mathematics, postponing until the end of this chapter the alternative approach that is necessary for study of general three-dimensional vector quantities. As already mentioned, each is valuable as a separate discipline, though, as would be expected, each has a separate notation and, generally, a quite different field of application.

The following introduction to complex numbers is based only on a knowledge of elementary trigonometric identities, and not until after more study of the exponential and trigonometric functions will we unify our treatment of these two topics.

The origin of complex numbers was the desire of eighteenth-century mathematicians always to be able to compute the roots of polynomials, even when they are of the form

$$x^2 = -1. \tag{4.1}$$

It was Leonhard Euler (1707–83) who first recognized that the real number system was deficient in respect of admitting solutions to all possible poly- nomials and, in connection with Eqn (4·1), he proposed that a new number i be introduced to extend the number system. In keeping with the mathematical beliefs of that period, he called i the *unit imaginary number* and related it to real numbers by requiring that

$$i^2 = -1. \tag{4.2}$$

If we allow the use of this new symbol, then $i = \sqrt{-1}$ is the positive square root of minus one, whence Eqn (4·1) may be seen to have the two roots $x = i$ and $x = -i$. That $x = i$ is a root follows from the definition of i, whilst $x = -i$ is also a root since $(-i)^2 = (-1)^2 . i^2 = 1 . i^2 = -1$. With the introduction of i, equations such as

$$x^2 = -k,$$

which are slightly more general than Eqn (4·1), can also be solved. The equation may be re-expressed in the form $x^2 = k . (-1)$, showing that its roots are $x = i\sqrt{k}$ and $x = -i\sqrt{k}$, where the positive square root is always taken. For example, if $x^2 = -9$, then the roots are $x = 3i$ and $x = -3i$.

The success of Euler's idea lies in the fact that only this one new number need be introduced to enable solutions to be found to all polynomials, irrespective of their degree. As a first step towards seeing this, consider the quadratic equation

$$ax^2 + bx + c = 0, \qquad (4\cdot3)$$

and suppose that $b^2 - 4ac < 0$. Then, setting $4ac - b^2 = m^2$, and formally applying the usual formula for the roots of a quadratic, we obtain

$$x = \frac{-b \pm \sqrt{-m^2}}{2a} \qquad \text{or} \qquad x = \left(\frac{-b}{2a}\right) \pm i\left(\frac{m}{2a}\right).$$

Hence, denoting the two roots by x_1 and x_2, they take the form

$$x_1 = \left(\frac{-b}{2a}\right) + i\left(\frac{m}{2a}\right) \qquad \text{and} \qquad x_2 = \left(\frac{-b}{2a}\right) - i\left(\frac{m}{2a}\right). \qquad (4\cdot4)$$

The numbers x_1 and x_2 are not ordinary numbers since each comprises the sum of a real number and a multiple of the unit imaginary number i. On this basis it is reasonable to conjecture that each root of any arbitrary polynomial will be of the same form and, should the multiplier of i be zero, that root will reduce to a real number.

This conjecture is correct, but before we may verify it, we must see how to perform arithmetic on numbers of this special type. These are the complex numbers already mentioned and, henceforth, we shall always refer to them by this name. Unless the exact form of a complex number is needed, it is useful to denote it by a single symbol, usually z, so that an arbitrary complex number z is of the form

$$z = x + iy, \qquad (4\cdot5)$$

where x and y are real numbers. We call Eqn (4·5) the *real–imaginary form* of a complex number, and refer to x as the *real part* of z, and to y as the *imaginary part* of z. In symbolic form we write

$$x = \operatorname{Re} z, \qquad y = \operatorname{Im} z. \qquad (4\cdot6)$$

Hence if $z = 4 - 7i$, then $\operatorname{Re} z = 4$ and $\operatorname{Im} z = -7$. We stress that $\operatorname{Re} z$ and $\operatorname{Im} z$ are real numbers. The *zero complex number* is denoted by 0 and represents the number $z = 0 + i \cdot 0$.

Already, and without proper justification, we have attributed some reasonable arithmetic properties to i. We have, for example, assumed results such as $\alpha i = i\alpha$ for all real α, and $\sqrt{-\alpha} = \sqrt{-1} \cdot \sqrt{\alpha} = i\sqrt{\alpha}$. To proceed logically and rigorously it would be necessary to define addition, subtraction, multiplication, and division for complex numbers and then to examine the applicability of the real number axioms of Chapter 1 in the case of complex numbers. This is necessary since whatever the arithmetic laws we now propose for complex numbers, they must obviously be in agreement with the real number axioms of Chapter 1, whenever the imaginary parts of complex numbers are zero. We shall not in fact justify the complex number axioms we now formulate, since this is a straightforward matter and provides good exercise for the student (see the problems at the end of the chapter). Instead, we simply summarize the results, pausing only to discuss in detail the most basic operations necessary for the manipulation of complex numbers.

4·2   Basic algebraic rules for complex numbers

First we shall agree to denote addition and subtraction of the complex numbers z_1 and z_2 in the usual manner by writing $z_1 + z_2$ and $z_1 - z_2$, respectively. Multiplication of the complex numbers z_1 and z_2 will be denoted by juxtaposition thus, $z_1 z_2$. Before going on, and in order to work with equations, we must define the meaning of equality between two complex numbers, and then we can define the operations of addition, subtraction, and multiplication. The following definitions are all phrased in terms of the arbitrary complex numbers $z_1 = a + ib$ and $z_2 = c + id$.

DEFINITION 4·1   We shall say that the two complex numbers z_1 and z_2 are *equal*, and will write $z_1 = z_2$ if, and only if, $a = c$ and $b = d$. That is if, and only if, their real parts and their imaginary parts are separately equal.

Example 4·1   Of the complex numbers z_1, z_2, and z_3 defined by $z_1 = 3 - 2i$, $z_2 = 1 + 3i$, and $z_3 = 3 - 2i$, it is obvious that $z_1 = z_3$ but that $z_1 \neq z_2$ and $z_3 \neq z_2$.

DEFINITION 4·2   By the *sum* $z_1 + z_2$ will be understood the single complex number which written in real–imaginary form has a real part that is the sum of the real parts of z_1 and z_2, and an imaginary part that is the sum of the imaginary parts of z_1 and z_2. Thus for the stated numbers z_1 and z_2 we have

$$z_1 + z_2 = (a + c) + i(b + d).$$

Example 4·2   If $z_1 = 2 + i$ and $z_2 = 1 - 3i$, then $z_1 + z_2 = 3 - 2i$.

DEFINITION 4·3   By the *difference* $z_1 - z_2$ will be understood the single complex number which written in real–imaginary form has a real part that is the difference of the real parts of z_1 and z_2 and an imaginary part that is the difference between the imaginary parts of z_1 and z_2. Thus for the stated numbers z_1 and z_2 we have

$$z_1 - z_2 = (a - c) + i(b - d).$$

Example 4·3   If $z_1 = 5 + 6i$ and $z_2 = 4 - 2i$, then $z_1 - z_2 = 1 + 8i$.

Using these definitions it is easily verified that axioms A·1 to A·5 of Chapter 1 also apply to complex numbers. To proceed to an examination of the other axioms we must define the operation of multiplication.

DEFINITION 4·4   The *product* z_1z_2, in which $z_1 = a + ib$ and $z_2 = c + id$, is a single complex number which may be written in real–imaginary form. The product is carried out algebraically as would be the ordinary product $(\alpha + \beta)(\gamma + \delta)$, and the final result is obtained by making the identifications $\alpha = a$, $\beta = ib$, $\gamma = c$, $\delta = id$ and using the result $i^2 = -1$ to combine the four terms that result into a real part and an imaginary part. Thus we have

$$z_1z_2 = (a + ib)(c + id) = ac + iad + ibc + i^2bd = (ac - bd) + i(ad + bc).$$

Example 4·4   If $z_1 = 2 + 3i$ and $z_2 = 1 - i$, then $z_1z_2 = 5 + i$. As a more difficult example let us express $(1 + i)^4 + (1 - i)^4$ in real–imaginary form.

Now $(1 + i)^4 = (1 + 4i + 6i^2 + 4i^3 + i^4)$ and $(1 - i)^4 = (1 - 4i + 6i^2 - 4i^3 + i^4)$, but as $i^2 = -1$, $i^3 = -i$, and $i^4 = 1$, these expressions become $(1 + i)^4 = -4$ and $(1 - i)^4 = -4$. Hence $(1 + i)^4 + (1 - i)^4 = -8$.

The definitions of addition, subtraction, and multiplication of complex numbers are used in the obvious manner for the solution of simple equations. Thus, if $2z - (2 + i) = 4 - 3i$, then adding $(2 + i)$ to both sides of the equation gives $2z + 0 = (4 - 3i) + (2 + i)$ or $2z = 6 - 2i$ whence $z = 3 - i$.

In all cases, the reader should memorize the method employed in the definitions, and not the quoted formulae.

With this definition of multiplication it is a simple matter to verify that axioms M·1 to M·4 and also axiom D·1 apply to complex numbers. When one of the numbers z_1 or z_2 reduces to a real number, then the real and imaginary parts of the other are both scaled by the same factor. If the scale factor is -1 the sign of the complex number is reversed. To discuss axiom M·5 and division we need to proceed more carefully.

As it stands, an expression such as $(a + ib)/c$ is well defined as a complex number, for we may regard $(1/c)$ as a multiplier of $(a + ib)$ and, provided $c \neq 0$, Definition 4·4 will give the result. In this case a and b are both scaled by the factor $(1/c)$. However, it is not clear that the more general expression

$$z_3 = \frac{z_1}{z_2} = \frac{a + ib}{c + id} \tag{4·7}$$

is reducible to a complex number expressible in real–imaginary form. The key to this problem is to be found in M·5 itself when we recall that division is really defined as the operation inverse to multiplication. Hence, we must rewrite Eqn (4·7) in the equivalent form

$$z_3(c + id) = a + ib, \tag{4·8}$$

and then try to determine z_3. Now it is easily verified that any complex number $\alpha + i\beta$ when multiplied by the associated complex number $\alpha - i\beta$ gives the real number $\alpha^2 + \beta^2$. Hence, if both sides of Eqn (4·8) are multiplied by $(c - id)$, the multiplier of z_3 will simply become the real number $c^2 + d^2$. Carrying out this operation, Eqn (4·8) takes the form

$$z_3(c^2 + d^2) = (a + ib)(c - id) \tag{4·9}$$

whence, dividing by the real number $(c^2 + d^2)$, we find that

$$z_3 = \frac{(ac + bd) + i(bc - ad)}{c^2 + d^2}. \tag{4·10}$$

Equation (4·10) is now in the real–imaginary form of a complex number and is the result of the quotient (4·7). Many books take expression (4·10) as the formal definition of the quotient (4·7). The definition we shall propose shortly is equivalent to Eqn (4·10) in all respects, but its form is much easier to memorize. The simplification is achieved by the introduction of a new and useful operation called forming the complex conjugate of a complex number.

DEFINITION 4·5 If $z = a + ib$ is an arbitrary complex number, then the complex number $\bar{z} = a - ib$ is the *complex conjugate* of z. The symbol $\bar{z}$ is read 'z bar'. Equivalently, we may state that the complex conjugate of a number is always obtained by changing the sign of the imaginary part of that number.

With this definition in mind it is easy to show that the following definition of the quotient z_1/z_2 is equivalent to Eqn (4·10).

DEFINITION 4·6 (division) The *quotient* z_1/z_2 of the two complex numbers z_1 and z_2 is the complex number $(z_1\bar{z}_2)/(z_2\bar{z}_2)$.

Using this definition it is a straightforward matter to verify axiom M·5 for complex numbers, provided only that $z_2 \neq 0$.

Example 4·5 We illustrate division by setting $z_1 = 2 + i$ and $z_2 = 3 - 2i$. Now $\bar{z}_2 = 3 + 2i$ and $z_1/z_2 = (z_1\bar{z}_2)/(z_2\bar{z}_2) = (2 + i)(3 + 2i)/(3 - 2i)(3 + 2i)$, whence $z_1/z_2 = (4 + 7i)/13$. By this same method, an equation of the form $2z(2 + i) = 1 + i$ is seen to have the solution $z = (1 + i)/(4 + 2i) = (3 + i)/10$.

On account of the fact that $\bar{z}$ is an ordinary complex number, its general properties are exactly the same as those of any other complex number. Hence the number axioms that apply to z, apply equally well to $\bar{z}$. The following specially useful results are easily proved, and are related to the arbitrary complex number $z = x + iy$, to its complex conjugate $\bar{z} = x - iy$ and to the real number $|z|$ associated with z and defined to be $|z| = (x^2 + y^2)^{\frac{1}{2}}$. (See Definition 4·7.)

$$z + \bar{z} = 2\,\mathrm{Re}\,z = 2x;$$

$$z - \bar{z} = 2i\,\mathrm{Im}\,z = 2iy;$$

$$z = \overline{(\bar{z})};$$

$$\frac{1}{\bar{z}} = \overline{\left(\frac{1}{z}\right)};$$

$$\overline{(z^n)} = (\bar{z})^n;$$

$$\left|\frac{\bar{z}_1}{\bar{z}_2}\right| = \frac{|\bar{z}_1|}{|\bar{z}_2|};$$

$$\overline{(z_1 + z_2 + \cdots + z_n)} = \bar{z}_1 + \bar{z}_2 + \cdots + \bar{z}_n;$$

$$\overline{z_1 z_2 \cdots z_n} = \bar{z}_1 \bar{z}_2 \cdots \bar{z}_n.$$

We now utilize some of these simple properties of the complex conjugate operation to prove an important theorem concerning the roots of a polynomial, and shall then deduce three very useful corollaries. In the process of doing so, we shall take as self-evident the fact that a polynomial $P(z)$ of degree n has n factors of the form $(z - \zeta)$. These are called *linear* factors because they are of degree 1. The numbers ζ may, or may not, be complex.

THEOREM 4·1   If the nth degree polynomial

$$P(z) \equiv a_0 z^n + a_1 z^{n-1} + \cdots + a_n$$

has its coefficients $a_0, a_1, \ldots, a_n$ real, then if $z = \zeta$ is a zero of $P(z)$, so also is $z = \bar{\zeta}$ a zero of $P(z)$.

Proof   Suppose that $z = \zeta$ is a zero of $P(z)$. Then by definition

$$a_0 \zeta^n + a_1 \zeta^{n-1} + \cdots + a_n = 0.$$

Hence, taking the complex conjugate of this equation we may write

$$\overline{(a_0 \zeta^n + a_1 \zeta^{n-1} + \cdots + a_n)} = 0.$$

However, the complex conjugate of a sum is the sum of the complex conjugates of the individual terms comprising the sum so that

$$\overline{(a_0 \zeta^n + a_1 \zeta^{n-1} + \cdots + a_n)} = \overline{a_0 \zeta^n} + \overline{a_1 \zeta^{n-1}} + \cdots + \bar{a}_n.$$

Now as the a_r, $r = 0, 1, \ldots, n$ are real, it follows that $\bar{a}_r = a_r$ and so

$$\overline{a_r \zeta^{n-r}} = a_r \overline{\zeta^{n-r}} = a_r (\bar{\zeta})^{n-r}, \text{ for } r = 0, 1, \ldots, n.$$

Hence,

$$a_0 \bar{\zeta}^n + a_1 \bar{\zeta}^{n-1} + \cdots + a_n = 0;$$

showing that $P(\bar{\zeta}) = 0$. Thus $z = \bar{\zeta}$ is also a zero of $P(z)$.

Paraphrased, Theorem 4·1 asserts that if a polynomial with real coefficients has complex zeros, then they must occur in complex conjugate pairs.

As any zero which is not complex must be real, it follows that we may formulate a Corollary to Theorem 4·1.

Corollary 4·1 (a)   If a polynomial has real coefficients, then those of its zeros that are not real, occur in complex conjugate pairs.

If $z = \zeta$ and $z = \bar{\zeta}$ represent any pair of complete conjugate zeros in Theorem 4·1, then $(z - \zeta)$ and $(z - \bar{\zeta})$ must both be factors of $P(z)$. Hence their product $(z - \zeta)(z - \bar{\zeta})$ must also be a factor. Now

$$(z - \zeta)(z - \bar{\zeta}) = z^2 - (\zeta + \bar{\zeta})z + \zeta\bar{\zeta},$$

and as $\zeta + \bar{\zeta} = 2\,\mathrm{Re}\,\zeta$ is a real number and $\zeta\bar{\zeta} = |\zeta|^2$ is also a real number, it follows that the pair of complex conjugate zeros correspond to a single quadratic factor with real coefficients. Hence Corollary 4·1 (a) may be re-phrased thus:

Corollary 4·1 (b)   Any polynomial with real coefficients may always be factorized into a set of factors which are linear or at most quadratic, each of which has real coefficients. Specifically, if the polynomial is of degree n and there are m pairs of complex conjugate zeros, then there will be $(n - 2m)$ linear factors with real coefficients and m quadratic factors with real coefficients.

Finally, as an obvious consequence of this last corollary:

Corollary 4·1 (c)   An odd degree polynomial with real coefficients must have at least one real zero.

The significance of these results is best illustrated by an example which shows how they may often be used to simplify a difficult problem to the point at which the solution may be determined by familiar methods.

Example 4·6   A polynomial $P(z)$ of degree 5 is defined by the relationship

$$P(z) \equiv z^5 + 5z^4 + 10z^3 + 10z^2 + 9z + 5.$$

Given that $z = i$ is a zero, deduce the remaining four zeros and use the result to express $P(z)$ as the simplest possible product of factors having real coefficients.

Solution   First, as the coefficients of $P(z)$ are all real, Theorem 4·1 is applicable. Hence if $z = i$ is a zero, then so also is $z = -i$. Thus $(z - i)$ and $(z + i)$ are factors, as is their product $(z - i)(z + i) = z^2 + 1$. Using ordinary long division to divide $P(z)$ by $(z^2 + 1)$ we find that

$$P(z)/(z^2 + 1) = z^3 + 5z^2 + 9z + 5.$$

Hence to find the remaining factors we must now factorize this cubic polynomial. As the degree is odd, and the coefficients are real, Corollary 4·1 (c) applies showing that it must have at least one real zero. At this point we have recourse to trial and error to find the real zero which for the purposes of this example has been made an integer.

Thus, setting

$$Q(z) = z^3 + 5z^2 + 9z + 5,$$

we must find a value $z = z_1$ such that $Q(z_1) = 0$. By inspection we see that $Q(-1) = 0$ showing that the real zero is $z = -1$. This corresponds to the linear factor with real coefficients $(z + 1)$. Removing the factor $(z + 1)$ from the cubic by long division, we then find that

$$\frac{P(z)}{(z^2 + 1)(z + 1)} = \frac{Q(z)}{(z + 1)} = z^2 + 4z + 5.$$

Finally we apply the standard formula for the roots of a quadratic to this expression to obtain the remaining two zeros. Completing the calculation, these are found to be $z = -2 - i$ and $z = -2 + i$. Thus the five zeros are $z = i$, $z = -i$, $z = -1$, $z = -2 - i$, and $z = -2 + i$. The required factorization is

$$P(z) \equiv (z + 1)(z^2 + 1)(z^2 + 4z + 5).$$

4·3   Complex numbers as vectors

So far we have discussed the basic arithmetic of complex numbers but have not mentioned their vector properties. To do this, and to give a geometrical representation of complex numbers, we plot them as points in a plane called the *complex plane* or, sometimes, the *z-plane*. Specifically, we shall use the real part of the complex number as its horizontal or x-coordinate and the imaginary part of the complex number as its vertical or y-coordinate. Thus to each complex number there corresponds just one point in the complex plane and, conversely, to each point in the complex plane there corresponds just one complex number. The relationship between points and complex numbers is one–one. In the complex plane, the x-axis is the real axis and the y-axis is the imaginary axis. Other accounts of this subject often refer to this geometrical representation of complex numbers as the Argand diagram, in honour of its inventor.

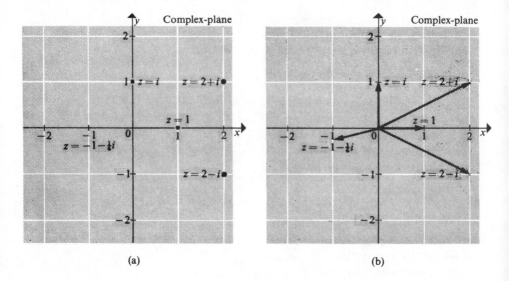

Fig. 4·1 Representation of complex numbers: (a) point representation; (b) vector representation.

In the complex plane, a complex number may either be considered as a point in the plane or, equivalently, as the directed straight line element from the origin to the point in question. We shall remember this dual relationship between points and vectors but, for simplicity, will usually speak only of points in the complex plane.

This duality between points and vectors is indicated in Fig. 4·1 where the complex numbers $z = 1$, $z = i$, $z = 2 + i$, $z = 2 - i$, and $z = -1 - \frac{1}{4}i$ have been represented as points (Fig. 4·1 (a)) and as vectors (Fig. 4·1 (b)). In the case of the vector representation, arrows have been added to show that the vector is drawn from the origin to the point in question.

Notice that if a number, together with its complex conjugate, are plotted in the complex plane, as for example $2 + i$ and $2 - i$ in Fig. 4·1 (a) and (b), then geometrically, in both the point and the vector representations, one is obtainable from the other by reflection in the x-axis as though it were a mirror.

Instead of adding and subtracting vectors analytically by use of Definitions 4·2 and 4·3, the same result may be achieved entirely geometrically as we now indicate. Consider the sum of the vectors $z_1 = 2 + i$ and $z_2 = 1 + 2i$. Analytically $z_1 + z_2 = 3 + 3i$, and Fig. 4·2 (a) shows this result. The same result may be obtained geometrically by the following construction. If we wish to add vector z_2 to z_1, then for the purposes of addition we shall imagine vector z_2 to be freed from the origin, so that it is capable of translation any-

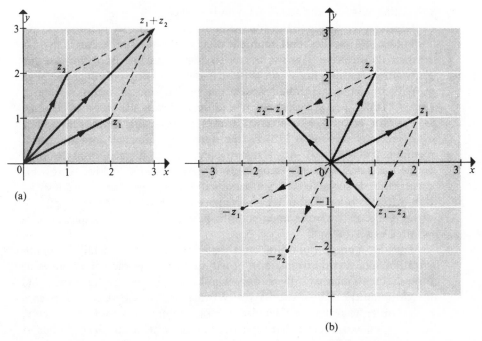

Fig. 4·2 Algebraic operations with complex numbers: (a) vector addition: $z_1 + z_2$; (b) vector subtraction: $z_1 - z_2$.

where in the complex plane, but we shall assume that wherever we re-locate it in the complex plane it will always be kept parallel to its original position, and its length and sense will be preserved. The result of adding z_2 to z_1 is then achieved by translating z_2 in the manner described until its origin is located at the tip of vector z_1. The two arrows of vectors z_1 and z_2 then point in the same direction, and the vector $z_1 + z_2$ is the line element directed from the origin 0 to the tip of the vector z_2 in its new position. In Fig. 4·2 (a) this construction is represented by the lower triangle comprising the parallelogram. Such triangles are *vector triangles*.

A vector not attached to a specific origin or one which, for the purposes of combination with another vector, is freed from its origin to be re-located in some other part of the complex plane will be called a *free* vector. This is in contrast to a vector that is attached to a definite origin which we shall call a *bound* vector. In the addition of z_2 to z_1 that we have just performed, z_1 was regarded as a bound vector and z_2 as a free vector.

Notice that by the same argument, z_1 may be freed and its origin translated to the tip of the bound vector z_2 to form the vector $z_2 + z_1$, which is the line element directed from the origin to the tip of vector z_1 in its new position. In Fig. 4·2 (a) this construction is represented by the upper triangle comprising the parallelogram. The fact that both constructions give rise to the same line representing on the one hand $z_1 + z_2$, and on the other $z_2 + z_1$, proves that vector addition is commutative, since $z_1 + z_2 = z_2 + z_1$.

Before proceeding with the discussion of subtraction, we first observe that Definition 4·4 implies that multiplication of the bound vector z by -1 reverses its direction. That is to say its origin remains fixed, but the line element representing the vector is rotated about the origin through the angle π. With this remark in mind we see that subtraction of vector z_2 from z_1 (Fig. 4·2 (b)), is just a special case of addition in which the vector to be added is $-z_2$. The vector $-z_2$ is obtained from z_2 by reversing the direction of z_2, as is indicated in Fig. 4·2 (b) by the dotted line directed into the fourth quadrant. The vector $z_1 - z_2$ is then the line element directed from the origin to the tip of the reversed vector z_2 in its new position. In Fig. 4·2 (b) this construction is shown in the right-hand half of the plane. The same construction, with the roles of z_1 and z_2 interchanged, is shown in the left-hand half of the plane and when compared with the first result proves that $z_1 - z_2 = -(z_2 - z_1)$. (Why?)

Thus far, complex numbers have been seen to obey the addition, multiplication, and distributive axioms of real numbers, and the reader might be forgiven for wondering if there is any significant difference between them and the real numbers. The answer is yes. Whereas real numbers can be given a natural order according to their size, complex numbers cannot. A glance at Fig. 4·1 (b) makes it clear that no natural order exists in the *field of complex numbers*, comprising all numbers in real–imaginary form, since even vectors of the same length may be differently directed, for instance the pairs of vectors 1 and i, and $2 + i$ and $2 - i$. Whereas it makes sense to order the lengths of vectors, since these are scalar quantities and may be so ordered, the vectors themselves have no natural order. To further our argument we now name the length of a vector and introduce a notation whereby it may be manipulated in equations.

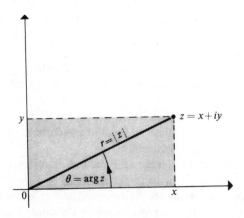

Fig. 4·3   Modulus and argument representation.

DEFINITION 4·7 (modulus of a vector)   The quantity

$$|z| = (x^2 + y^2)^{1/2}$$

is called the *modulus* of the vector $z = x + iy$. It is the length of the line element drawn from the origin to the point (x, y) in the complex plane (see Fig. 4·3).

Example 4·7   If $z = 3 + 4i$, then $|z| = (3^2 + 4^2)^{1/2} = 5$.

Notice that in the special case Im $z = 0$, $|z|$ reduces to the absolute value of a real number since, as always, the positive square root is involved in the definition. The following useful results are easily verified:

$$z\bar{z} = |z|^2; \qquad |z_1 z_2| = |z_1| \cdot |z_2|.$$

If either the upper or lower triangles comprising the parallelogram in Fig. 4·2 (a) are considered, then clearly, when expressed in terms of the modulus, the Euclidean theorem 'the sum of the lengths of any two sides of a triangle exceeds the length of the third side' becomes the following inequality relating moduli:

$$|z_1| + |z_2| \geq |z_1 + z_2|. \tag{4·11}$$

Equality will occur only when z_1 and z_2 are collinear. For obvious reasons Eqn (4·11) is called the *triangle inequality*, and it has already been encountered in simple form when we discussed the absolute value of the sum of two real numbers. An analytic proof of result (4·11) is set as a problem at the end of the chapter.

Another useful inequality relating the moduli of the complex numbers z_1 and z_2 is

$$|z_1 + z_2| \geq ||z_1| - |z_2||, \tag{4·12}$$

where again equality occurs only when z_1 and z_2 are collinear. The proof of this is also left to the reader as a problem.

Example 4·8   If $z_1 = 3 + 4i$ and $z_2 = 4 + 3i$, then $z_1 + z_2 = 7 + 7i$. Hence $|z_1| = (3^2 + 4^2)^{1/2} = 5$, $|z_2| = (4^2 + 3^2)^{1/2} = 5$, and $|z_1 + z_2| = (7^2 + 7^2)^{1/2} = \sqrt{98}$, so that $|z_1| + |z_2| = 10$ and $||z_1| - |z_2|| = 0$. We have thus verified inequalities (4·11) and (4·12) in this special case, for they demand that for any z_1 and z_2

$$||z_1| - |z_2|| \leq |z_1 + z_2| \leq |z_1| + |z_2|$$

which in this case corresponds to the valid inequality

$$0 < \sqrt{98} < 10.$$

4·4   Modulus–argument form of complex numbers

Referring again to Fig. 4·3, we see that the complex number z need not be specified in the standard form for it may equally well be specified by giving both the value of $|z|$ and the angle θ which, by convention, is always measured positively in an anti-clockwise direction from the x-axis to the line of the vector z. The angle θ is the *argument* of z and we shall write $\theta = \arg z$. The argument of z is indeterminate with respect to multiples of 2π, because angles θ and $\theta + 2k\pi$, where k is any integer, will give rise to the same line on Fig. 4·3. Later we shall see that this indeterminacy in θ plays an important role in the determination of the roots of complex numbers. When $\theta = \arg z$ is restricted to the interval $-\pi < \theta \leq \pi$, it will be termed the *principal value* of the argument.

If we define the real number r by the equation $r = |z|$, and still set $\theta = \arg z$, then the ordered number pair (r, θ) describes the *polar coordinates* of the point z in Fig. 4·3. That is, the radial distance of a point from the origin together with its bearing measured from a fixed line through the origin. The relationship between the Cartesian coordinates (x, y) and the polar coordinates (r, θ) of the same complex number z is immediate, since from Fig. 4·3 we have

$$x = r \cos \theta \qquad y = r \sin \theta \tag{4·13}$$

or, equivalently,

$$r = (x^2 + y^2)^{1/2} \qquad \cos \theta = \frac{x}{(x^2 + y^2)^{1/2}} \qquad \sin \theta = \frac{y}{(x^2 + y^2)^{1/2}} \tag{4·14}$$

Thus the complex number, or vector, $z = x + iy$ may also be written in the *modulus–argument* form

$$z = r(\cos \theta + i \sin \theta). \tag{4·15}$$

Because $\arg z$ is indeterminate up to an angle $2k\pi$, we must phrase our definition of equality between two complex numbers carefully when it is to refer to complex numbers expressed in modulus–argument form.

DEFINITION 4·8   The two numbers $z_1 = r(\cos \theta + i \sin \theta)$ and $z_2 = \rho(\cos \phi + i \sin \phi)$ expressed in modulus–argument form will be said to be equal if, and only if, $r = \rho$ and $\theta = \phi + 2k\pi$.

Equations (4·13) and (4·14) enable immediate interchange between the modulus–argument and the real–imaginary forms of z, as the following examples indicate.

Example 4·9

(a) Express $z = -4\sqrt{3} + 4i$ in modulus–argument form;

(b) Express $z = 2 + 5i$ in modulus–argument form;

(c) If $|z| = 3$ and arg $z = -\pi/10$, express z in real–imaginary form.

Solution (a)   From Eqn (4·14), $r = |z| = [(-4\sqrt{3})^2 + 4^2]^{1/2} = 8$, whilst $\cos \theta = -(4\sqrt{3})/8 = -(\sqrt{3})/2$ and $\sin \theta = 4/8 = \frac{1}{2}$, from which we deduce that the principal value of θ must lie in the second quadrant with $\theta = \arg z = 5\pi/6$. Hence, in modulus–argument form

$$z = 8\left(\cos \frac{5\pi}{6} + i \sin \frac{5\pi}{6}\right).$$

Notice that although we could have written $\theta = \arg z = \arctan (-1/\sqrt{3})$, it would not then have been clear in which quadrant θ must lie, and, consequently, we shall always specify $\sin \theta$ and $\cos \theta$ separately.

Solution (b)   Again from Eqn (4·14), $r = |z| = (2^2 + 5^2)^{1/2} = \sqrt{29}$, whilst this time $\cos \theta = 2/\sqrt{29}$ and $\sin \theta = 5/\sqrt{29}$, from which we deduce that the principal value of θ must lie in the first quadrant with $\theta = \arg z = 1\cdot1903$ rad. Hence, in modulus–argument form

$$z = \sqrt{29}(\cos 1\cdot1903 + i \sin 1\cdot1903).$$

Solution (c)   The result is immediate, since Eqn (4·15) gives

$$z = 3\left\{\cos \left(-\frac{\pi}{10}\right) + i \sin \left(-\frac{\pi}{10}\right)\right\}$$

$$= 2\cdot8533 - 0\cdot9270i.$$

We now examine the consequences of multiplication and division for complex numbers expressed in modulus–argument form. Let z_1 and z_2 be the two complex numbers:

$$z_1 = r_1(\cos \theta_1 + i \sin \theta_1) \quad \text{and} \quad z_2 = r_2(\cos \theta_2 + i \sin \theta_2). \quad (4\cdot16)$$

Then by direct multiplication we find that

$$z_1z_2 = r_1r_2[(\cos \theta_1 \cos \theta_2 - \sin \theta_1 \sin \theta_2)$$
$$+ i(\sin \theta_1 \cos \theta_2 + \cos \theta_1 \sin \theta_2)],$$

and using the trigonometric identities for $\cos (\theta_1 + \theta_2)$ and $\sin (\theta_1 + \theta_2)$ this may be written as

$$z_1z_2 = r_1r_2[\cos (\theta_1 + \theta_2) + i \sin (\theta_1 + \theta_2)]. \quad (4\cdot17)$$

We have thus proved that the result of the product z_1z_2 is a complex number with modulus $|z_1z_2| = r_1r_2$ and argument $\arg (z_1z_2) = \theta_1 + \theta_2 = \arg z_1 + \arg z_2$. Thus the result of multiplying two complex numbers is to produce a complex number whose modulus is the product of the two separate moduli and whose argument is the sum of the two separate arguments (see Fig. 4·4). A special case results if we write

$$i = \cos \tfrac{1}{2}\pi + i \sin \tfrac{1}{2}\pi. \tag{4·18}$$

It follows that in the z-plane, multiplication by i corresponds geometrically to an anti-clockwise rotation through $\tfrac{1}{2}\pi$ without any change of size. To illustrate this, the vectors iz_1 and iz_2 have been added to Fig. 4·4.

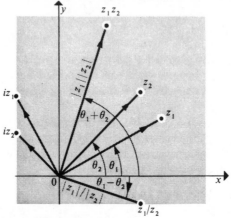

Fig. 4·4 Multiplication and division; z_1z_2, z_1/z_2.

By repeated application of Eqn (4·17) it is easily proved that if $z_m = r_m(\cos \theta_m + i \sin \theta_m)$ for $m = 1, 2, \ldots, n$, then

$$z_1z_2 \cdots z_n = r_1r_2 \cdots r_n[\cos (\theta_1 + \theta_2 + \cdots + \theta_n) \\ + i \sin (\theta_1 + \theta_2 + \cdots + \theta_n)]. \tag{4·19}$$

An argument essentially similar to that which gave rise to Eqn (4·17), but this time using the trigonometric identities for $\cos (\theta_1 - \theta_2)$ and $\sin (\theta_1 - \theta_2)$, establishes that whenever $z_2 \neq 0$, then with the same notation we have

$$\frac{z_1}{z_2} = \frac{r_1}{r_2} [\cos (\theta_1 - \theta_2) + i \sin (\theta_1 - \theta_2)]. \tag{4·20}$$

Obviously $|z_1/z_2| = r_1/r_2 = |z_1|/|z_2|$ and $\arg (z_1/z_2) = \theta_1 - \theta_2 = \arg z_1 - \arg z_2$. Expressed in words, this says that the result of dividing two complex numbers is to produce a complex number whose modulus is the quotient of the separate moduli and whose argument is the difference of the two separate arguments.

A most important special case of Eqn (4·19) occurs when all the $z_1, z_2, \ldots, z_n$ are equal to the same complex number $z = r(\cos \theta + i \sin \theta)$, say. The result then becomes

$$z^n = r^n(\cos n\theta + i \sin n\theta).$$

Substituting for z and cancelling a real factor r^n, we obtain the following important theorem.

THEOREM 4·2 (de Moivre's Theorem)

$(\cos \theta + i \sin \theta)^n = \cos n\theta + i \sin n\theta.$

A more subtle argument would have yielded the fact that this remarkable result is true for all real values of n, and not just for the integral values utilized in our proof. This will be undertaken later when the complex exponential function has been discussed.

Theorem 4·2 provides a simple method by which certain forms of trigonometric identity may be established. One typical example is enough to illustrate this.

Example 4·10   Let us relate $\sin 4\theta$ and $\cos 4\theta$ to sums of powers of $\sin \theta$ and $\cos \theta$. Set $n = 4$ in Theorem 4·2 and expand the left-hand side by the binomial theorem, using the fact that $i^2 = -1$, $i^3 = -i$, $i^4 = 1$, etc., to obtain

$$\cos^4 \theta + 4i \cos^3 \theta \sin \theta - 6 \cos^2 \theta \sin^2 \theta - 4i \cos \theta \sin^3 \theta + \sin^4 \theta$$
$$= \cos 4\theta + i \sin 4\theta.$$

Then, recalling that equality of complex numbers means equality of their real and imaginary parts considered separately, we have the two results:

equality of real parts

$$\cos^4 \theta - 6 \cos^2 \theta \sin^2 \theta + \sin^4 \theta = \cos 4\theta,$$

and

equality of imaginary parts

$$4(\cos^3 \theta \sin \theta - \cos \theta \sin^3 \theta) = \sin 4\theta.$$

These are the desired results. It is characteristic of complex numbers that any single complex equality implies two real equalities, and even if only one is sought the other will be generated automatically. The same method works for any positive integral value of n when it will connect $\sin n\theta$ and $\cos n\theta$ with sums of powers of $\sin \theta$ and $\cos \theta$.

We shall return to this idea in connection with the exponential function, and show that it is possible to use de Moivre's theorem to express $\sin^n \theta$ and $\cos^n \theta$ in terms of sums involving $\sin r\theta$ and $\cos r\theta$.

Sometimes Theorem 4·2 can be used to reduce the labour of computation as now shown.

Example 4·11   We shall evaluate z^{10} where $z = 1 + i$. Rather than making repeated multiplications, or applying the binomial theorem, we write z in modulus–argument form as $z = \sqrt{2}(\cos \pi/4 + i \sin \pi/4)$, when we have $z^{10} = (\sqrt{2})^{10}(\cos \pi/4 + i \sin \pi/4)^{10}$. By de Moivre's theorem this becomes

$$z^{10} = 2^5 \left(\cos \frac{5\pi}{2} + i \sin \frac{5\pi}{2} \right) = 32i.$$

4·5   Roots of complex numbers

When performing algebra on real numbers the idea of the root of a number plays a fundamental part. The same is true when manipulating complex numbers, and we now discuss the general ideas involved in determining their roots.

Let p/q be any rational number, where p and q are integers with q supposed positive. We shall assume that p and q have no common factor.

DEFINITION 4·9   We define $z^{p/q}$ by saying that:

$$w = z^{p/q} \Leftrightarrow w^q = z^p.$$

Let

$$w = \rho(\cos \phi + i \sin \phi) \quad \text{and} \quad z = r(\cos \theta + i \sin \theta). \tag{4·21}$$

Then from Definition 4·9 and de Moivre's theorem we have

$$\rho^q(\cos q\phi + i \sin q\phi) = r^p(\cos p\theta + i \sin p\theta). \tag{4·22}$$

Now from Definition 4·8 it follows that

$$\rho^q = r^p \quad \text{and} \quad q\phi = p\theta + 2k\pi, \tag{4·23}$$

and so

$$\rho = r^{p/q} \quad \text{and} \quad \phi = \frac{p\theta + 2k\pi}{q}. \tag{4·24}$$

The expressions $w = z^{p/q}$ thus have the general form

$$w = z^{p/q} = r^{p/q} \left[\cos \left(\frac{p\theta + 2k\pi}{q} \right) + i \sin \left(\frac{p\theta + 2k\pi}{q} \right) \right]$$

$$\text{with } k \text{ an integer.} \tag{4·25}$$

It is easily seen that only q different values $w_0, w_1, w_2, \ldots, w_{q-1}$ of w will result from Eqn (4·25) as the integer k increases through successive integral values. It is usual to give k the q successive values $k = 0, 1, 2, \ldots, q - 1$. If k is allowed to increase beyond the value $q - 1$, then the numbers $w_0, w_1, \ldots, w_{q-1}$ will simply be generated again because of the periodicity properties of the sine and cosine functions.

Example 4·12   We illustrate the use of Eqn (4·25) by determining the n numbers w satisfying the equation $w = (1)^{1/n}$. For obvious reasons these are called the *nth roots of unity*. Comparing this equation with the general expression $w = z^{p/q}$ that has just been discussed we see that we must make the identifications $z = 1$, $p = 1$, and $q = n$. To proceed further we must write the number unity in its modulus–argument form

$$1 = 1 . (\cos 0 + i \sin 0),$$

so that comparing this with z in Eqn (4·21) we see that the further identifications $r = 1$ and $\theta = 0$ must be made. Substitution of these quantities into Eqn (4·25) then gives the result

$$w_k = \cos \frac{2k\pi}{n} + i \sin \frac{2k\pi}{n} \qquad \text{with } k = 0, 1, 2, \ldots, n - 1.$$

The result of this calculation with $n = 5$, for example, is to generate the fifth roots of unity. In Fig. 4·5 these roots are plotted as the numbers w_0, w_1, . . ., w_4 in the complex plane. They are uniformly distributed around the unit circle centred on the origin. By making use of the vector properties of complex numbers we shall usually represent this circle by the convenient notation $|z| = 1$. (Why?)

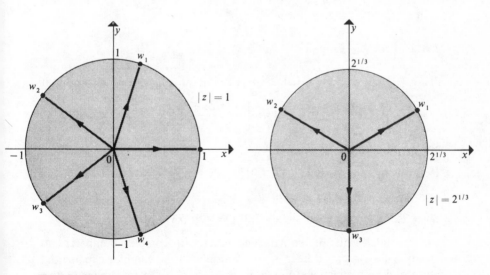

Fig. 4·5 Fifth roots of unity.   **Fig. 4·6** Roots of $\omega = (1 + i)^{2/3}$.

Example 4·13 As a slightly more general example we now determine $z^{2/3}$, when $z = 1 + i$. In this case $p = 2$, $q = 3$, and in modulus–argument form, $z = \sqrt{2}(\cos \pi/4 + i \sin \pi/4)$ showing that $r = \sqrt{2}$ and $\theta = \pi/4$. Substitution into Eqn (4·25) gives

$$w = 2^{1/3} \left[\cos \left(\frac{1 + 4k}{6} \right) \pi + i \sin \left(\frac{1 + 4k}{6} \right) \pi \right] \qquad \text{with } k = 0, 1, 2.$$

The three roots w_0, w_1, and w_2 are thus:

$$(k = 0): w_0 = 2^{1/3} \left(\cos \frac{\pi}{6} + i \sin \frac{\pi}{6} \right) = 2^{1/3} \left(\frac{\sqrt{3}}{2} + \frac{i}{2} \right),$$

$$(k = 1): w_1 = 2^{1/3} \left(\cos \frac{5\pi}{6} + i \sin \frac{5\pi}{6} \right) = 2^{1/3} \left(-\frac{\sqrt{3}}{2} + \frac{i}{2} \right),$$

$$(k = 2): w_2 = 2^{1/3} \left(\cos \frac{3\pi}{2} + i \sin \frac{3\pi}{2} \right) = -2^{1/3} i.$$

These are plotted in the complex plane in Fig. 4·6, where they are seen to be uniformly distributed around the circle $|z| = 2^{1/3}$.

Example 4·14   As a final example let us find the roots of the equation

$$w = i^{-1/3}.$$

In terms of the notation of Eqns (4·21) and (4·25), and recalling that we have agreed always to take q as positive, we have $p = -1$, $q = 3$, and $z = i$. Now in modulus–argument form

$$i = 1 \cdot \left(\cos \frac{\pi}{2} + i \sin \frac{\pi}{2} \right),$$

so that $r = 1$ and $\theta = \pi/2$. Hence, substituting into Eqn (4·25), we find that

$$w = \cos \left[\frac{(-\pi/2) + 2k\pi}{3} \right] + i \sin \left[\frac{(-\pi/2) + 2k\pi}{3} \right] \qquad \text{with } k = 0, 1, 2.$$

Hence the three roots w_0, w_1, and w_2 are:

$(k = 0): w_0 = (\cos \pi/6 - i \sin \pi/6) = \frac{1}{2}(\sqrt{3} - i),$

$(k = 1): w_1 = (\cos \pi/2 + i \sin \pi/2) = i,$

$(k = 2): w_2 = (\cos 7\pi/6 + i \sin 7\pi/6) = -\frac{1}{2}(\sqrt{3} + i).$

This completes our preliminary encounter with complex numbers, and our study will be resumed later in connection with the complex exponential function and with functions of a complex variable. The remainder of this chapter is devoted to developing the foundations of our study of general vectors.

4·6   Introduction to space vectors

It is clear that any set of vector quantities that do not all lie in a plane cannot be represented vectorially in the form of complex numbers. For example, even the vectors describing the velocity of a vehicle as it is driven at constant speed past fixed points on a winding hill could not be so represented. Pairwise these velocity vectors define planes, and so could be represented by complex numbers in those planes, though different pairs of vectors would define different planes, thereby making any general representation impossible in terms of complex numbers. The trouble here is not hard to find. It is that complex numbers just happen to be capable of representation as planar vectors with their own appropriate descriptive language, and they were not developed with general vector representation in mind. In short, they are complex numbers first and vectors second; not the other way around.

To overcome this limitation and to be able to describe arbitrary vector quantities we must preserve the idea of a vector as a directed length, but re-think its description. This is best achieved using a diagram, so consider Fig. 4·7 which depicts the mutually perpendicular Cartesian axes $O\{x, y, z\}$ with origin O. In more mathematical terms we describe these axes as being mutually *orthogonal*. This is a technical term that in a geometrical context has the same meaning as perpendicular, though it is often used in a wider sense, when the word perpendicular would be inappropriate. Henceforth we shall almost always use the term orthogonal.

The manner of identification of the x, y, and z coordinate axes is not

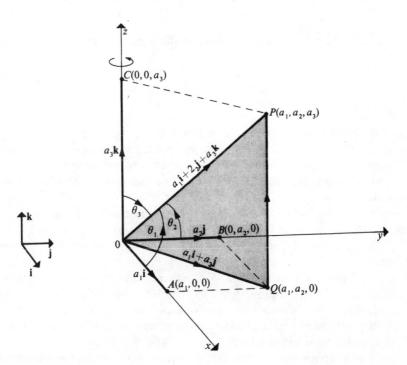

Fig. 4·7   Right-handed Cartesian axes.

arbitrary, but is made in such a manner that they form a right-handed system of axes. By this we mean that having assigned axes for the variables x and y, together with the directions in which they increase positively, the direction of positive z is then chosen to be that in which a right-handed screw would advance were it aligned with the third axis and rotated in the sense x to y. This sense of rotation is indicated in Fig. 4·7 by means of a directed spiral about the z-axis. In the diagram the y- and z-axes are supposed to lie in the plane of the paper with the x-axis pointing out of the paper towards the viewer. Later we shall refer to this right-handed property in connection with axes which are not orthogonal, when right-handedness is still to be interpreted in exactly the same sense as above.

This right-handed property of the system of axes is shared by each pair of axes in turn, provided the senses of rotation are appropriately defined. The following table describes the convention that is always adopted.

Table 4·1   Right-handed axes

Rotate From	To	R-H screw advances in direction of positive
x	y	z
y	z	x
z	x	y

The table can easily be remembered in the concise form

$$x \quad y \quad z$$
$$y \quad z \quad \cdot x$$
$$z \quad x \quad y$$

where the entry in any row is obtained from the entry in the row above by transferring the first letter of that entry to the last position. These entries are called *cyclic permutations* of the letters x, y, and z, and further cyclic permutations will simply regenerate the table. These rules describe the right-handed symmetry of the $O\{x, y, z\}$ axes. If any two letters in an entry are interchanged, then by the same rule, the negative direction of the third axis is defined. Hence the set of letters $y\,x\,z$ are to be interpreted 'rotate from y to x to make a right-handed screw aligned with the z-axis advance in the direction of negative z'.

If in the above argument a right-handed screw motion had been replaced by a left-handed screw motion, then a left-handed system of axes would have resulted. Although a left-handed system of axes is in all respects equivalent to a right-handed system for the purposes of vector representation, it is customary to work with right-handed systems.

Let P be the point with coordinates $x = a_1$, $y = a_2$, and $z = a_3$ illustrated in Fig. 4·7. We shall denote it by the more concise notation (a_1, a_2, a_3) where

the first, second, and third entries in this ordered number triple represent the
x, y, and z coordinates, respectively. Then from the point of view of coordinate
geometry it is the point P that is of interest, whereas from the point of view
of vectors it is the directed line element from O to P that is of interest. To
signify that it is the vector quantity that interests us here we shall write $\overline{OP}$.
Notice that by this convention the vector $\overline{PO}$ is the directed line from P to O
and is opposite in sense to $\overline{OP}$. In future we will denote the length of the vector
$\overline{OP}$ by $|\overline{OP}|$, which is a scalar, and by definition this length will always be
positive.

In Fig. 4·7 the lengths OA $= a_1$, OB $= a_2$, and OC $= a_3$ are called the
orthogonal projections of $\overline{OP}$ onto the x-, y-, and z-axes, and a simple application of Pythagoras' theorem gives the result

$$|\overline{OP}|^2 = (OA)^2 + (OB)^2 + (OC)^2$$

or,

$$|\overline{OP}|^2 = a_1^2 + a_2^2 + a_3^2.$$

Dividing by $|\overline{OP}|^2$ this becomes

$$1 = \left(\frac{a_1}{|\overline{OP}|}\right)^2 + \left(\frac{a_2}{|\overline{OP}|}\right)^2 + \left(\frac{a_3}{|\overline{OP}|}\right)^2,$$

which can then be rewritten in terms of the angles θ_1, θ_2, θ_3 as

$$1 = \cos^2 \theta_1 + \cos^2 \theta_2 + \cos^2 \theta_3. \tag{4·26}$$

If the numbers l, m, and n are defined by the relations

$$l = \cos \theta_1, \qquad m = \cos \theta_2, \qquad n = \cos \theta_3, \tag{4·27}$$

then Eqn (4·26) becomes

$$1 = l^2 + m^2 + n^2. \tag{4·28}$$

For obvious reasons l, m, and n are called the *direction cosines* of $\overline{OP}$ with
respect to the axes O$\{x, y, z\}$ and it is often convenient to write them in the
form of an ordered number triple as $\{l, m, n\}$. The angles θ_1, θ_2, and θ_3 are
indeterminate to within a multiple of 2π and, by convention, they will always
be taken to lie in the interval $[0, \pi]$.

Consider the direction cosines l, m, n as defining a point P' in space with
coordinates $x = l$, $y = m$, and $z = n$, then, by Pythagoras' theorem and
Eqn (4·28), the vector $\overline{OP'}$ must have unit length. The direction and sense of
$\overline{OP'}$ are the same as those of $\overline{OP}$; only the lengths are different. Vectors of
unit length in given directions prove to be extremely useful in vector analysis
so they are appropriately called *unit vectors*.

Now by definition, the direction cosines l, m, n are proportional to the coordinates a_1, a_2, a_3 of the point P and consequently the numbers a_1, a_2, and a_3 are often called the *direction ratios* of OP. To convert direction ratios to direction cosines it is necessary to normalize them by dividing by the square root of the sum of the squares of the direction ratios. This is, of course, equivalent to division by the quantity we have agreed to denote by $|\overline{OP}|$.

Example 4·15 Find the direction ratios, the direction cosines and the angles θ_1, θ_2, and θ_3 of the vector $\overline{OP}$, where P is the point $(1, -2, 4)$.

Solution The direction ratios are 1, -2, 4, and $|\overline{OP}|$, which is the square root of the sum of the squares of the direction ratios, is

$$|\overline{OP}| = (1^2 + (-2)^2 + 4^2)^{1/2} = \sqrt{21}.$$

Hence the direction cosines of $\overline{OP}$ are $l = 1/\sqrt{21}$, $m = -2/\sqrt{21}$, and $n = 4/\sqrt{21}$, from which the angles θ_1, θ_2, and θ_3 are seen to be $1·351$, $2·022$, and $0·509$ radians, respectively. Unless otherwise stated we shall always express angles in terms of radians, as here.

Example 4·16 Determine the angles of inclination θ_1, θ_2, and θ_3 of a vector to the x-, y-, and z-axes, respectively, given that its direction cosines are:

(a) $\{\frac{1}{2}, -\sqrt{3}/2, 0\}$,

(b) $\{\frac{1}{2}, \frac{1}{4}, \sqrt{11}/4\}$.

Solution (a) Here $l = \cos \theta_1 = 1/2$, $m = \cos \theta_2 = -\sqrt{3}/2$, $n = \cos \theta_3 = 0$, so that $\theta_1 = \pi/3$, $\theta_2 = 5\pi/6$, and $\theta_3 = \pi/2$. Hence in this case the vector lies entirely in the (x, y)-plane.

Solution (b) In this case, $l = \cos \theta_1 = 1/2$, $m = \cos \theta_2 = 1/4$, $n = \cos \theta_3 = \sqrt{11}/4$, so that $\theta_1 = \pi/3$, $\theta_2 = 1·318$, and $\theta_3 = 0·593$.

Example 4·17 If a vector has direction cosines $\{\frac{1}{2}, m, \frac{1}{2}\}$ deduce the possible values of m. If, in addition, it is stated that the vector makes an obtuse angle θ_2 with the y-axis determine the value of θ_2.

Solution We use Eqn (4·28), setting $l = \frac{1}{2}$ and $n = \frac{1}{2}$ to obtain

$$(\tfrac{1}{2})^2 + m^2 + (\tfrac{1}{2})^2 = 1.$$

Whence, $m^2 = 1/2$ or $m = \pm 1/\sqrt{2}$. These values of m correspond to $\theta_2 = \pi/4$ for $m = 1/\sqrt{2}$, and to $\theta_2 = 3\pi/4$ for $m = -1/\sqrt{2}$. As the angle θ_2 is required to be obtuse we must select $\theta_2 = 3\pi/4$.

The idea of a fixed origin is fundamental to coordinate geometry though it proves to be rather too restrictive in vector analysis. This is because it is only the magnitude, direction, and sense of a vector that usually matter, and not the choice of origin and coordinate system in which the vector is represented. For example, when specifying a wind velocity it is normally sufficient to say 20 ft/s due East, without identifying the particular points in space at which the air has this velocity.

In vector work this ambiguity as to the location of a vector in space is allowed by considering as equivalent, any two vectors that may be repre-

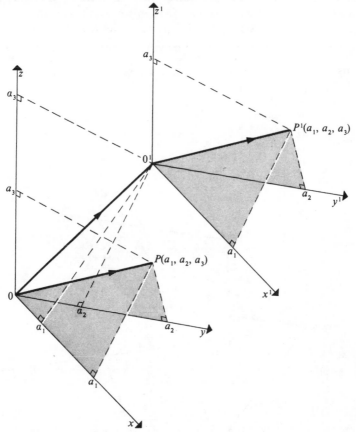

Fig. 4·8 Translation of axes without rotation.

sented by directed line elements of equal length which are parallel, and have the same sense. In Fig. 4·8 we have depicted two vectors $\overline{OP}$ and $\overline{O'P'}$ that are equivalent in the sense just defined. Another way to define this equivalence is to require that when the axes $O\{x, y, z\}$ are translated, without rotation, to the position $O'\{x', y', z'\}$, the coordinates of P' with respect to the axes through O' are the same as those of P with respect to the axes through O. That is, if P is the point (a_1, a_2, a_3) in the system of axes $O\{x, y, z\}$, then P' is the point (a_1, a_2, a_3) in the system of axes $O'\{x', y', z'\}$. Do not get confused by this. If O′ is the point (α_1, α_2, α_3) with respect to O$\{x, y, z\}$, then coordinates in the unprimed system are related to those in the primed system by the equations $x = \alpha_1 + x'$, $y = \alpha_2 + y'$, and $z = \alpha_3 + z'$.

This freedom to translate vectors now enables us to give direction cosines to any vector in space and not just to those having their base at O. Suppose, for example, that we require the length and direction cosines of the vector $\overline{AB}$, where A is the point (a_1, a_2, a_3) and B is the point (b_1, b_2, b_3) when expressed relative to some set of axes O$\{x, y, z\}$. Then we see at once that the lengths of the projections of $\overline{AB}$ on the x, y, and z axes are ($b_1 - a_1$), ($b_2 - a_2$), and ($b_3 - a_3$), respectively. Accordingly, by translating the vector $\overline{AB}$ until A in its new position A′ coincides with O, we see that the tip B in its new

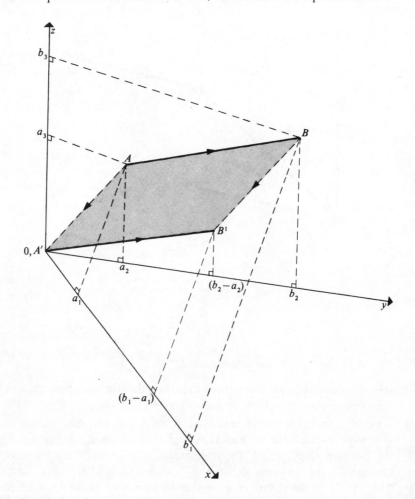

Fig. 4·9 Translation of a vector.

position B' must be the point $((b_1 - a_1), (b_2 - a_2), (b_3 - a_3))$ (see Fig. 4·9). Hence $|AB|$, that is the length of AB, is

$$|AB| = [(b_1 - a_1)^2 + (b_2 - a_2)^2 + (b_3 - a_3)^2]^{1/2}. \tag{4·29}$$

The direction cosines of AB then follow as before and are

$$l = \frac{b_1 - a_1}{|AB|}, \qquad m = \frac{b_2 - a_2}{|AB|}, \qquad n = \frac{b_3 - a_3}{|AB|}. \tag{4·30}$$

Example 4·18 Find $|AB|$ and the direction cosines of the vector AB, if A has coordinates $(1, 2, 3)$ and B the coordinates $(4, 3, 6)$.

Solution From Eqn (4·29) we see that $|AB| = [(4 - 1)^2 + (3 - 2)^2 + (6 - 3)^2]^{1/2} = \sqrt{19}$, whilst from Eqn (4·30) it follows that $l = 3/\sqrt{19}$, $m = 1/\sqrt{19}$, and $n = 3/\sqrt{19}$.

It is now convenient to introduce a triad of unit vectors, denoted by **i**, **j**, and **k**, that are parallel to and are directed in the positive senses of the *x*-, *y*-, and *z*-axes, respectively. Here we remind the reader that these are called unit vectors because they are each of unit length on the *x*-, *y*-, and *z*-length scales. Notice that the term right-handed that was applied to the system of axes O{*x, y, z*} also applies to the triad of vectors **i**, **j**, **k** when taken in this order. We shall use this idea again later.

An arbitrary vector in any one of the **i**, **j**, or **k** directions may then be obtained by scaling the length of the appropriate unit vector by a multiplication factor μ. Thus a vector three times the size of the unit vector **i** will be written 3**i**, whilst a vector twice the size of the unit vector **k**, but oppositely directed, will be written $-2\mathbf{k}$.

Returning to Fig. 4·7 we see that in terms of **i**, **j**, and **k**, the vectors OA, OB, and OC may be written as

$$OA = a_1\mathbf{i}, \qquad OB = a_2\mathbf{j}, \qquad OC = a_3\mathbf{k}.$$

From our ideas of vector addition in a plane the vector OQ lying in the (x, y)-plane is $OQ = OA + AQ$ or, because vectors may be translated, $OQ = OA + AB$. Now in terms of our unit vector notation this may be written $OQ = a_1\mathbf{i} + a_2\mathbf{j}$. Turning attention to the plane containing points O, Q, and P, we see that by the same argument $OP = OQ + QP$. Again, because vectors may be translated, $QP = OC$ so that finally, on substituting for OQ and QP in the equation $OP = OQ + QP$, we obtain

$$OP = a_1\mathbf{i} + a_2\mathbf{j} + a_3\mathbf{k}. \tag{4·31}$$

For ease of notation, arbitrary vectors, like unit vectors, will usually be denoted by a single symbol such as α, **a**, or **r**. Thus a general point P in space with coordinates (x, y, z) will often be written

$$\mathbf{r} = x\mathbf{i} + y\mathbf{j} + z\mathbf{k}. \tag{4.32}$$

The almost universally accepted convention which we adopt here is to denote vector quantities by bold face type and scalar quantities by italic type.

Because a vector such as that in Eqn (4·32) identifies a point P in space it is called a *position vector*. In the vector representation Eqn (4·31) the numbers a_1, a_2, and a_3 are called the *components* of $\overline{OP}$.

Two vectors will only be said to be *equal* if, when written in the form of Eqn (4·31), their corresponding components are equal. The vector **a** $= a_1\mathbf{i} + a_2\mathbf{j} + a_3\mathbf{k}$ will be said to be a scalar multiple λ of vector **b** $= b_1\mathbf{i} + b_2\mathbf{j} + b_3\mathbf{k}$, and we will write $\mathbf{a} = \lambda\mathbf{b}$ if, and only if, $a_1 = \lambda b_1$, $a_2 = \lambda b_2$, and $a_3 = \lambda b_3$. In the special case $\lambda = -1$ we have $\mathbf{a} = -\mathbf{b}$, showing that $|\mathbf{a}| = |\mathbf{b}|$, but that the senses of **a** and **b** are opposite. Thus in Fig. 4·7 we have $\overline{OP} = -\overline{PO}$.

The *zero* or *null* vector $\mathbf{0}$ is the vector whose three components are each identically zero. It is often denoted by 0 instead of **0**, since confusion is unlikely to arise on account of this simplification of the notation. Following on from our first ideas of vectors, and in accordance with the derivation of Eqn (4·31), we now define the operations of addition and subtraction of vectors.

DEFINITION 4·10 Let **a** and **b** be arbitrary vectors with components (a_1, a_2, a_3) and (b_1, b_2, b_3), respectively, so that they may be written $\mathbf{a} = a_1\mathbf{i} + a_2\mathbf{j} + a_3\mathbf{k}$ and $\mathbf{b} = b_1\mathbf{i} + b_2\mathbf{j} + b_3\mathbf{k}$. Then we define the *sum* $\mathbf{a} + \mathbf{b}$ of the two vectors **a** and **b** to be the vector $(a_1 + b_1)\mathbf{i} + (a_2 + b_2)\mathbf{j} + (a_3 + b_3)\mathbf{k}$. The *difference* $\mathbf{a} - \mathbf{b}$ of the two vectors **a** and **b** is defined to be the vector $(a_1 - b_1)\mathbf{i} + (a_2 - b_2)\mathbf{j} + (a_3 - b_3)\mathbf{k}$.

Because real numbers are commutative with respect to addition, it follows directly from this definition that the operation of vector addition is commutative. That is we have $\mathbf{a} + \mathbf{b} = \mathbf{b} + \mathbf{a}$ for all vectors **a** and **b**. When the subtraction operation is considered the properties of real numbers imply the result $\mathbf{a} - \mathbf{b} = -(\mathbf{b} - \mathbf{a})$ for all vectors **a** and **b**.

Example 4·19 If $\mathbf{a} = \mathbf{i} + \mathbf{j} + 2\mathbf{k}$ and $\mathbf{b} = 3\mathbf{i} - 3\mathbf{j} + \mathbf{k}$, then $\mathbf{a} + \mathbf{b}$ $= (1 + 3)\mathbf{i} + (1 - 3)\mathbf{j} + (2 + 1)\mathbf{k}$, showing that $\mathbf{a} + \mathbf{b} = 4\mathbf{i} - 2\mathbf{j} + 3\mathbf{k}$. Reversal of the order of the sum followed by the same argument proves the commutative property $\mathbf{a} + \mathbf{b} = \mathbf{b} + \mathbf{a}$ for these particular vectors. In the case of subtraction we have $\mathbf{a} - \mathbf{b} = (1 - 3)\mathbf{i} + (1 - (-3))\mathbf{j} + (2 - 1)\mathbf{k}$, showing that $\mathbf{a} - \mathbf{b} = -2\mathbf{i} + 4\mathbf{j} + \mathbf{k}$. It is easily established that $\mathbf{a} - \mathbf{b}$ $= -(\mathbf{b} - \mathbf{a})$.

Although these particular results could be illustrated diagrammatically, the vector triangles involved would look essentially the same as those used earlier in connection with addition and subtraction of complex numbers and would be arrived at by the same reasoning. Rather than illustrate this specific case, we present in Fig. 4·10 the results of addition and subtraction of arbitrary vectors **a** and **b**. Because a geometrical projection method is necessary to illustrate three-dimensional problems on a sheet of paper, such diagrams are much less useful as a tool than was the case in a plane. Accordingly, we shall usually concentrate on an analytical approach to vectors, using diagrams

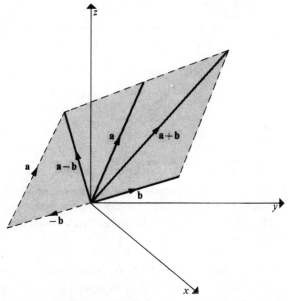

Fig. 4·10   Addition and subtraction of vectors.

only when they seem likely to be helpful.

Two terms worthy of note that are applied to vectors are the names parallel and anti-parallel. Two vectors will be said to be *parallel* when their lines of action are parallel and their senses are the same. Conversely, two vectors will be said to be *anti-parallel* when their lines of action are parallel but their senses are opposite. Thus if **a** is a vector and μ is a scalar, the vectors **a** and μ**a** are parallel if $\mu > 0$ and are anti-parallel if $\mu < 0$. It follows that two vectors will be parallel if their corresponding direction cosines are equal and they will be anti-parallel if their corresponding direction cosines are equal in magnitude but opposite in sign.

Example 4·20   The vectors **a** = **i** + 2**j** − 4**k** and **b** = 3**i** + 6**j** − 12**k** are such that we may write 3**a** = **b**. Since the scalar 3 > 0 it follows that **a** and **b** are parallel. However the vectors **c** = **i** − 3**j** + **k** and **d** = −2**i** + 6**j** − 2**k**

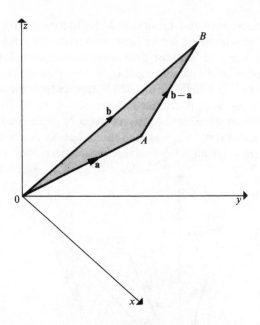

Fig. 4·11 Position vectors defining the vector $\underline{AB}$.

are such that we may write $-2\mathbf{c} = \mathbf{d}$ and, as the scalar $-2 < 0$, it follows that $\mathbf{c}$ and $\mathbf{d}$ are anti-parallel. By the same argument, the two vectors $\mathbf{p} = 3\mathbf{i} - \mathbf{j} + 2\mathbf{k}$ and $\mathbf{q} = 6\mathbf{i} + 2\mathbf{j} + 4\mathbf{k}$ are neither parallel nor anti-parallel, since for no scalar μ is it true that $\mu\mathbf{p} = \mathbf{q}$.

The length of the vector $\underline{AB}$ which we have already denoted by $|\underline{AB}|$ is a useful quantity and, as with complex numbers, is called the modulus of the vector $\underline{AB}$. Its formal definition follows.

DEFINITION 4·11 The *modulus* $|\mathbf{a}|$ of the vector $\mathbf{a} = a_1\mathbf{i} + a_2\mathbf{j} + a_3\mathbf{k}$ is the positive square root

$$|\mathbf{a}| = (a_1{}^2 + a_2{}^2 + a_3{}^2)^{1/2}.$$

It is an immediate consequence of this definition that any vector $\mathbf{r}$ with direction cosines $\{l, m, n\}$ may be written in the form

$$\mathbf{r} = |\mathbf{r}|(l\mathbf{i} + m\mathbf{j} + n\mathbf{k}). \tag{4·33}$$

The proof of this is obvious for by definition, $l|\mathbf{r}|$ is the x-component of $\mathbf{r}$, $m|\mathbf{r}|$ is the y-component, and $n|\mathbf{r}|$ is the z-component. The form of Eqn (4·33) shows that any vector may be expressed as the product of a scalar (its modulus) and a unit vector defining its direction and sense.

When it is necessary to define an arbitrary vector AB in space, this may easily be accomplished by using position vectors **a** and **b** to identify its end points A and B. This is illustrated in Fig. 4·11 from which, by the rules of vector addition, we may write

$$\overline{OA} + \overline{AB} = \overline{OB}$$

or,

$$\overline{AB} = \overline{OB} - \overline{OA} = \mathbf{b} - \mathbf{a}.$$

Examination of this simple but useful result suggests that an accurate name for the vector $\overline{AB}$ would be the 'position vector of B relative to A', since in this role it is $\overline{A}$ that plays the part of the origin. This more exact name is seldom used since the symbol AB is sufficiently clear as it stands.

Example 4·21 Let points A and B be identified by the position vectors $\mathbf{a} = -2\mathbf{i} - 3\mathbf{j} + \mathbf{k}$ and $\mathbf{b} = 3\mathbf{i} - \mathbf{j} + 4\mathbf{k}$, respectively. Find the vector $\overline{AB}$ together with its modulus and direction cosines.

Solution The diagram in Fig. 4·11 can be taken to represent this situation showing that vector $\overline{AB} = \mathbf{b} - \mathbf{a}$. Substituting for the values of **a** and **b**, we find $\overline{AB} = (3\mathbf{i} - \mathbf{j} + 4\mathbf{k}) - (-2\mathbf{i} - 3\mathbf{j} + \mathbf{k})$, whence $\overline{AB} = 5\mathbf{i} + 2\mathbf{j} + 3\mathbf{k}$. Then $|\overline{AB}| = (5^2 + 2^2 + 3^2)^{1/2} = \sqrt{38}$ after which the usual argument establishes that $l = 5/\sqrt{38}$, $m = 2/\sqrt{38}$, and $n = 3/\sqrt{38}$.

By considering the plane containing the vectors **a**, **b**, and **b** − **a** in Fig. 4·11, the arguments that established the triangle inequalities for complex numbers also establish them for arbitrary space vectors. Hence for arbitrary vectors **a** and **b** we have

$$||\mathbf{a}| - |\mathbf{b}|| \le |\mathbf{a} + \mathbf{b}| \le |\mathbf{a}| + |\mathbf{b}|. \tag{4·34}$$

Finally, to close this section, let us find the angle θ between two vectors **a** and **b** with the direction cosines $\{l_1, m_1, n_1\}$ and $\{l_2, m_2, n_2\}$, respectively. When the lines of action of the vectors intersect the angle θ is well defined and, by convention, is always chosen to lie in the interval $[0, \pi]$. If the lines of action of two vectors do not intersect then they are merely translated until they do, when the angle θ is defined as above. It will suffice to consider the angle between two unit vectors directed along **a** and **b** since the length of the vectors will obviously not influence the angle between them. From Eqn (4·33), these unit vectors are seen to be $(l_1\mathbf{i} + m_1\mathbf{j} + n_1\mathbf{k})$ and $(l_2\mathbf{i} + m_2\mathbf{j} + n_2\mathbf{k})$. These are shown in Fig. 4·12. They have their tips P and Q at the respective points (l_1, m_1, n_1) and (l_2, m_2, n_2).

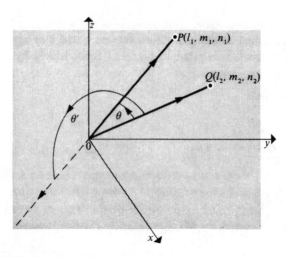

Fig. 4·12 Angle between two lines.

Now, by the cosine rule

$$|PQ|^2 = |OP|^2 + |OQ|^2 - 2|OP| \cdot |OQ| \cos \theta, \qquad (4\cdot35)$$

but $|OP| = |OQ| = 1$, and by Eqn (4·29), $|PQ|^2 = (l_2 - l_1)^2 + (m_2 - m_1)^2 + (n_2 - n_1)^2$, whilst by Eqn (4·28), $l_1{}^2 + m_1{}^2 + n_1{}^2 = l_2{}^2 + m_2{}^2 + n_2{}^2 = 1$. Consequently, substituting into Eqn (4·35) and simplifying, we find the desired result

$$\cos \theta = l_1 l_2 + m_1 m_2 + n_1 n_2. \qquad (4\cdot36)$$

The angle of inclination θ follows directly from this equation. The restriction of the angle between the vectors to the interval $[0, \pi]$ means that in Fig. 4·12, it is the angle θ that is selected, and not the angle θ'.

As a particular case, if

$$l_1 l_2 + m_1 m_2 + n_1 n_2 = 0, \qquad (4\cdot37)$$

then the two vectors **a** and **b** must be orthogonal.

Example 4·22 Find the angle of inclination θ between the vectors $\mathbf{a} = \mathbf{i} + 2\mathbf{j} + 3\mathbf{k}$ and $\mathbf{b} = 2\mathbf{i} - \mathbf{j} - \mathbf{k}$.

Solution Here $|\mathbf{a}| = \sqrt{14}$, $|\mathbf{b}| = \sqrt{6}$, so that the direction cosines $\{l_1, m_1, n_1\}$ of **a** are $l_1 = 1/\sqrt{14}$, $m_1 = 2/\sqrt{14}$, $n_1 = 3/\sqrt{14}$ whilst the direction cosines $\{l_2, m_2, n_2\}$ of **b** are $l_2 = 2/\sqrt{6}$, $m_2 = -1/\sqrt{6}$, $n_2 = -1/\sqrt{6}$. Hence by Eqn (4·36), the angle θ is the solution of the equation

$$\cos \theta = \left(\frac{1}{\sqrt{14}}\right)\left(\frac{2}{\sqrt{6}}\right) + \left(\frac{2}{\sqrt{14}}\right)\left(\frac{-1}{\sqrt{6}}\right) + \left(\frac{3}{\sqrt{14}}\right)\left(\frac{-1}{\sqrt{6}}\right),$$

or

$$\theta = \arccos\left(\frac{-3}{2\sqrt{21}}\right).$$

On account of the restriction of θ to the interval $[0, \pi]$ it finally follows that $\theta = 1{\cdot}905$ rad.

4·7   Scalar and vector products

If $\mathbf{a} = a_1\mathbf{i} + a_2\mathbf{j} + a_3\mathbf{k}$ is an arbitrary vector and λ is a scalar, then we have already defined the product $\lambda\mathbf{a}$ to be the vector $\lambda\mathbf{a} = \lambda a_1\mathbf{i} + \lambda a_2\mathbf{j} + \lambda a_3\mathbf{k}$. Hence the effect of multiplying a vector by a scalar is to magnify the vector without changing its direction. The result of this product is to generate a vector. We must now discuss the multiplication of two vectors.

Here three-dimensional vector algebra differs radically from the vector algebra of complex numbers. With complex numbers there is only one multiplication operation defined, and the product of two complex numbers is always a complex number. In the case of vectors we shall see that two multiplication operations are defined for a pair of vectors. One operation called a scalar product generates a scalar, whereas the other operation called a vector product generates a vector. The operation of division is not defined for vectors.

The scalar product of two vectors is a generalization of the notion of the orthogonal projection of a line element onto another line and is suggested by Eqn (4·36). Its definition follows.

DEFINITION 4·12   The *scalar product* of the two vectors $\mathbf{a} = a_1\mathbf{i} + a_2\mathbf{j} + a_3\mathbf{k}$ and $\mathbf{b} = b_1\mathbf{i} + b_2\mathbf{j} + b_3\mathbf{k}$ is written $\mathbf{a} . \mathbf{b}$ and is defined to be the scalar quantity

$$\mathbf{a} . \mathbf{b} = a_1 b_1 + a_2 b_2 + a_3 b_3.$$

Because of the notation used, a scalar product is often colloquially called the *dot product*. Some books favour the notation $(\mathbf{a}, \mathbf{b})$ for the scalar product when it is then usually called the *inner product* of vectors $\mathbf{a}$ and $\mathbf{b}$. To exhibit the relation of $\mathbf{a} . \mathbf{b}$ to Eqn (4·36) we first divide $\mathbf{a} . \mathbf{b}$ by the product of the moduli $|\mathbf{a}||\mathbf{b}|$ to get

$$\frac{\mathbf{a} . \mathbf{b}}{|\mathbf{a}||\mathbf{b}|} = \left(\frac{a_1}{|\mathbf{a}|}\right)\left(\frac{b_1}{|\mathbf{b}|}\right) + \left(\frac{a_2}{|\mathbf{a}|}\right)\left(\frac{b_2}{|\mathbf{b}|}\right) + \left(\frac{a_3}{|\mathbf{a}|}\right)\left(\frac{b_3}{|\mathbf{b}|}\right).$$

Then, from the definition of direction cosines, we recognize that this may be written

$$\frac{\mathbf{a} . \mathbf{b}}{|\mathbf{a}||\mathbf{b}|} = l_1 l_2 + m_1 m_2 + n_1 n_2, \tag{4·38}$$

where $\{l_1, m_1, n_1\}$ are the direction cosines of $\mathbf{a}$ and $\{l_2, m_2, n_2\}$ are the direction cosines of **b**. If θ is the angle of inclination between **a** and **b** then, by virtue of Eqn (4·36), expression (4·38) becomes

$$\mathbf{a} \cdot \mathbf{b} = |\mathbf{a}||\mathbf{b}| \cos \theta. \tag{4·39}$$

This may be taken as an alternative definition of the scalar product **a** . **b**.

ALTERNATIVE DEFINITION 4·13   The *scalar product* of the two vectors **a** and **b** is written **a** . **b** and is defined to be the scalar quantity

$$\mathbf{a} \cdot \mathbf{b} = |\mathbf{a}||\mathbf{b}| \cos \theta,$$

where θ is the angle between the vectors.

Notice that it is a direct consequence of the definition that the scalar product of two vectors is commutative. That is, we have **a** . **b** = **b** . **a** for any two vectors **a** and **b**.

Because of this property we shall sometimes, and without confusion, write $\mathbf{a}^2$ with the understanding that $\mathbf{a}^2 \equiv \mathbf{a} \cdot \mathbf{a}$. In practice Definition 4·12 is most used to find the scalar product since it relates the scalar product directly to the components of the vectors involved. The alternative form set out in Definition 4·13 is used to find the angle between the two vectors once the scalar product is known.

Example 4·23   Find the scalar product of the vectors $\mathbf{a} = -2\mathbf{i} - 3\mathbf{j} + \mathbf{k}$ and $\mathbf{b} = -\mathbf{i} + \mathbf{j} + 3\mathbf{k}$ and use the result to find the angle between **a** and **b**.

Solution   From Definition 4·12 we have $\mathbf{a} \cdot \mathbf{b} = (-2)(-1) + (-3)(1) + (1)(3) = 2$. Now $|\mathbf{a}| = \sqrt{14}$ and $|\mathbf{b}| = \sqrt{11}$, so that substituting in Definition 4·13 we have $2 = \sqrt{14} \cdot \sqrt{11} \cos \theta$ and hence $\cos \theta = 2/\sqrt{154}$, or $\theta = \text{arc cos } (2/\sqrt{154})$.

Consider the scalar products of the unit vectors **i**, **j**, and **k**. Since these are mutually orthogonal the angle between any two is $\pi/2$. It follows from Definition 4·13 that the scalar product of any two different unit vectors from this triad is zero. As each of the vectors **i**, **j**, and **k** is parallel to itself, when forming the scalar product of one of these vectors with itself we must set $\theta = 0$. Thus as $|\mathbf{i}| = |\mathbf{j}| = |\mathbf{k}| = 1$, it follows from Definition 4·13 that **i** . **i** = **j** . **j** = **k** . **k** = 1. In summary we have these important results, which should be memorized since they are fundamental to everything that follows:

$$\mathbf{i} \cdot \mathbf{i} = \mathbf{j} \cdot \mathbf{j} = \mathbf{k} \cdot \mathbf{k} = 1,$$

$$\mathbf{i} \cdot \mathbf{j} = \mathbf{j} \cdot \mathbf{i} = 0,$$

$$\mathbf{i} \cdot \mathbf{k} = \mathbf{k} \cdot \mathbf{i} = 0,$$

$$\mathbf{j} \cdot \mathbf{k} = \mathbf{k} \cdot \mathbf{j} = 0.$$

These results are conveniently combined in Table 4·2. Each entry is to be interpreted as the scalar product of the vector at the left of the row of the entry, with the vector at the top of the column of the entry.

Table 4·2 Table of scalar products of **i**, **j**, and **k**

First member	Second member		
	i	**j**	**k**
i	1	0	0
j	0	1	0
k	0	0	1

The scalar product of two vectors may be deduced using Table 4·2 by simple algebraic manipulation without the use of Definition 4·12. To see this consider the vectors $\mathbf{a} = a_1\mathbf{i} + a_2\mathbf{j} + a_3\mathbf{k}$ and $\mathbf{b} = b_1\mathbf{i} + b_2\mathbf{j} + b_3\mathbf{k}$. First form their scalar product

$$\mathbf{a} \cdot \mathbf{b} = (a_1\mathbf{i} + a_2\mathbf{j} + a_3\mathbf{k}) \cdot (b_1\mathbf{i} + b_2\mathbf{j} + b_3\mathbf{k}),$$

and then expand the right-hand side as though ordinary algebraic quantities were involved to obtain

$$\mathbf{a} \cdot \mathbf{b} = (a_1\mathbf{i}) \cdot (b_1\mathbf{i}) + (a_1\mathbf{i}) \cdot (b_2\mathbf{j}) + (a_1\mathbf{i}) \cdot (b_3\mathbf{k}) + (a_2\mathbf{j}) \cdot (b_1\mathbf{i})$$
$$+ (a_2\mathbf{j}) \cdot (b_2\mathbf{j}) + (a_2\mathbf{j}) \cdot (b_3\mathbf{k}) + (a_3\mathbf{k}) \cdot (b_1\mathbf{i}) + (a_3\mathbf{k}) \cdot (b_2\mathbf{j})$$
$$+ (a_3\mathbf{k}) \cdot (b_3\mathbf{k}).$$

Next, recognizing that the scalars a_i, b_i may be taken to the front of each scalar product involved, rewrite the result thus:

$$\mathbf{a} \cdot \mathbf{b} = a_1b_1\mathbf{i} \cdot \mathbf{i} + a_1b_2\mathbf{i} \cdot \mathbf{j} + a_1b_3\mathbf{i} \cdot \mathbf{k} + a_2b_1\mathbf{j} \cdot \mathbf{i} + a_2b_2\mathbf{j} \cdot \mathbf{j}$$
$$+ a_2b_3\mathbf{j} \cdot \mathbf{k} + a_3b_1\mathbf{k} \cdot \mathbf{i} + a_3b_2\mathbf{k} \cdot \mathbf{j} + a_3b_3\mathbf{k} \cdot \mathbf{k}.$$

Finally, using Table 4·2, this reduces to the desired result

$$\mathbf{a} \cdot \mathbf{b} = a_1b_1 + a_2b_2 + a_3b_3.$$

In practice the intermediate working is always omitted and the result of a scalar product is written on sight by retaining only the products involving $\mathbf{i} \cdot \mathbf{i}$, $\mathbf{j} \cdot \mathbf{j}$, and $\mathbf{k} \cdot \mathbf{k}$.

Example 4·24 Determine the scalar products of these pairs of vectors:

(a) $\mathbf{a} = \mathbf{i} - 3\mathbf{j} + \mathbf{k}$, $\mathbf{b} = -\mathbf{i} + \mathbf{j} - 3\mathbf{k}$;

(b) $\mathbf{a} = 2\mathbf{i} + \mathbf{j} - \mathbf{k}$, $\mathbf{b} = -\mathbf{i} + \mathbf{j} - \mathbf{k}$;

(c) $\mathbf{a} = 2\mathbf{i} - \mathbf{j} + 3\mathbf{k}$, $\mathbf{b} = -2\mathbf{i} + \mathbf{j} - 3\mathbf{k}$;

(d) $\mathbf{a} = \mathbf{i} + 2\mathbf{j} - \mathbf{k}$, $\mathbf{b} = \mathbf{i} + 2\mathbf{j} - \mathbf{k}$.

Solutions To show the application of scalar products of unit vectors we shall retain the notation $\mathbf{i} \cdot \mathbf{i}$, $\mathbf{j} \cdot \mathbf{j}$, and $\mathbf{k} \cdot \mathbf{k}$ in the first part of each calculation to indicate the origin of the terms involved. The terms involving products such as $\mathbf{i} \cdot \mathbf{j}$, $\mathbf{i} \cdot \mathbf{k}$, . . ., will be omitted as these scalar products are zero. The result will usually be written down on sight without any intermediate working.

(a) $\mathbf{a} \cdot \mathbf{b} = (\mathbf{i} - 3\mathbf{j} + \mathbf{k}) \cdot (-\mathbf{i} + \mathbf{j} - 3\mathbf{k})$
$$= (1)(-1)\mathbf{i} \cdot \mathbf{i} + (-3)(1)\mathbf{j} \cdot \mathbf{j} + (1)(-3)\mathbf{k} \cdot \mathbf{k}$$
$$= -1 - 3 - 3 = -7.$$

(b) $\mathbf{a} \cdot \mathbf{b} = (2\mathbf{i} + \mathbf{j} - \mathbf{k}) \cdot (-\mathbf{i} + \mathbf{j} - \mathbf{k})$
$$= (2)(-1)\mathbf{i} \cdot \mathbf{i} + (1)(1)\mathbf{j} \cdot \mathbf{j} + (-1)(-1)\mathbf{k} \cdot \mathbf{k}$$
$$= -2 + 1 + 1 = 0.$$

Thus $\mathbf{a}$ and $\mathbf{b}$ are orthogonal.

(c) $\mathbf{a} \cdot \mathbf{b} = (2\mathbf{i} - \mathbf{j} + 3\mathbf{k}) \cdot (-2\mathbf{i} + \mathbf{j} - 3\mathbf{k})$
$$= (2)(-2)\mathbf{i} \cdot \mathbf{i} + (-1)(1)\mathbf{j} \cdot \mathbf{j} + (3)(-3)\mathbf{k} \cdot \mathbf{k}$$
$$= -4 - 1 - 9 = -14.$$

(d) $\mathbf{a} \cdot \mathbf{b} = (\mathbf{i} + 2\mathbf{j} - \mathbf{k}) \cdot (\mathbf{i} + 2\mathbf{j} - \mathbf{k})$
$$= (1)(1)\mathbf{i} \cdot \mathbf{i} + (2)(2)\mathbf{j} \cdot \mathbf{j} + (-1)(-1)\mathbf{k} \cdot \mathbf{k}$$
$$= 1 + 4 + 1 = 6.$$

Example (d) above is a special case of the scalar product of a vector with itself and either from Definition 4·12 or 4·13 we see that for an arbitrary vector $\mathbf{a}$,

$$\mathbf{a} \cdot \mathbf{a} = |\mathbf{a}|^2. \tag{4·40}$$

In words, 'the scalar product of a vector with itself is equal to the square of the modulus of that vector'. This simple result is often valuable when finding a unit vector parallel to a given arbitrary vector $\boldsymbol{\alpha}$. To see how this comes about, if we divide $\boldsymbol{\alpha}$ by its modulus $|\boldsymbol{\alpha}|$ to form the vector $\hat{\boldsymbol{\alpha}} = \boldsymbol{\alpha}/|\boldsymbol{\alpha}|$, then result (4·40) shows that $\hat{\boldsymbol{\alpha}} \cdot \hat{\boldsymbol{\alpha}} = 1$ and so $\hat{\boldsymbol{\alpha}}$ is a unit vector.

Example 4·25 Find a unit vector $\hat{\boldsymbol{\alpha}}$ parallel to the vector $\boldsymbol{\alpha} = 3\mathbf{i} - \mathbf{j} - 2\mathbf{k}$. Use the result to determine the projection of the vector $\mathbf{b} = 2\mathbf{i} + 3\mathbf{j} + \mathbf{k}$ in the direction of $\boldsymbol{\alpha}$.

Solution Here $|\boldsymbol{\alpha}| = \sqrt{14}$ so that the desired unit vector $\hat{\boldsymbol{\alpha}} = \boldsymbol{\alpha}/\sqrt{14}$ $= (3/\sqrt{14})\mathbf{i} - (1/\sqrt{14})\mathbf{j} - (2/\sqrt{14})\mathbf{k}$. Now the projection of vector $\mathbf{b}$ along $\boldsymbol{\alpha}$ is by definition the length l of vector $\mathbf{b}$ when projected normally onto the line determined by $\boldsymbol{\alpha}$. Thus it is $l = |\mathbf{b}| \cos \theta$, where θ is the angle between $\mathbf{b}$ and $\boldsymbol{\alpha}$. Since $|\hat{\boldsymbol{\alpha}}| = 1$ we may write this as $l = |\mathbf{b}||\hat{\boldsymbol{\alpha}}| \cos \theta$ or, by Definition 4·13, as $l = \mathbf{b} \cdot \hat{\boldsymbol{\alpha}}$. Hence in this problem $l = (2\mathbf{i} + 3\mathbf{j} + \mathbf{k}) \cdot \hat{\boldsymbol{\alpha}} = 1/\sqrt{14}$.

It follows from the definition of a scalar product of two vectors and from the properties of real numbers, that if **a**, **b**, and **c** are three arbitrary vectors, then

$$\mathbf{a} . (\mathbf{b} + \mathbf{c}) = \mathbf{a} . \mathbf{b} + \mathbf{a} . \mathbf{c}.$$

This is the distributive law for the scalar product of vectors.

Expressions of the form **a . b . c**, **a . b . c . d**, . . ., are meaningless since the scalar product is only defined between a pair of vectors. Note also that division by vectors is not defined, since although we may write $\mathbf{a} . \mathbf{b} = n$, it makes no sense to write either $\mathbf{a} = n/. \mathbf{b}$ or $\mathbf{a} . = n/\mathbf{b}$.

The other form of product of two vectors is the vector product. We shall denote the vector product of vectors **a** and **b** by **a** × **b**. Again because of the notation this is often colloquially called the *cross product* of two vectors. Other notations in use for the vector product are [**a**, **b**] and **a** ∧ **b**. In preparation for the definition of **a** × **b** we now introduce a unit vector **n̂** that is normal (i.e. orthogonal) to the plane defined by the vectors **a** and **b**, and whose sense is such that **a**, **b**, and **n̂**, in this order, form a right-handed set of vectors. Here, although **a**, **b**, and **n̂** are not necessarily mutually orthogonal, we use right-handedness exactly as was defined at the start of Section 4·6.

DEFINITION 4·14   The *vector product* of vectors **a** and **b** will be written **a** × **b** and is defined to be the vector quantity

$$\mathbf{a} \times \mathbf{b} = |\mathbf{a}||\mathbf{b}| \sin \theta \mathbf{\hat{n}},$$

where θ is the angle between vectors **a** and **b** with $\sin \theta \geq 0$, and **n̂** is a unit vector normal to the plane of **a** and **b** such that **a**, **b**, and **n̂**, in this order, form a right-handed set of vectors.

This shows that the vector **a** × **b** is normal to both **a** and **b** and has magnitude $|\mathbf{a}||\mathbf{b}| \sin \theta$. The first interesting and unusual feature of this form of product is that it is not commutative. If **a**, **b**, **n̂**, in this order, form a right-handed set for the definition of **a** × **b**, then for the definition of **b** × **a** it is necessary to take for the right-handed set the vectors **b**, **a**, −**n̂**, in the stated order. The immediate consequence is the important general result that if **a** and **b** are arbitrary vectors, then

$$\mathbf{a} \times \mathbf{b} = -(\mathbf{b} \times \mathbf{a}). \tag{4·41}$$

In contrast with the scalar product, it is easily seen that the vector product of parallel vectors is identically zero, whereas the vector product of orthogonal vectors is non-zero. A simple calculation gives Table 4·3 of vector products of the unit vectors **i**, **j**, and **k**. The left-hand column identifies the first member of the vector product and the top row identifies the second member of the vector product. The corresponding entry in the table gives the result of the vector product. The entries along the diagonal are all seen to be the zero or null vector.

Table 4·3   Table of vector products of **i**, **j**, and **k**

First member	Second member		
	i	j	k
i	0	k	j
j	−k	0	i
k	j	−i	0

If we take, for example, the first element in the left-hand column and the last element in the top row, we see that $\mathbf{i} \times \mathbf{k} = -\mathbf{j}$. In many respects it is easier to memorize these three results:

$$\mathbf{i} \times \mathbf{j} = \mathbf{k}, \qquad \mathbf{j} \times \mathbf{k} = \mathbf{i}, \qquad \mathbf{k} \times \mathbf{i} = \mathbf{j}, \tag{4·42}$$

and then to use property (4·41), than to remember Table 4·3 complete. The order of the vectors occurring in these key relations can be remembered by making the cyclic permutations

i j k

j k i

k i j

As with scalar products, this table of vector products may be used to calculate the vector product of any two vectors expressed in component form. Consider the vector product $\mathbf{a} \times \mathbf{b}$ where $\mathbf{a} = a_1\mathbf{i} + a_2\mathbf{j} + a_3\mathbf{k}$ and $\mathbf{b} = b_1\mathbf{i} + b_2\mathbf{j} + b_3\mathbf{k}$. Proceeding as though ordinary algebraic quantities were involved we write

$$\mathbf{a} \times \mathbf{b} = (a_1\mathbf{i} + a_2\mathbf{j} + a_3\mathbf{k}) \times (b_1\mathbf{i} + b_2\mathbf{j} + b_3\mathbf{k})$$
$$= (a_1\mathbf{i}) \times (b_1\mathbf{i}) + (a_1\mathbf{i}) \times (b_2\mathbf{j}) + (a_1\mathbf{i}) \times (b_3\mathbf{k})$$
$$+ (a_2\mathbf{j}) \times (b_1\mathbf{i}) + (a_2\mathbf{j}) \times (b_2\mathbf{j}) + (a_2\mathbf{j}) \times (b_3\mathbf{k})$$
$$+ (a_3\mathbf{k}) \times (b_1\mathbf{i}) + (a_3\mathbf{k}) \times (b_2\mathbf{j}) + (a_3\mathbf{k}) \times (b_3\mathbf{k}),$$

working on the assumption that vector multiplication is distributive over addition. Next we recognize that the scalars a_i, b_j may be taken out in front of each vector product that is involved so that the expression becomes

$$\mathbf{a} \times \mathbf{b} = a_1b_1\mathbf{i} \times \mathbf{i} + a_1b_2\mathbf{i} \times \mathbf{j} + a_1b_3\mathbf{i} \times \mathbf{k} + a_2b_1\mathbf{j} \times \mathbf{i} + a_2b_2\mathbf{j} \times \mathbf{j}$$
$$+ a_2b_3\mathbf{j} \times \mathbf{k} + a_3b_1\mathbf{k} \times \mathbf{i} + a_3b_2\mathbf{k} \times \mathbf{j} + a_3b_3\mathbf{k} \times \mathbf{k}.$$

Finally, using Table 4·3 and collecting together the **i**, **j**, and **k** terms, we obtain

$$\mathbf{a} \times \mathbf{b} = (a_2b_3 - a_3b_2)\mathbf{i} + (a_3b_1 - a_1b_3)\mathbf{j} + (a_1b_2 - a_2b_1)\mathbf{k}. \qquad (4·43)$$

This is often taken as the definition of the vector product **a** × **b** in place of our Definition 4·14. Expression (4·43) may be considerably simplified if the concept of a determinant is used. Before showing this we must digress slightly to define this term.

DEFINITION 4·15   Let a, b, c, and d be any four real numbers. Consider the two-row by two-column array of these numbers

$$\begin{matrix} a & b \\ c & d. \end{matrix} \qquad (A)$$

Define the expression

$$\begin{vmatrix} a & b \\ c & d \end{vmatrix} \qquad (B)$$

that is associated with this array by the identity

$$\begin{vmatrix} a & b \\ c & d \end{vmatrix} \equiv (ad - cb). \qquad (C)$$

We define the *second-order determinant* associated with the array (A) to be the number represented in symbols by (B) and having the value defined by (C). The process of expressing the left-hand side of (C) in the form of the right-hand side is called *expanding* the determinant.

Example 4·26   Evaluate the second-order determinants

$$(a) \begin{vmatrix} 1 & 7 \\ 3 & 9 \end{vmatrix}; \qquad (b) \begin{vmatrix} 0 & -1 \\ 4 & 2 \end{vmatrix}; \qquad (c) \begin{vmatrix} 2 & 6 \\ 1 & 3 \end{vmatrix}.$$

Solution   The values of the determinants follow directly from the definition:

$$(a) \begin{vmatrix} 1 & 7 \\ 3 & 9 \end{vmatrix} = (1)(9) - (3)(7) = 9 - 21 = -12;$$

$$(b) \begin{vmatrix} 0 & -1 \\ 4 & 2 \end{vmatrix} = (0)(2) - (4)(-1) = 0 + 4 = 4;$$

$$(c) \begin{vmatrix} 2 & 6 \\ 1 & 3 \end{vmatrix} = (2)(3) - (1)(6) = 6 - 6 = 0.$$

DEFINITION 4·16   Let a_i, b_i, and c_i with $i = 1$, 2, 3 be any set of nine real numbers. Consider the three-row by three-column array of these numbers

$$a_1 \quad a_2 \quad a_3$$
$$b_1 \quad b_2 \quad b_3 \tag{A}$$
$$c_1 \quad c_2 \quad c_3.$$

Define the expression

$$\begin{vmatrix} a_1 & a_2 & a_3 \\ b_1 & b_2 & b_3 \\ c_1 & c_2 & c_3 \end{vmatrix} \tag{B}$$

that is associated with this array to be the single number that is determined by the identity

$$\begin{vmatrix} a_1 & a_2 & a_3 \\ b_1 & b_2 & b_3 \\ c_1 & c_2 & c_3 \end{vmatrix} \equiv a_1 \begin{vmatrix} b_2 & b_3 \\ c_2 & c_3 \end{vmatrix} - a_2 \begin{vmatrix} b_1 & b_3 \\ c_1 & c_3 \end{vmatrix} + a_3 \begin{vmatrix} b_1 & b_2 \\ c_1 & c_2 \end{vmatrix}. \tag{C}$$

We define the *third-order determinant* associated with the array (A) to be the number represented in symbols by (B) and having the value defined by (C).

Example 4·27   Evaluate the third order determinant

$$\Delta = \begin{vmatrix} 3 & -2 & -7 \\ 2 & 1 & 2 \\ 2 & 1 & 1 \end{vmatrix}.$$

Solution   From the definition,

$$\begin{vmatrix} 3 & -2 & -7 \\ 2 & 1 & 2 \\ 2 & 1 & 1 \end{vmatrix} = (3) \begin{vmatrix} 1 & 2 \\ 1 & 1 \end{vmatrix} - (-2) \begin{vmatrix} 2 & 2 \\ 2 & 1 \end{vmatrix} + (-7) \begin{vmatrix} 2 & 1 \\ 2 & 1 \end{vmatrix}.$$

Expanding the three second-order determinants and adding, we obtain the desired result

$$\Delta = 3(1 - 2) + 2(2 - 4) - 7(2 - 2) = -7.$$

It is helpful to classify determinants in some simple way, which the next definition achieves.

DEFINITION 4·17   We define the *order* of a determinant to be the number of terms that lie on a diagonal drawn from the top left-hand corner to the bottom right-hand corner. The values of these terms are immaterial.

Thus in Example 4·26 the determinants are second-order, whereas in Example 4·27 the determinant is third-order, and is evaluated in terms of three second-order determinants.

We are now able to give the promised alternative definition of a vector product.

ALTERNATIVE DEFINITION 4·18 We define the *vector product* $\mathbf{a} \times \mathbf{b}$ of the two vectors $\mathbf{a} = a_1\mathbf{i} + a_2\mathbf{j} + a_3\mathbf{k}$ and $\mathbf{b} = b_1\mathbf{i} + b_2\mathbf{j} + b_3\mathbf{k}$ to be the formal expansion of the determinant

$$\mathbf{a} \times \mathbf{b} \equiv \begin{vmatrix} \mathbf{i} & \mathbf{j} & \mathbf{k} \\ a_1 & a_2 & a_3 \\ b_1 & b_2 & b_3 \end{vmatrix}.$$

In this definition we have used the word 'formal' because, although the a_i and b_i are real numbers, the $\mathbf{i}$, $\mathbf{j}$, and $\mathbf{k}$ are unit vectors. Aside from this the expansion of the third-order determinant is performed exactly as in Example 4·27.

Example 4·28 Determine the vector product $\mathbf{a} \times \mathbf{b}$ where $\mathbf{a} = \mathbf{i} + \mathbf{j} - 2\mathbf{k}$ and $\mathbf{b} = -2\mathbf{i} + 3\mathbf{j} + \mathbf{k}$.

Solution To apply Definition 4·18 we first notice that the components a_1, a_2, and a_3 of $\mathbf{a}$ are 1, 1, and -2 whilst the components b_1, b_2, and b_3 of $\mathbf{b}$ are -2, 3, and 1. Hence

$$\mathbf{a} \times \mathbf{b} = \begin{vmatrix} \mathbf{i} & \mathbf{j} & \mathbf{k} \\ 1 & 1 & -2 \\ -2 & 3 & 1 \end{vmatrix} = \mathbf{i}\begin{vmatrix} 1 & -2 \\ 3 & 1 \end{vmatrix} - \mathbf{j}\begin{vmatrix} 1 & -2 \\ -2 & 1 \end{vmatrix} + \mathbf{k}\begin{vmatrix} 1 & 1 \\ -2 & 3 \end{vmatrix}$$

and so

$$\mathbf{a} \times \mathbf{b} = 7\mathbf{i} + 3\mathbf{j} + 5\mathbf{k}.$$

This effectively demonstrates that for most practical purposes Definition 4·18 involves the least manipulation.

It is easily proved that the vector product is distributive, so that for any three vectors $\mathbf{a}$, $\mathbf{b}$, and $\mathbf{c}$ we always have

$$\mathbf{a} \times (\mathbf{b} + \mathbf{c}) = \mathbf{a} \times \mathbf{b} + \mathbf{a} \times \mathbf{c}.$$

Indeed this is implied by the way in which Eqn (4·43) was derived.

With the introduction of the vector product, mixed products of the form $\mathbf{a} \cdot (\mathbf{b} \times \mathbf{c})$ become possible. This type of product is known as a *triple scalar product* and as it involves the scalar product of **a** with (**b** × **c**) it is seen to be a scalar. If $\mathbf{a} = a_1\mathbf{i} + a_2\mathbf{j} + a_3\mathbf{k}$, $\mathbf{b} = b_1\mathbf{i} + b_2\mathbf{j} + b_3\mathbf{k}$, and $\mathbf{c} = c_1\mathbf{i} + c_2\mathbf{j} + c_3\mathbf{k}$ then by combination of Definitions 4·12 and 4·18 we have

$$\mathbf{a} \cdot (\mathbf{b} \times \mathbf{c}) = (a_1\mathbf{i} + a_2\mathbf{j} + a_3\mathbf{k}) \cdot \begin{vmatrix} \mathbf{i} & \mathbf{j} & \mathbf{k} \\ b_1 & b_2 & b_3 \\ c_1 & c_2 & c_3 \end{vmatrix}$$

or,

$$\mathbf{a} \cdot (\mathbf{b} \times \mathbf{c}) = a_1(b_2c_3 - c_2b_3) - a_2(b_1c_3 - c_1b_3) + a_3(b_1c_2 - c_1b_2).$$

The terms on the right-hand side of this expression are the result of expanding (C) in Definition 4·16, so that they may be re-combined into a determinant to give the general result

$$\mathbf{a} \cdot (\mathbf{b} \times \mathbf{c}) = \begin{vmatrix} a_1 & a_2 & a_3 \\ b_1 & b_2 & b_3 \\ c_1 & c_2 & c_3 \end{vmatrix}. \tag{4·44}$$

By interchanging rows of the determinant it is readily shown that the dot . and the cross × in a triple scalar product may be interchanged so that

$$\mathbf{a} \cdot (\mathbf{b} \times \mathbf{c}) = (\mathbf{a} \times \mathbf{b}) \cdot \mathbf{c}. \tag{4·45}$$

Example 4·29 Evaluate the triple scalar product $\mathbf{a} \cdot (\mathbf{b} \times \mathbf{c})$ given that $\mathbf{a} = 2\mathbf{i} + \mathbf{k}$, $\mathbf{b} = \mathbf{i} + \mathbf{j} + 2\mathbf{k}$, and $\mathbf{c} = -\mathbf{i} + \mathbf{j}$.

Solution The components of **a**, **b**, and **c** are, respectively, $(2, 0, 1)$, $(1, 1, 2)$, and $(-1, 1, 0)$. Hence

$$\mathbf{a} \cdot (\mathbf{b} \times \mathbf{c}) = \begin{vmatrix} 2 & 0 & 1 \\ 1 & 1 & 2 \\ -1 & 1 & 0 \end{vmatrix} = 2 \cdot (-2) - 0 \cdot (2) + 1 \cdot (2) = -2.$$

As our next generalization, we notice that vector products of more than two vectors are defined provided the order in which these products are to be carried out is specified by bracketing. As a special case we have the triple vector product $\mathbf{a} \times (\mathbf{b} \times \mathbf{c})$ of the three vectors **a**, **b**, and **c** which differs from the triple vector product $(\mathbf{a} \times \mathbf{b}) \times \mathbf{c}$. The first expression signifies the vector product of **a** and (**b** × **c**), whilst the second signifies the vector product of (**a** × **b**) and **c**, and in general these are different vectors.

A straightforward application of Definition 4·18 establishes the following useful identity from which some interesting results may be derived

$$\mathbf{a} \times (\mathbf{b} \times \mathbf{c}) = (\mathbf{a} \cdot \mathbf{c})\mathbf{b} - (\mathbf{a} \cdot \mathbf{b})\mathbf{c}. \tag{4·46}$$

The details of the proof are left to the reader.

Example 4·30   Demonstrate the difference between the triple vector products $\mathbf{a} \times (\mathbf{b} \times \mathbf{c})$ and $(\mathbf{a} \times \mathbf{b}) \times \mathbf{c}$ by making the identifications $\mathbf{a} = \mathbf{i}$, $\mathbf{b} = \mathbf{i} + \mathbf{j}$, $\mathbf{c} = \mathbf{k}$.

Solution   By direct substitution we find that $\mathbf{a} \times (\mathbf{b} \times \mathbf{c}) = \mathbf{i} \times [(\mathbf{i} + \mathbf{j}) \times \mathbf{k}]$ and so expanding this result by using Eqn (4·42) gives $\mathbf{a} \times (\mathbf{b} \times \mathbf{c}) = \mathbf{i} \times [-\mathbf{j} + \mathbf{i}] = -\mathbf{k}$. Similarly, in the second case, $(\mathbf{a} \times \mathbf{b}) \times \mathbf{c} = [\mathbf{i} \times (\mathbf{i} + \mathbf{j})] \times \mathbf{k} = \mathbf{k} \times \mathbf{k} = 0$.

4·8   Geometrical applications

This section illustrates something of the application of vectors to elementary geometry, and gives some simple but useful results. First we consider the representation of a straight line in vector form, and then show how the single vector equation may be reduced to the more familiar set of three Cartesian equations.

The straight line

Consider the problem of determining the equation of a straight line given that it passes through the point A with position vector $\mathbf{a}$ relative to O, and is parallel to vector $\mathbf{b}$. We shall denote the position vector of a general point P on the line by $\mathbf{r}$ as shown in Fig. 4·13.

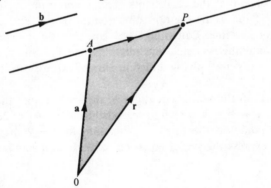

Fig. 4.13   Straight line through A parallel to **b**.

By the rules of vector addition we have

$$\overline{OP} = \overline{OA} + \overline{AP}$$

or,

$$\mathbf{r} = \mathbf{a} + \overline{AP}.$$

However, as the straight line through A is parallel to the free vector **b**, it follows that for any point P on the line there is a scalar λ such that we can write $\underline{AP} = \lambda\mathbf{b}$. Applying this result to the equation above we see that the vector equation for the straight line becomes

$$\mathbf{r} = \mathbf{a} + \lambda\mathbf{b}. \tag{4·47}$$

The scalar λ in this equation is simply a parameter, and different values of λ will determine different points on the line. To express this result in Cartesian form, set $\mathbf{r} = x\mathbf{i} + y\mathbf{j} + z\mathbf{k}$, $\mathbf{a} = a_1\mathbf{i} + a_2\mathbf{j} + a_3\mathbf{k}$ and $\mathbf{b} = b_1\mathbf{i} + b_2\mathbf{j} + b_3\mathbf{k}$, when Eqn (4·47) reduces to

$$x\mathbf{i} + y\mathbf{j} + z\mathbf{k} = a_1\mathbf{i} + a_2\mathbf{j} + a_3\mathbf{k} + \lambda(b_1\mathbf{i} + b_2\mathbf{j} + b_3\mathbf{k}).$$

This vector equation implies three scalar equations by virtue of the equality of its $\mathbf{i}$, $\mathbf{j}$, and $\mathbf{k}$ components. Hence we arrive at the three scalar equations

$$x = a_1 + \lambda b_1 \quad \text{(\textbf{i}-component)}$$
$$y = a_2 + \lambda b_2 \quad \text{(\textbf{j}-component)}$$
$$z = a_3 + \lambda b_3 \quad \text{(\textbf{k}-component)}.$$

If these are each solved for λ and equated, we obtain the more familiar result

$$\frac{x - a_1}{b_1} = \frac{y - a_2}{b_2} = \frac{z - a_3}{b_3} = \lambda. \tag{4·48}$$

Equations (4·48) are the standard Cartesian form for the equations of a straight line. Notice that the coefficients of x, y, and z in Eqn (4·48) are all unity; that b_1, b_2, and b_3 are then the direction ratios of $\mathbf{b}$ and a_1, a_2, and a_3 define a point on the line. Equations (4·48) are sometimes expressed in the form of three simultaneous equations relating x and y, x and z, and y and z. This follows by cross-multiplying different pairs of expressions in Eqn (4·48).

Example 4·31 Find the vector equation of the line through the point with position vector $\mathbf{i} + 3\mathbf{j} - \mathbf{k}$ which is parallel to the vector $2\mathbf{i} + 3\mathbf{j} + 4\mathbf{k}$. Determine the point on the line corresponding to $\lambda = 2$ in the resulting equation. Also express the vector equation of the line in standard Cartesian form.

Solution From Eqn (4·47) we have

$$\mathbf{r} = (\mathbf{i} + 3\mathbf{j} - \mathbf{k}) + \lambda(2\mathbf{i} + 3\mathbf{j} + 4\mathbf{k})$$

or,

$$\mathbf{r} = (1 + 2\lambda)\mathbf{i} + 3(1 + \lambda)\mathbf{j} + (4\lambda - 1)\mathbf{k}.$$

This is the vector equation of the line, and setting $\lambda = 2$ determines the point $\mathbf{r} = 5\mathbf{i} + 9\mathbf{j} + 7\mathbf{k}$. To express the equation of the line in Cartesian form we appeal to Eqns (4·48) and use the fact that $\mathbf{a} = \mathbf{i} + 3\mathbf{j} - \mathbf{k}$ and

$\mathbf{b} = 2\mathbf{i} + 3\mathbf{j} + 4\mathbf{k}$. Hence $a_1 = 1$, $a_2 = 3$, $a_3 = -1$, and $b_1 = 2$, $b_2 = 3$, and $b_3 = 4$, so that the desired Cartesian equations are

$$\frac{x-1}{2} = \frac{y-3}{3} = \frac{z+1}{4} = \lambda.$$

As a check we can also use these equations to determine the point corresponding to $\lambda = 2$. We must solve the three equations

$$\frac{x-1}{2} = 2, \qquad \frac{y-3}{3} = 2, \qquad \frac{z+1}{4} = 2,$$

which give $x = 5$, $y = 9$, and $z = 7$. These are of course the coordinates of the tip of the position vector $\mathbf{r} = 5\mathbf{i} + 9\mathbf{j} + 7\mathbf{k}$ which confirms our previous result.

The same approach may be used if the line is required to pass through the two points A and B with position vectors $\boldsymbol{\alpha}$ and $\boldsymbol{\beta}$, respectively. For then the line passes through $\boldsymbol{\alpha}$ and is parallel to the vector $\boldsymbol{\beta} - \boldsymbol{\alpha}$ which is just a segment of the line itself. Hence we identify $\mathbf{a}$ with $\boldsymbol{\alpha}$ and $\mathbf{b}$ with $\boldsymbol{\beta} - \boldsymbol{\alpha}$, after which the argument proceeds as before.

In the next example we illustrate how the non-standard Cartesian equations of a straight line may be re-interpreted in vector form.

Example 4·32   The equations

$$\frac{2x-1}{3} = \frac{y+2}{3} = \frac{-z+4}{2}$$

determine a straight line. Express them in vector form and find the direction ratios of the line.

Solution   To express the equations in standard Cartesian form we must first make the coefficients of x, y, and z each equal to unity. Hence we rewrite the equations:

$$\frac{x-\frac{1}{2}}{(3/2)} = \frac{y+2}{3} = \frac{z-4}{(-2)}.$$

The vector $\mathbf{a}$ then has components $a_1 = \frac{1}{2}$, $a_2 = -2$, $a_3 = 4$ and the vector $\mathbf{b}$ has the components $b_1 = 3/2$, $b_2 = 3$, $b_3 = -2$. These latter three numbers are the desired direction ratios. The vector equation of the straight line itself is

$$\mathbf{r} = \tfrac{1}{2}(1 + 3\lambda)\mathbf{i} + (3\lambda - 2)\mathbf{j} + 2(2 - \lambda)\mathbf{k}.$$

(Why?)

On occasion it is necessary to determine the perpendicular distance p from a point C with position vector $\mathbf{c}$ to the line L with equation $\mathbf{r} = \mathbf{a} + \lambda\mathbf{b}$.

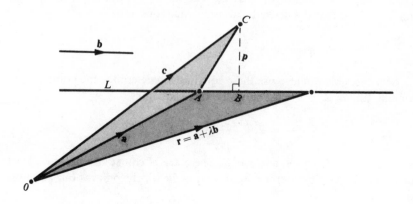

Fig. 4·14 Perpendicular distance of point from line.

This can be done by applying Pythagoras' theorem in Fig. 4·14.
We have the obvious result

$$p^2 = (AC)^2 - (AB)^2$$

but $\overline{AC} = \mathbf{c} - \mathbf{a}$ so that $(AC)^2 = |AC|^2 = (\mathbf{c} - \mathbf{a}) . (\mathbf{c} - \mathbf{a})$, whilst length $\overline{AB}$ is the projection of $\overline{AC}$ onto the line L. Now the unit vector along L is $\mathbf{b}/|\mathbf{b}|$ so that $AB = (\mathbf{c} - \overline{\mathbf{a}}) . \mathbf{b}/|\mathbf{b}|$ and thus

$$(AB)^2 = \left(\frac{(\mathbf{c} - \mathbf{a}) . \mathbf{b}}{|\mathbf{b}|}\right)^2.$$

Combining these results gives

$$p^2 = (\mathbf{c} - \mathbf{a}) . (\mathbf{c} - \mathbf{a}) - \left(\frac{(\mathbf{c} - \mathbf{a}) . \mathbf{b}}{|\mathbf{b}|}\right)^2, \tag{4·49}$$

from which p may be deduced.

Example 4·33 Find the distance of the point with position vector $\mathbf{i} + \mathbf{j} + \mathbf{k}$ from the line $\mathbf{r} = (\mathbf{i} + 2\mathbf{j} + \mathbf{k}) + \lambda(\mathbf{i} - 2\mathbf{j} + \mathbf{k})$.

Solution In the notation leading to Eqn (4·49) we have $\mathbf{a} = \mathbf{i} + 2\mathbf{j} + \mathbf{k}$, $\mathbf{b} = \mathbf{i} - 2\mathbf{j} + \mathbf{k}$, and $\mathbf{c} = \mathbf{i} + \mathbf{j} + \mathbf{k}$. Hence $\mathbf{c} - \mathbf{a} = -\mathbf{j}$ and thus $(\mathbf{c} - \mathbf{a}) . (\mathbf{c} - \mathbf{a}) = (-\mathbf{j}) . (-\mathbf{j}) = 1$. Also $(\mathbf{c} - \mathbf{a}) . \mathbf{b} = -\mathbf{j} . (\mathbf{i} - 2\mathbf{j} + \mathbf{k}) = 2$ so that $((\mathbf{c} - \mathbf{a}) . \mathbf{b})^2 = 4$, whilst $|\mathbf{b}|^2 = 6$. Hence

$$\left(\frac{(\mathbf{c} - \mathbf{a}) . \mathbf{b}}{|\mathbf{b}|}\right)^2 = \frac{4}{6} = \frac{2}{3},$$

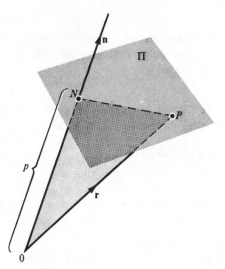

Fig. 4·15   Vector equation of a plane $\mathbf{n} \cdot \mathbf{r} = |\mathbf{n}|p$.

and so from Eqn (4·49), $p^2 = 1 - \frac{2}{3} = \frac{1}{3}$ or $p = 1/\sqrt{3}$ as p is essentially positive.

The plane

The equation of a plane is easily determined once it is recognized that a plane Π is specified when one point on it is known, together with any vector perpendicular to it. Such a vector, when normalized, is a *unit-normal* to the plane II and is unique except for its sign. The ambiguity as to the sign of the normal is, of course, because a plane has no preferred side. To derive its equation consider Fig. 4·15.

Let $\mathbf{r}$ be the position vector relative to O of a point P on the plane Π, and $\mathbf{n}$ be a vector normal to the plane directed through the plane away from O so that the corresponding unit normal is $\hat{\mathbf{n}} = \mathbf{n}/|\mathbf{n}|$. Further, let the perpendicular distance ON from the origin O to the plane be p. Then for all points P we have $(OP) \cos \theta = p$. In terms of vectors this is

$$\frac{\mathbf{r} \cdot \mathbf{n}}{|\mathbf{n}|} = p, \tag{4.50}$$

which is just the *vector equation* of a *plane*. If the number p in Eqn (4·50) is positive then the plane lies on the side of the origin towards which $\mathbf{n}$ is directed, otherwise it lies on the opposite side.

To express result (4·50) in Cartesian form let $\mathbf{r} = x\mathbf{i} + y\mathbf{j} + z\mathbf{k}$ and the unit normal $\hat{\mathbf{n}} = \mathbf{n}/|\mathbf{n}| = l\mathbf{i} + m\mathbf{j} + n\mathbf{k}$, where of course $l^2 + m^2 + n^2 = 1$. Equation (4·50) becomes

$$lx + my + nz = p. \tag{4.51}$$

This is the standard Cartesian form of the equation of a plane. Any equation of this form represents a plane having for its unit normal the vector $l\mathbf{i} + m\mathbf{j} + n\mathbf{k}$ and lying at a perpendicular distance p from the origin. If $p = 0$ the plane passes through the origin.

Example 4·34 Find the Cartesian equation of the plane containing the point $(1, 2, 3)$ which is normal to the vector $\mathbf{i} + 2\mathbf{j} + 2\mathbf{k}$.

Solution First we use Eqn (4·50) to determine p. Since the point $(1, 2, 3)$ lies in the plane, $\mathbf{r} = \mathbf{i} + 2\mathbf{j} + 3\mathbf{k}$ is the position vector of a point in the plane. The vector normal to the plane in this case is $\mathbf{n} = \mathbf{i} + 2\mathbf{j} + 2\mathbf{k}$, so that $|\mathbf{n}| = 3$ and the unit normal $\hat{\mathbf{n}} = \mathbf{n}/|\mathbf{n}| = (\mathbf{i} + 2\mathbf{j} + 2\mathbf{k})/3$. This shows that $l = \frac{1}{3}, m = \frac{2}{3}, n = \frac{2}{3}$. Hence, substituting into Eqn (4·50),

$$p = \frac{(\mathbf{i} + 2\mathbf{j} + 3\mathbf{k}) \cdot (\mathbf{i} + 2\mathbf{j} + 2\mathbf{k})}{3},$$

or $p = 11/3$. As $p > 0$, the plane must lie on the side of the origin towards which $\hat{\mathbf{n}}$ is directed. Substituting in Eqn (4·51) we find the desired Cartesian form of the equation of the plane:

$$\tfrac{1}{3}x + \tfrac{2}{3}y + \tfrac{2}{3}z = \tfrac{11}{3}.$$

This equation could equally well be written in the non-standard Cartesian form $x + 2y + 2z = 11$, though then the constant on the right-hand side is no longer the perpendicular distance of the plane from the origin.

Simple geometrical considerations similar to those set out above, when coupled with the scalar and vector product, enable various useful results to be derived very quickly. For example, as the angle θ between two planes is defined to be the angle between their unit normals $\hat{\mathbf{n}}_1$ and $\hat{\mathbf{n}}_2$ it follows that θ may be obtained from the scalar product $\hat{\mathbf{n}}_1 \cdot \hat{\mathbf{n}}_2 = \cos \theta$. Also the line of intersection of these two planes is perpendicular to both normals $\hat{\mathbf{n}}_1$ and $\hat{\mathbf{n}}_2$ and so is parallel to the vector $\mathbf{t}$ determined by the vector product $\mathbf{t} = \hat{\mathbf{n}}_1 \times \hat{\mathbf{n}}_2$. Rather than elaborate on these ideas here, a number of problems are given at the end of the chapter.

The sphere

Consider a sphere of radius R with its centre at the point A with the position vector $\mathbf{a}$. Then if $\mathbf{r}$ is the position vector of any point on the surface of the sphere, the modulus of the vector $\mathbf{r} - \mathbf{a}$ must equal R. In terms of vectors the equation of the sphere is

$$|\mathbf{r} - \mathbf{a}| = R$$

or, alternatively,

$(\mathbf{r} - \mathbf{a}) \cdot (\mathbf{r} - \mathbf{a}) = R^2.$ (4·52)

If, now, we expand this equation to get

$\mathbf{r} \cdot \mathbf{r} - 2\mathbf{r} \cdot \mathbf{a} = R^2 - \mathbf{a} \cdot \mathbf{a},$

and then set $\mathbf{r} = x\mathbf{i} + y\mathbf{j} + z\mathbf{k}$, $\mathbf{a} = a_1\mathbf{i} + a_2\mathbf{j} + a_3\mathbf{k}$ and $R^2 - \mathbf{a} \cdot \mathbf{a} = q$, we obtain the standard Cartesian form of the equation of a sphere

$$x^2 + y^2 + z^2 - 2a_1 x - 2a_2 y - 2a_3 z = q.$$ (4·53)

Example 4·35 Find the Cartesian form of equation of the sphere of radius 2 having its centre at $\mathbf{a} = \mathbf{i} + \mathbf{j} + 2\mathbf{k}$.

Solution As $\mathbf{r} = x\mathbf{i} + y\mathbf{j} + z\mathbf{k}$ and $\mathbf{a} = \mathbf{i} + \mathbf{j} + 2\mathbf{k}$ we have $\mathbf{r} - \mathbf{a}$ $= (x - 1)\mathbf{i} + (y - 1)\mathbf{j} + (z - 2)\mathbf{k}$, whilst $R = 2$. Hence Eqn (4·52) becomes

$$(x - 1)^2 + (y - 1)^2 + (z - 2)^2 = 4,$$

which is the desired Cartesian form of the equation.

4·9 Applications to mechanics

This section briefly introduces some of the many situations in mechanics that are best described vectorially. First is one of the simplest applications of vectors, that will already be familiar to the reader.

Polygon of forces—resultant

It is known from experiment that when forces $\mathbf{F}_1, \mathbf{F}_2, \ldots, \mathbf{F}_n$ act on a rigid body through a single point O, their combined effect is equivalent to that of a single force $\mathbf{R}$, their *resultant*, which acts through the same point O and is equal to their vector sum. Such a system of forces acting through a single point is a *concurrent* system of forces. Thus we have

$\mathbf{R} = \mathbf{F}_1 + \mathbf{F}_2 + \cdots + \mathbf{F}_n.$ (4·54)

These forces are often represented in the form of a *vector polygon* of forces as shown in Fig. 4·16, in which the senses of the forces $\mathbf{F}_i$ are all similarly directed and are opposite to the sense of $\mathbf{R}$.

Conversely, the vector polygon shows that the vector $-\mathbf{R}$ is the additional force that is required to act through O in order to maintain the system of forces in equilibrium.

Example 4·36 Forces $\mathbf{F}_1, \mathbf{F}_2$, and $\mathbf{F}_3$ have magnitudes $3\sqrt{3}$, $\sqrt{14}$, and $2\sqrt{6}$ lb and act concurrently through a point O along the lines of the vector $\mathbf{i} + \mathbf{j} + \mathbf{k}$, $3\mathbf{i} - \mathbf{j} + 2\mathbf{k}$, and $-\mathbf{i} + 2\mathbf{j} + \mathbf{k}$, respectively. Find force $\mathbf{Q}$ that must act through O for the system to remain in equilibrium.

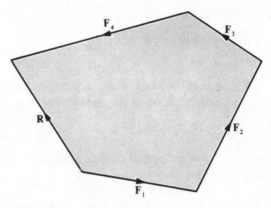

Fig. 4·16   Vector polygon.

Solution   This is a direct application of the last remark about the vector polygon of forces, and the only problem is one of scaling. Let us agree that a vector of unit modulus represents a force of 1 lb. From the conditions of the question we see that $\mathbf{F}_1$, $\mathbf{F}_2$, and $\mathbf{F}_3$ are respectively directed along the unit vectors

$$\hat{\mathbf{f}}_1 = \frac{1}{\sqrt{3}}\,(\mathbf{i} + \mathbf{j} + \mathbf{k}),$$

$$\hat{\mathbf{f}}_2 = \frac{1}{\sqrt{14}}\,(3\mathbf{i} - \mathbf{j} + 2\mathbf{k}),$$

$$\hat{\mathbf{f}}_3 = \frac{1}{\sqrt{6}}\,(-\mathbf{i} + 2\mathbf{j} + \mathbf{k}).$$

Using the scale factor we can use these to write

$$\mathbf{F}_1 = 3\sqrt{3}\hat{\mathbf{f}}_1 = 3\mathbf{i} + 3\mathbf{j} + 3\mathbf{k},$$

$$\mathbf{F}_2 = \sqrt{14}\hat{\mathbf{f}}_2 = 3\mathbf{i} - \mathbf{j} + 2\mathbf{k},$$

$$\mathbf{F}_3 = 2\sqrt{6}\hat{\mathbf{f}}_3 = -2\mathbf{i} + 4\mathbf{j} + 2\mathbf{k}.$$

Hence the resultant $\mathbf{R} = \mathbf{F}_1 + \mathbf{F}_2 + \mathbf{F}_3 = 4\mathbf{i} + 6\mathbf{j} + 7\mathbf{k}$. The force necessary for equilibrium is $\mathbf{Q} = -\mathbf{R}$ showing that $\mathbf{Q} = -4\mathbf{i} - 6\mathbf{j} - 7\mathbf{k}$.

As $|\mathbf{Q}| = \sqrt{101}$, it follows immediately that the desired force is $\sqrt{101}$ lbs and acts in the direction of the unit vector $\hat{\mathbf{q}}$, where

$$\hat{\mathbf{q}} = \frac{-1}{\sqrt{101}}\,(4\mathbf{i} + 6\mathbf{j} + 7\mathbf{k}).$$

In many problems of statics the centroid or the centre of mass of a system of particles is of importance. We now define this concept in terms of vectors.

DEFINITION 4·19 The *centre of mass* of the system of masses m_1, m_2, ..., m_n whose position vectors are $\mathbf{a}_1$, $\mathbf{a}_2$, ..., $\mathbf{a}_n$ is at the point G, where G has the position vector $\mathbf{g}$ determined by

$$\mathbf{g} = \frac{m_1\mathbf{a}_1 + m_2\mathbf{a}_2 + \cdots + m_n\mathbf{a}_n}{m_1 + m_2 + \cdots + m_n}.$$

Next we discuss simple problems about relative motions, and relative velocity.

Relative velocity

Problems involving the motion of one point relative to another, which is itself moving, occur frequently in mechanics and easily lend themselves to vector treatment. They are best illustrated by example but first we define relative velocity.

DEFINITION 4·20 The relative velocity of a point P with velocity $\mathbf{u}$, relative to the point Q with velocity $\mathbf{v}$, is defined to be the velocity $\mathbf{u} - \mathbf{v}$.

Example 4·37 A man walks due east at 4 mile/h and his dog runs northeast at 12 mile/h. Find the velocity and speed of the man relative to his dog.

Solution Let a unit vector denote a velocity of magnitude 1 mile/h and take $\mathbf{j}$ pointing due north and $\mathbf{i}$ pointing due east.

Unit vectors in the directions of motion of the man and dog are then $\mathbf{i}$ and $(\mathbf{i} + \mathbf{j})/\sqrt{2}$. The velocity $\mathbf{u}$ of the man is thus $\mathbf{u} = 4\mathbf{i}$ and the velocity $\mathbf{v}$ of the dog is $\mathbf{v} = 6\sqrt{2}(\mathbf{i} + \mathbf{j})$. Hence the velocity of the man relative to his dog is

$$\mathbf{u} - \mathbf{v} = 2(2 - 3\sqrt{2})\mathbf{i} - 6\sqrt{2}\mathbf{j}.$$

His relative speed is $|\mathbf{u} - \mathbf{v}| = (160 - 48\sqrt{2})^{1/2}$ mile/h.

Work done by a force

The scalar product can be used to give a convenient representation of the work W done by a force $\mathbf{F}$ that produces a displacement d of the particle on which it acts. The work done by a force of magnitude $|\mathbf{F}|$ when it displaces a particle through a distance $|\mathbf{d}|$ is defined as the product of the distance moved and the component of force in the direction of the displacement.

Hence, as W is positive we have

$$W = |\mathbf{F}||\mathbf{d}||\cos\theta|,$$

where θ is the angle of inclination between $\mathbf{F}$ and $\mathbf{d}$. So the final result is:

$$W = |\mathbf{F} \cdot \mathbf{d}|. \tag{4·55}$$

Example 4·38 Calculate the work W done by a force $\mathbf{F}$ of 12 lbs whose line of action is parallel to $2\mathbf{i} + 3\mathbf{j} - 2\mathbf{k}$ when it moves its point of application through a displacement $\mathbf{d}$ of 4 ft in a direction parallel to $-2\mathbf{i} + \mathbf{j} - 3\mathbf{k}$.

Solution The unit vectors parallel to the force $\mathbf{F}$ and displacement $\mathbf{d}$ are $\hat{\mathbf{f}} = (2\mathbf{i} + 3\mathbf{j} - 2\mathbf{k})/\sqrt{17}$ and $\hat{\mathbf{d}} = (-2\mathbf{i} + \mathbf{j} - 3\mathbf{k})/\sqrt{14}$, respectively. Let $\hat{\mathbf{f}}$ denote a force of 1 lb and $\hat{\mathbf{d}}$ a displacement of 1 ft so that $\mathbf{F} = 12\hat{\mathbf{f}}$ $= (24\mathbf{i} + 36\mathbf{j} - 24\mathbf{k})/\sqrt{17}$ and $\mathbf{d} = 4\hat{\mathbf{d}} = (-8\mathbf{i} + 4\mathbf{j} - 12\mathbf{k})/\sqrt{14}$. Then the work W that is done is

$$W = |\mathbf{F} . \mathbf{d}| \text{ ft lbs}$$
$$= (24)(-8) + (36)(4) + (-24)(-12) = 240 \text{ ft lbs}.$$

We now turn to applications of the vector product. One of the easiest occurs in the determination of the angular velocity of a point rotating about a fixed axis.

Angular velocity

Consider a rigid body rotating with a constant spin Ω rad/s about a fixed axis L. Fig. 4·17 represents a point P in such a body, having the position vector $\mathbf{d}$ relative to a point O on the spin axis L. Point Q is the foot of the perpendicular from P to the line L.

The vector $\mathbf{\Omega}$ parallel to L with magnitude Ω and sense determined by a right-hand screw rule with respect to L and the direction of the spin Ω is called the *angular velocity* of the body. The instantaneous linear velocity $\mathbf{v}$ of point P with position vector $\mathbf{d}$ is obviously Ω . (QP) in a direction tangent to the dotted circle in Fig. 4·17. It is easily seen that we may rewrite this as

$$|\mathbf{v}| = |\mathbf{\Omega}||\mathbf{d}| \sin \theta$$

or as

$$\mathbf{v} = \mathbf{\Omega} \times \mathbf{d}. \tag{4·56}$$

The final two applications of the vector product involve the concept of the moment of a vector which is first defined and they require the use of a bound vector.

DEFINITION 4·21 We define $\mathbf{M} = \mathbf{d} \times \mathbf{Q}$ to be the *moment of vector* $\mathbf{Q}$ about the point O, where $\mathbf{d}$ is the position vector relative to O of any point on the line of action of the bound vector $\mathbf{Q}$.

This definition is illustrated in Fig. 4·18 in which the plane Π contains the vectors $\mathbf{d}$ and $\mathbf{Q}$ and, by virtue of the definition of the moment, $\mathbf{M}$ is normal to Π.

The natural mechanical applications of this definition are to the moment of a force and to the moment of momentum about a fixed point. In both

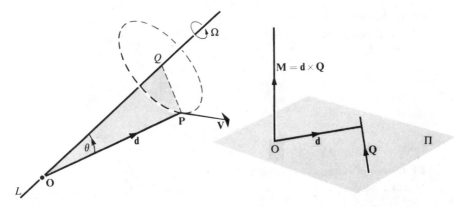

Fig. 4·17 Angular velocity.　　　　**Fig. 4·18** Moment of a vector about O.

situations the line of action of the vector whose moment is to be found is important, as is its point of application in some circumstances.

If **Q** is identified with a force **F**, then the expression

$$\mathbf{M} = \mathbf{d} \times \mathbf{F} \tag{4·57}$$

is the moment or *torque* of the force **F** about O. If the force is expressed in ·lb and the displacement vector in ft, the units of torque are lb-ft. Similarly, if **Q** is identified with the *momentum m*v of a particle of mass m moving with velocity **v**, then the vector

$$\mathbf{M} = \mathbf{d} \times (m\mathbf{v})$$
$$= m\mathbf{d} \times \mathbf{v} \tag{4·58}$$

is the *moment of momentum* or the *angular momentum* of the particle about O.

PROBLEMS

Section 4·1

4·1 By setting $x^2 = w$, reduce the following quartic equations to quadratic equations, and hence obtain their roots:

(a) $x^4 + x^2 - 2 = 0$;　(b) $x^4 + 5x^2 + 6 = 0$;　(c) $x^4 - 5x^2 + 6 = 0$.

4·2 Find the real and imaginary parts of each of these complex numbers:

(a) $z = 9 - 6i$;　(b) $z = 32$;　(c) $z = 14 + 2i$;　(d) $z = 17i$;　(e) $z = -3 + i$.

Section 4·2

4·3 Form the sums $z_1 + z_2$ given that:

(a) $z_1 = 3 - i, z_2 = 4 + 7i$;
(b) $z_1 = -2 - 4i, z_2 = 2 + 3i$;
(c) $z_1 = 5 + 6i, z_2 = -5 - 6i$.

4·4 Form the differences $z_1 - z_2$ given that:
(a) $z_1 = 2 + 6i$, $z_2 = 4 + 2i$;
(b) $z_1 = -2 + i$, $z_2 = -2 + 2i$;
(c) $z_1 = 4 + 7i$, $z_2 = 2 + 7i$.

4·5 Form the products $z_1 z_2$ given that:
(a) $z_1 = 1 + i$, $z_2 = 2 + 3i$;
(b) $z_1 = 3 - 5i$, $z_2 = 3 + 5i$.
(c) $z_1 = i$, $z_2 = 4 - 3i$.

4·6 Evaluate $(1 + i)^5 - (1 - i)^5$.

4·7 Form the quotients z_1/z_2 given that:
(a) $z_1 = 3 + 2i$, $z_2 = 1 - i$;   (b) $z_1 = 9 + 3i$, $z_2 = 3 + i$;
(c) $z_1 = 8 + 4i$, $z_2 = 2 - 4i$.

4·8 Given that $z = i$ is a root of the polynomial

$$P(z) = z^5 - 2z^4 + 10z^3 - 20z^2 + 9z - 18,$$

deduce the values of the other four roots. Factorize $P(z)$ into linear and quadratic factors with real coefficients.

Section 4·3

4·9 Plot the following vectors z_1 and z_2 in the complex-plane and use geometrical methods to form their sum $z_1 + z_2$ and their difference $z_1 - z_2$:
(a) $z_1 = 2 + 3i$, $z_2 = -1 + 2i$;   (b) $z_1 = 3$, $z_2 = 4 - i$;
(c) $z_1 = 4i$, $z_2 = 3 - 4i$;   (d) $z_1 = -1 - 2i$, $z_2 = -1 + 2i$.

4·10 Find the modulus of each of these vectors:
(a) $4 - 3i$;   (b) $-2 + 3i$;   (c) $2 - 3i$;   (d) $3 + 4i$;   (e) $5i$.

4·11 Use the properties of the complex conjugate operation to prove that for any two complex numbers z_1 and z_2,

$$z_1 \bar{z}_2 + \bar{z}_1 z_2 = 2 \operatorname{Re} z_1 \bar{z}_2.$$

Then, using this result together with the obvious inequality

$$| \operatorname{Re} z_1 \bar{z}_2 | \le | z_1 \bar{z}_2 |$$

and the identity

$$| z_1 + z_2 |^2 = (z_1 + z_2)(\bar{z}_1 + \bar{z}_2),$$

prove the triangle inequality,

$$| z_1 + z_2 | \le | z_1 | + | z_2 |.$$

4·12 Use the same form of argument as in Problem 4·29 together with the obvious inequality $\operatorname{Re} z_1 \bar{z}_2 \ge - | z_1 \bar{z}_2 |$ to prove

$$|| z_1 | - | z_2 || \le | z_1 + z_2 |.$$

Section 4·4

4·13 Express these numbers in modulus–argument form:
(a) $z = -3 + 4i$;   (b) $z = -3 - 4i$;
(c) $z = -3 + 3i$;   (d) $z = 2\sqrt{3} - 2i$.

4·14 Express the following numbers z in real–imaginary form given that:

(a) $|z| = 4$, $\arg z = \dfrac{\pi}{3}$;

(b) $|z| = 2$, $\arg z = \dfrac{-\pi}{6}$;

(c) $|z| = 6$, $\arg z = \dfrac{3\pi}{2}$;

(d) $|z| = 3$, $\arg z = \dfrac{4\pi}{3}$.

4·15 Form the products $z_1 z_2$ and the quotients z_1/z_2 of the following numbers expressed in modulus–argument form:

(a) $z_1 = 3(\cos \tfrac{1}{6}\pi + i \sin \tfrac{1}{6}\pi)$; $z_2 = \tfrac{1}{2}(\cos \tfrac{1}{3}\pi + i \sin \tfrac{1}{3}\pi)$;

(b) $z_1 = 4(\cos \tfrac{1}{3}\pi - i \sin \tfrac{1}{3}\pi)$; $z_2 = 2(\cos \tfrac{1}{4}\pi + i \sin \tfrac{1}{4}\pi)$.

Section 4·5

4·16 Use de Moivre's theorem to express $\sin 11\theta$ and $\cos 11\theta$ in terms of powers of $\sin \theta$ and $\cos \theta$.

4·17 Evaluate z^{20} when $z = \sqrt{3} + i$.

4·18 Find the roots of the equation $w = (-i)^{2/3}$.

4·19 Find the roots of the equation $w = (1 + i\sqrt{3})^{1/4}$.

Section 4·6

4·20 Determine the lengths $|OP|$ of the vectors OP given that O is the origin and the points P are:

(a) $(1, 1, 1)$; (b) $(-2, 1, 3)$; (c) $(-1, -1, -1)$; (d) $(3, -2, -4)$.

4·21 Find the lengths $|OP|$, the direction cosines and the angles θ_1, θ_2, θ_3 of the vectors OP, where the points P are:

(a) $(2, -1, -1)$; (b) $(4, 0, 2)$; (c) $(-1, 2, 1)$.

4·22 Find the direction ratios, the direction cosines and the angles θ_1, θ_2, θ_3 of the vectors OP, where the points P are:

(a) $(1, 1, 1)$; (b) $(-1, 1, 1)$; (c) $(2, 1, -1)$.

4·23 Determine the angles θ_1, θ_2, θ_3 for the vectors with the direction cosines:

(a) $\left\{ \dfrac{\sqrt{3}}{2}, 0, \dfrac{1}{2} \right\}$; (b) $\left\{ \dfrac{1}{\sqrt{3}}, \dfrac{1}{\sqrt{3}}, \dfrac{1}{\sqrt{3}} \right\}$; (c) $\left\{ \dfrac{1}{3}, -\dfrac{1}{3}, \dfrac{\sqrt{7}}{3} \right\}$.

4·24 Determine the lengths $|AB|$ of the vectors AB, given that the end points A and B are:

(a) $A = (1, 1, 1)$, $B = (2, 0, 1)$;

(b) $A = (2, -1, 1)$, $B = (-2, 2, 2)$.

Use your results to determine the direction cosines for each of these vectors.

4·25 Write down the position vectors OP in terms of the unit vectors $\mathbf{i}$, $\mathbf{j}$, $\mathbf{k}$ given that O is the origin and the points P are:

(a) $(1, 1, 1)$; (b) $(-2, 3, 7)$; (c) $(3, -1, 11)$; (d) $(0, 1, 0)$.

4·26 Determine the values of α, β, and γ in order that:

$(1 - \alpha)\mathbf{i} + \beta(1 - \alpha^2)\mathbf{j} + (\gamma - 2)\mathbf{k} = \tfrac{1}{2}\mathbf{i} + 3\mathbf{j} + 2\mathbf{k}$.

4·27 Form the sum $\mathbf{a} + \mathbf{b}$ and difference $\mathbf{a} - \mathbf{b}$ of the vectors:
(a) $\mathbf{a} = 3\mathbf{i} - 2\mathbf{j} + \mathbf{k}$, $\mathbf{b} = -\mathbf{i} - 2\mathbf{j} + 3\mathbf{k}$;
(b) $\mathbf{a} = -\mathbf{i} + 2\mathbf{j} - \mathbf{k}$, $\mathbf{b} = 2\mathbf{i} - 4\mathbf{j} + 2\mathbf{k}$.

4·28 State which of the following pairs of vectors $\mathbf{a}$ and $\mathbf{b}$ are parallel and which are anti-parallel:
(a) $\mathbf{a} = \mathbf{i} - 3\mathbf{j} + \mathbf{k}$, $\mathbf{b} = -4\mathbf{i} + 12\mathbf{j} - 4\mathbf{k}$;
(b) $\mathbf{a} = -2\mathbf{i} + 3\mathbf{j} - \mathbf{k}$, $\mathbf{b} = 2\mathbf{i} - 3\mathbf{j} + \mathbf{k}$;
(c) $\mathbf{a} = 4\mathbf{i} - \mathbf{j} - 3\mathbf{k}$, $\mathbf{b} = 8\mathbf{i} - 2\mathbf{j} - 6\mathbf{k}$.

Section 4·7

4·29 Express the following vectors $\mathbf{a}$ as the product of a scalar and a unit vector:

(a) $\mathbf{a} = 2\mathbf{i} - \mathbf{j} + 3\mathbf{k}$;   (b) $\mathbf{a} = 3\mathbf{i} - 3\mathbf{j} + \mathbf{k}$;   (c) $\mathbf{a} = -\dfrac{\sqrt{71}}{9}\mathbf{i} + \dfrac{1}{3}\mathbf{j} - \dfrac{1}{9}\mathbf{k}$.

4·30 Find the vectors $\underline{AB}$, and their direction cosines given that A and B have position vectors $\mathbf{a}$ and $\mathbf{b}$, respectively, where
(a) $\mathbf{a} = 3\mathbf{i} - 3\mathbf{j} + 5\mathbf{k}$, $\mathbf{b} = \mathbf{i} + 2\mathbf{j} - \mathbf{k}$;
(b) $\mathbf{a} = 2\mathbf{i} + 2\mathbf{j} + \mathbf{k}$, $\mathbf{b} = \mathbf{i} + 3\mathbf{j} + 2\mathbf{k}$.

4·31 Find the scalar products $\mathbf{a} \cdot \mathbf{b}$ and hence find the angle between the vectors $\mathbf{a}$ and $\mathbf{b}$ given that:
(a) $\mathbf{a} = 7\mathbf{i} - 3\mathbf{j} + \mathbf{k}$, $\mathbf{b} = -\mathbf{i} + 2\mathbf{j} + 2\mathbf{k}$;
(b) $\mathbf{a} = 2\mathbf{i} - 2\mathbf{j} + \mathbf{k}$, $\mathbf{b} = -3\mathbf{i} - 3\mathbf{j} + 4\mathbf{k}$;
(c) $\mathbf{a} = \mathbf{i} + 2\mathbf{j} + 3\mathbf{k}$, $\mathbf{b} = -2\mathbf{i} - 4\mathbf{j} - 6\mathbf{k}$.

4·32 Find unit vectors parallel to the vectors $\mathbf{a}$ where:

(a) $\mathbf{a} = 2\mathbf{i} - 2\mathbf{j} + \mathbf{k}$;   (b) $\mathbf{a} = -3\mathbf{i} + \mathbf{j} + 2\mathbf{k}$;   (c) $\mathbf{a} = 7\mathbf{i} - 2\mathbf{j} - 3\mathbf{k}$.

4·33 Prove the distributive law for the scalar product by using either definition of the scalar product.

4·34 Form the vector products $\mathbf{a} \times \mathbf{b}$ if:
(a) $\mathbf{a} = \mathbf{i} - 2\mathbf{j} - 4\mathbf{k}$, $\mathbf{b} = 2\mathbf{i} - 2\mathbf{j} + 3\mathbf{k}$;
(b) $\mathbf{a} = -\mathbf{i} + 4\mathbf{j} - \mathbf{k}$, $\mathbf{b} = 3\mathbf{i} + 2\mathbf{j} + 4\mathbf{k}$;
(c) $\mathbf{a} = -2\mathbf{i} + 4\mathbf{k}$, $\mathbf{b} = 3\mathbf{j} - 2\mathbf{k}$.

4·35 Evaluate the determinants:

(a) $\begin{vmatrix} 2 & 1 \\ 4 & 6 \end{vmatrix}$;   (b) $\begin{vmatrix} 4 & 16 \\ -2 & 6 \end{vmatrix}$;   (c) $\begin{vmatrix} 2 & 0 \\ 0 & 16 \end{vmatrix}$;   (d) $\begin{vmatrix} 3 & 9 \\ 1 & 3 \end{vmatrix}$.

4·36 For what values of λ, if any, do these determinants vanish:

(a) $\begin{vmatrix} \lambda & 2 \\ 3 & 1 \end{vmatrix}$;   (b) $\begin{vmatrix} \lambda & 2 \\ 3 & 2\lambda \end{vmatrix}$;   (c) $\begin{vmatrix} 3 & \lambda \\ 0 & 2 \end{vmatrix}$;   (d) $\begin{vmatrix} 3\lambda & 4 \\ 2 & -\lambda \end{vmatrix}$.

4·37 Evaluate the determinants:

(a) $\begin{vmatrix} 2 & 1 & 1 \\ 1 & 2 & 1 \\ 1 & 1 & 1 \end{vmatrix}$;   (b) $\begin{vmatrix} 3 & 4 & 5 \\ 2 & 2 & 1 \\ 1 & 0 & 2 \end{vmatrix}$;   (c) $\begin{vmatrix} 3 & 4 & 5 \\ 3 & 1 & 2 \\ 6 & 5 & 7 \end{vmatrix}$.

4·38 Evaluate the vector products $\mathbf{b} \times \mathbf{a}$ given that:

    (a) $\mathbf{a} = 2\mathbf{i} - \mathbf{j} + 2\mathbf{k}$, $\mathbf{b} = -3\mathbf{i} + 2\mathbf{j} + \mathbf{k}$;

    (b) $\mathbf{a} = -\mathbf{i} + \mathbf{j} + \mathbf{k}$, $\mathbf{b} = 4\mathbf{i} + 2\mathbf{j} + 3\mathbf{k}$;

    (c) $\mathbf{a} = -\mathbf{i} - \mathbf{j} - \mathbf{k}$, $\mathbf{b} = 2\mathbf{i} + 2\mathbf{j} + 2\mathbf{k}$.

4·39 Determine unit vectors that are normal to both vectors $\mathbf{a}$ and $\mathbf{b}$ when:

    (a) $\mathbf{a} = 3\mathbf{i} + 5\mathbf{j} - 2\mathbf{k}$, $\mathbf{b} = \mathbf{i} + \mathbf{j} + \mathbf{k}$;

    (b) $\mathbf{a} = -4\mathbf{i} + 2\mathbf{k}$, $\mathbf{b} = \mathbf{j} - 3\mathbf{k}$.

State whether the results are unique and, if not, in what way are they indeterminate.

4·40 Evaluate the triple scalar products $\mathbf{a} \cdot (\mathbf{b} \times \mathbf{c})$ and $(\mathbf{b} \times \mathbf{a}) \cdot \mathbf{c}$ given that:

    (a) $\mathbf{a} = 2\mathbf{i} - \mathbf{j} - 3\mathbf{k}, \mathbf{b} = 3\mathbf{k}, \mathbf{c} = \mathbf{i} + 2\mathbf{j} + 2\mathbf{k}$;

    (b) $\mathbf{a} = \mathbf{i} + 2\mathbf{j} + \mathbf{k}, \mathbf{b} = 2\mathbf{i} + \mathbf{j} + \mathbf{k}, \mathbf{c} = 4\mathbf{i} + 2\mathbf{j} + 2\mathbf{k}$.

4·41 Prove that if $\mathbf{a}$, $\mathbf{b}$, and $\mathbf{c}$ form three edges of a parallelepiped all meeting at a common point, then the volume of this solid figure is given by $|\mathbf{a} \cdot (\mathbf{b} \times \mathbf{c})|$. Deduce that the vanishing of the triple scalar product implies that the vectors $\mathbf{a}$, $\mathbf{b}$, and $\mathbf{c}$ are co-planar (that is, all lie in a common plane).

4·42 Determine the vector products $\mathbf{a} \times (\mathbf{b} \times \mathbf{c})$ given that:

    (a) $\mathbf{a} = 2\mathbf{i} - \mathbf{j} - 3\mathbf{k}, \mathbf{b} = 3\mathbf{i} + \mathbf{j} + \mathbf{k}, \mathbf{c} = -\mathbf{i} + \mathbf{j} + \mathbf{k}$;

    (b) $\mathbf{a} = -\mathbf{i} + \mathbf{j} - \mathbf{k}, \mathbf{b} = 2\mathbf{i} - 2\mathbf{j} + 2\mathbf{k}, \mathbf{c} = \mathbf{i} + \mathbf{k}$.

Section 4·8

4·43 Find the vector equation of the line through the points A and B with position vectors $\mathbf{a} = 2\mathbf{i} + \mathbf{j} - \mathbf{k}$ and $\mathbf{b} = -\mathbf{i} + \mathbf{j} + 2\mathbf{k}$. Determine the direction cosines of this line.

4·44 The equations

$$\frac{3x + 3}{2} = \frac{-2y + 1}{7} = \frac{2z + 6}{3}$$

determine a straight line. Express them in vector form and find the direction cosines of the line.

4·45 Find the perpendicular distance of the point $2\mathbf{i} + \mathbf{j} + \mathbf{k}$ from the line $\mathbf{r} = (1 + \lambda)\mathbf{i} + (2 - 3\lambda)\mathbf{j} + (2 + \lambda)\mathbf{k}$.

4·46 Find the Cartesian equation of the plane containing the point $3\mathbf{i} - \mathbf{k}$ and also containing the two vectors $\boldsymbol{\alpha}$, $\boldsymbol{\beta}$ where $\boldsymbol{\alpha} = \mathbf{i} + 2\mathbf{j} + \mathbf{k}$ and $\boldsymbol{\beta} = -\mathbf{i} + 2\mathbf{j} + 2\mathbf{k}$.

4.47 Find the angle between the two planes

$$\frac{x - 1}{2} = \frac{y + 2}{1} = \frac{z - 3}{3}$$

and

$$\frac{2x - 1}{2} = \frac{y - 1}{3} = \frac{2z + 1}{3}.$$

4·48 A line may be uniquely determined as the intersection of two planes

    $\mathbf{r} \cdot \mathbf{n}_1 = p_1$ and $\mathbf{r} \cdot \mathbf{n}_2 = p_2$ (A)

where $\mathbf{n}_1$ and $\mathbf{n}_2$ are not necessarily unit vectors. The direction of the line is normal to both n_1 and n_2 and so is parallel to $n_1 \times n_2$. Hence the line has the equation $r = a + \lambda(n_1 \times n_2)$ where λ is a parameter and a is some point common to the two planes in (A). Apply these arguments to obtain the vector equation of the line determined by the planes

$$x + 2y - z = 3 \quad \text{and} \quad 2x + y + 2z = 1.$$

4·49 The inward drawn normal to a sphere of radius 2 at the point (1, 1, 2) on its surface is $n = 2i - j + k$. Deduce its equation in Cartesian form.

Section 4·9

4·50 Forces F_1, F_2, F_3, and F_4 have magnitudes $2\sqrt{6}$, $3\sqrt{5}$, 3, and 15 lb and act concurrently through a point O along the lines of the vectors $-i + 2j - k$, $2i + k$, $2j$, and $4i + 3j$, respectively. Find the resultant of these forces and determine its magnitude in lb.

4·51 Forces 1, 2, and 3 act at one corner of a cube along the diagonals of the faces meeting at that corner. Find the magnitude of their resultant, and its inclination to the edges of the cube.

4·52 Find the centre of mass of the masses 1, 3, 4, and 2 lb situated at points with the respective position vectors $3i - j + k$, $2i + 2j + 2k$, $-i + 7j - k$, and $4i - 10k$.

4·53 Prove that the centre of mass of a system of masses is independent of the choice of origin.
    (Hint: Choose a new origin O′ with position vector b relative to the original origin O and apply the definition of centre of mass.)

4·54 The velocity of a boat relative to the water is represented by $4i + 3j$, and that of the water relative to the earth by $2i - j$. What is the velocity of the boat relative to the earth if i and j represent velocities of 1 mile/h to the east and north, respectively?

4·55 The point of application of the force $9i + 6j + 7k$ moves a distance 5 ft in the direction of the vector $3i + j + 4k$. If the modulus of the force vector is equal to the magnitude of the force in lb, find the work done.

4·56 A body spins about a line through the origin parallel to the vector $2i - j + k$ at 15 rad/s. Find the angular velocity vector Ω for the body and find the instantaneous linear velocity of a point in the body with position vector $i + 2j + 3k$.

4·57 Find the torque of a force represented by $3i + 6j + k$ about point O given that it acts through the point with position vector $-i + j + 2k$ relative to O.

4·58 Masses 1, 3, and 2 units at the points specified by the position vectors $3i - k$, $2i - 3j + k$, and $i + j + k$ relative to point O have velocities represented by $2j + k$, $3i + j + 2k$, and $i - j + k$, respectively. Determine the vector sum of the moments of momentum of each of these masses about O.

5 Differentiation of functions of one or more real variables

5·1 The derivative

The important branch of mathematics known as the calculus is concerned with two basic operations called differentiation and integration. These operations are related and both rely for their definition on the use of limits.

The calculus was founded jointly, and independently, by Newton in England, and by his contemporary Leibnitz in Germany to whom we owe the essentials of our present day notation. In introducing the ideas underlying a derivative we shall make use of a simple dynamical problem in very much the same way that Newton did when first formulating his early ideas on differentiation. However we have the advantage of understanding the nature of a limit more clearly than was the case in his day, so that after presenting our heuristic argument, we shall quickly formalize it in terms of the ideas set down in Chapter 3.

We shall consider how to define and determine the instantaneous speed of a point P moving in a non-uniform manner along a straight line. To be precise, we shall suppose that a fixed point O on the line has been selected, and that the distance s of point P from O at time t is determined by the equation

$$s = f(t),$$

where $f(t)$ is some suitable continuous function of t defined on some interval $\mathscr{I}$. Thus we know the position of P at a general time t, and are required to use this information to define and find the speed of P at any given instant of time. When the motion of P is uniform, so that its displacement is proportional to the elapsed time, the familiar definition of speed as distance per unit time can be used. However if the motion is non-uniform we must consider the situation more carefully. We shall use intuition here and first consider the *difference quotient*

$$\frac{f(t_2) - f(t_1)}{t_2 - t_1} \tag{5·1}$$

in which t_1 and t_2 are two different times belonging to $\mathscr{I}$.

It seems reasonable to suppose that if t_2 were to be taken sufficiently close to t_1 then expression (5·1), which is the quotient of the finite distance travelled and the elapsed time, would in some sense provide a measure of the average speed of P in the small time interval $t_2 - t_1$. Even better would be the idea that we compute the difference quotient (5·1) not for one time t_2 close to t_1, but for a monotonic sequence $\{\tau_i\}$ of times having for its limit the time t_1 which is not a member of the sequence. This last condition is necessary because Eqn (5·1) is not defined if $t_2 = t_1$. Then if the sequence of difference quotients corresponding to Eqn (5·1) has a limit we propose to call the value of this limit the *instantaneous speed* $u(t_1)$ of P at time t_1.

Expressed in the symbolic form of Chapter 3 we may write

$$u(t_1) = \lim_{i \to \infty} \left[\frac{f(\tau_i) - f(t_1)}{\tau_i - t_1} \right]. \tag{5·2}$$

This definition is obviously consistent with the case of the uniform motion of P, for then every difference quotient involved in the determination of the limit (5·2) would give the same constant value u, say. We will call this value u the constant speed of P.

As the function $f(t)$ is continuous it is clearly desirable that we define not in terms of the discrete variable τ_i but in terms of a continuous variable τ. Fortunately we can do this easily, for the conditions of the connecting Theorem 3·5 are satisfied and allow us to rewrite Eqn (5·2) thus:

$$u(t) = \lim_{\tau \to t} \left[\frac{f(\tau) - f(t)}{\tau - t} \right]. \tag{5·3}$$

We have now dropped the suffix 1 since t_1 was not specific and represented any value of the time t belonging to $\mathcal{I}$.

It should be appreciated that the limit $u(t)$ in Eqn (5·3) is a number and not a ratio of quantities as were the members of the sequence used to define the limit. The instantaneous speed $u(t)$ can be interpreted as the distance through which P would move in unit time if, during that time, it were to move at a constant speed equal to the value $u(t)$. Because Eqn (5·3) is consistent with the notion of a constant speed, it is customary to omit the adjective 'instantaneous' and to speak only of the speed of P.

The limit involved in Eqn (5·3) is of the indeterminate type and it will be our object to devise techniques for evaluating such limits for a wide class of functions $f(t)$. In trivial cases these may be determined by simple algebraic considerations as this example shows.

Example 5·1 Suppose that the distance of a point P from a fixed origin at time t is determined by the equation $f(t) = kt^3$, where k is a constant with dimensions $(\text{Length})(\text{Time})^{-3}$. Find the functional form of the speed $u(t)$ at time t, and determine its value when $t = 4$.

Solution We are here required to evaluate the limit

$$u(t) = \lim_{\tau \to t} \left[\frac{k(\tau^3 - t^3)}{\tau - t} \right]$$

which is the form assumed by Eqn (5·3) when $f(t) \equiv kt^3$.

Using the identity $\tau^3 - t^3 = (\tau - t)(\tau^2 + \tau t + t^2)$ we may write

$$u(t) = \lim_{\tau \to t} \left[\frac{k(\tau - t)(\tau^2 + \tau t + t^2)}{(\tau - t)} \right]$$

$$= \lim_{\tau \to t} k(\tau^2 + \tau t + t^2)$$

$$= 3kt^2.$$

Thus the functional form of the speed is $u(t) = 3kt^2$, so that at $t = 4$ the speed has the value $u(4) = 48k$.

It is often helpful to check the form of a result by means of *dimensional analysis*. This is achieved by representing the fundamental quantities of mass, length, and time occurring in expressions and equations by the symbols M, L, and T, and ignoring any purely numerical multipliers that may be involved. The equations then become identities between expressions of the form $L^p M^r T^s$, where p, r, and s are real numbers. Quantities other than length, mass, and time are represented as suitable combinations of these fundamental quantities. Thus speed and acceleration would be written LT^{-1} and LT^{-2}, respectively, with no account being taken of their magnitudes. We illustrate this approach with Example 5·1. By supposition k has dimensions LT^{-3}, so that from the form of the solution we see that $u(t)$ must have the dimensions $kT^2 = (LT^{-3})T^2 = LT^{-1}$, which are the dimensions of speed, as required.

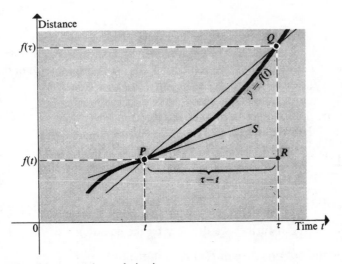

Fig. 5·1   Speed interpreted as a derivative.

There is a valuable graphical interpretation of the limit (5·3) shown in Fig. 5·1 which is the graph of a function $f(t)$ together with the chord PQ, where P is the point $(t, f(t))$ and Q the point $(\tau, f(\tau))$.

The difference quotient within the brackets of Eqn (5·3) before the limit is taken is the tangent of the angle $Q\hat{P}R$. In the limit as $\tau \to t$, so the point Q approaches the point P and the chord PQ approaches the tangent PS to the curve $y = f(t)$ at P. The value $u(t)$ arrived at by considering the limit of the difference quotient (5·3) is thus the tangent of the angle $S\hat{P}R$ and so is equal to the *gradient* or slope of the curve $y = f(t)$ at P. The number $u(t_1)$ evaluated at any specific time $t = t_1$ is the *derivative* of $f(t)$ with respect to t at $t = t_1$. The limit $u(t)$ as a function of t is simply called the derivative of $f(t)$ with respect to t and the operation of computing the derivative of a function is called *differentiation*. A function that possesses a derivative at each point of an interval is said to be *differentiable* in that interval. Hence in Example 5·1, the derivative of kt^3 with respect to t at $t = 4$ is $48k$, whereas the derivative of kt^3 with respect to t is the function $3kt^2$. The function kt^3 is obviously differentiable in any finite interval.

This heuristic approach has served to introduce the limiting arguments underlying the concept of a derivative, and we must now carefully reformulate these arguments and express them in general terms. We shall use the following key definitions.

DEFINITION 5·1   A function $f(x)$ of the real variable x will be said to be *differentiable* at x_0 if, and only if,

$$\lim_{x \to x_0} \frac{f(x) - f(x_0)}{x - x_0}$$

exists and is independent of the side from which x approaches x_0. More generally, $f(x)$ will be said to be *differentiable* in an interval $\mathscr{I}$ if it is differentiable at each point of $\mathscr{I}$. At any points of $\mathscr{I}$ for which the limit is not defined the function $f(x)$ will be said to be *non-differentiable*.

DEFINITION 5·2   If $f(x)$ is a differentiable function of the real variable x at x_0, then the value of the expression

$$\lim_{x \to x_0} \frac{f(x) - f(x_0)}{x - x_0}$$

will be denoted by $f'(x_0)$ or $\left. \dfrac{df}{dx} \right|_{x=x_0}$, and we shall say that it is the *derivative* of $f(x)$ at $x = x_0$. If further we define y by the equation $y = f(x)$, then we can also write the derivative of $f(x)$ at x_0 in the form $\left. \dfrac{dy}{dx} \right|_{x=x_0}$

These definitions merely express in a more sophisticated way, what is usually put as follows.

Let $y = f(x)$. Then if δy is the increment in y occasioned by an increment

δx in x, we have $y + \delta y = f(x + \delta x)$ and hence

$$\frac{\delta y}{\delta x} = \frac{f(x + \delta x) - f(x)}{\delta x}.$$

Thus at $x = x_0$,

$$\frac{\delta y}{\delta x} = \frac{f(x_0 + \delta x) - f(x_0)}{\delta x}$$

and so

$$\left.\frac{dy}{dx}\right|_{x=x_0} = \lim_{\delta x \to 0} \frac{f(x_0 + \delta x) - f(x_0)}{\delta x}.$$

To obtain the formulation of Definition 5·2 above, first write h in place of δx to obtain

$$\left.\frac{dy}{dx}\right|_{x=x_0} = \lim_{h \to 0} \frac{f(x_0 + h) - f(x_0)}{h},$$

and then write x in place of $x_0 + h$, so that $h = x - x_0$.

What does the requirement, that $\lim_{x \to x_0}\{[f(x) - f(x_0)]/(x - x_0)\}$ should exist, actually mean? It is this. There is a number $f'(x_0)$ such that the left- and right-hand limits of the function $\varphi(x) = [f(x) - f(x_0)]/(x - x_0)$ as x approaches x_0 exist and are both equal to $f'(x_0)$. The function $\varphi(x)$ itself is defined near but not at $x = x_0$ but has the property that $\lim_{x \to x_0} \varphi(x) = f'(x_0)$. We shall use this idea together with Theorem 3·4 when we discuss the general properties of derivatives of combinations of functions.

If in Definition 5·2 we write $x_0 + h$ in place of x, and replace x_0 by x in the subsequent result, we may formulate this definition.

DEFINITION 5·3   If $y = f(x)$ is a differentiable function of the real variable x at all points of an interval $\mathscr{I}$, then the *derivative* of $f(x)$ in $\mathscr{I}$ is the function denoted either by $f'(x)$ or dy/dx and defined by

$$f'(x) = \frac{dy}{dx} = \lim_{h \to 0} \frac{f(x + h) - f(x)}{h}.$$

The operation of computing the derivative of a function is *differentiation*.

Let us now apply exactly the same arguments to Fig. 5·2 as were used in connection with the speed at a point of the particle trajectory in Fig. 5·1. This time the graph represents any function $y = f(x)$ satisfying the conditions of Definition 5·3. Then if P is any point in the interval within which $f(x)$ is differentiable, and Q is an adjacent point, the chord PQ is, in some sense, an approximation to the tangent line to the curve PR at P. The limiting position

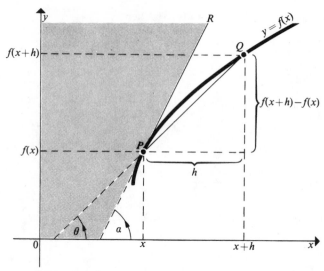

Fig. 5·2   Derivative interpreted as a gradient.

of the chord PQ will lie along the tangent line to the curve at P and in terms of angles we have $\lim_{Q \to P} \theta = \alpha$. However,

$$\frac{f(x + h) - f(x)}{h} = \tan \theta$$

so that

$$\lim_{h \to 0} \frac{f(x + h) - f(x)}{h} = \lim_{h \to 0} \tan \theta$$

whence, finally,

$$f'(x) = \tan \alpha, \tag{5·4}$$

or, equivalently,

$$\frac{dy}{dx} = \tan \alpha. \tag{5·5}$$

This result shows that we may interpret the derivative of a differentiable function at a point as the *gradient* of the tangent line drawn to the curve at that point. It is implicit in the definition that the tangent line so defined should be independent of whether Q approaches P from the left or right.

The geometrical interpretation of a derivative allows us to see quite clearly that in addition to the function needing to be continuous in the neighbourhood of a point at which it is required to be differentiable, it also needs a special kind of smoothness. Specifically, the left- and right-hand tangents to the curve at the point in question must be one and the same. Indeed, we could re-phrase our definition of differentiability in terms of the

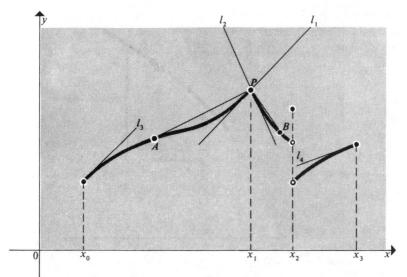

Fig. 5·3   Non-differentiable function at $x = x_1$ and $x = x_2$.

equality of the left- and right-hand tangents at a point on the curve, just as we did when dealing with continuity.

Consider the function $f = f(x)$ shown in Fig. 5·3 and defined on the interval $[x_0, x_3]$, but only continuous in the semi-open intervals $[x_0, x_2)$ and $(x_2, x_3]$.

Then, despite the fact that the function $f(x)$ is continuous in $[x_0, x_2)$ and $(x_2, x_3]$, it is only possible to assert that tangent lines in the sense implied by Definition 5·3 can be constructed for points in the open intervals (x_0, x_1), (x_1, x_2), and (x_2, x_3). No tangent line can be constructed at x_2 because of the discontinuity; two tangent lines l_1 and l_2 can be constructed at point P according as A and B approach P from the left and the right; whilst only right- and left-hand tangents l_3 and l_4 can be constructed at the end points x_0 and x_3 because the function $f(x)$ is not defined outside $[x_0, x_3]$.

We close this section by deducing the derivatives of some important elementary functions, and stating them as theorems.

THEOREM 5·1   The derivative of a constant function is zero.

Proof   Let k be any constant and consider the function $f(x)$ where $f(x) = k$ for all x. Then

$$\frac{f(x + h) - f(x)}{h} = \frac{k - k}{h} = 0 \text{ for all } x.$$

Hence

$$\lim_{h \to 0} \frac{f(x + h) - f(x)}{h} = 0 \text{ for all } x.$$

THEOREM 5·2 If n is any positive integer, then the real function $y = x^n$ is differentiable everywhere and has the derivative $dy/dx = nx^{n-1}$. If m is any negative integer, then the function $y = x^m$ is differentiable everywhere except at the origin and has the derivative $dy/dx = mx^{m-1}$.

Proof We must first consider the limit of the difference quotient $[(x + h)^n - x^n]/h$. By the binomial theorem we have

$$\frac{(x + h)^n - x^n}{h}$$

$$= \frac{x^n + nx^{n-1}h + \frac{n(n - 1)}{2!} x^{n-2}h^2 + \cdots + \binom{n}{r} x^{n-r}h^r + \cdots + h^n - x^n}{h}$$

$$= nx^{n-1} + \frac{n(n - 1)}{2!} x^{n-2}h + \cdots + \binom{n}{r} x^{n-r}h^{r-1} + \cdots + h^{n-1}.$$

Now $\lim_{h \to 0} h = 0$ so $\lim_{h \to 0} h^r = 0$ for $1 \leq r \leq n - 1$ and so

$$\lim_{h \to 0} \binom{n}{r} a^{n-r}h^{r-1} = 0.$$

Consequently,

$$\lim_{h \to 0} \frac{(x + h)^n - x^n}{h} = nx^{n-1}.$$

This is defined for all finite x including $x = 0$ and so proves the first part of the theorem. Next let $m = -n$. Then

$$\frac{(x + h)^m - x^m}{h} = \frac{(x + h)^{-n} - x^{-n}}{h} = \left(\frac{x^n - (x + h)^n}{h} \right) \frac{1}{x^n(x + h)^n}.$$

Now from our result above

$$\lim_{h \to 0} \frac{x^n - (x + h)^n}{h} = -nx^{n-1}$$

whilst

$$\lim_{h \to 0} (x + h) = x \quad \text{and so} \quad \lim_{h \to 0} (x + h)^n = x^n.$$

If $x \neq 0$,

$$\lim_{h \to 0} \frac{1}{x^n(x + h)^n} = \frac{1}{\lim_{h \to 0} x^n . \lim_{h \to 0} (x + h)^n} = \frac{1}{x^{2n}}.$$

Thus

$$\lim_{h \to 0} \frac{(x + h)^m - x^m}{h} = -nx^{n-1} . \frac{1}{x^{2n}} = -nx^{-n-1} = mx^{m-1}.$$

Hence we have proved that

$$\frac{dy}{dx}\bigg|_{x=x_0} = \frac{d}{dx}(x^n)\bigg|_{x=x_0} = nx_0^{n-1} \text{ for all } x_0$$

if n is a positive integer, and for all non-zero x_0 if n is a negative integer. This can easily be proved for all real n so, henceforth, we shall use the result freely, irrespective of the value of n.

THEOREM 5·3   The functions $\sin \alpha x$ and $\cos \alpha x$ of the real variable x, where α is any real number, are differentiable everywhere and

$$\frac{d}{dx}(\sin \alpha x) = \alpha \cos \alpha x \qquad \frac{d}{dx}(\cos \alpha x) = -\alpha \sin \alpha x.$$

Proof   These results follow by applying Definition 5·3 and then using limits (3·9) and (3·10). Thus we have

$$\frac{d}{dx}(\sin \alpha x) = \lim_{h \to 0} \frac{\sin \alpha(x+h) - \sin \alpha x}{h}$$

$$= \lim_{h \to 0} \left[\frac{\sin \alpha x \cos \alpha h + \cos \alpha x \sin \alpha h - \sin \alpha x}{h} \right]$$

$$= \sin \alpha x \lim_{h \to 0} \left(\frac{\cos \alpha h - 1}{h} \right) + \cos \alpha x \lim_{h \to 0} \left(\frac{\sin \alpha h}{h} \right)$$

$$= 0 + \alpha \cos \alpha x.$$

As this function is defined for all finite x, the first part of the required result has been established. The remainder of the proof follows exactly similar lines, and so will be omitted.

Example 5·2   Find the derivatives of the following functions stating any point at which they are not differentiable.

(a) $f(x) = \begin{cases} 3 \text{ for } -\infty < x < 1 \\ 2 \text{ for } 1 \leq x \leq \infty. \end{cases}$

(b) $f(x) = x^5$ for all x.

(c) $f(x) = \begin{cases} x^{-3} \text{ for } x \neq 0 \\ 1 \text{ for } x = 0. \end{cases}$

(d) $f(x) = \sin 4x$.

(e) $f(x) = \cos 7x$.

Solution   (a) By virtue of Theorem 5·1, the function $f(x)$ has a zero derivative for all x except at the point $x = 1$ where it is not defined.

(b) From Theorem 5·2 we have $dy/dx = 5x^4$ for all x.

(c) From Theorem 5·2 we have $dy/dx = -3x^{-4}$ for $x \neq 0$, and the derivative is not defined at $x = 0$.

(d) and (e) From Theorem 5·3 we have

$$\frac{d}{dx}(\sin 4x) = 4 \cos 4x \qquad \frac{d}{dx}(\cos 7x) = -7 \sin 7x \quad \text{for all } x.$$

By now it is obvious that Definition 5·3 is a working definition that can be used. However, some better method than its direct application is obviously needed to compute derivatives of complicated functions. This requirement will be systematically pursued in the next section.

5·2   Rules of differentiation

The complicated functions that occur in mathematical and physical studies are invariably the result of forming sums, products, and quotients of simple algebraic and trigonometric functions. This suggests that our next task should comprise a general study of the operation of differentiation when applied to sums, products, and quotients of arbitrary differentiable functions. We will present our results in the form of basic theorems which must become thoroughly familiar to the reader.

THEOREM 5·4 (differentiation of a sum) If $f(x)$ and $g(x)$ are real valued functions of x, differentiable at x_0, and k_1 and k_2 are constants, then the linear combination $k_1f(x) + k_2g(x)$ is also differentiable at x_0. Furthermore,

$$\frac{d}{dx}(k_1f(x) + k_2g(x))\Big|_{x=x_0} = k_1f'(x_0) + k_2g'(x_0).$$

Proof Here we must apply Definition 5·3 to the linear combination $k_1f(x) + k_2g(x)$. We obtain

$$\frac{d}{dx}(k_1f(x) + k_2g(x))\Big|_{x=x_0}$$

$$= \lim_{h \to 0} \frac{k_1f(x_0 + h) + k_2g(x_0 + h) - [k_1f(x_0) + k_2g(x_0)]}{h}$$

$$= k_1\lim_{h \to 0} \frac{f(x_0 + h) - f(x_0)}{h} + k_2\lim_{h \to 0} \frac{g(x_0 + h) - g(x_0)}{h}$$

$$= k_1f'(x_0) + k_2g'(x_0).$$

If f and g are both differentiable in some common interval $\mathscr{I}$, then the above argument when applied to each point of $\mathscr{I}$ yields the result

$$\frac{d}{dx}[k_1f(x) + k_2g(x)] = k_1f'(x) + k_2g'(x),$$

where x is any point of $\mathscr{I}$. The constants k_1 and k_2 are often absorbed into the functions f and g, when the result could be expressed 'the derivative of a sum of functions is equal to the sum of their derivatives'. The task of showing that this result is true for a linear combination of an arbitrary number of differentiable functions is left to the reader as an exercise involving proof by induction.

Example 5·3   Let us use Theorem 5·4 to compute the derivative of $f(x) = \sin^2 x$.

Solution   As it stands we cannot differentiate $f(x)$. However by a well known trigonometric identity we may transform $f(x)$ to the form

$$f(x) = \tfrac{1}{2}(1 - \cos 2x),$$

when Theorem 5·4 becomes applicable. Then, using our earlier results concerning the differentiation of a constant and of $\cos \alpha x$ we find that

$$\frac{d}{dx}(\sin^2 x) = \frac{d}{dx}\{\tfrac{1}{2}(1 - \cos 2x)\}$$

$$= \frac{d}{dx}(\tfrac{1}{2}) - \frac{d}{dx}(\tfrac{1}{2}\cos 2x)$$

$$= 0 - \tfrac{1}{2}\frac{d}{dx}(\cos 2x)$$

$$= -\tfrac{1}{2}.(-2)\sin 2x$$

$$= 2\sin x \cos x.$$

THEOREM 5·5 (differentiation of a product)   If $f(x)$ and $g(x)$ are differentiable real valued functions at x_0, then so also is the product function $f(x)g(x)$. Furthermore,

$$\frac{d}{dx}(f(x)g(x))\bigg|_{x \to x_0} = f'(x_0)g(x_0) + f(x_0)g'(x_0).$$

Proof   Again we consider a difference quotient but this time, for economy of expression, use the form of limit given in Definition 5·2. We have the identity

$$\frac{f(x)g(x) - f(x_0)g(x_0)}{x - x_0} \equiv \left(\frac{f(x) - f(x_0)}{x - x_0}\right)g(x) + \left(\frac{g(x) - g(x_0)}{x - x_0}\right)f(x_0). \quad \text{(I)}$$

Now we wish to show that $\lim_{x \to x_0} f(x) = f(x_0)$. This would be true if $f(x)$ were continuous but we only know that it is differentiable and as yet do not know that this implies continuity. We shall prove that it does. As $f(x)$ is differentiable at $x = x_0$ we must have

$$\frac{f(x) - f(x_0)}{x - x_0} = f'(x_0) + \varepsilon \qquad \text{as } x \to x_0,$$

where $h = x - x_0$ and $\varepsilon \to 0$ as $x \to x_0$. Hence

$$f(x) - f(x_0) = (x - x_0)[f'(x_0) + \varepsilon], \quad \text{where } \varepsilon \to 0 \text{ as } x \to x_0.$$

This implies that if x is taken sufficiently close to x_0 then the difference $f(x) - f(x_0)$ can be made arbitrarily small. This is just our definition of continuity and so we have proved that differentiability of $f(x)$ at x_0 implies its continuity at that point. Thus we are permitted to write

$$\lim_{x \to x_0} f(x) = f(x_0)$$

and, similarly,

$$\lim_{x \to x_0} g(x) = g(x_0).$$

Now

$$\lim_{x \to x_0} \left(\frac{f(x) - f(x_0)}{x - x_0} \right) = f'(x_0), \qquad \lim_{x \to x_0} \left(\frac{g(x) - g(x_0)}{x - x_0} \right) = g'(x_0),$$

so, finally, taking the limit of (I) as $x \to x_0$, we obtain the result

$$\frac{\mathrm{d}}{\mathrm{d}x} (f(x)g(x)) \Big|_{x = x_0} = f'(x_0)g(x_0) + f(x_0)g'(x_0).$$

Again, if f and g are both differentiable in some common interval $\mathscr{I}$ then, as before, we obtain the more general result

$$\frac{\mathrm{d}}{\mathrm{d}x} (f(x)g(x)) = f'(x)g(x) + f(x)g'(x) \qquad \text{for } x \in \mathscr{I}.$$

As an incidental detail of this proof we have shown that differentiability at a point implies continuity. This result is worth stating formally.

THEOREM 5·6   If a real valued function $f(x)$ is differentiable at the point x_0, then it is also continuous there. The converse result is not true.

Proof   It only remains to prove that the converse result is not true: namely, that continuity does not imply differentiability. This has already been seen in connection with Fig. 5·3, but let us give a specific example. Our final assertion in Theorem 5·6 will be valid even if we can produce only one example of a function that is continuous at a point but is not differentiable there. Such an example used to prove the falsity of an assertion is a *counterexample*, and in this case we choose the function $f(x) = |x|$. This is known to be continuous at $x = 0$, but the derivative as defined in Definition 5·3 is not defined at the origin so the function is not differentiable at that point.

Example 5·4 Differentiate the function $f(x) = \sin^2 x$ and compute $f'(\tfrac{1}{4}\pi)$.

Solution We express the function as a product and use Theorem 5·5.

$$\frac{d}{dx}(\sin^2 x) = \frac{d}{dx}(\sin x \cdot \sin x)$$

$$= \left[\frac{d}{dx}(\sin x)\right] \sin x + \sin x \left[\frac{d}{dx}(\sin x)\right]$$

$$= 2 \sin x \left[\frac{d}{dx}(\sin x)\right]$$

$$= 2 \sin x \cos x.$$

As would be expected, this verifies the result of Example 5·4. Finally, using this expression we compute

$$\frac{d}{dx}(\sin^2 x)\bigg|_{x=\frac{1}{4}\pi} = 2 \sin \frac{\pi}{4} \cos \frac{\pi}{4} = 1.$$

Our next theorem is important and concerns the rule for differentiating a composite function or, more simply, the rule for the differentiation of a function of a function.

THEOREM 5·7 (differentiation of composite functions) If $g(x)$ is a real valued differentiable function at $x = x_0$ and $f(u)$ is a real valued differentiable function at $u = g(x_0)$, then $f[g(x)]$ is differentiable at $x = x_0$. Furthermore,

$$\frac{d}{dx}\{f[g(x)]\}\bigg|_{x=x_0} = f'[g(x_0)] \cdot g'(x_0).$$

Proof We have the obvious result

$$\frac{f[g(x)] - f[g(x_0)]}{x - x_0} = \frac{f[g(x)] - f[g(x_0)]}{g(x) - g(x_0)} \cdot \frac{g(x) - g(x_0)}{x - x_0}.$$

Since $g(x)$ is differentiable at x_0 it is continuous there, and so $g(x) \to g(x_0)$ as $x \to x_0$. So, writing $g(x) = u$, $g(x_0) = a$ we have

$$\frac{f[g(x)] - f[g(x_0)]}{x - x_0} = \frac{f(u) - f(a)}{u - a} \cdot \frac{g(x) - g(x_0)}{x - x_0}. \tag{A}$$

Now for ease of argument we shall assume the behaviour of $g(x)$ to be strictly monotonic in some neighbourhood of x_0, so that $g(x) = g(x_0)$ only when $x = x_0$. In these circumstances the difference quotients on the right-hand side of (A) are well defined as $x \to x_0$ so that we may take limits and obtain

$$\frac{d}{dx}\{f[g(x)]\}\bigg|_{x=x_0} = \lim_{x \to x_0} \frac{f[g(x)] - f[g(x_0)]}{x - x_0}$$

$$= \lim_{u \to a} \left[\frac{f(u) - f(a)}{u - a} \right] \cdot \lim_{x \to x_0} \left[\frac{g(x) - g(x_0)}{x - x_0} \right]$$

$$= f'(a) \cdot g'(x_0)$$

$$= f'[g(x_0)] \cdot g'(x_0). \tag{B}$$

It is not difficult to show that the theorem is still true when $g(x)$ is not monotonic in some neighbourhood of x_0 and an infinite sequence of points $\{x_i\}$ exist with limit point x_0 at all of which $g(x_i) = g(x_0)$.

All that is necessary here is to observe that if $x \to x_0$ through the successive values x_i of this sequence, then $g(x_i) - g(x_0) = 0$ and so

$$\frac{g(x_i) - g(x_0)}{x_i - x_0} = 0 \quad \text{for every } i.$$

Hence, by Theorem 3·5, it follows that

$$\frac{d}{dx} \{g(x)\} \bigg|_{x=x_0} = 0.$$

However, by the same argument,

$$\frac{f[g(x_i)] - f[g(x_0)]}{x_i - x_0} = 0 \quad \text{for every } i,$$

showing that

$$\frac{d}{dx} \{f[g(x)]\} \bigg|_{x=x_0} = 0,$$

and so result (B) is also valid in this case.

If (B) is true at each point of some interval $\mathscr{I}$, then we have the general result

$$\frac{d}{dx} \{f[g(x)]\} = f'[g(x)] \cdot g'(x).$$

When the substitution $u = g(x)$ is made, this result can be written:

$$\frac{d}{dx} [f(u)] = \frac{df}{du} \cdot \frac{du}{dx}. \tag{5·6}$$

In this form the theorem is known as the *chain rule* for differentiation, and it is this result that is most often found in textbooks. By repeated application, the chain rule readily extends to enable the differentiation of more complicated composite functions such as the triple composite function $f\{g[h(x)]\}$, always provided the functions f, g, and h have suitable differentiability properties. In this case, setting $v = h(x)$ and $u = g(v)$ result (5·6) takes the form

$$\frac{d}{dx}[f(u)] = \frac{df}{du} \cdot \frac{du}{dv} \cdot \frac{dv}{dx}. \tag{5·7}$$

Further extensions of the same kind are obviously possible and are left to the reader.

Example 5·5 Differentiate the following functions and find the values of their derivatives at $x = 1$:

(a) $\sin(x^2 + 3)$;

(b) $(x^3 + x + 1)^{1/3}$;

(c) $\sin \sqrt{(1 + x^2)}$.

Solution (a) Set $u = x^2 + 3$ so that

$$\frac{d}{dx}[\sin(x^2 + 3)] = \frac{d}{dx}(\sin u).$$

From the chain rule:

$$\frac{d}{dx}[\sin(x^2 + 3)] = \frac{d}{du}(\sin u) \cdot \frac{du}{dx}.$$

Now $(d/du)(\sin u) = \cos u$, $du/dx = 2x$ so that

$$\frac{d}{dx}[\sin(x^2 + 3)] = (\cos u) \cdot 2x$$

$$= 2x \cos(x^2 + 3).$$

Hence at $x = 1$,

$$\frac{d}{dx}[\sin(x^2 + 3)]\Big|_{x=1} = 2 \cos 4.$$

(b) This time set $u = x^3 + x + 1$,

$$\frac{d}{dx}[(x^3 + x + 1)^{1/3}] = \frac{d}{dx}(u^{1/3}).$$

From the chain rule:

$$\frac{d}{dx}[(x^3 + x + 1)^{1/3}] = \frac{d}{du}(u^{1/3}) \cdot \frac{du}{dx}.$$

Hence as $(d/du)(u^{1/3}) = \frac{1}{3}u^{-2/3}$, $du/dx = 3x^2 + 1$ we obtain

$$\frac{d}{dx}[(x^3 + x + 1)^{1/3}] = (\tfrac{1}{3}u^{-2/3}) \cdot (3x^2 + 1)$$

$$= \tfrac{1}{3}(3x^2 + 1) \cdot (x^3 + x + 1)^{-2/3}.$$

Thus when $x = 1$,

$$\frac{d}{dx}[(x^3 + x + 1)^{1/3}]\bigg|_{x=1} = \frac{4}{3^{5/3}}.$$

(c) We must use the extension of the chain rule given in Eqn (5·7). Set $v = 1 + x^2$ when $\sin\sqrt{(1 + x^2)} = \sin\sqrt{v}$, and $u = \sqrt{v}$ when $\sin\sqrt{(1 + x^2)} = \sin u$.

Then

$$\frac{d}{dx}[\sin\sqrt{(1 + x^2)}] = \frac{d}{dx}(\sin u)$$

$$= \left[\frac{d}{du}(\sin u)\right]\frac{du}{dv}\cdot\frac{dv}{dx}$$

$$= \cos u\,\frac{du}{dv}\cdot\frac{dv}{dx}.$$

However,

$$\frac{dv}{dx} = 2x \quad \text{and} \quad \frac{du}{dv} = \tfrac{1}{2}v^{-1/2} = \frac{1}{2\sqrt{(1 + x^2)}}$$

so that, combining all the results,

$$\frac{d}{dx}[\sin\sqrt{(1 + x^2)}] = \frac{x\cos\sqrt{(1 + x^2)}}{\sqrt{(1 + x^2)}}.$$

Whence at $x = 1$,

$$\frac{d}{dx}[\sin\sqrt{(1 + x^2)}]\bigg|_{x=1} = \frac{\cos\sqrt{2}}{\sqrt{2}}.$$

THEOREM 5·8 (differentiation of a quotient) If $f(x)$ and $g(x)$ are real valued differentiable functions at x_0 and $g(x_0) \neq 0$, then the quotient $f(x)/g(x)$ is differentiable at x_0. Furthermore

$$\frac{d}{dx}\left[\frac{f(x)}{g(x)}\right]\bigg|_{x=x_0} = \frac{g(x_0)f'(x_0) - g'(x_0)f(x_0)}{[g(x_0)]^2}.$$

Proof If we consider the quotient $f(x)/g(x)$ to be the product of the two functions $f(x)$ and $1/g(x)$, we have by Theorem 5·5

$$\frac{d}{dx}\left[\frac{f(x)}{g(x)}\right]\bigg|_{x=x_0} = \frac{1}{g(x)}\cdot f'(x) + f(x)\frac{d}{dx}\left[\frac{1}{g(x)}\right]\bigg|_{x=x_0}$$

Now we must compute $(d/dx)(1/g)$. We set $g(x) = u$ when, from the chain rule,

$$\frac{d}{dx}\left[\frac{1}{g(x)}\right]\bigg|_{x=x_0} = \frac{d}{dx}\left[\frac{1}{u}\right]\bigg|_{x=x_0}$$

$$= -\frac{1}{u^2}\frac{du}{dx}\bigg|_{x=x_0}$$

$$= \frac{-g'(x_0)}{[g(x_0)]^2}.$$

Hence, combining our results, we obtain the desired result

$$\frac{d}{dx}\left[\frac{f(x)}{g(x)}\right]\bigg|_{x=x_0} = \frac{g(x_0)f'(x_0) - g'(x_0)f(x_0)}{[g(x_0)]^2}.$$

As in the other cases the general result follows when the conditions of the theorem are satisfied throughout some interval $\mathscr{I}$. It has the obvious form

$$\frac{d}{dx}\left[\frac{f(x)}{g(x)}\right] = \frac{g(x)f'(x) - g'(x)f(x)}{[g(x)]^2}.$$

Example 5·6 Differentiate $(3x + 1)/(x^2 - 2)$ and determine the values of x for which the derivative is not defined.

Solution Set $f(x) = 3x + 1$ and $g(x) = x^2 - 2$. Then $f'(x) = 3$ and $g'(x) = 2x$ for all x, whilst $g(x) = 0$ for $x = \pm\sqrt{2}$. Hence applying Theorem 5·8 we have

$$\frac{d}{dx}\left[\frac{3x + 1}{x^2 - 2}\right] = \frac{(x^2 - 2) \cdot 3 - (2x)(3x + 1)}{(x^2 - 2)^2}$$

$$= -\left[\frac{3x^2 + 2x + 6}{(x^2 - 2)^2}\right],$$

provided $x \neq \pm\sqrt{2}$.

To complete this section, Table 5·1 summarizes the results of differentiating the trigonometric functions. Unfamiliar results may be deduced by directly applying Theorem 5·8 to the definitions of the functions concerned.

Table 5·1   Derivatives of trigonometric functions

$\frac{d}{dx}(\sin x) = \cos x$	$\frac{d}{dx}(\cos x) = -\sin x$	$\frac{d}{dx}(\tan x) = \sec^2 x$
$\frac{d}{dx}(\operatorname{cosec} x) = -\operatorname{cosec} x \cot x$	$\frac{d}{dx}(\sec x) = \sec x \tan x$	$\frac{d}{dx}(\cot x) = -\operatorname{cosec}^2 x$

5·3   Some important consequences of differentiability

We preface this section by proving a result that belongs more properly to Chapter 3 since it depends for its validity only on the property of continuity. Our sole reason for discussing it here is to present it in the context in which it will first be used. It is usually known by the name of the intermediate value theorem and we shall now show that the idea underlying it is extremely simple.

Consider the situation in which a recording thermometer attached to some piece of equipment records its temperature at pre-assigned times. Suppose, for instance, that at times t_1 and t_2 the temperatures recorded were T_1 and T_2, respectively. Then although there is no record of the variation of the temperature $T(t)$ at times t between t_1 and t_2, it may be safely inferred that the temperature will pass at least once through each intermediate value between T_1 and T_2. It is quite possible for the temperature to assume values that do not lie between T_1 and T_2, but no assertion can be made about such an event. The situation is illustrated in Fig. 5·4 where T^* is a typical tempera-

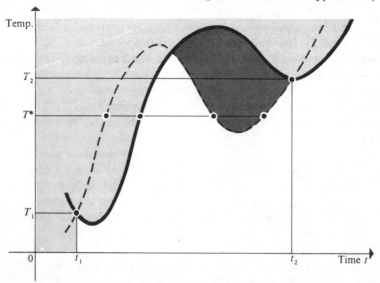

Fig. 5·4   Physical illustration of intermediate value theorem.

ture intermediate between T_1 and T_2, and the dotted and solid lines represent two possible temperature variations with time.

This physical situation is an example of the operation of the intermediate value theorem in everyday life, and we are able to make our assertion because we know from experience that however rapidly a temperature may change, it can never undergo an abrupt jump. In mathematical terms we are saying that temperature change must be a continuous process. Expressed like this the result seems obvious, but how may we prove it? Our simple proof relies on the postulate of Section 3·2, which asserts that every bounded monotonic sequence tends to a limit, but first we state the formal result.

THEOREM 5·9 (intermediate value theorem)   Let the real valued function $f(x)$ be continuous on the closed interval $[a, b]$ and such that $f(a) \neq f(b)$. Then if y^* is any number intermediate between $f(a)$ and $f(b)$, there exists a number x^* between a and b such that $y^* = f(x^*)$.

Proof   Although a diagram is not essential for this proof, the representative situation shown in Fig. 5·5 will be of help.

First set $x_1 = \frac{1}{2}(a + b)$, then if $f(x_1) = y^*$ the result is proved. If not consider the intervals (a, x_1), (x_1, b). Then in one of these two intervals, y^* will lie between the functional values occurring at either end of the interval. Call this interval I_1 and let it be represented by the open interval (a_1, b_1). Thus in Fig. 5·5, I_1 is the right-hand interval and so in that case $a_1 = \frac{1}{2}(a + b)$, $b_1 = b$.

Next set $x_2 = \frac{1}{2}(a_1 + b_1)$. If $f(x_2) = y^*$ the result is proved. If not consider the intervals (a_1, x_2), (x_2, b_1). Then in one of these two intervals, y^* will lie between the functional values occurring at either end of the interval. Call this interval I_2 and let it be represented by the open interval (a_2, b_2). in Fig. 5·5 the interval I_2 is the left-hand sub-interval of I_1, so that $a_2 = a_1$, $b_2 = \frac{1}{2}(a_1 + b_1)$.

We either prove the result directly for some x_n or we define an infinite sequence of open intervals $I_1 \supset I_2 \supset I_3 \supset \ldots ..$. Because each interval is

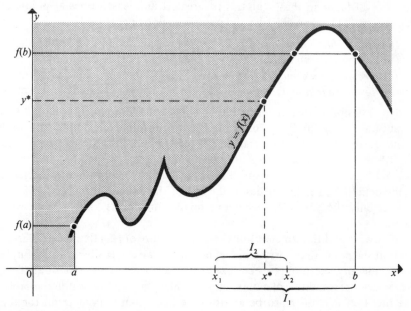

Fig. 5·5   Intermediate value theorem.

contained by all its predecessors it then follows that the sequence of numbers $a_1, a_2, a_3, \ldots$ is monotonic increasing and bounded above whilst the sequence of numbers $b_1, b_2, b_3, \ldots$ is monotonic decreasing and bounded below. Hence by the postulate of Section 3·2, the sequences $\{a_i\}$ and $\{b_i\}$ both tend to a limit. That they both tend to the same limit follows from the fact that the length of the nth interval I_n is $(b - a)/2^n$, which tends to zero as $n \to \infty$. Letting the common value of these two limits be denoted by x^* we have $\lim_{n \to \infty} |f(a_n) - f(x^*)| = 0$, thereby showing the existence of the required number x^*.

The following is an obvious consequence of the intermediate value theorem:

Corollary 5·9 Every function that is continuous in a closed interval attains both its greatest and least values at points of that interval. These values may occur at the end points of the interval.

5·3 (a) Maxima and minima

One of the most familiar and useful applications of differentiation is to the problem of determining those points in some interval $[a, b]$ at which a function $f(x)$ assumes its maximum and minimum values. Collectively these values are known as the *extrema* of the function $f(x)$ on the interval $[a, b]$ and they are of various types as this definition indicates.

DEFINITION 5·4 (extrema)   Let $f(x)$ be a continuous function defined on the interval $[a, b]$ so that it attains its greatest and least values at points of that interval. Then we say that the point x_0 belonging to $[a, b]$ is:

    (a) an *absolute maximum* if $f(x_0) \geq f(x)$ for all points x in $[a, b]$;
    (b) an *absolute minimum* if $f(x_0) \leq f(x)$ for all points x in $[a, b]$;
    (c) a *relative maximum* if $f(x_0 + h) - f(x_0) \leq 0$ for $|h|$ sufficiently small;
    (d) a *relative minimum* if $f(x_0 + h) - f(x_0) \geq 0$ for $|h|$ sufficiently small.

No assumption of differentiability has been made when formulating this definition so that in Fig. 5·6, point P is an absolute maximum and both points R and T are relative maxima. Point Q is an absolute minimum and point S a relative minimum. Although the functional value at U lies intermediate between those at Q and S, it is not a relative minimum in the sense of the definition, because it lies at the end of the domain of definition $[a, b]$ so that only the one-sided behaviour of the function is known there with respect to h.

If now, in addition to continuity, we also require of $f(x)$ that it be differentiable at the point x_0 occurring in Definition 5·4, we can easily devise a simple test to identify the points where extrema must occur. Consider point P in Fig. 5·6 as representative of a maximum at which the function is differentiable. The fact that P happens to be an absolute maximum is immaterial for the subsequent argument.

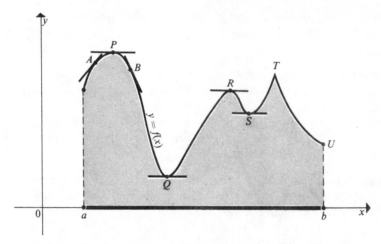

Fig. 5·6   Extrema of a function on $[a, b]$.

By supposition, if f is differentiable at P, the expression

$$f'(x_0) = \lim_{x \to x_0} \left[\frac{f(x) - f(x_0)}{x - x_0} \right]$$

must be independent of the manner of approach of x to x_0. Now for maxima of types (a) and (c) we have $f(x) - f(x_0) \leq 0$, and hence it follows that when $x < x_0$, $f'(x_0)$ is the limit of an essentially positive function; whereas when $x > x_0$, $f'(x_0)$ is the limit of an essentially negative function. Clearly this is only possible if $f'(x_0) = 0$. We have thus proved that if f is differentiable at x_0, then a necessary condition that f should have a maximum at x_0 is $f'(x_0) = 0$.

Similar reasoning establishes that the condition $f'(x_0) = 0$ is also a necessary condition for the differentiable function f to have a minimum at x_0. To show that the vanishing of the derivative f' at a point is not a sufficient condition for that point to be an extremum, we appeal to a counter-example. The function $f = x^3$ has a continuous derivative $f' = 3x^2$ which vanishes at the origin. Nevertheless, f is negative for $x < 0$ and f is positive for $x > 0$, thereby showing that despite the vanishing of the derivative, neither a maximum nor a minimum of the function can occur at the origin. Later we shall identify behaviour of this nature as typical of a *point of inflection* with a horizontal tangent. Generally speaking, a point of inflection is a demarcation point on the graph of a differentiable function separating a region of convexity from a region of concavity. Collectively the points at which the derivative vanishes, regardless of whether or not they are maxima, minima, or points of inflection are called *critical* points or *stationary* points of the function.

Combining the previous results, and recalling that the condition that f be differentiable at x_0 precludes behaviour of the type encountered at point T in Fig. 5·6, we are able to formulate the following general result.

THEOREM 5·10 Let f be a real valued differentiable function on some interval $[a, b]$. Then the *stationary points* of f are the numbers ξ for which $f'(\xi) = 0$.

Once the stationary points of a function have been determined it is necessary to examine the functional behaviour in the vicinity of each one in order to determine the nature of the point involved. An absolute maximum is identified from amongst the relative maxima by direct comparison of the functional values at the stationary points in question. A similar process identifies an absolute minimum.

Example 5·7 Without appealing to graphical ideas, find the location and nature of the extrema of the following two functions and determine if they are differentiable at these points:

(a) $f(x) = \frac{1}{3}x^3 + 2x^2 + 3x + 1$;

(b) $f(x) = (2x - 5)x^{2/3}$.

Solution (a) The stationary points are determined by finding those values $x = \xi$ for which the derivative f' vanishes.

Now $f' = x^2 + 4x + 3$ and so the desired stationary points are given by the roots of the equation

$$\xi^2 + 4\xi + 3 = 0.$$

These roots are $\xi = -1$ and $\xi = -3$, and the functional values at the respective points are $f(-1) = -\frac{1}{3}$ and $f(-3) = 1$. As the derivative f' is the sum of continuous functions it is everywhere continuous, so that no cusp-like behaviour with associated extrema as typified by point T in Fig. 5·6 can arise. So the two points $\xi = -1$ and $\xi = -3$ are the only ones at which stationary values can occur. An examination of the behaviour of the function near these points will determine if these stationary values correspond to maxima, minima, or points of inflection.

A sketch graph would quickly show that in fact $\xi = -3$ corresponds to a local maximum and $\xi = -1$ to a local minimum, but we are specifically required to establish these results by analytical means. How then can we do this? The solution lies in a direct application of Definition 5·4, and we illustrate the argument by considering the stationary point $\xi = -1$. To find the behaviour of f close to $\xi = -1$ we shall set $x = -1 + h$, where h is small, and substitute in $f(x)$ to obtain

$$f(-1 + h) = \frac{1}{3}(-1 + h)^3 + 2(-1 + h)^2 + 3(-1 + h) + 1,$$

whence,

$$f(-1 + h) = -\frac{1}{3} + h^2 + \frac{h^3}{3}.$$

Now $f(-1) = -\frac{1}{3}$ so that we may also write this result in the form

$$f(-1 + h) - f(-1) = h^2\left(1 + \frac{h}{3}\right).$$

Clearly for $|h|$ small, the right-hand side is essentially positive, and so we have succeeded in showing that close to $\xi = -1$,

$$f(\xi + h) - f(\xi) > 0,$$

and so by Definition 5·4 (d) the stationary point $\xi = -1$, at which $f(\xi) = -\frac{1}{3}$, is seen to be a local minimum. An exactly similar argument will establish that the stationary point $\xi = -3$, at which $f(\xi) = 1$, is a local maximum. These are only local extrema because it is possible to find values of x for which $f > 1$ and $f < -\frac{1}{3}$.

Solution (b)   This case is more complicated. We have

$$\frac{df}{dx} = 2x^{2/3} + \frac{2(2x - 5)}{3x^{1/3}}$$

showing that the stationary points of f are determined by the roots of the equation

$$0 = 2\xi^{2/3} + \frac{2(2\xi - 5)}{3\xi^{1/3}}.$$

This has the single root $\xi = 1$ at which $f(1) = -3$, showing that the function has only one stationary point. To determine the nature of this point let us set $x = 1 + h$, where $|h|$ is small, and substitute into $f(x)$ to find

$$f(1 + h) = (2h - 3)(1 + h)^{2/3}.$$

Next we expand the factor $(1 + h)^{2/3}$ by the binomial theorem as far as terms involving h^2 to obtain

$$f(1 + h) = (2h - 3)(1 + \tfrac{2}{3}h - \tfrac{1}{9}h^2 + O(h^3))$$

or,

$$f(1 + h) = -3 + \tfrac{4}{3}h^2 + O(h^3).$$

Using the fact that $f(1) = -3$ this becomes

$$f(1 + h) - f(1) = \tfrac{4}{3}h^2 + O(h^3)$$

showing that close to $\xi = 1$, $f(\xi + h) - f(\xi) > 0$. Hence by Definition 5·4 (d), the stationary point $\xi = 1$ is seen to correspond to a local minimum. Again, it is only a local minimum because for large negative x we have $f < -3$.

We now observe that f' is defined for all x other than for $x = 0$, at which point $f(0) = 0$. The behaviour of the function in the vicinity of the origin needs examination since, as it is not differentiable there, Theorem 5·10 can provide no information about that point. Set $x = h$, where h is small, and substitute in f to get

$$f(h) = (2h - 5)h^{2/3}.$$

Now $f(0) = 0$, so that we may rewrite this as

$$f(h) - f(0) = (2h - 5)h^{2/3},$$

thereby showing that as the right-hand side is essentially negative for suitably small h, close to $\xi = 0$ we have $f(\xi + h) - f(\xi) < 0$. From Definition 5·4 (c) we now see that the origin is a local maximum, despite the fact that f is not differentiable at that point. It is only a local maximum because for large positive x we have $f > f(0)$. For reference purposes the function is shown in Fig. 5·7.

The method of classification of stationary points that we have just illustrated is always applicable, though it provides more information than is often required. This is so because not only does it discriminate between maxima and minima, but it also provides the approximate behaviour of the function close to the point in question. We shall return to this problem later to provide much simpler criteria by which the nature of stationary points may be identified.

5·3 (b) Rolle's theorem
One form of Rolle's theorem may be stated as follows.

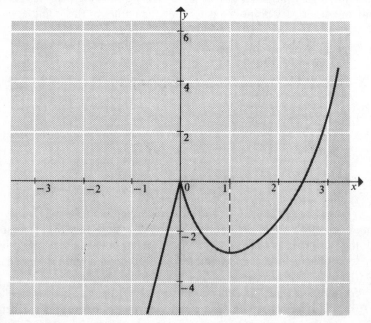

Fig. 5·7 $y = (2x - 5)x^{2/3}$.

THEOREM 5·11   Let f be a real valued function that is continuous on the closed interval $[a, b]$ and differentiable at all points of the open interval (a, b). Then if $f(a) = f(b)$ there is at least one point $x = \xi$ interior to (a, b) at which $f'(\xi) = 0$.

Proof   We know from Corollary 5·9 that a continuous function $f(x)$ defined on the closed interval $[a, b]$ must attain its maximum value M and its minimum value m at points of $[a, b]$. Then if $m = M$ on $[a, b]$, the function $f(x) = $ constant, and since the derivative of a constant is zero, the point $x = \xi$ at which $f'(\xi) = 0$ may be taken anywhere within the interval.

If $f(x)$ is not a constant function then $m \neq M$, and as $f(a) = f(b)$ it follows that at least one of the numbers m, M must differ from the value $f(a)$. We shall suppose that $M \neq f(a)$. Then clearly the value M must be attained at some point $x = \xi$ interior to (a, b). As f is assumed to be differentiable in (a, b) it follows that Theorem 5·10 must be applicable showing that $f'(\xi) = 0$. A similar argument applies if $m \neq f(a)$. Geometrically this theorem simply asserts that the graph of any function satisfying the conditions of the theorem must have at least one point in the interval $[a, b]$ at which the tangent to the curve is horizontal.

If f is not differentiable at even one interior point of (a, b) then Rolle's theorem cannot be applied. Our counter-example in this instance is the simple function $f(x) = |x|$ with $-1 \leq x \leq 1$. This function is everywhere continuous, and is differentiable at all points other than at the origin, but there is certainly no point $x = \xi$ on $[-1, 1]$ at which $f' = 0$. The graph of this function is shown in Fig. 5·8, with one of a function $g(x)$ not satisfying

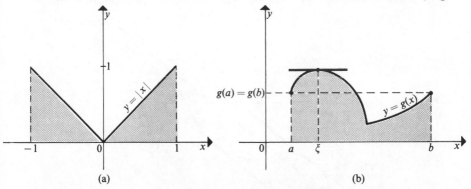

Fig. 5·8   Counter examples for Rolle's theorem: (a) Rolle's theorem does not apply—no point ξ for which $f'(\xi) = 0$; (b) $g'(\xi) = 0$, but Rolle's theorem does not apply.

the conditions of the theorem but for which the result happens to be true.

5·3 (c)   Mean value theorems for derivatives

Our most important application of Rolle's theorem will be in the proof of the mean value theorem for derivatives. In a first account of the subject it is difficult to indicate just how valuable and powerful this deceptively simple theorem really is as an analytical tool. However something of its utility will, perhaps, be appreciated after studying the remainder of this chapter. First let us present an intuitive approach to the theorem.

Consider Fig. 5·9 which represents a graph of a differentiable function $f(x)$ on the open interval (a, b). Then as P and S are the points $(a, f(a))$ and $(b, f(b))$, the gradient m of the line PS is

$$m = \frac{f(b) - f(a)}{b - a}.$$

Now we may identify points Q and R, with respective x-coordinates ξ and η interior to (a, b), at which the tangent lines l_1 and l_2 to the graph are parallel to PS, and so must also have the same gradient m. Then because of the geometrical interpretation of the derivative f' as the gradient of the tangent line, at either P or Q we may equate m and f'. If we confine attention to point Q we have

$$\frac{f(b) - f(a)}{b - a} = f'(\xi),$$

where $a < \xi < b$. This is the form in which the mean value theorem for derivatives, also known as the *law of the mean*, is usually quoted. In geometrical terms the theorem asserts that there is always a point $(\xi, f(\xi))$ on the graph of the function, with $a < \xi < b$, at which the tangent to the curve is parallel to the secant line PS. The fact that the precise value of ξ is not usually known is, generally speaking, unimportant in the application of this

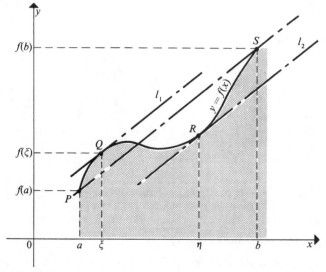

Fig. 5·9 Illustration of the mean value theorem.

theorem. This is because it is often used with some limiting argument in which $b \to a$, so that $\xi \to a$ also. A formal statement of the theorem is as follows.

THEOREM 5·12 (mean value theorem for derivatives)   If $f(x)$ is a real valued function that is continuous in $[a, b]$ and differentiable in (a, b), then there exists a point ξ interior to (a, b) such that

$$\frac{f(b) - f(a)}{b - a} = f'(\xi).$$

The existence of more than one point ξ in (a, b) at which this result is true is not precluded. This is so because it is only asserted that such a point exists, and not that there is necessarily only one such point. Such is the case, for example, in Fig. 5·9 since as was remarked, $f'(\xi) = f'(\eta)$ with $\xi \neq \eta$, though both points ξ and η are interior to (a, b).

Many people would regard the argument above as proof enough of the mean value theorem, but for the more critical reader we now offer the promised proof based on Rolle's theorem.

Proof As with the proofs of many mathematical theorems, our result is established more easily by a somewhat artificial approach than by a direct method. Here we shall utilize the intuitively obtained result above to suggest the form of a special function $F(x)$ to which Rolle's theorem can be applied, thereby yielding the desired result.

Specifically, since by implication the result depends on $f(x)$ and x, we shall try to find the simplest function $F(x)$ that depends on $f(x)$ and x, that is continuous in $[a, b]$ and is differentiable in (a, b), and is such that $F(a) = F(b)$. The value of $F(a)$ may be assigned arbitrarily and $F(x)$ will still satisfy Rolle's theorem, so to simplify slightly the working we shall assume that $F(a) = F(b) = 0$.

We consider the obvious function

$$F(x) = A + Bx + f(x)$$

which clearly satisfies the continuity and differentiability conditions of Rolle's theorem. The constants A and B must be chosen in order that $F(a) = F(b) = 0$.

Thus

$$0 = A + Ba + f(a)$$

and

$$0 = A + Bb + f(b)$$

from which it follows that,

$$A = -f(a) + a\left(\frac{f(b) - f(a)}{b - a}\right), \qquad B = \frac{f(b) - f(a)}{a - b}.$$

Hence $F(x)$ has the form

$$F(x) = f(x) - f(a) + \left[\frac{f(b) - f(a)}{b - a}\right](a - x).$$

Thus we have succeeded in finding a function $F(x)$ with the desired properties which satisfies Rolle's theorem. Differentiating $F(x)$ we obtain

$$F'(x) = f'(x) - \left[\frac{f(b) - f(a)}{b - a}\right].$$

Now by Rolle's theorem there exists a point ξ, with $a < \xi < b$, such that $F'(\xi) = 0$ and so we have our desired result

$$\frac{f(b) - f(a)}{b - a} = f'(\xi).$$

Since we may write $\xi = a + \theta(b - a)$, where $0 < \theta < 1$, this result is sometimes expressed in the following form attributable to Cauchy,

$$f(b) - f(a) = (b - a)f'[a + \theta(b - a)] \qquad \text{with } 0 < \theta < 1.$$

By applying the same arguments to a suitably constructed function $\varphi(x)$, analogous to $F(x)$, it is a simple matter to prove the following extension of the mean value theorem due to Cauchy. (See Problem 5·15.)

Corollary 5·12 If $g'(x) = h'(x)$ at all points of $[a, b]$ then $g(x) = h(x) +$ constant in $[a, b]$.

Proof Set $f = g - h$ in Theorem 5·12 applied to the interval $[a, x]$. Then $g(x) - h(x) = g(a) - h(a) = $ constant and the result follows.

THEOREM 5·13 (Cauchy extended mean value theorem) If $f(x)$ and $g(x)$ are real valued functions that are continuous in $[a, b]$ and differentiable in (a, b) and $g'(x) \neq 0$ in (a, b), then there exists a point ξ interior to (a, b) such that

$$\frac{f(b) - f(a)}{g(b) - g(a)} = \frac{f'(\xi)}{g'(\xi)}.$$

5·3 (d) Indeterminate forms—L'Hospital's rule

Limits such as $\lim_{x \to 0} (\sin \alpha x)/x$ which apparently tend to the form $0/0$ have already been encountered and given meaning in special cases. A closely related problem is that of giving meaning to the limit of a quotient which apparently tends to ∞/∞. These limit problems are both called *indeterminate forms.* One of the most obvious applications of the extended mean value theorem is to resolve the value of the limit in either of these situations, and we now prove the simplest statement of a useful result generally known as *L'Hospital's Rule.*

THEOREM 5·14 (first form of L'Hospital's rule)   If $f(x)$ and $g(x)$ are real valued differentiable functions at $x = x_0$ and,

(a) $f(x_0) = g(x_0) = 0$,

(b) $\lim\limits_{x \to x_0} \dfrac{f'(x)}{g'(x)} = \lambda$, where λ is either a real number or infinity,

then

$$\lim_{x \to x_0} \frac{f(x)}{g(x)} = \lim_{x \to x_0} \frac{f'(x)}{g'(x)} = \lambda.$$

Proof   Apply the extended mean value theorem to the functions $f(x)$ and $g(x)$ defined on the interval $[x, x_0]$ and use condition (a) to obtain

$$\frac{f(x)}{g(x)} = \frac{f'(\xi)}{g'(\xi)},$$

where $x < \xi < x_0$.

Now $x \to x_0$ implies that $\xi \to x_0$, so that by condition (b) we have the desired result

$$\lim_{x \to x_0} \frac{f(x)}{g(x)} = \lim_{\xi \to x_0} \frac{f'(\xi)}{g'(\xi)} = \lambda.$$

The fact that the variable ξ appears in the second limit in place of the x stated in the theorem is unimportant. Its function is simply that of a variable and the symbol used to denote it is immaterial.

In general, when the symbol used to denote a variable is unimportant because it only appears in some intermediate calculation, the details of which do not concern us, we shall call it a *dummy variable.*

A useful extension of L'Hospital's rule is contained in the following corollary which allows examination of limits which tend to the form ∞/∞.

Corollary 5·14   If $\varphi(x)$ and $\psi(x)$ are real valued differentiable functions at $x = x_0$ and,

(a) $\lim\limits_{x \to x_0} \varphi(x) \to \pm\infty$, $\lim\limits_{x \to x_0} \psi(x) \to \pm\infty$,

(b) $\lim\limits_{x \to x_0} \dfrac{\varphi'(x)}{\psi'(x)} = \lambda$, where λ is either a real number or infinity,

then

$$\lim_{x \to x_0} \frac{\varphi(x)}{\psi(x)} = \lim_{x \to x_0} \frac{\varphi'(x)}{\psi'(x)} = \lambda.$$

Proof Apply the extended mean value theorem to the quotient $\varphi(x)/\psi(x)$ in the open interval (x, x_1) with $x_0 < x < x_1$, and write the result in the form

$$\frac{\varphi(x)}{\psi(x)} = \left[\frac{1 - \dfrac{\psi(x_1)}{\psi(x)}}{1 - \dfrac{\varphi(x_1)}{\varphi(x)}} \right] \cdot \frac{\varphi'(\xi)}{\psi'(\xi)},$$

where $x < \xi < x_1$. Then, taking x_1 fixed and arbitrarily close to x_0 so that $\xi \to x_0$, allow $x \to x_0$. The first factor on the right-hand side then approaches arbitrarily close to unity thereby giving rise to the stated result. A modification of this argument shows that the result is also true if $x_0 \to \infty$.

Example 5·8 Determine the value of the following indeterminate forms using L'Hospital's rule and Corollary 5·14:

(a) $\displaystyle\lim_{x \to 0} \frac{\sin \alpha x}{x}$;

(b) $\displaystyle\lim_{x \to 1} \frac{x^3 + 3x^2 - 2x - 2}{2x^2 - x - 1}$;

(c) $\displaystyle\lim_{x \to 0} \frac{\sin 3x}{x^3}$;

(d) $\displaystyle\lim_{x \to \frac{1}{4}\pi} \frac{\tan 3x}{\tan x}$;

(e) $\displaystyle\lim_{x \to 0} \frac{\left(\dfrac{a}{x}\right)}{\cot bx}$.

Solution (a) This is of the form $\lim f/g \to 0/0$ with $f(x) = \sin \alpha x$ and $g(x) = x$. As $f'(x) = \alpha \cos \alpha x$ and $g'(x) = 1$ it follows that

$$\lim_{x \to 0} \frac{\sin \alpha x}{x} = \lim_{x \to 0} \frac{\alpha \cos \alpha x}{1} = \alpha.$$

This confirms the limit that was obtained by a different method in Chapter 3.

(b) This is also of the form $\lim f/g \to 0/0$ but this time with $f(x) = x^3 + 3x^2 - 2x - 2$ and $g(x) = 2x^2 - x - 1$. It follows that $f'(x) = 3x^2 + 6x - 2$ and $g'(x) = 4x - 1$ so that

$$\lim_{x \to 1} \frac{x^3 + 3x^2 - 2x - 2}{2x^2 - x - 1} = \lim_{x \to 1} \frac{3x^2 + 6x - 2}{4x - 1} = \frac{7}{3}.$$

(c) This is again of the form $\lim f/g \to 0/0$ with $f(x) = \sin 3x$ and $g(x) = x^3$. Hence $f'(x) = 3 \cos 3x$ and $g'(x) = 3x^2$ so that

$$\lim_{x \to 0} \frac{\sin 3x}{x^3} = \lim_{x \to 0} \frac{\cos 3x}{x^2} \to +\infty.$$

(d) This is of the form $\lim f/g \to \infty/\infty$ with $f(x) = \tan 3x$ and $g(x) = \tan x$. Hence $f'(x) = 3 \sec^2 3x$ and $g'(x) = \sec^2 x$ and by Corollary 5·14,

$$\lim_{x \to \frac{1}{2}\pi} \frac{\tan 3x}{\tan x} = \lim_{x \to \frac{1}{2}\pi} \frac{3 \sec^2 3x}{\sec^2 x} = 3 \lim_{x \to \frac{1}{2}\pi} \frac{\cos^2 x}{\cos^2 3x}.$$

This is again an indeterminate form, but now of the type 0/0. Applying Theorem 5·14 we have

$$3 \lim_{x \to \frac{1}{2}\pi} \frac{\cos^2 x}{\cos^2 3x} = 3 \lim_{x \to \frac{1}{2}\pi} \frac{2 \sin x \cos x}{6 \sin 3x \cos 3x} = \lim_{x \to \frac{1}{2}\pi} \left(\frac{\sin x}{\sin 3x} \right) \cdot \lim_{x \to \frac{1}{2}\pi} \left(\frac{\cos x}{\cos 3x} \right)$$

and hence

$$\lim_{x \to \frac{1}{2}\pi} \frac{\tan 3x}{\tan x} = - \lim_{x \to \frac{1}{2}\pi} \frac{\cos x}{\cos 3x}.$$

This last result is yet again an indeterminate form of the type 0/0 so that a further application of Theorem 5·14 finally gives

$$\lim_{x \to \frac{1}{2}\pi} \frac{\tan 3x}{\tan x} = - \lim_{x \to \frac{1}{2}\pi} \frac{\sin x}{3 \sin 3x} = \frac{1}{3}.$$

(e) This is of the form $\lim f/g \to \infty/\infty$ but it is easily seen that an application of Corollary 5·14 will not simplify the limit to be evaluated. Instead, we rewrite the limit in the form

$$\lim_{x \to 0} \frac{\left(\dfrac{a}{x} \right)}{\cot bx} = \lim_{x \to 0} a \frac{\tan bx}{x}$$

when it is seen that the alternative form is of the type $\lim f/g \to 0/0$ with $f(x) = a \tan bx$ and $g(x) = x$. Now $f'(x) = ab \sec^2 bx$ and $g'(x) = 1$ so that by Theorem 5·14,

$$\lim_{x \to 0} \frac{\left(\dfrac{a}{x} \right)}{\cot bx} = \lim_{x \to 0} \frac{ab \sec^2 x}{1} = ab.$$

5·3 (e)   Identification of extrema

We return to the topic of extrema and, in particular, to the identification of functional behaviour at stationary values by means of the mean value theorem.

Suppose that a real valued function $f(x)$ is differentiable in the interval (a, b) and has a maximum at an interior point x_0 of (a, b).

Then if h is assumed to be positive and we consider the interval $[x_0 - h, x_0]$ to the left of x_0, by the mean value theorem

$$\frac{f(x_0) - f(x_0 - h)}{h} = f'(\xi),$$

where $x_0 - h < \xi < x_0$.

Now by supposition $h > 0$ and as x_0 is a maximum, the numerator of this expression will also be positive showing that $f'(\xi) > 0$. Hence by allowing h to tend to zero, it follows that $\xi \to x_0$ and we have shown that to the immediate left of the maximum we must have $f' > 0$.

To the right of the maximum, and in the interval $[x_0, x_0 + h]$, the same argument shows that

$$\frac{f(x_0 + h) - f(x_0)}{h} = f'(\eta),$$

where $x_0 < \eta < x_0 + h$. This numerator is negative so that to the immediate right of the maximum we must have $f' < 0$.

Similar arguments applied to a minimum and a point of inflection with a horizontal tangent yield the following useful theorem, illustrated in Fig. 5·10.

THEOREM 5·15 (identification of extrema using first derivative)   If $f(x)$ is a real valued differentiable function in the neighbourhood of a point x_0 at which $f'(x_0) = 0$ then:

(a) the function has a maximum at x_0 if $f'(x) > 0$ to the left of x_0 and $f'(x) < 0$ to the right of x_0;

(b) the function has a minimum at x_0 if $f'(x) < 0$ to the left of x_0 and $f'(x) > 0$ to the right of x_0;

(c) the function has a point of inflection with zero gradient at x_0 if $f'(x)$ has the same sign to the left and right of x_0.

In many books these results are regarded as intuitively obvious deductions from the geometrical interpretation of a derivative in conjunction with the behaviour of the graph of the function. However we have discussed them formally here as an illustration of an important consequence of the mean value theorem.

Example 5·9   We again consider the functions of Example 5·7.

Case (a) $f(x) = \frac{1}{3}x^3 + 2x^2 + 3x + 1$ with stationary points $x = \xi$ at $\xi = -1$ and $\xi = -3$. As $f'(x) = x^2 + 4x + 3$ it follows that to the immediate left of $\xi = -1$ we have $f' < 0$, whilst to the immediate right $f' > 0$

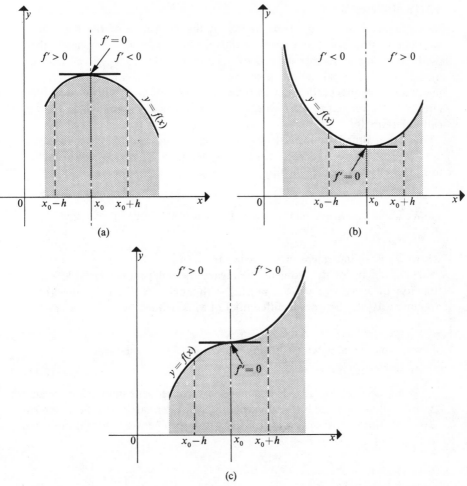

Fig. 5·10   Stationary values of $y = f(x)$: (a) local maximum; (b) local minimum; (c) point of inflection with zero gradient.

showing that $\xi = -1$ corresponds to a minimum. A similar argument shows that $\xi = -3$ corresponds to a maximum.

Case (b) $f(x) = (2x - 5)x^{2/3}$ with the one stationary point $x = \xi$ at $\xi = 1$. As $f'(x) = 2x^{2/3} + 2(2x - 5)/3x^{1/3}$ it follows that $f' < 0$ to the immediate left of $\xi = 1$ and $f' > 0$ to the immediate right. Hence $\xi = 1$ corresponds to a minimum. As Theorem 5·15 stands, since f is not differentiable at the origin, the maximum that occurs there must be identified as in Example 5·7. However a trivial modification of the proof would show that results (a) and (b) of the theorem are still valid if f is not differentiable at x_0.

5·3 (f) Differentials

In using the notation dy/dx to represent the derivative of the dependent variable y with respect to x we have thus far been careful to emphasize that dy/dx is simply a number defined by a limit. Although suggestive of increments, dy and dx taken separately have as yet no individual meaning. In many applications, particularly in differential equations which we encounter later, it is convenient to work with actual quantities dy and dx which we will call differentials.

However differentials must obviously be defined in a manner consistent with the notation dy/dx when it is used to denote the derivative with respect to x of the function y defined by

$$y = f(x). \tag{5·8}$$

We achieve this by defining dy, the *first-order* differential of y, by

$$dy = f'(x) \,.\, \Delta x, \tag{5·9}$$

where Δx is an increment in x of arbitrary size.

However, if, for the moment, we regard the independent variable x as a function of x we can write $x = g(x)$ with $g(x) = x$. Then by the above argument dx, the first-order differential of x, is defined by

$$dx = 1 \,.\, \Delta x, \tag{5·10}$$

showing that we may with meaning write Eqn (5·9) in the form

$$dy = f'(x)dx. \tag{5·11}$$

When needed, the actual increment in y consequent upon an increment Δx in x will be denoted by Δy. In general the differential dy and the increment Δy are distinct quantities and the interrelationship between them is indicated in Fig. 5·11.

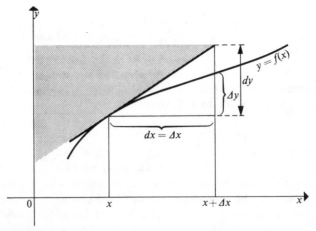

Fig. 5·11 Differentials dx and dy.

In more advanced treatments the use of differentials is strictly avoided on account of logical difficulties encountered with their definition. However they are so useful that we shall ignore these objections and use them freely whenever necessary.

It is an immediate consequence of this that if

$$y = k_1 f(x) + k_2 g(x)$$

then by Theorem 5·4,

$$dy = k_1 f'(x)dx + k_2 g'(x)dx$$

or, equivalently, in symbolic notation

$$d(k_1 f + k_2 g) = k_1 df + k_2 dg. \tag{5·12}$$

If we have

$$y = f(x)g(x)$$

then by Theorem 5·5,

$$dy = g(x)f'(x)dx + f(x)g'(x)dx$$

or, equivalently, in symbolic notation

$$d(fg) = g df + f dg. \tag{5·13}$$

Finally, if

$$y = f(x)/g(x)$$

then by Theorem 5·8,

$$dy = \frac{g(x)f'(x)dx - f(x)g'(x)dx}{g^2(x)}$$

or, equivalently, in symbolic notation

$$d\left(\frac{f}{g}\right) = \frac{g df - f dg}{g^2}. \tag{5·14}$$

Example 5·10   If $f(x) = \sin(x^2 + 4)$ and $g(x) = x^3$ find the differentials:

(a) $d(3f + g)$;

(b) $d(fg)$;

(c) $d\left(\dfrac{f}{g}\right)$.

Solution

(a) $d(3f + g) = d[3 \sin(x^2 + 4) + x^3]$

$\qquad\qquad\quad = 3 \cos(x^2 + 4)d(x^2 + 4) + 3x^2 dx$

$\qquad\qquad\quad = 6x \cos(x^2 + 4)dx + 3x^2 dx.$

(b) $d(fg) = d[x^3 \sin (x^2 + 4)]$

$\qquad = 3x^2 \sin (x^2 + 4)dx + x^3 \cos (x^2 + 4)d(x^2 + 4)$

$\qquad = 3x^2 \sin (x^2 + 4)dx + 2x^4 \cos (x^2 + 4)dx.$

(c) $d\left(\dfrac{f}{g}\right) = d\left[\dfrac{\sin (x^2 + 4)}{x^3}\right]$

$\qquad = \dfrac{x^3 \cos (x^2 + 4)d(x^2 + 4) - 3x^2 \sin (x^2 + 4)dx}{x^6}$

$\qquad = \dfrac{2x^2 \cos (x^2 + 4)dx - 3 \sin (x^2 + 4)dx}{x^4}.$

For small values of dx, the differential dy is obviously a reasonable approximation to the actual increment Δy. This simple observation is often utilized to relate small changes in dependent and independent variables as the next example shows.

Example 5·11 The pressure p of a polytropic gas is related to the density ρ by the expression

$$p = A\rho^\gamma,$$

where A is a constant. Deduce the relationship connecting the differentials dp and dρ. Given that $\gamma = 3/2$ and $\rho = 4$, and taking dp as an approximation to the actual pressure change Δp, compute the approximate new pressure if ρ is increased by 0·1. Compare the approximate and exact results.

Solution In this case $p = f(\rho)$ with $f(\rho) = A\rho^\gamma$. Hence $f'(\rho) = \gamma A\rho^{\gamma-1}$ and thus the desired differential relation is

$$dp = \gamma A\rho^{\gamma-1}d\rho.$$

When $\gamma = 3/2$ and $\rho = 4$ it follows from the stated pressure–density law that the initial pressure p_0 is

$$p_0 = 4^{3/2}A = 8A.$$

Using the differential relation to compute the approximate pressure increase represented by the differential dp we find

$$dp = (3/2) \cdot A \cdot 4^{1/2} \cdot (0\cdot1) = 0\cdot3A.$$

Hence the approximate new pressure $p_0 + dp = 8\cdot3A$.

The exact new pressure $p_0 + \Delta p$ may be computed from the pressure–density law by setting $\rho = 4\cdot1$ to obtain

$$p_0 + \Delta p = (4\cdot1)^{3/2}A = 8\cdot308A.$$

This shows that in this case the differential relation gives a good approximation to the pressure increase.

5·4   Higher derivatives—applications

We have seen how differentiation applied to a suitable function $f(x)$ yields as a result another function $f'(x)$, the derivative of $f(x)$ with respect to x. If the function $f'(x)$ is itself differentiable then a repetition of differentiation will result in a further function that we shall denote by $f''(x)$ and will call the *second derivative* of $f(x)$ with respect to x. We may usefully employ the dynamical problem that served to introduce the notion of a derivative to give meaning to the notion of a second derivative, for if $f'(x)$ represents a velocity, then $f''(x)$ represents an acceleration. If the function $f''(x)$ is itself differentiable then it is customary to denote the *third derivative* of $f(x)$ by $f'''(x)$ after which, if necessary, further derivatives are conventionally denoted by the use of superscript roman numerals. Hence the sixth derivative of a suitably differentiable function $f(x)$ would be written $f^{\text{vi}}(x)$.

A better notation than this is needed for general purposes and the two most often used because of their versatility are

$$\frac{d^n y}{dx^n} \quad \text{or} \quad D^n y.$$

These both represent the nth derivative with respect to x of $y = f(x)$ and for their determination require the successive application of differentiation n times. The number n is the *order* of the derivative and the symbol D symbolizes the operation of differentiation. Computationally the definition of the nth derivative of y with respect to x is equivalent to using either of these two equivalent algorithms

$$\frac{d}{dx}\left(\frac{d^{n-1}y}{dx^{n-1}}\right) = \frac{d^n y}{dx^n} \quad \text{or} \quad D[D^{n-1}y] = D^n y. \tag{5·15}$$

These expressions are, of course, only meaningful when n is an integer and we shall agree to the convention $D^0 y \equiv y$.

Geometrically, the function $d^n y/dx^n$ bears to the graph of $d^{n-1}y/dx^{n-1}$, the same relationship as does the function dy/dx to the graph of y. Namely $d^n y/dx^n$ at $x = x_0$ is the gradient of the graph of $d^{n-1}y/dx^{n-1}$ as a function of x at the same point $x = x_0$.

Example 5·12 Determine dy/dx, d^2y/dx^2, and d^3y/dx^3 given that $y = f(x)$ with:

(a) $f(x) = \cos mx$;

(b) $f(x) = \tan x$;

(c) $f(x) = 1/(1 + x)$.

If possible make deductions about the nth derivative.

Solution

(a) $\dfrac{dy}{dx} = f'(x) = \dfrac{d}{dx}\,(\cos mx) = -m \sin mx,$

$\dfrac{d^2y}{dx^2} = \dfrac{d}{dx}\left(\dfrac{dy}{dx}\right) = \dfrac{d}{dx}\,[-m \sin mx] = -m^2 \cos mx,$

$\dfrac{d^3y}{dx^3} = \dfrac{d}{dx}\left(\dfrac{d^2y}{dx^2}\right) = \dfrac{d}{dx}\,[-m^2 \cos mx] = m^3 \sin mx.$

An inductive argument easily shows that the nth derivative $(d^n/dx^n)(\cos mx)$ $= m^n \cos [mx + (n\pi/2)]$.

In respect of the function $y = \cos mx$, it is of importance to notice that the simple algebraic equation

$$\frac{d^2y}{dx^2} + m^2 y = 0$$

connects the function and its second derivative. Because this equation involves derivatives it is a *differential equation*. Such equations are very important in both mathematics and the mathematical sciences; the last three chapters of this book provide an introductory study of them.

(b) $\dfrac{dy}{dx} = f'(x) = \dfrac{d}{dx}\,(\tan x) = \sec^2 x,$

$\dfrac{d^2y}{dx^2} = \dfrac{d}{dx}\left(\dfrac{dy}{dx}\right) = \dfrac{d}{dx}\,(\sec^2 x)\,2 = \sec^2 x \tan x,$

$\dfrac{d^3y}{dx^3} = \dfrac{d}{dx}\left(\dfrac{d^2y}{dx^2}\right) = \dfrac{d}{dx}\,(2 \sec^2 x \tan x) = 2 \sec^2 x(2 \tan^2 x + \sec^2 x).$

There is no simple rule by which $(d^n/dx^n)(\tan x)$ may be computed.

(c) $\dfrac{dy}{dx} = f'(x) = \dfrac{d}{dx}\left(\dfrac{1}{1+x}\right) = \dfrac{-1}{(1+x)^2},$

$\dfrac{d^2y}{dx^2} = \dfrac{d}{dx}\left(\dfrac{dy}{dx}\right) = \dfrac{d}{dx}\left[\dfrac{-1}{(1+x)^2}\right] = \dfrac{2}{(1+x)^3},$

$\dfrac{d^3y}{dx^3} = \dfrac{d}{dx}\left(\dfrac{d^2y}{dx^2}\right) = \dfrac{d}{dx}\left[\dfrac{2}{(1+x)^3}\right] = \dfrac{-3!}{(1+x)^4}.$

It follows by induction that

$$\frac{d^n}{dx^n}\left(\frac{1}{1+x}\right) = \frac{(-1)^n n!}{(1+x)^{n+1}}.$$

In general, functions are not capable of differentiation an indefinite number of times, and at some stage they usually become non-differentiable. A simple example of a function that is not differentiable an indefinite number of times, though for a different reason from the above, is x^n, with n an integer. The nth derivative of x^n is the constant number $n!$ so that the $(n + 1)$th and all subsequent derivatives are identically zero.

5·4 (a) Leibnitz's theorem

This useful theorem is a consequence of Theorem 5·5 and facilitates the computation of high-order derivatives of the product $f(x)g(x)$ of the two functions $f(x)$ and $g(x)$, in terms of the derivatives of the individual functions $f(x)$ and $g(x)$ themselves.

The result is, perhaps, best expressed in terms of the symbolic differentiation operator D, and for our starting point we now re-express the result of Theorem 5·5 in terms of the operator D.

$$D(fg) = fDg + gDf.$$

Assuming functions $f(x)$ and $g(x)$ are suitably differentiable, a further application of the operator D together with Theorem 5·5 yields

$$D^2(fg) = D(fDg + gDf)$$
$$= Df \cdot Dg + fD^2g + Dg \cdot Df + gD^2f.$$

However

$$Df \cdot Dg = \frac{df}{dx} \cdot \frac{dg}{dx} = \frac{dg}{dx}\frac{df}{dx} = Dg \cdot Df,$$

so that

$$D^2(fg) = fD^2g + 2Df \cdot Dg + gD^2f. \tag{5·16}$$

A repetition of the same argument shows that

$$D^3(fg) = fD^3g + 3Df \cdot D^2g + 3D^2f \cdot Dg + gD^3f. \tag{5·17}$$

The coefficients involved in Eqns (5·16) and (5·17) are seen to belong to the general pattern of binomial coefficients in the expansion of $(a + b)^n$, namely to the rows of numbers

$(n = 0)$ $\quad \binom{0}{0}$

$(n = 1)$ $\quad \binom{1}{0} \quad \binom{1}{1}$

$(n = 2)$ $\quad \binom{2}{0} \quad \binom{2}{1} \quad \binom{2}{2}$

$(n = 3)$ $\quad \binom{3}{0} \quad \binom{3}{1} \quad \binom{3}{2} \quad \binom{3}{3}$

· · · · · · · · · · ·

or, equivalently, to the rows

$(n = 0)$	1			
$(n = 1)$	1	1		
$(n = 2)$	1	2	1	
$(n = 3)$	1	3	3	1

. . . .

This suggests that in evaluating $D^n(fg)$, the coefficients arising should belong to the $(n + 1)$th row of either of these arrays, which are *Pascal triangles*. That this is so can be proved fairly easily, using an inductive argument similar to that used to prove the binomial theorem. We shall not give the details, preferring simply to state the theorem.

THEOREM 5·16 (Leibnitz's theorem)   If $f(x)$ and $g(x)$ are n times differentiable real valued functions in the interval (a, b), then

$$D^n(fg) = \sum_{k=0}^{n} \binom{n}{k} D^{n-k}f \cdot D^k g.$$

The value and power of this is best shown by an application.

Example 5·13   Use Leibnitz's theorem to evaluate $(d^3/dx^3)(x^6 \sin x)$.

Solution   Setting $n = 3$ in the general result gives

$$D^3(fg) = gD^3f + 3D^2f \cdot Dg + 3Df \cdot D^2g + fD^3g.$$

This is, of course, result (5·17) differently expressed. Now we make the identifications $f(x) = x^6$ and $g(x) = \sin x$ when it follows that $Df = 6x^5$, $D^2f = 30x^4$, $D^3f = 120x^3$, and $Dg = \cos x$, $D^2g = -\sin x$, $D^3g = -\cos x$. Hence substitution into the above result gives

$$D^3(x^6 \sin x) = 120x^3 \sin x + 90x^4 \cos x - 18x^5 \sin x - x^6 \cos x.$$

5·4 (b)   Identification of extrema by second derivatives

An important application of the second derivative of a function $f(x)$ is to the identification of the nature of its extrema. Let us suppose that $f(x)$ is twice differentiable and that $f'(x_0) = 0$ and $f''(x_0) = L < 0$.

Then from Definition 5·2 and the notion of a second derivative we must have that

$$f''(x_0) = \lim_{x \to x_0} \frac{f'(x) - f'(x_0)}{x - x_0} = L < 0.$$

By supposition $f'(x_0) = 0$, so that

$$f''(x_0) = \lim_{x \to x_0} \frac{f'(x)}{x - x_0} = L < 0.$$

This limit must be independent of the manner in which x approaches x_0 so that we must consider separately the cases that x lies to the left or to the right of x_0.

If x lies to the left of x_0 then $x - x_0 < 0$. Consequently, as the value L of the limit is negative, the expression defining $f''(x_0)$ implies that to the immediate left of x_0 it must be true that $f'(x) > 0$.

If x lies to the right of x_0 then $x - x_0 > 0$. Consequently, as the value L of the limit is negative, the expression defining $f''(x_0)$ implies that to the immediate right of x_0 it must be true that $f'(x) < 0$.

These results, in conjunction with Theorem 5·15 (a) prove that at a stationary value x_0, for which $f''(x_0) < 0$, the function $f(x)$ attains a *maximum* value. An exactly similar argument proves that at a stationary value x_0, for which $f''(x_0) > 0$, the function $f(x)$ attains a *minimum* value.

To complete the argument, consider the situation in which $f''(x_0) = 0$. It might be conjectured that this corresponds to a point of inflection; and to establish the correctness of our intuition let us appeal to the geometrical interpretation of a derivative as a gradient.

Suppose that x_0 corresponds to a point of inflection with zero gradient. Then as x increases through the value x_0, either

(a) $f'(x)$ is initially positive and decreases to a minimum value $f'(x_0) = 0$, thereafter increasing again (cf. Fig. 5·10 (c));

or,

(b) $f'(x)$ is initially negative and increases to a maximum value $f'(x_0) = 0$, thereafter decreasing again.

In each case x_0 is a stationary value of the first derivative $f'(x)$, so that by an application of Theorem 5·10 to the function $f'(x)$ we find that $f''(x_0) = 0$ at a point of inflection.

We have thus proved the following theorem.

THEOREM 5·17 (identification of extrema using second derivatives) Let $f(x)$ be a real valued twice differentiable function in (a, b) with a stationary point x_0 in (a, b), so that $f'(x_0) = 0$. Then, if

(a) $f''(x_0) < 0$ the function $f(x)$ has a maximum at x_0,

(b) $f''(x_0) > 0$ the function $f(x)$ has a minimum at x_0,

(c) $f''(x_0) = 0$ the function $f(x)$ has a point of inflection at x_0 with zero gradient provided that the sign of $f'(x)$ is the same to the immediate left and right of x_0.

The proof of this theorem shows clearly what was asserted earlier; namely that a point of inflection on the graph of a function separates a region of convexity from a region of concavity. There is, of course, no necessity that this point should have associated with it a zero gradient.

Following this argument to its logical conclusion we see that the proof of

(c) above need only involve the sign of $f'(x)$ to the left and right of x_0 when $f'(x_0) = 0$, for then such arguments are needed to distinguish between an extremum and a point of inflection. If $f'(x_0) \neq 0$ such problems do not arise and it is sufficient to look for those values ξ for which $f''(\xi) = 0$. We have thus proved the following general result.

THEOREM 5·18 (location of points of inflection) If $f(x)$ is a real valued twice differentiable function then its points of inflection, if any, occur at the numbers ξ for which $f''(\xi) = 0$ provided that $f'(\xi) \neq 0$. If however this is not so, and $f'(\xi) = 0$, then ξ corresponds to a point of inflection provided that the sign of $f'(x)$ is the same to the immediate left and right of ξ.

It is left to the reader as an exercise to prove that when $f'(x_0) = f''(x_0) = 0$, then provided $f'''(x_0)$ exists, our condition on $f'(x)$ may be replaced by the requirement $f'''(x_0) \neq 0$. The proof is essentially similar to that given for Theorem 5·17 though this time the starting point is the definition of $f''(x_0)$ expressed as a limit. We give this result as a corollary.

Corollary 5·18 If $f(x)$ is a real valued thrice differentiable function and $f'(\xi) = f''(\xi) = 0$, then $f(x)$ has a point of inflection at $x = \xi$ if $f'''(\xi) \neq 0$.

Example 5·14 Locate and identify the stationary values of the following functions. Find any points of inflection they may have, together with the gradient of the tangent line at such points:

(a) $f(x) = x^3 - 12x + 1$ in $[-10, 10]$;
(b) $f(x) = \tan x$ in $[-\frac{1}{4}\pi, \frac{1}{4}\pi]$;
(c) $f(x) = (x - 1)^3$ in $(-\infty, \infty)$.

Solution (a) The stationary values are those numbers ξ for which $f'(\xi) = 0$. Hence as $f'(x) = 3x^2 - 12$, the stationary values are determined by the equation

$$3\xi^2 - 12 = 0.$$

This has roots $\xi = 2$, $\xi = -2$ which both lie in $[-10, 10]$ and are the desired stationary values. As $f''(x) = 6x$, it follows that $f''(2) = 12 > 0$ and $f''(-2) = -12 < 0$. Hence by Theorem 5·17, the point $\xi = 2$ is a minimum and the point $\xi = -2$ is a maximum. Since the function has no other stationary value there can be no point of inflection at which the tangent line has zero gradient. However $f''(x) = 6x$ vanishes when $x = 0$, so that by Theorem 5·18 we see that $x = 0$ must correspond to a point of inflection. The gradient at $x = 0$ is $f'(0) = -12$ which is the gradient of the desired tangent line to the graph at the point of inflection.

(b) Here we have $f'(x) = \sec^2 x$ and clearly, since $\sec^2 x = 1 + \tan^2 x$, it follows that $f'(x) \neq 0$ in $[-\frac{1}{4}\pi, \frac{1}{4}\pi]$. The function $f(x) = \tan x$ thus has no stationary values in $[-\frac{1}{4}\pi, \frac{1}{4}\pi]$, though it assumes its greatest value at $\frac{1}{4}\pi$

and its least value at $-\frac{1}{4}\pi$. We have $f''(x) = 2\sec^2 x \tan x$ which vanishes for $x = 0$. Hence by Theorem 5·18, the function $\tan x$ has a point of inflection at the origin at which the gradient of the tangent to the graph has the value $f'(0) = 1$.

(c) We see that $f'(x) = 3(x - 1)^2$ and so the condition $f'(\xi) = 0$ yields $\xi = 1$ as the single stationary value. However, $f''(x) = 6(x - 1)$ which shows that we also have $f''(1) = 0$. Appealing to the last part of Theorem 5·18 we see that, as $f'(x) = 3(x - 1)^2 > 0$ to both the left and right of $x = 1$, it follows that $f(x) = (x - 1)^3$ has a point of inflection at that point. The tangent line to the graph there has a zero gradient. Alternatively, as $f'''(x) = 6 \neq 0$, the result also follows from Corollary 5·18.

5·5   Partial differentiation

The notion of continuity has already been extended so that it is meaningful in the context of functions of several independent variables. It is now appropriate to extend the notion of a derivative in a similar fashion. For simplicity of argument we shall work with the function $f(x, y)$ of two independent variables, and in order to visualize its behaviour geometrically we will define a dependent variable by the equation

$$u = f(x, y). \tag{5·18}$$

The function may then be represented as a surface in three dimensional space.

A typical surface generated by a function of the form of Eqn (5·18) is shown in Fig. 5·12 and, unlike functions of one independent variable, it is necessary to define more than one first-order derivative. The idea involved is simple: by holding one of the independent variables in f constant at some value of interest, the function f then becomes a function of the single remaining independent variable. We may then differentiate f as though it were a function only of that one variable. By holding first x and then y constant in this manner, two different derivatives may be defined which, because of their manner of computation, will be called *partial derivatives* to distinguish them from our earlier use of the term derivative. We shall now express these ideas formally as a definition and set down the standard notation to be used.

DEFINITION 5·5 (partial derivatives)   Let $f(x, y)$ be a function defined near (x_0, y_0). Suppose that

$$\lim_{x \to x_0} \frac{f(x, y_0) - f(x_0, y_0)}{x - x_0} \tag{A}$$

exists and is independent of the direction of approach of x to x_0. Then f is differentiable *partially* with respect to x at (x_0, y_0). The value of the limit is denoted by $f_x(x_0, y_0)$ or by $\partial f / \partial x|_{(x_0, y_0)}$ and called the first-order partial derivative of f with respect to x at (x_0, y_0).

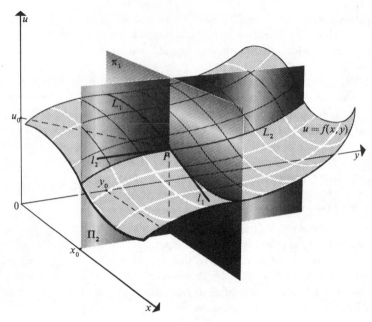

Fig. 5·12 Geometrical interpretation of partial derivatives.

Similarly, suppose that

$$\lim_{y \to y_0} \frac{f(x_0, y) - f(x_0, y_0)}{y - y_0}$$
(B)

exists and is independent of the direction of approach of y to y_0. Then f is differentiable partially with respect to y at (x_0, y_0). The limit is denoted by $f_y(x_0, y_0)$ or by $\partial f/\partial y|_{(x_0, y_0)}$ and called the first-order partial derivative of f with respect to y at (x_0, y_0).

By analogy with ordinary derivatives, if $f(x, y)$ is differentiable partially with respect to x and y at all points of some region in the (x, y)-plane and these derivatives are continuous, then we say f is differentiable in that region. The operations of partial differentiation with respect to x and y are usually denoted by the differentiation operators $\partial/\partial x$ and $\partial/\partial y$, respectively.

Let us now interpret these definitions in terms of Fig. 5·12. The function $f(x, y_0)$ occurring in the numerator of limit (A) in Definition 5·5 is represented in that figure by the intersection of the surface $u = f(x, y)$ with the plane $y = y_0$ which has been labelled Π_1. It is the curve L_1. The number $f_x(x_0, y_0)$ defined by limit (A) is the gradient of the tangent line l_1 to this curve at point P. By requiring the limit to be independent of the direction of approach of x to x_0, we have ensured that the tangent lines drawn to the curve at P, whether from the left or the right, will have the same gradient. In simpler terms this ensures that the curve L_1 is smooth and has no kink at P.

The number $f(x_0, y)$ occurring in the numerator of limit (B) in the definition is represented in Fig. 5·12 by the intersection of the surface $u = f(x, y)$ with the plane $x = x_0$ which has been labelled Π_2. It is the curve L_2. The number $f_y(x_0, y_0)$ defined by limit (B) is the gradient of the tangent line l_2 to this curve at point P.

Thus by differentiating partially we mean that, during the process of differentiation, the other independent variable is to be regarded as a constant. In consequence, all the rules of differentiation developed for functions of a single variable are also rules of partial differentiation, provided only that the functions involved are suitably differentiable. On account of this when, for example, the operator $\partial/\partial x$ acts on a function only of y, say $g(y)$, that function is to be regarded as a constant with respect to this operator and so $(\partial/\partial x)[g(y)] \equiv 0$. Similarly $(\partial/\partial y)[h(x)] \equiv 0$.

Example 5·15 In each of the following cases compute f_x and f_y as functions of x and y. Use the result to determine the numerical value of these derivatives at the stated points:

(a) $f(x, y) = x^3 + 2xy + 2y^2$; $(1, 2)$;

(b) $f(x, y) = x \sin xy + 3$; $(1, \frac{1}{2}\pi)$;

(c) $f(x, y) = x/(x^2 + y^2)$; $(1, 0)$.

Solution

(a) $f_x = \dfrac{\partial}{\partial x} [x^3 + 2xy + 2y^2]$

$$= \frac{\partial}{\partial x} [x^3] + 2y \frac{\partial}{\partial x} [x] + 2y^2 \frac{\partial}{\partial x} [1],$$

whence

$$\frac{\partial f}{\partial x} = 3x^2 + 2y.$$

At the point $(1, 2)$ we find that $\partial f/\partial x|_{(1,2)} = 7$. Similarly,

$$f_y = \frac{\partial}{\partial y} [x^3 + 2xy + 2y^2]$$

$$= x^3 \frac{\partial}{\partial y} [1] + 2x \frac{\partial}{\partial y} [y] + 2 \frac{\partial}{\partial y} [y^2]$$

whence

$$\frac{\partial f}{\partial y} = 2x + 4y.$$

At the point $(1, 2)$ we find that $\partial f/\partial y|_{(1,2)} = 10$.

(b) $f_x = \dfrac{\partial}{\partial x}[x \sin xy + 3]$

$= x \dfrac{\partial}{\partial x}[\sin xy] + \sin xy \dfrac{\partial}{\partial x}[x] + \dfrac{\partial}{\partial x}[3]$

whence

$\dfrac{\partial f}{\partial x} = xy \cos xy + \sin xy.$

At the point $(1, \frac{1}{2}\pi)$ we find that $\partial f/\partial x|_{(1,\frac{1}{2}\pi)} = 1$. Similarly,

$f_y = \dfrac{\partial}{\partial y}[x \sin xy + 3]$

$= x \dfrac{\partial}{\partial y}[\sin xy] + \dfrac{\partial}{\partial y}[3]$

whence

$\dfrac{\partial f}{\partial y} = x^2 \cos xy$

and

$\dfrac{\partial f}{\partial y}\bigg|_{(1,\frac{1}{2}\pi)} = 0.$

(c) $f_x = \dfrac{\partial}{\partial x}\left[\dfrac{x}{x^2 + y^2}\right]$

$= \dfrac{1}{x^2 + y^2}\dfrac{\partial}{\partial x}[x] + x\dfrac{\partial}{\partial x}[(x^2 + y^2)^{-1}]$

$= \dfrac{1}{x^2 + y^2} - \dfrac{x}{(x^2 + y^2)^2}\dfrac{\partial}{\partial x}[x^2 + y^2],$

whence

$\dfrac{\partial f}{\partial x} = \dfrac{1}{x^2 + y^2} - \dfrac{2x^2}{(x^2 + y^2)^2} = \dfrac{y^2 - x^2}{(x^2 + y^2)^2}.$

At the point $(1, 0)$ we find that $\partial f/\partial x|_{(1,0)} = -1$. Similarly,

$f_y = \dfrac{\partial}{\partial y}\left[\dfrac{x}{x^2 + y^2}\right]$

$= x\dfrac{\partial}{\partial y}[(x^2 + y^2)^{-1}]$

$$= \frac{-x}{(x^2 + y^2)^2} \frac{\partial}{\partial y} [x^2 + y^2],$$

whence

$$\frac{\partial f}{\partial y} = \frac{-2xy}{(x^2 + y^2)^2}$$

and so

$$\frac{\partial f}{\partial y}\bigg|_{(1,0)} = 0.$$

The notion of partial differentiation extends to functions of more than two independent variables in an obvious manner. Suppose that the function $f(x, y, z)$ is defined near the point (x_0, y_0, z_0) then, provided the limits exist, we define the three first-order partial derivatives f_x, f_y, and f_z by the expressions

$$\frac{\partial f}{\partial x}\bigg|_{(x_0, y_0, z_0)} = \lim_{x \to x_0} \frac{f(x, y_0, z_0) - f(x_0, y_0, z_0)}{x - x_0},$$

$$\frac{\partial f}{\partial y}\bigg|_{(x_0, y_0, z_0)} = \lim_{y \to y_0} \frac{f(x_0, y, z_0) - f(x_0, y_0, z_0)}{y - y_0},$$

$$\frac{\partial f}{\partial z}\bigg|_{(x_0, y_0, z_0)} = \lim_{z \to z_0} \frac{f(x_0, y_0, z) - f(x_0, y_0, z_0)}{z - z_0}.$$

Clearly a function of n independent variables will have n different first-order partial derivatives; one with respect to each of the independent variables. The actual computation of these partial derivatives is carried out exactly as before.

Example 5·16 Find the first-order partial derivatives of

$$f(x, y, z) = x^3 y^2 + 3 \sin yz + 2.$$

Solution This function has three independent variables so we must obtain three first-order partial derivatives. Namely, f_x, f_y, and f_z. First we have

$$\frac{\partial f}{\partial x} = \frac{\partial}{\partial x} [x^3 y^2 + 3 \sin yz + 2]$$

$$= y^2 \frac{\partial}{\partial x} [x^3] + 3 \sin yz \frac{\partial}{\partial x} [1] + \frac{\partial}{\partial x} [2],$$

so

$$\frac{\partial f}{\partial x} = 3x^2 y^2.$$

Next,

$$\frac{\partial f}{\partial y} = \frac{\partial}{\partial y} [x^3y^2 + 3 \sin yz + 2]$$

$$= x^3 \frac{\partial}{\partial y} [y^2] + 3 \frac{\partial}{\partial y} [\sin yz] + \frac{\partial}{\partial y} [2],$$

so

$$\frac{\partial f}{\partial y} = 2x^3y + 3z \cos yz.$$

Finally,

$$\frac{\partial f}{\partial z} = \frac{\partial}{\partial z} [x^3y^2 + 3 \sin yz + 2]$$

$$= x^3y^2 \frac{\partial}{\partial z} [1] + 3 \frac{\partial}{\partial z} [\sin yz] + \frac{\partial}{\partial z} [2],$$

so

$$\frac{\partial f}{\partial z} = 3y \cos yz.$$

5·6   Total differential

The idea of a differential, that was useful in ordinary differentiation, may also be developed to advantage in connection with partial differentiation. . We first approach this problem from the geometrical standpoint, and then indicate how an analytical counterpart of these arguments can be produced.

Let us consider Eqn (5·18) and its geometrical representation in Fig. 5·12. The conditions for differentiability at P ensure that the surface has a tangent plane Π at that point (why?), and it is to this plane that we now confine our attention. An element of this tangent plane defined by the lines l_1 and l_2 through P is depicted in Fig. 5·13. Obviously points on Π close to P must also be close to those points on the surface $u = f(x, y)$ that lie vertically below them. This suggests that for such points, the element of plane Π neighbouring P represents a good approximation to the element of the curved surface defining the function u near to P. Thus variations of u close to P may, with propriety, be approximated by the variations of the corresponding points on Π.

Since we are interested in variations of u about the point P at which $u_0 = f(x_0, y_0)$, we shall start by translating our coordinate axes without rotation to the point P. In this position the new x, y, and u coordinate axes will be denoted by x', y', and u', respectively, as shown in Fig. 5·14.

If, relative to P, the x' and y' coordinates of a point P′ are Δx and Δy, then it is obvious from Fig. 5·14 that the increment du must be

$$\mathrm{d}u = \Delta x \tan \alpha + \Delta y \tan \beta,$$

where α and β are the angles between the lines l_1 and l_2 and the x'- and y'-axes, respectively.

However, by the definition of f_x and f_y, we have

$$f_x(x_0, y_0) = \tan \alpha, \qquad f_y(x_0, y_0) = \tan \beta,$$

so that

$$\mathrm{d}u = f_x(x_0, y_0)\Delta x + f_y(x_0, y_0)\Delta y. \tag{5·19}$$

We now define differentials dx and dy in the independent variables x and y

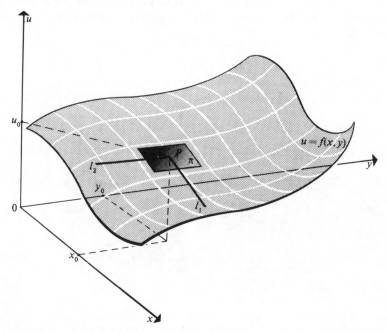

Fig. 5·13   Tangent plane Π to surface $u = f(x, y)$ at point P.

by setting $\mathrm{d}x = \Delta x$ and $\mathrm{d}y = \Delta y$. Expression (5·19) then becomes

$$\mathrm{d}u = f_x(x_0, y_0)\mathrm{d}x + f_y(x_0, y_0)\mathrm{d}y, \tag{5·20}$$

which is the relationship by which we define the *total* differential du of the function $u = f(x, y)$. This is so called because it takes account of the total effect, on u, of the changes dx in x and dy in y. The additive effect of these changes is clearly apparent in Fig. 5·14 and results from using a tangent plane approximation to the surface near P. As before, when dx and dy are suitably small, du is a reasonable approximation to the true change Δu given by

$$\Delta u = f(x_0 + dx, y_0 + dy) - f(x_0, y_0). \tag{5·21}$$

An analytic rather than geometric justification of the tangent plane approximation used to define du in Eqn (5·20) can be based on Theorem 5·12.

Equation (5·21), which is exact, is taken to be the starting point and by addition and subtraction of a term $f(x_0, y_0 + \Delta y)$, is written

$$\Delta u = [f(x_0 + \Delta x, y_0 + \Delta y) - f(x_0, y_0 + \Delta y)]$$
$$+ [f(x_0, y_0 + \Delta y) - f(x_0, y_0)],$$

where the first bracket is a function only of x and the second bracket is a function only of y.

Then Theorem 5·12 expressed in the Cauchy form may be applied to the first bracket with respect to x and to the second bracket with respect to y to yield

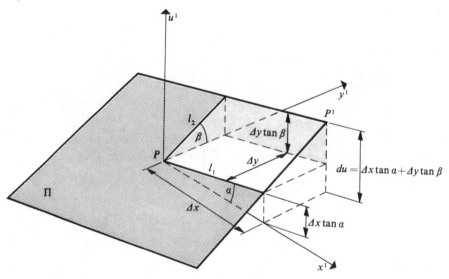

Fig. 5·14 Element of tangent plane.

$$\Delta u = \Delta x f_x(x_0 + \xi \Delta x, y_0 + \Delta y) + \Delta y f_y(x_0, y_0 + \eta \Delta y), \tag{5·22}$$

where $0 < \xi < 1$ and $0 < \eta < 1$. Partial derivatives have been used here because, although in the first bracket it is only x that varies whilst in the second bracket it is only y that varies, both brackets are nevertheless functions of x and y.

Result (5·20) then follows by letting Δx and Δy become small. The continuity of $f_x(x_0 + \xi \Delta x, y_0 + \Delta y)$ allows it to be approximated by

$f_x(x_0, y_0)$ with an error ε_1 and, similarly, the continuity of $f_y(x_0, y_0 + \eta\Delta y)$ allows it to be approximated by $f_y(x_0, y_0)$ with an error ε_2. Then, as Δx, $\Delta y \to 0$, so also do ε_1 and ε_2. It is left as an exercise for the reader to supply the details necessary to make this argument rigorous. If Eqn (5·20) is defined for all points (x_0, y_0) of some region in the (x, y)-plane, theh the suffix zero may be discarded and Eqn (5·20) can then be regarded as a functional relationship rather than a result that is true only near one point.

We have thus proved a special case of the following more general result whose proof differs in no significant detail.

THEOREM 5·19 (total differential)   Let $f(x_1, x_2, \ldots, x_n)$ be a real valued function of n real variables and let its first-order partial derivatives exist and be continuous in some region $\mathscr{R}$. Then the total differential du of the function $u = f(x_1, x_2, \ldots, x_n)$ in the region $\mathscr{R}$ is given by

$$du = \frac{\partial f}{\partial x_1}\, dx_1 + \frac{\partial f}{\partial x_2}\, dx_2 + \cdots + \frac{\partial f}{\partial x_n}\, dx_n.$$

If we consider the surface generated by setting $u = $ constant, then on that surface $du \equiv 0$. Theorem 5·19 then takes the form

$$0 = \frac{\partial f}{\partial x_1}\, dx_1 + \frac{\partial f}{\partial x_2}\, dx_2 + \cdots + \frac{\partial f}{\partial x_n}\, dx_n, \tag{5·23}$$

showing that the differentials $dx_1, dx_2, \ldots, dx_n$ are no longer independent since this *constraint* condition has been imposed on them. This is of course to be expected, since we have imposed the single condition $f(x_1, x_2, \ldots, x_n)$ = constant on the independent variables $u_1, u_2, \ldots, u_n$ so that we are no longer free to change them arbitrarily. Indeed, if differentials $dx_1, dx_2, \ldots, dx_{n-1}$ are chosen arbitrarily, then the remaining differential dx_n is uniquely determined by Eqn (5·23). If we call the number of independent variables the number of *degrees of freedom* associated with the equation $u = f(x_1, x_2, \ldots, x_n)$, then Eqn (5·23) implies the loss of a single degree of freedom.

Example 5·17   In thermodynamics, the pressure p of an ideal gas, its volume V, its absolute temperature T and the gas constant R are related by the *ideal gas law* $pV = RT$. Find the expression relating the total differential dp and the differentials dV and dT.

Solution   We have $p = RT/V$, and so $p = f(T, V)$ with $f(T, V) = RT/V$. Hence $\partial f/\partial T = R/V$ and $\partial f/\partial V = -RT/V^2$. Now interpreting Theorem 5·19 in this case we find

$$dp = \left(\frac{\partial f}{\partial T}\right) dT + \left(\frac{\partial f}{\partial V}\right) dV, \tag{*}$$

and so

$$dp = \frac{R}{V}\,dT - \frac{RT}{V^2}\,dV.$$

Notice that the use of the symbol f in the total differential relation (*) to bring it into accord with the notation of Theorem 5·19 is not strictly necessary since $p \equiv f$. We could equally well have written equation (*) as

$$dp = \left(\frac{\partial p}{\partial T}\right)dT + \left(\frac{\partial p}{\partial V}\right)dV,$$

and used the immediately obvious result that

$$\frac{\partial p}{\partial T} = \frac{R}{V} \quad \text{and} \quad \frac{\partial p}{\partial V} = -\frac{RT}{V^2}.$$

Let us now consider the function $u = f(x, y)$ and, as a special case, set $u = 0$ so that the equation

$$f(x, y) = 0$$

defines y implicitly in terms of x. How then may we compute the derivative dy/dx without solving for y in terms of x? The solution to this problem is provided by Eqn (5·23), which in this case takes the form

$$0 = \frac{\partial f}{\partial x}\,dx + \frac{\partial f}{\partial y}\,dy.$$

We saw in connection with the definition of the differentials dy and dx in Eqn (5·11), that the function (dy/dx), called the derivative of y with respect to x, is the ratio $dy : dx$ of the differentials. Hence dividing by the differential dx, assuming that $\partial f/\partial y \neq 0$, and rearranging gives the result

$$\frac{dy}{dx} = \frac{-(\partial f/\partial x)}{(\partial f/\partial y)}.$$

We state this as a corollary to Theorem 5·19.

Corollary 5·19   If the real variables x and y are related implicitly by the equation $f(x, y) = 0$, and the partial derivatives $\partial f/\partial x$ and $\partial f/\partial y$ exist and are continuous, then

$$\frac{dy}{dx} = -\left(\frac{\partial f}{\partial x}\right)\bigg/\left(\frac{\partial f}{\partial y}\right),$$

whenever $\partial f/\partial y \neq 0$. Insistence on this latter condition may be avoided by writing the result in the alternative form

$$\left(\frac{\partial f}{\partial y}\right)\frac{dy}{dx} + \frac{\partial f}{\partial x} = 0.$$

5·7   The chain rule and its consequences

If, in Theorem 5·19, the variables x_1, x_2, . . ., x_n are specified in terms of a parameter t, say, then the result requires slight modification. Suppose that

$$x_1 = x_1(t), x_2 = x_2(t), . . ., x_n = x_n(t),$$

which are all differentiable functions of t. Then the variable u becomes a function of the single real variable t for we may write

$$u = F(t),   \tag{5·24}$$

where $F(t) = f(x_1(t), x_2(t), . . ., x_n(t))$.

Hence by an obvious adaptation of Eqn (5·11) defining differentials we may write

$$du = F'(t)dt,   \tag{5·25}$$

where, of course, $F'(t) = du/dt$ the derivative of u with respect to t.

However by a further application of Eqn (5·11) to each of the variables $x_1 = x_1(t)$, $x_2 = x_2(t)$, . . ., $x_n = x_n(t)$ we have the result

$$dx_1 = \left(\frac{dx_1}{dt}\right) dt, dx_2 = \left(\frac{dx_2}{dt}\right) dt, . . ., dx_n = \left(\frac{dx_n}{dt}\right) dt.   \tag{5·26}$$

Substituting these expressions for the differentials dx_i in terms of the differential dt into the statement of Theorem 5·19 gives

$$du = \left(\frac{\partial f}{\partial x_1}\frac{dx_1}{dt} + \frac{\partial f}{\partial x_2}\frac{dx_2}{dt} + \cdots + \frac{\partial f}{\partial x_n}\frac{dx_n}{dt}\right) dt.   \tag{5·27}$$

Finally, a comparison of Eqns (5·30) and (5·32) shows that

$$F'(t) = \frac{\partial f}{\partial x_1}\frac{dx_1}{dt} + \frac{\partial f}{\partial x_2}\frac{dx_2}{dt} + \cdots + \frac{\partial f}{\partial x_n}\frac{dx_n}{dt}.$$

As $F'(t) = du/dt$, this result facilitates the calculation of du/dt without the need for formal substitution into $u = f(x_1, x_2, . . ., x_n)$ of the values $x_1 = x_1(t)$, $x_2 = x_2(t)$, . . ., $x_n = x_n(t)$.

We have proved the following useful result.

THEOREM 5·20 (chain rule for partial derivatives)   Let $u = f(x_1, x_2, . . ., x_n)$ be a real valued function of n real variables and let its first-order partial derivatives exist and be continuous. Further, let each of the variables x_1, x_2, . . ., x_n be a differentiable function of the single real variable t so that we may write

$$x_1 = x_1(t), x_2 = x_2(t), . . ., x_n = x_n(t).$$

Then the total derivative of u with respect to t is given by

$$\frac{du}{dt} = \frac{\partial f}{\partial x_1}\frac{dx_1}{dt} + \frac{\partial f}{\partial x_2}\frac{dx_2}{dt} + \cdots + \frac{\partial f}{\partial x_n}\frac{dx_n}{dt}.$$

Two special cases of this theorem are of sufficient importance to merit recording as corollaries. The first arises when f is a function of only two variables between which an explicit relationship exists, and the parameter t is identified with one of these variables.

As only two variables are involved we shall avoid the use of numerical suffixes by agreeing to write $x_1 = x$ and $x_2 = y$ where, by supposition, $y = y(x)$ is some known explicit relation. The statement of Theorem 5·20 then becomes

$$\frac{du}{dt} = \frac{\partial f}{\partial x}\frac{dx}{dt} + \frac{\partial f}{\partial y}\frac{dy}{dt}.$$

If, now, we identify t with x, then $t = x$ and $dx/dt = 1$, $dy/dt = dy/dx$ so that the above result becomes

$$\frac{du}{dx} = \frac{\partial f}{\partial x} + \frac{\partial f}{\partial y}\frac{dy}{dx}.$$

The expression on the right-hand side is the total derivative of u with respect to x. The first term on the right takes account of the change directly due to x whilst the second term takes account of the fact that y is itself a function of x. This result enables du/dx to be obtained without needing to substitute $y = y(x)$ in the relation $u = f(x, y)$.

Corollary 5·20 (a)   If $u = f(x, y)$ is a real valued function of the real variables x and y with continuous first-order derivatives and y is related to x by the explicit equation $y = y(x)$, then

$$\frac{du}{dx} = \frac{\partial f}{\partial x} + \frac{\partial f}{\partial y}\frac{dy}{dx}.$$

More generally, suppose that $u = f(x, y)$ whilst x and y are related implicitly by the equation

$$g(x, y) = 0.$$

How must we modify our previous argument in order that we may compute the total derivative du/dx? The result of Corollary 5·20 (a) is still true but obviously dy/dx now depends on the form of g. To find the form of dy/dx we can use Corollary 5·19, writing $f = g$, to see that

$$\frac{dy}{dx} = -\left(\frac{\partial g}{\partial x}\right) \bigg/ \left(\frac{\partial g}{\partial y}\right),$$

showing that

$$\frac{du}{dx} = \frac{\partial f}{\partial x} - \left(\frac{\partial f}{\partial y}\right)\left(\frac{\partial g}{\partial x}\right) \bigg/ \left(\frac{\partial g}{\partial y}\right),$$

provided $\partial g/\partial y \neq 0$. We state this as our next result.

Corollary 5·20 (b)   If $u = f(x, y)$ is a real valued function of the real variables x and y with continuous first-order derivatives, and y is related implicitly to x by the equation $g(x, y) = 0$, then

$$\frac{du}{dx} = \frac{\partial f}{\partial x} - \left(\frac{\partial f}{\partial y}\right)\left(\frac{\partial g}{\partial x}\right) \bigg/ \left(\frac{\partial g}{\partial y}\right),$$

provided $\partial g/\partial y \neq 0$.

Example 5·18   Determine the derivative du/dt given that

$$u = \sin(x^2 + y^2) \qquad \text{with } x = 3t, \ y = 1/(1 + t^2).$$

Solution   We must apply Theorem 5·20 making the identifications $x_1 = x$, $x_2 = y$, and $f(x, y) = \sin(x^2 + y^2)$ with $x = 3t$ and $y = 1/(1 + t^2)$. Hence

$$\frac{\partial f}{\partial x} = 2x \cos(x^2 + y^2) \qquad \frac{\partial f}{\partial y} = 2y \cos(x^2 + y^2)$$

whilst

$$\frac{dx}{dt} = 3, \qquad \frac{dy}{dt} = \frac{-2t}{(1 + t^2)^2}.$$

Substituting in Theorem 5·20,

$$\frac{du}{dt} = 2x \cos(x^2 + y^2) \cdot (3) + 2y \cos(x^2 + y^2) \cdot \left[\frac{-2t}{(1 + t^2)^2}\right]$$

or

$$\frac{du}{dt} = 2 \cos(x^2 + y^2)\left[3x - \frac{2yt}{(1 + t^2)^2}\right].$$

Using the known relationships between x, y, and t, the derivative du/dt can thus be computed for any desired value of t. The details are left to the reader.

Example 5·19   Determine the total derivative du/dx in each case:
  (a) $u = x \cos y + y \cos x$ when $y = 1 + x + x^3$;
  (b) $u = x^2 + 2xy - y^2$ when $x^2 + y^2 + \cos xy = 0$.

Solution (a) This requires an application of Corollary 5·20 (a). We set

$$f(x, y) = x \cos y + y \cos x \quad \text{and} \quad y = 1 + x + x^3$$

so that

$$\frac{\partial f}{\partial x} = \cos y - y \sin x, \quad \frac{\partial f}{\partial y} = -x \sin y + \cos x$$

and

$$\frac{dy}{dx} = 1 + 3x^2.$$

Hence, substituting into Corollary 5·20 (a),

$$\frac{du}{dx} = \cos y - y \sin x + (\cos x - x \sin y)(1 + 3x^2).$$

(b) In this case we use Corollary 5·20 (b), with

$$f(x, y) = x^2 + 2xy - y^2 \quad \text{and} \quad g(x, y) = x^2 + y^2 + \cos xy.$$

Hence

$$\frac{\partial f}{\partial x} = 2x + 2y, \quad \frac{\partial f}{\partial y} = 2x - 2y,$$

$$\frac{\partial g}{\partial x} = 2x - y \sin xy \quad \frac{\partial g}{\partial y} = 2y - x \sin xy.$$

Finally, applying Corollary 5·21 (b),

$$\frac{du}{dx} = 2(x + y) - \frac{2(x - y)(2x - y \sin xy)}{(2y - x \sin xy)}.$$

5·8 Change of variable

This section discusses a somewhat more complicated situation than that covered by Theorem 5·20, namely, the implications on partial differentiation of changing the independent variables in a function $u = f(x_1, x_2, \ldots, x_n)$ that is to be differentiated. This situation commonly occurs as a result of changing coordinate systems to suit physical problems as the following example illustrates. Suppose that $p = p(x, y, z)$ is the pressure in a fluid flowing parallel to the z-axis. Then $\partial p/\partial z$ is the pressure gradient along·the direction of flow and $\partial p/\partial x$, $\partial p/\partial y$ are the transverse pressure gradients in the plane $z = $ constant.

Now, if the flow takes place in a rectangular duct with sides described by $x = $ constant, $y = $ constant, then the Cartesian coordinates $O\{x, y, z\}$ are obviously the natural ones to use. However, if the flow takes place in a cylindrical pipe, then the z-axis is still convenient as it can be aligned with the axis of the pipe, but the x-, y-axes are now less useful since the wall of the pipe becomes the curve $x^2 + y^2 =$ constant. Clearly, a more sensible coordinate system would be the cylindrical polar coordinates r, θ, z' in which r and θ define a point in the plane $z' =$ constant. Figure 5·15 illustrates this idea.

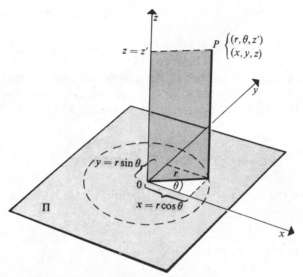

Fig. 5·15 Cylindrical polar coordinates.

Plane $z = z' = 0$ in both the O$\{x, y, z\}$ and O$\{r, \theta, z'\}$ systems of axes, and is denoted by Π. Relative to these two systems the point P has the coordinates O$\{x, y, z\}$ and O$\{r, \theta, z'\}$, respectively, where

$$x = r \cos \theta, \qquad y = r \sin \theta, \qquad z = z'. \tag{5.28}$$

How can the pressure gradients described by the partial derivatives $\partial p/\partial r$, $\partial p/\partial \theta$, and $\partial p/\partial z'$ be determined from Eqn (5.28), and the known functions $\partial p/\partial x$, $\partial p/\partial y$, and $\partial p/\partial z$. The rest of this section is devoted to solving this type of problem. Notice that from the definition of partial differentiation, $\partial p/\partial z$ and $\partial p/\partial z'$ have essentially the same meaning, whereas $\partial p/\partial r$ is the derivative of p computed along a radius with θ and z' held constant, whilst $\partial p/\partial \theta$ is the derivative of p tangential to a circle $r =$ constant drawn on the plane $z' =$ constant.

Although the replacement of coordinate variables in this manner involves replacing a set of n independent variables by a new set also comprising n in number ($n = 3$ above), we shall first prove a more general result. Specifically, consider the implication of the situation in which

$$u = f(x_1, x_2, \ldots, x_n), \tag{5.29}$$

when the independent variables $x_1, x_2, \ldots, x_n$ are themselves differentiable functions of another set of variables which we denote by $\alpha_1, \alpha_2, \ldots, \alpha_m$. It is not necessary that m should equal n. Thus we have

$$
\begin{aligned}
x_1 &= x_1(\alpha_1, \alpha_2, \ldots, \alpha_m), \\
x_2 &= x_2(\alpha_1, \alpha_2, \ldots, \alpha_m), \\
&\quad\cdot\quad\cdot\quad\cdot\quad\cdot\quad\cdot\quad\cdot \\
x_n &= x_n(\alpha_1, \alpha_2, \ldots, \alpha_m),
\end{aligned}
\tag{5.30}
$$

If the variables $x.$ in Eqn (5.29) were to be replaced by the equivalent functions (5.30) involving the variables $\alpha.$, then f would become some function $F(\alpha_1, \alpha_2, \ldots, \alpha_m)$ of $\alpha_1, \alpha_2, \ldots, \alpha_m$ so that by Theorem 5·19 we could write

$$
du = \frac{\partial F}{\partial \alpha_1}\, d\alpha_1 + \frac{\partial F}{\partial \alpha_2}\, d\alpha_2 + \cdots + \frac{\partial F}{\partial \alpha_m}\, d\alpha_m.
\tag{5.31}
$$

Next, observe that by applying this same theorem to the equation for x_i in Eqn (5.30) we obtain

$$
dx_i = \frac{\partial x_i}{\partial \alpha_1}\, d\alpha_1 + \frac{\partial x_i}{\partial \alpha_2}\, d\alpha_2 + \cdots + \frac{\partial x_i}{\partial \alpha_m}\, d\alpha_m,
\tag{5.32}
$$

for $i = 1, 2, \ldots, n$.

Substituting these expressions into the statement of Theorem 5·19 then gives

$$
du = \frac{\partial f}{\partial x_1}\left[\frac{\partial x_1}{\partial \alpha_1}\, d\alpha_1 + \frac{\partial x_1}{\partial \alpha_2}\, d\alpha_2 + \cdots + \frac{\partial x_1}{\partial \alpha_m}\, d\alpha_m\right] + \cdots
$$
$$
+ \frac{\partial f}{\partial x_n}\left[\frac{\partial x_n}{\partial \alpha_1}\, d\alpha_1 + \frac{\partial x_n}{\partial \alpha_2}\, d\alpha_2 + \cdots + \frac{\partial x_n}{\partial \alpha_m}\, d\alpha_m\right].
\tag{5.33}
$$

On re-arrangement this becomes

$$
du = \left[\frac{\partial f}{\partial x_1}\frac{\partial x_1}{\partial \alpha_1} + \frac{\partial f}{\partial x_2}\frac{\partial x_2}{\partial \alpha_1} + \cdots + \frac{\partial f}{\partial x_n}\frac{\partial x_n}{\partial \alpha_1}\right] d\alpha_1 + \cdots
$$
$$
+ \left[\frac{\partial f}{\partial x_1}\frac{\partial x_1}{\partial \alpha_m} + \frac{\partial f}{\partial x_2}\frac{\partial x_2}{\partial \alpha_m} + \cdots + \frac{\partial f}{\partial x_n}\frac{\partial x_n}{\partial \alpha_m}\right] d\alpha_m.
\tag{5.34}
$$

Since $f(x_1, x_2, \ldots, x_n) = F(\alpha_1, \alpha_2, \ldots, \alpha_m)$, it follows by a direct comparison of the ith terms of Eqns (5.31) and (5.34) that

$$\frac{\partial f}{\partial \alpha_i} = \frac{\partial f}{\partial x_1}\frac{\partial x_1}{\partial \alpha_i} + \frac{\partial f}{\partial x_2}\frac{\partial x_2}{\partial \alpha_i} + \cdots + \frac{\partial f}{\partial x_n}\frac{\partial x_n}{\partial \alpha_i} \qquad (5.35)$$

for $i = 1, 2, \ldots, m$.

We state this result in the form of a general theorem.

THEOREM 5·21 (change of variable) Let $f(x_1, x_2, \ldots, x_n)$ be a real valued function of the real variables $x_1, x_2, \ldots, x_n$ whose first-order derivatives exist and are continuous. Further, let $x_1 = x_1(\alpha_1, \alpha_2, \ldots, \alpha_m), x_2 = x_2(\alpha_1, \alpha_2, \ldots, \alpha_m),$ $\ldots, x_n = x_n(\alpha_1, \alpha_2, \ldots, \alpha_m)$ be differentiable functions of the real variables $\alpha_1, \alpha_2, \ldots, \alpha_m$, then

$$\frac{\partial f}{\partial \alpha_1} = \frac{\partial f}{\partial x_1}\frac{\partial x_1}{\partial \alpha_1} + \frac{\partial f}{\partial x_2}\frac{\partial x_2}{\partial \alpha_1} + \cdots + \frac{\partial f}{\partial x_n}\frac{\partial x_n}{\partial \alpha_1}$$

$$\frac{\partial f}{\partial \alpha_2} = \frac{\partial f}{\partial x_1}\frac{\partial x_1}{\partial \alpha_2} + \frac{\partial f}{\partial x_2}\frac{\partial x_2}{\partial \alpha_2} + \cdots + \frac{\partial f}{\partial x_n}\frac{\partial x_n}{\partial \alpha_2}$$

$$\cdots \cdots \cdots \cdots \cdots$$

$$\frac{\partial f}{\partial \alpha_m} = \frac{\partial f}{\partial x_1}\frac{\partial x_1}{\partial \alpha_m} + \frac{\partial f}{\partial x_2}\frac{\partial x_2}{\partial \alpha_m} + \cdots + \frac{\partial f}{\partial x_n}\frac{\partial x_n}{\partial \alpha_m}.$$

Example 5·20 Express $\partial f/\partial r$, $\partial f/\partial \theta$, and $\partial f/\partial z'$ in terms of $\partial f/\partial x$, $\partial f/\partial y$, and $\partial f/\partial z$ given that $x = r \cos \theta$, $y = r \sin \theta$, $z = z'$. Find their values given that

$$f(x, y, z) = x^2 + 3xy + y^2 + z^2.$$

Solution We must apply Theorem 5.21 with $m = n = 3$ by making the identifications $x_1 = x$, $x_2 = y$, $x_3 = z$ and $\alpha_1 = r$, $\alpha_2 = \theta$, $\alpha_3 = z'$. Our first result is

$$\frac{\partial f}{\partial r} = \frac{\partial f}{\partial x}\frac{\partial x}{\partial r} + \frac{\partial f}{\partial y}\frac{\partial y}{\partial r} + \frac{\partial f}{\partial z}\frac{\partial z}{\partial r},$$

$$\frac{\partial f}{\partial \theta} = \frac{\partial f}{\partial x}\frac{\partial x}{\partial \theta} + \frac{\partial f}{\partial y}\frac{\partial y}{\partial \theta} + \frac{\partial f}{\partial z}\frac{\partial z}{\partial \theta},$$

$$\frac{\partial f}{\partial z'} = \frac{\partial f}{\partial x}\frac{\partial x}{\partial z'} + \frac{\partial f}{\partial y}\frac{\partial y}{\partial z'} + \frac{\partial f}{\partial z}\frac{\partial z}{\partial z'}.$$

However,

$$\frac{\partial x}{\partial r} = \cos \theta, \qquad \frac{\partial x}{\partial \theta} = -r \sin \theta, \qquad \frac{\partial x}{\partial z'} = 0, \qquad \frac{\partial y}{\partial r} = \sin \theta,$$

$$\frac{\partial y}{\partial \theta} = r \cos \theta, \quad \frac{\partial y}{\partial z'} = 0, \quad \frac{\partial x}{\partial z'} = \frac{\partial y}{\partial z'} = 0, \quad \frac{\partial z'}{\partial z} = 1.$$

Hence, substituting these values into the above transformation equations shows that

$$\frac{\partial f}{\partial r} = \frac{\partial f}{\partial x} \cos \theta + \frac{\partial f}{\partial y} \sin \theta,$$

$$\frac{\partial f}{\partial \theta} = - \frac{\partial f}{\partial x} r \sin \theta + \frac{\partial f}{\partial y} r \cos \theta,$$

$$\frac{\partial f}{\partial z'} = \frac{\partial f}{\partial z}.$$

Next, using the fact that $f(x, y, z) = x^2 + 3xy + y^2 + z^2$ we see that

$$\frac{\partial f}{\partial x} = 2x + 3y, \quad \frac{\partial f}{\partial y} = 3x + 2y, \quad \frac{\partial f}{\partial z} = 2z,$$

so that

$$\frac{\partial f}{\partial r} = (2x + 3y) \cos \theta + (3x + 2y) \sin \theta.$$

However, as $r^2 = x^2 + y^2$ and $\cos \theta = x/(x^2 + y^2)^{1/2}$, $\sin \theta = y/(x^2 + y^2)^{1/2}$, this result simplifies to

$$\frac{\partial f}{\partial r} = \frac{2x^2 + 6xy + 2y^2}{(x^2 + y^2)^{1/2}}.$$

A similar calculation shows that

$$\frac{\partial f}{\partial \theta} = 3(x^2 - y^2), \quad \frac{\partial f}{\partial z'} = 2z.$$

5·9 Implicit functions

We have already used implicit functions when discussing various consequences of total differentials, and will now examine these ideas more closely. Consider the equation $f(x, y) = 0$. Often the argument is used that from this implicit function of x and y we can, in principle, solve for y, and as y depends on x, we are entitled to express y in the explicit form $y = \varphi(x)$.

Suppose that $f(x, y) = x^2 + y^2 + 1$. Then *no* real values of x and y satisfy the implicit equation $f(x, y) = 0$, so certainly in this case one cannot solve for y. Thus a necessary condition that we may solve for y near to some point P with coordinates (x_0, y_0) is that there are real numbers x_0, y_0 such that $f(x_0, y_0) = 0$.

Now let $u = f(x, y)$ be the graph of $f(x, y)$, and assume that f_x and f_y

exist and are continuous so that the graph will be a smooth surface of the type shown in Fig. 5.16. Then $f(x, y) = 0$ is the curve of the section of this surface by the plane $u = 0$. In general the curve of the section will be similar to the smooth curve L shown in the figure and can be described by an equation of the form $y = \varphi(x)$. This will obviously be the case provided firstly, that the surface $u = f(x, y)$ and the plane $u = 0$ intersect and secondly, that they are nowhere tangential. The curve L will be smooth, and the function $\varphi(x)$ differentiable, because the assumed continuity of the derivatives f_x and f_y will ensure that the surface $u = f(x, y)$ is itself smooth, and so will generate a smooth curve of section. This is, of course, the assertion made in Corollary 5·19. Let P be a representative point on L with coordinates (x_0, y_0) in the $u = 0$ plane, and line l be drawn tangential to the surface $u = f(x, y)$ at P in the plane $x = x_0$. Then by Definition 5·5, the angle α between line l and the plane $u = 0$ is such that $\tan \alpha = \partial f / \partial y |_{(x_0, y_0)}$.

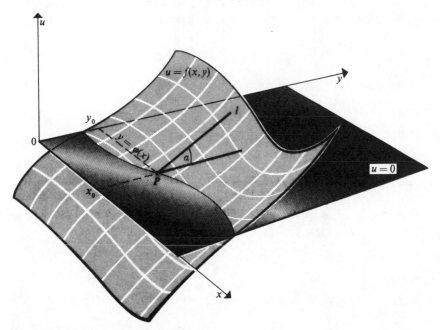

Fig. 5·16 The function $y = \phi(x)$ defined by the intersection of $u = f(x, y)$ and the plane $u = 0$.

Hence the condition that the surface $u = f(x, y)$ and the plane $u = 0$ should not be tangential at P is seen to be $f_y(x_0, y_0) \neq 0$. Collecting our results we now formulate them as the following theorem.

THEOREM 5·22 (implicit function theorem) Let $f(x, y)$ be **differentiable and have continuous first-order partial derivatives near to (x_0, y_0) at which**

$f(x_0, y_0) = 0$ and $f_y(x_0, y_0) \neq 0$. Then, near (x_0, y_0), it is possible to solve the implicit equation $f(x, y) = 0$ uniquely for y in the explicit form $y = \varphi(x)$, where $\varphi(x)$ is differentiable. That is, near to (x_0, y_0), $f(x, \varphi(x)) = 0$.

Notice that this theorem is only of the existence type in that it ensures that an explicit representation $y = \varphi(x)$ exists, but gives no information on how such a representation may be found in any specific case.

As a corollary to this theorem, consider the relationship between the derivatives of a function and its inverse. Let $F(x, y) = y - f(x)$, so that $F(x, y) = 0$ implies the relationship $y = f(x)$. Suppose that at some point (x_0, y_0) we have $f'(x_0) \neq 0$ and $y_0 = f(x_0)$. Then, noticing that $\partial F/\partial x = (\partial/\partial x)[-f(x)] = (d/dx)[-f(x)] = -f'(x)$ and $\partial F/\partial y = 1$, it follows from Theorem 5·22 that close to (x_0, y_0) we may solve for x as a function of y to obtain an inverse function $x = \varphi(y)$. That is, $F(\varphi(y), y) = y - f[\varphi(y)] = 0$.

Furthermore, applying Corollary 5·19 to $F(x, y) = 0$ and regarding y as the independent variable and x as the dependent variable, we have

$$1 - f'(x)\frac{dx}{dy} = 0,$$

so that provided $f'(x) \neq 0$, we have

$$\frac{dx}{dy} = 1/f'(x) \quad \text{or} \quad \varphi'(y) = 1/f'(x),$$

which is the desired result.

Corollary 5·22 Let $y = f(x)$ be a real valued differentiable function of x close to some point (x_0, y_0) at which $y_0 = f(x_0)$. Let $x = \varphi(y)$ be the function inverse to it close to the same point (x_0, y_0) so that $x_0 = \varphi(y_0)$, and let $f'(x_0) \neq 0$. Then close to (x_0, y_0), we have

$$\varphi'(y) = 1/f'(x)$$

or, equivalently,

$$\frac{dx}{dy} = 1 \bigg/ \left(\frac{dy}{dx}\right).$$

This corollary has two important applications which we mention next. The first application of Corollary 5·22 is to the differentiation of inverse circular functions. In Section 2·2, we agreed to write

$$y = \arcsin x \quad \text{when} \quad x = \sin y \quad \text{and} \quad -\pi/2 \leq y \leq \pi/2.$$

Now,

$$\frac{d}{dy}(\sin y) = \cos y \neq 0 \quad \text{for } -\pi/2 < y < \pi/2;$$

that is, for $-1 < x < 1$ and so, by Corollary 5·22,

$$\frac{dy}{dx} = 1 \Big/ \left(\frac{dx}{dy}\right) = \frac{1}{\cos y} = \frac{1}{\sqrt{(1 - \sin^2 y)}} = \frac{1}{\sqrt{(1 - x^2)}}.$$

The *positive* square root has been taken here because the principal branch of the function $y = \text{arc sin } x$ is a monotonic *increasing* function of x in its domain of definition $-1 \le x \le 1$. By this same argument, the *negative* square root is taken when differentiating the principal branch of the function $y = \text{arc cos } x$ which is a monotonic *decreasing* function of x in its domain of definition $-1 \le x \le 1$. Thus

$$\frac{d}{dx} (\text{arc sin } x) = \frac{1}{\sqrt{(1 - x^2)}} \qquad \text{for } -1 < x < 1.$$

Similar arguments establish Table 5·2. In the entries for the derivatives of arc cosec and arc sec, the term $|x|$ has been introduced to take account of the two separate cases that need consideration when deriving these results; namely, when $x > a$ and when $x < -a$. These same ideas will be encountered again in the next chapter in connection with Table 6·3, when they will be discussed in more detail.

Table 5·2   Derivatives of inverse circular functions

$$\frac{d}{dx}(\text{arc sin } x/a) = \frac{1}{\sqrt{(a^2 - x^2)}} \qquad\qquad \frac{d}{dx}(\text{arc cos } x/a) = \frac{-1}{\sqrt{(a^2 - x^2)}}$$
$$\text{for } -a < x < a \qquad\qquad\qquad\qquad \text{for } -a < x < a$$

$$\frac{d}{dx}(\text{arc tan } x/a) = \frac{a}{a^2 + x^2} \qquad\qquad \frac{d}{dx}(\text{arc cosec } x/a) = \frac{-a}{|x| \sqrt{(x^2 - a^2)}}$$
$$\text{for all } x \qquad\qquad\qquad\qquad\qquad \text{for } |x| > a$$

$$\frac{d}{dx}(\text{arc sec } x/a) = \frac{a}{|x| \sqrt{(x^2 - a^2)}} \qquad \frac{d}{dx}(\text{arc cot } x/a) = \frac{-a}{a^2 + x^2}$$
$$\text{for } |x| > a \qquad\qquad\qquad\qquad \text{for all } x$$

In Chapter 2 we saw that curves may be described parametrically thus:

$$x = X(t), \qquad y = Y(t),$$

where t is a parameter defined in some interval $\mathscr{I}$. The question that now arises is how may we find dy/dx in terms of the functions $X(t)$ and $Y(t)$.

Let us suppose that $X(t)$ and $Y(t)$ are differentiable functions of t with continuous derivatives and that $X'(t) \ne 0$. Then by Theorem 5·22, we may solve $x = X(t)$ in the form $t = f(x)$, say, so that then $y = Y[f(x)]$. From Theorem 5·7 on the differentiation of composite functions we have

$$\frac{dy}{dx} = \frac{d}{dx}\{Y[f(x)]\} = \frac{dY}{df}\frac{df}{dx}$$

or, equivalently,

$$\frac{dy}{dx} = \frac{dy}{dt} \cdot \frac{dt}{dx}. \tag{5·36}$$

However, by Corollary 5·22, $dt/dx = 1/(dx/dt)$ so that

$$\frac{dy}{dx} = \frac{dy}{dt}\bigg/\frac{dx}{dt}. \tag{5·37}$$

Hence, like x and y, the derivative dy/dx is now also known parametrically in terms of t.

This result is best remembered in symbolic operator form:

$$\frac{d}{dx} = \frac{1}{(dx/dt)}\frac{d}{dt}. \tag{5·38}$$

Higher order derivatives with respect to x may be found either by a repetition of the argument leading to Eqn (5.37), or by successive applications of Eqn (5·38).

Thus, using Eqn (5·38), we have

$$\frac{d^2y}{dx^2} = \frac{d}{dx}\left(\frac{dy}{dx}\right) = \frac{1}{(dx/dt)}\left[\frac{d}{dt}\left(\frac{dy}{dt}\bigg/\frac{dx}{dt}\right)\right]$$

or, denoting differentiation with respect to t by a dot,

$$\frac{d^2y}{dx^2} = \frac{d}{dx}\left(\frac{dy}{dx}\right) = \frac{1}{\dot{x}}\frac{d}{dt}\left(\frac{dy}{dx}\right).$$

Using the fact that $dy/dx = \dot{y}/\dot{x}$ and performing the indicated differentiations gives

$$\frac{d^2y}{dx^2} = \frac{\dot{x}\ddot{y} - \ddot{x}\dot{y}}{\dot{x}^3}. \tag{5·39}$$

It is recommended that the reader remembers the arguments leading to the operator rule (5·38) together with the rule itself, rather than remembering results of the form (5·39).

Example 5·21 If $x = t + 2\sin t$, $y = \cos t$ determine dy/dx and d^2y/dx^2 and hence deduce their values when $t = 0$.

Solution We have

$$\frac{dx}{dt} = 1 + 2\cos t, \qquad \frac{dy}{dt} = -\sin t,$$

so that by Eqn (5·46)

$$\frac{dy}{dx} = \frac{dy}{dt} \bigg/ \frac{dx}{dt} = \frac{-\sin t}{1 + 2 \cos t}.$$

When $t = 0$ we have $x = 0$, $y = 1$ and

$$\frac{dy}{dx}\bigg|_{x=0} = \frac{-\sin t}{1 + 2 \cos t}\bigg|_{t=0} = 0.$$

Next, as

$$\frac{d^2y}{dx^2} = \frac{1}{(dx/dt)} \frac{d}{dt}\left(\frac{dy}{dx}\right)$$

we have

$$\frac{d^2y}{dx^2} = \frac{1}{1 + 2 \cos t} \frac{d}{dt}\left[\frac{-\sin t}{1 + 2 \cos t}\right].$$

Thus, performing the differentiation and simplifying,

$$\frac{d^2y}{dx^2} = -\left[\frac{2 + \cos t}{(1 + 2 \cos t)^3}\right]$$

and so

$$\frac{d^2y}{dx^2}\bigg|_{x=0} = -\left[\frac{2 + \cos t}{(1 + 2 \cos t)^3}\right]\bigg|_{t=0} = -\frac{1}{9}.$$

5·10   Higher order partial derivatives

If the function $f(x, y)$ is differentiable with continuous first-order derivatives f_x and f_y, then it can also happen that these partial derivatives which are functions of x and y are themselves differentiable. Thus we are led to consider the further partial derivatives

$$\frac{\partial}{\partial x}(f_x), \quad \frac{\partial}{\partial y}(f_x), \quad \frac{\partial}{\partial x}(f_y), \quad \text{and} \quad \frac{\partial}{\partial y}(f_y).$$

These functions, when they exist, are *second-order* partial derivatives of f and are respectively denoted by

$$\frac{\partial^2 f}{\partial x^2}, \quad \frac{\partial^2 f}{\partial y \partial x}, \quad \frac{\partial^2 f}{\partial x \partial y}, \quad \text{and} \quad \frac{\partial^2 f}{\partial y^2}.$$

Using an alternative notation we often write these same derivatives as

$$f_{xx}, f_{xy}, f_{yx}, \text{ and } f_{yy}.$$

In this notation the first suffix signifies the partial derivative of f that is to be differentiated partially with respect to the second suffix. The centre pair of derivatives are *mixed* second-order partial derivatives and it is conventional that the order of x and y in corresponding mixed derivatives in the two notations is interchanged. Thus we have,

$$\frac{\partial}{\partial y}(f_x) = \frac{\partial^2 f}{\partial y \partial x} = f_{xy} \quad \text{but} \quad \frac{\partial}{\partial x}(f_y) = \frac{\partial^2 f}{\partial x \partial y} = f_{yx}.$$

It is important to notice that the double operations of partial differentiation that lead to the mixed derivatives f_{xy} and f_{yx} are performed in different orders. Consequently we have no right to expect that the derivatives that result will be equal to one another. To emphasize this point we now write out in full the limiting operations involved in arriving at f_{xy} and f_{yx}:

$$f_{xy}(x_0, y_0) = \frac{\partial}{\partial y}\left[f_x(x, y)\right]\Big|_{(x_0, y_0)}$$

$$= \lim_{k \to 0} \frac{1}{k}\left[\lim_{h \to 0} \frac{f(x_0 + h, y_0 + k) - f(x_0, y_0 + k)}{h}\right.$$

$$\left. - \lim_{h \to 0} \frac{f(x_0 + h, y_0) - f(x_0, y_0)}{h}\right]$$

and so, writing

$$g(x_0, y_0, h, k) = f(x_0 + h, y_0 + k) - f(x_0, y_0 + k) - f(x_0 + h, y_0) \\ + f(x_0, y_0),$$

we obtain the result

$$f_{xy}(x_0, y_0) = \lim_{k \to 0} \lim_{h \to 0} \frac{1}{hk} g(x_0, y_0, h, k), \tag{5.40}$$

where the inner limit with respect to h is to be taken first. Exactly similar reasoning gives the corresponding result

$$f_{yx}(x_0, y_0) = \lim_{h \to 0} \lim_{k \to 0} \frac{1}{hk} g(x_0, y_0, h, k). \tag{5.41}$$

Here it is the inner limit with respect to k that is to be taken first.

The double limits used in Eqns (5·40) and (5·41) are called *iterated limits* on account of the fact that they are taken sequentially so that their order is important. They are not to be confused with the simple double limit of Definition 3·7 into which questions of order do not enter.

Let us now explore the consequence of requiring one of the mixed derivatives, say f_{xy}, to be continuous. This is, of course, the usual situation, Definitions 3·7 and 3·8 imply that if f_{xy} is continuous at (x_0, y_0), then a limit $L = f_{xy}(x_0, y_0)$ exists with the property that

$$L = \lim_{\substack{h \to 0 \\ k \to 0}} f_{xy}(x_0 + h, y_0 + k), \tag{5.42}$$

where the question of the order in which the limits are to be taken does not occur. Hence, as $f_{xy}(x_0, y_0)$ is also defined by Eqn (5.40) in which an iterated limit is involved, the equating of these two results implies that if f_{xy} is continuous, then the order of the iterated limits in Eqn (5.40) is immaterial. Thus, under the stated conditions, expressions (5·40) and (5·41) become identical and the continuity of f_{xy} implies not only the existence of f_{yx}, but also that $f_{xy} = f_{yx}$. This establishes our next result.

THEOREM 5·23 (equality of mixed derivatives)   Let $f(x, y)$ be a real valued function of the real variables x, y, and let f_x, f_y, f_{xy} exist and be continuous in the neighbourhood of the point (x_0, y_0). Then f_{yx} also exists at (x_0, y_0) and

$$\frac{\partial^2 f}{\partial x \partial y}\bigg|_{(x_0, y_0)} = \frac{\partial^2 f}{\partial y \partial x}\bigg|_{(x_0, y_0)}.$$

Still higher-order derivatives can be defined by an obvious extension of the notation. Thus, for a suitably differentiable function f we may define the third-order partial derivatives

$$f_{xxx}, f_{yyx}, f_{xyx}, f_{yyy}, \text{ etc.}$$

If the higher-order derivatives involved are continuous then, by an obvious extension of Theorem 5·23, the order of performing differentiations may be disregarded. In the case of the mixed third-order partial derivative f_{xyx} this would imply that

$$f_{xyx} = \frac{\partial}{\partial x}\left[\frac{\partial}{\partial y}(f_x)\right] = \frac{\partial}{\partial y}\left[\frac{\partial}{\partial x}(f_x)\right] = f_{xxy}.$$

Hence, under these conditions, it is proper to extend the ∂ notation by writing

$$\frac{\partial^3 f}{\partial x^3}, \frac{\partial^3 f}{\partial x \partial y^2}, \frac{\partial^3 f}{\partial x^2 \partial y}, \frac{\partial^3 f}{\partial y^3}, \text{ etc.}$$

Example 5·22   If $f(x, y) = x^4 + 2x^2 y^2 + xy^4$ find the second- and third-order partial derivatives of f.

Solution   First-order derivatives:

$$f_x = 4x^3 + 4xy^2 + y^4, \qquad f_y = 4x^2 y + 4xy^3.$$

Second-order derivatives:

$$f_{xx} = 12x^2 + 4y^2, \qquad f_{yy} = 4x^2 + 12xy^2,$$

$$f_{xy} = \frac{\partial}{\partial y}(f_x) = 8xy + 4y^3.$$

This mixed derivative is continuous, and so $f_{xy} = f_{yx}$. As a check in this case we compute f_{yx} directly:

$$f_{yx} = \frac{\partial}{\partial x}(f_y) = 8xy + 4y^3.$$

Third-order derivatives:

$$f_{xxx} = 24x, \qquad f_{yyy} = 24xy, \qquad f_{xyy} = \frac{\partial}{\partial y}(f_{xy}) = 8x + 12y^2,$$

$$f_{xxy} = \frac{\partial}{\partial y}(f_{xx}) = 8y.$$

The continuity of the third-order derivatives we have computed ensures the existence and equality of the other corresponding third-order derivatives that may be defined. Thus, for example, as $f_{xxy} = 8y$ is continuous, there is no need to compute f_{xyx}, since it exists and is equal to f_{xxy}.

Example 5·23 Define the function f by the requirement

$$f(x, y) = \begin{cases} \dfrac{xy(x^2 - y^2)}{x^2 + y^2} & \text{if either } x \neq 0, \text{ or } y \neq 0 \\ 0 & \text{if both } x = 0 \text{ and } y = 0 \end{cases}$$

Deduce the value of each of the mixed derivatives at the origin.

Solution   We shall use definitions (5·40) and (5·41) for this purpose by setting $x_0 = 0$, $y_0 = 0$ so that

$$g(0, 0, h, k) = \frac{hk(h^2 - k^2)}{h^2 + k^2}.$$

Then, from Eqn (5·40),

$$f_{xy}(0, 0) = \lim_{k \to 0} \lim_{h \to 0} \frac{1}{hk}\left\{ \frac{hk(h^2 - k^2)}{h^2 + k^2} \right\}$$

$$= \lim_{k \to 0} \lim_{h \to 0} \frac{h^2 - k^2}{h^2 + k^2} = \lim_{k \to 0}\left(\frac{-k^2}{k^2} \right) = -1.$$

However, because the order of the iterated limits are reversed in Eqn (5·41), the same argument also shows that

$$f_{yx}(0, 0) = \lim_{h \to 0} \lim_{k \to 0} \frac{h^2 - k^2}{h^2 + k^2} = \lim_{h \to 0}\left(\frac{h^2}{h^2} \right) = 1.$$

Thus $f_{xy}(0, 0) = -1$ whereas $f_{yx}(0, 0) = 1$. This occurs because the functions f_{xy} and f_{yx} are not continuous at $(0, 0)$ as may be checked by direct calculation.

PROBLEMS

Section 5·1

5·1 Give examples of four physical quantities that are essentially defined in terms of a derivative.

5·2 Use Definitions 5·1 and 5·2 to prove that the following functions are differentiable in the stated intervals and to compute their derivatives. Evaluate these derivatives for the stated values:

(a) $f(x) = 3x^2$ in $[0, 3]$, find $f'(2)$;
(b) $f(x) = 2x^3 + x + 1$ in $[-1, 4]$, find $f'(3)$;
(c) $f(x) = |x|$ in $(0, \infty)$, find $f'(1)$;
(d) $f(x) = |x|$ in $(-\infty, 0)$, find $f'(-3)$;
(e) $f(x) = 1/x$ in $[1, 5]$, find $f'(4)$;
(f) $f(x) = x^{1/4}$ in $(0, \infty)$, find $f'(2)$.

5·3 Deduce the gradients of the functions $f(x)$ to the immediate left and right of $x = 1$ given that:

(a) $f(x) = \begin{cases} x^3 + x + 1 \text{ for } x \geq 1 \\ 5 - x - x^2 \text{ for } x < 1; \end{cases}$

(b) $f(x) = \begin{cases} x^3 - x + 3 \text{ for } x \geq 1 \\ 2x + 1 \qquad \text{ for } x < 1. \end{cases}$

5·4 At which points in the stated intervals, if any, are the following functions $f(x)$ non-differentiable:

(a) $f(x) = x + \sin 2x$ for $0 \leq x \leq \pi$;

(b) $f(x) = \begin{cases} x + 1/x \text{ for } x \neq 0 \\ 0 \qquad \text{ for } x = 0 \end{cases}$ in the interval $[-1, 1]$;

(c) $f(x) = \begin{cases} 1 \text{ for } x \text{ rational} \\ 0 \text{ for } x \text{ irrational} \end{cases}$ in the interval $[0, 1]$.

Section 5·2

5·5 By assuming Theorem 5·2 is also valid for rational n where necessary, find the derivatives of the following functions, stating at which points in their domains of definition, if any, they are non-differentiable:

(a) $f(x) = \begin{cases} x^{1/3} + \cos 3x, \text{ for } x \neq 0 \\ 0, \qquad \text{ for } x = 0 \end{cases}$ in the interval $-\tfrac{1}{2}\pi \leq x \leq \pi$;

(b) $f(x) = x \sin 2x + x^{5/3}$ for $-1 \leq x \leq 3$;

(c) $f(x) = |\cos x|$ for $0 \leq x \leq \pi$.

5·6 Differentiate the following functions by making a repeated application of Theorem 5·5:

(a) $y = (1 + x^2) \sin 7x \cos 4x$;
(b) $y = (1 + 2x^2 + x^4)^3$;
(c) $y = \cos^3 2x$.

5·7 Differentiate these composite functions:

(a) $y = (x^2 + 2x + 1)^{3/2}$;

(b) $y = (a + bx^3)^{1/3}$;

(c) $y = (2 + 3 \sin 2x)^5$;

(d) $y = \sin (1 + 2x^3)$.

5·8 Differentiate these quotients:

(a) $y = (x^2 + 3x + 7)/(x^4 + 1)$;

(b) $y = \dfrac{\sin (1 + x^2)}{x^4 + 2x^2 + 6}$;

(c) $y = \dfrac{1}{3 \cos^3 x} - \dfrac{1}{\cos x}$.

5·9 Differentiate these functions:

(a) $y = \dfrac{1}{(1 - 3 \cos x)^2}$;

(b) $y = \dfrac{x}{a^2 \sqrt{(b^2 + x^2)}}$;

(c) $y = \dfrac{\tan (1 + x^2 + x^4)}{\sin (1 + x^2)}$.

Section 5·3

5·10 The function $f(x) = \tfrac{1}{3}x^3 - x + 2$ which is defined in the interval $(-\infty, \infty)$ has extrema at the points $x = 1, x = -1$. Identify their nature by considering the behaviour of the function close to these points. Are they relative or absolute extrema?

5·11 By considering the behaviour of $f(x) = \sin \tfrac{1}{2}x \cos \tfrac{1}{2}x$ in the neighbourhood of $x = \tfrac{1}{2}\pi$, show that the function attains an absolute maximum at that point.

5·12 Find the critical point of the function $f(x) = (x - 1)x^{2/3}$ and identify its nature. Do the points $x = -1, x = 0$ correspond to extrema of the function and, if so, of what type are they?

5·13 Find the critical points of the function $f(x) = x^2(3 - x)^2$.

5·14 Identify the critical points and extrema of the function

$$f(x) = \begin{cases} x^2 - 3x + 2 & \text{for } 0 \le x \le 2\cdot5 \\ x^2 - 7x + 12 & \text{for } 2\cdot5 < x \le 5. \end{cases}$$

5·15 Let the functions $f(x)$ and $g(x)$ be continuous in $[a, b]$ and differentiable in (a, b), with $g'(x)$ non-zero in (a, b). Show that under these conditions Rolle's theorem may be applied to the function $F(x)$ defined by $F(x) = f(a)g(a) - f(b) g(a) + [g(a) - g(b)]f(x) - [f(a) - f(b)]g(x)$, for $a \le x \le b$. Hence establish the Cauchy extended mean value theorem.

5·16 By repeatedly applying L'Hospital's rule where necessary, evaluate the following indeterminate forms of the type 0/0:

(a) $\lim\limits_{x \to 0} \dfrac{\tan \alpha x}{x}$;

(b) $\lim\limits_{x \to 0} \dfrac{x \cos x - \sin x}{x^3}$;

(c) $\lim\limits_{x \to 0} \dfrac{\tan x - \sin x}{x - \sin x}$;

(d) $\lim\limits_{x \to 1} \dfrac{x^3 - 2x^2 - x + 2}{x^3 - 7x + 6}$.

5·17 Evaluate the following indeterminate forms which are of the type ∞/∞:

(a) $\lim\limits_{x\to 0} (\pi/x)/\cot \pi x/2$;

(b) $\lim\limits_{x\to\frac{1}{2}\pi} \tan x/\tan 5x$;

(c) $\lim\limits_{x\to\infty} \dfrac{3x^2 + x - 1}{x^2 + 2}$;

(d) $\lim\limits_{x\to 0} \dfrac{\cot x}{x - \cot x}$.

5·18 Explain the fallacy in this argument. The limit

$$\lim_{x\to\infty} \frac{x^2 + x \sin x + \sin x}{x^2}$$

does not exist because, applying Corollary 5·14 to L'Hospital's rule gives

$$\lim_{x\to\infty} \frac{x^2 + x \sin x + \sin x}{x^2} = \lim_{x\to\infty} \frac{2x + \sin x + x \cos x + \cos x}{2x}$$

$$= \lim_{x\to\infty} \left[1 + \tfrac{1}{2}\cos x + \frac{\sin x + \cos x}{2x} \right] = 1 + \tfrac{1}{2} \lim_{x\to\infty} \cos x.$$

What is the true value of this limit?

5·19 Indeterminate limits of the form $\infty - \infty$, $0 \cdot \infty$ can be reduced to the types $0/0$ or ∞/∞ by means of the following simple devices. If the limit is of the type $0 \cdot \infty$ set $\lim\limits_{x\to a} f(x) = 0$ and $\lim\limits_{x\to a} g(x) \to \infty$, then

$$\lim_{x\to a} [f(x) g(x)] = \lim_{x\to a} [f(x)/(1/g(x))] \quad \text{(type 0/0)}$$

$$= \lim_{x\to a} [g(x)/(1/f(x))] \quad \text{(type } \infty/\infty\text{)}.$$

If the limit is of the type $\infty - \infty$ set $\lim\limits_{x\to a} f(x) = 0$, $\lim\limits_{x\to a} g(x) = 0$, then

$$\lim_{x\to a} \left[\frac{1}{f(x)} - \frac{1}{g(x)} \right] = \lim_{x\to a} \left[\frac{g(x) - f(x)}{f(x) g(x)} \right] \quad \text{(type 0/0)}$$

$$= \lim_{x\to a} \left[\frac{1/f(x) g(x)}{1/(g(x) - f(x))} \right] \quad \text{(type } \infty/\infty\text{)}.$$

Apply these results to evaluate the following limits:

(a) $\lim\limits_{x\to 0} \left(\dfrac{1}{\sin x} - \dfrac{1}{x} \right)$;

(b) $\lim\limits_{x\to 3} \left(\dfrac{1}{x - 3} - \dfrac{5}{x^2 - x - 6} \right)$;

(c) $\lim\limits_{x\to 0} (1 - \cos x) \cot x$;

(d) $\lim\limits_{x\to\infty} x \sin \dfrac{3}{x}$;

(e) $\lim\limits_{x\to 1} (1 - x) \tan \dfrac{\pi x}{2}$;

(f) $\lim\limits_{x\to\frac{1}{2}\pi} \left(\dfrac{x}{\cot x} - \dfrac{\pi}{2 \cos x} \right)$.

5·20 Apply to Problem 5·14 the modification to Theorem 5·15 indicated at the end of Example 5·9 (b) to identify the behaviour of the function at the origin.

5·21 Metals A and B have coefficients of linear expansion α and β, respectively.

That is to say, when the temperature changes by an amount t from the ambient value T_0, the linear dimensions of metal A change by a factor $(1 + \alpha t)$, whilst those of metal B change by a factor $(1 + \beta t)$. Suppose that a block of metal A contains a cylindrical cavity of height H_0 and radius R_0 at temperature T_0 which is empty apart from a cylinder of metal B which has height h_0 and radius r_0 at that same temperature. Obtain an approximate expression for the small volume change dV of the cavity between the cylinders consequent upon a small change of temperature dt.

Section 5·4

5·22 Compute the first and second derivatives of the functions $f(x)$ listed below:

(a) $f(x) = \tan x$;

(b) $f(x) = x^2 \sin x$;

(c) $f(x) = (1 + x)(3 \sin x + \cos 2x)$;

(d) $f(x) = (x^2 + 1)^{1/2}$.

5·23 Show that the function $f(x)$ defined below is continuous and has a continuous first derivative at $x = 1$, but that it has a discontinuous second derivative at that point:

$$f(x) = \begin{cases} x^4 - x^2 - x + 1 & \text{for } x \leq 1 \\ 2x^3 - x^2 + x & \text{for } x > 1. \end{cases}$$

5·24 Use Leibnitz's theorem to evaluate the third derivatives of the following functions:

(a) $f(x) = \dfrac{x^7}{1 + x}$;

(b) $f(x) = (x^7 - 1) \tan x$;

(c) $f(x) = \sin^2 x$;

(d) $f(x) = x^3 \sec 2x$.

5·25 Apply Theorems 5·17 and 5·18 to locate and identify the extrema and points of inflection of the following functions, using your results to determine the gradients at the points of inflection:

(a) $f(x) = 2x^3 + 3x^2 - 12x + 5$;

(b) $f(x) = \dfrac{x^3}{x^2 + 3}$;

(c) $f(x) = x^2(x - 12)^2$.

5·26 Determine the values of a and b in order that $f(x) = x^3 + ax^2 + bx + 1$ should have a point of inflection at $x = 2$ at which the gradient of the tangent to the graph is -3.

Section 5·5

5·27 Compute the derivatives f_x and f_y given that:

(a) $f(x, y) = x^2/y$;

(b) $f(x, y) = 3x^2y + (x + y)^2x + 1$;

(c) $f(x, y) = \sin(x^2 + y^2)$.

5·28 Given that

$$f(x, y) = x^3 + 3x^2y + 4xy^2 + 2y^3$$

prove that $xf_x + yf_y = 3f$.

5·29 Compute the derivatives f_x, f_y, f_z given that:

(a) $f(x, y, z) = x^2yz + \dfrac{1}{xyz^2}$;

(b) $f(x, y, z) = x \cos yz + y \cos xz + z \cos xy$;

(c) $f(x, y, z) = \cos (x^2 + xy + yz)$.

5·30 Show that if

$$f(x, y, z) = \frac{x}{(x^2 + y^2 + z^2)^{3/2}}$$

then $xf_x + yf_y + zf_z = -2f$.

Section 5·6

5·31 Find the total differential du given that $u = f(x, y, z)$, where:

(a) $f(x, y, z) = \dfrac{1}{x^2yz} + xyz$;

(b) $f(x, y, z) = x \sin (y^2 + z^2)$;

(c) $f(x, y, z) = (1 - x^2 - y^2 - z^2)^{3/2}$.

5·32 Compute dy/dx from the following implicit relationships:
(a) $x^2 + y^2 = 4$;
(b) $x \sin xy = 1$;
(c) $x^2y + 2xy^2 + y^3 = 2$.

Section 5·7

5.33 Find du/dt given that:
(a) $u = xy + \sin (x^2 + y^2)$ with $x = 2t$, $y = (1 + t^2)^{1/2}$;
(b) $u = (1 + x^2 + y^2)^{3/2}$ with $x = t(1 + t)$, $y = t^3$;

(c) $u = \dfrac{z}{(x^2 + y^2)^{1/2}}$ with $x = 3 \cos t$, $y = 3 \sin t$, $z = t^2$.

5·34 If $u = f(x, y)$, compute du/dx given that:

(a) $f(x, y) = (1 + xy + x^2)$ where $y = \tan \left(\dfrac{x}{2}\right)$;

(b) $f(x, y) = (1 + x^2 - y^2)^{3/2}$ where $y = \cos 3x$;
(c) $f(x, y) = x \cos y + y \cos x - 1$ where $y = 1 + \sin^2 x$.

5·35 If $u = f(x, y)$ and $g(x, y) = 0$ are differentiable functions, compute du/dx given that:
(a) $f(x, y) = x^3 + 3xy + y^3$ and $g(x, y) = x \cos y + y \cos x - 2$;
(b) $f(x, y) = x^2y^2 + \sin xy$ and $g(x, y) = x^2 - 2y^2 - 3$.

Section 5·8

5·36 Given that $f(x, y, z) = x^2 + xy + \sin yz$, compute $\partial f/\partial r$, $\partial f/\partial \theta$, and $\partial f/\partial z$, where (r, θ, z') are the cylindrical polar coordinates corresponding to the point (x, y, z).

5·37 If u and v are functions of x and y which satisfy $u^2 - v^2 + 2x + 3y = 0$ and $uv + x - y = 0$, find $\partial u/\partial x$, $\partial u/\partial y$, $\partial v/\partial x$, and $\partial v/\partial y$ in terms of u and v.

5·38 Use Theorem 5·21 with $n = 2$, $m = 3$ to determine $\partial f/\partial u$, $\partial f/\partial v$, and $\partial f/\partial w$, given that:

$$f = x^2 + \tfrac{1}{2}y^2$$

where $x = u^2 + v + w$ and $y = uvw$.

5·39 Show that if $u = 1/r^n$, where $r^2 = x^2 + y^2 + z^2$, then

$$\frac{\partial^2 u}{\partial x^2} + \frac{\partial^2 u}{\partial y^2} + \frac{\partial^2 u}{\partial z^2} = \frac{n(n-1)}{r^{n+2}}.$$

Section 5·9

5·40 Compute dx/dy for each of the following relationships:

(a) $y = 1 + x^2 + x \sin x$;
(b) $y = (1 - x + x^2)^{1/2}$;
(c) $y = x + \tan x$.

5·41 Differentiate these functions:

(a) $f(x) = x^2 \text{ arc sec } (x/a)$;
(b) $f(x) = (x^2 + x + 1)/\text{arc sin } (x^2 - 2)$;
(c) $f(x) = (1 + x + \text{arc cos } 2x)^{3/2}$.

5·42 Compute dy/dx and d^2y/dx^2 for each of the following parametrically defined curves:

(a) $x = t - 1$, $y = t^3$;
(b) $x = \cos^3 t$, $y = 2 \sin^3 t$;
(c) $x = \text{arc cos } \dfrac{1}{\sqrt{(1 + t^2)}}$, $y = \text{arc sin } \dfrac{t}{\sqrt{(1 + t^2)}}$;
(d) $x = 2(\cos t + t \sin t)$, $y = 2(\sin t - t \cos t)$.

5·43 Compute dy/dx and d^2y/dx^2 at $t = \tfrac{1}{2}\pi$ if $x = t - \sin t$ and $y = 2(1 - \cos t)$.

Section 5·10

5·44 Compute $\partial^2 z/\partial x^2$, $\partial^2 z/\partial x \partial y$, $\partial^2 z/\partial y \partial x$, and $\partial^2 z/\partial y^2$ for each of the following functions and hence show that $\partial^2 z/\partial x \partial y = \partial^2 z/\partial y \partial x$:

(a) $z = (x^2 + y^2)^{1/2}$;
(b) $z = x \cos y + y \cos x$;
(c) $z = \text{arc tan } (y/x)$.

5·45 Compute $f_{xx}(1, 1)$, $f_{xy}(1, 1)$, and $f_{yy}(1, 1)$ given that

$$f(x, y) = (1 + x)^4(1 + y)^3.$$

Is $\partial^2 f/\partial x \partial y = \partial^2 f/\partial y \partial x$? Give reasons for your answer.

5·46 Given that

$$f(x, y) = \begin{cases} \dfrac{xy}{x^2 + y^2} & \text{for } x \neq 0, y \neq 0 \\ 1 & \text{for } x = 0, y = 0 \end{cases}$$

compute $\partial^2 f/\partial x \partial y$ stating, with reasons, when it is equal to $\partial^2 f/\partial y \partial x$. Is there any point at which this result is not true and, if so, what property of the function invalidates the result? [Hint: Consider limits taken along the line $y = mx$.]

5·47 Show that if $w = \arctan(x/y)$, then $\partial^2 w/\partial x^2 + \partial^2 w/\partial y^2 + \partial^2 w/\partial z^2 = 0$.

6 Exponential, hyperbolic, and logarithmic functions

6·1 The exponential function

This chapter will be concerned primarily with the exponential function, first introduced in connection with limits in Section 3·3 and, thereafter, with a number of related functions. This time our approach will be to utilize both geometrical ideas and the elementary calculus to produce a more useful form of definition than that contained in Eqn (3·6).

Let us seek a function $E(x)$ equal to its own derivative and such that $E(0) = 1$. Specifically, we must solve the equation

$$E'(x) = E(x) \tag{6·1}$$

which, because it involves the unknown function $E(x)$ together with its derivative, is called a *differential* equation. This differential equation has the following simple geometrical interpretation: if the graph of the function $E(x)$ is drawn, then the gradient of the graph at the point $(x, E(x))$ is equal to the functional value of $E(x)$ itself.

Suppose, for the moment, that there is a unique function $E(x)$ defined by our requirements, and consider the new function $F(x)$, where

$$F(x) = E(x)E(a - x). \tag{6·2}$$

Then,

$$F'(x) = E(x)\frac{\mathrm{d}}{\mathrm{d}x}[E(a - x)] + E(a - x)\frac{\mathrm{d}}{\mathrm{d}x}[E(x)]^{\cdot}$$

which, using the defining property (6·1), becomes

$$F'(x) = -E(x)E(a - x) + E(a - x)E(x) = 0.$$

Consequently, $F(x) = $ constant but, as $F(0) = E(0)E(a) = E(a)$, it follows at once that $F(x) = F(0) = E(a)$ for all x, and thus Eqn (6·2) takes the form

$$E(x)E(a - x) = E(a).$$

Alternatively, by replacing a by $a + b$ and x by b this may be written

$$E(a + b) = E(a)E(b). \tag{6·3}$$

Hence, if n is a positive integer,

$$E(n) = E(n - 1)E(1) = E(n - 2)(E(1))^2 = \cdots = (E(1))^n. \tag{6·4}$$

If, now, we denote $E(1)$ by the symbol e, then Eqn (6·4) is equivalent to

$$E(n) = e^n. \tag{6·5}$$

The fact that $E(0) = 1$ taken together with Eqn (6·1) implies $E(1) > 1$, also implies, via Eqn (6·5), that $\lim_{n \to \infty} e^n \to \infty$.

Again,

$$E(-n)E(n) = E(0) = 1,$$

so that

$$E(-n) = \frac{1}{E(n)} = \frac{1}{e^n} = e^{-n}. \tag{6·6}$$

Now we must extend this notation to take account of rational and irrational x. Let us consider $E(x)$ for rational x, so that $x = p/q$ with p, q integers. Then, using Eqn (6·5), we may write

$$\left[E\left(\frac{p}{q}\right)\right]^q = E\left(\frac{pq}{q}\right) = E(p) = e^p,$$

and so

$$E\left(\frac{p}{q}\right) = e^{p/q}. \tag{6·7}$$

A similar argument using Eqn (6·6) shows that

$$E\left(\frac{-p}{q}\right) = e^{-p/q}. \tag{6·8}$$

Thus we have shown that for all rational x

$$E(x) = e^x. \tag{6·9}$$

To extend the definition of $E(x)$ to all the real numbers x and not just to the rationals, it only remains to add that for any irrational number ξ, we define $E(\xi)$ by the equation $E(\xi) = e^\xi$.

We now seek a series solution to our function $E(x)$ of the form

$$y = \sum_{r=0}^{\infty} a_r x^r \tag{6·10}$$

where, for simplicity, we have set $y = E(x)$ so that Eqn (6·1) now becomes

$$\frac{dy}{dx} = y, \tag{6·11}$$

with $y(0) = 1$.

Assuming that this infinite series may be differentiated termwise, we have

$$\frac{dy}{dx} = \sum_{r=0}^{\infty} ra_r x^{r-1},$$

so that substituting for y and dy/dx in Eqn (6·11) yields

$$\sum_{r=0}^{\infty} ra_r x^{r-1} = \sum_{r=0}^{\infty} a_r x^r$$

or, equivalently,

$$\sum_{r=0}^{\infty} (r + 1)a_{r+1} x^r = \sum_{r=0}^{\infty} a_r x^r. \tag{6·12}$$

For this result to be unconditionally true for all x, as it must be to satisfy our definition of $E(x)$, it follows that it must be an identity in x. This can only be possible if the coefficients of the corresponding powers of x on each side of Eqn (6·12) are identical. Hence, equating the coefficients of the general term involving x^r, we find that

$$(r + 1)a_{r+1} = a_r \tag{6·13}$$

for $r = 0, 1, 2, \ldots$.

As we require that $y(0) = 1$, it follows by setting $x = 0$ in Eqn (6·10) that $a_0 = 1$. Using this result together with Eqn (6·13), which defines the coefficients a_r recursively, it is easily seen that

$$a_0 = 1, a_1 = 1, a_2 = \frac{1}{2!}, a_3 = \frac{1}{3!}, \ldots, a_r = \frac{1}{r!}, \ldots.$$

Substitution of these coefficients into Eqn (6·10) then shows that

$$E(x) = 1 + x + \frac{x^2}{2!} + \frac{x^3}{3!} + \cdots + \frac{x^n}{n!} + \cdots \tag{6·14}$$

whatever this expression may mean.

We have already remarked that the sum of an infinite series is to be interpreted as the limit of the partial sums of the series, so let us now consider the nth partial sum

$$S_n = 1 + x + \frac{x^2}{2!} + \cdots + \frac{x^{n-1}}{(n-1)!} \tag{6·15}$$

of the function $E(x)$.

If $x > 0$ then $S_{n+1} - S_n = x^n/n! > 0$, so that $\{S_n\}$ is increasing. Is $\{S_n\}$ bounded? Let R be an integer greater than $2x$, then $x/r < \frac{1}{2}$ for $r \geq R$, and so

$$\frac{x^r}{r!} = \frac{x}{1} \cdot \frac{x}{2} \cdots \frac{x}{R-1} \cdot \frac{x}{R} \cdots \frac{x}{r} < \frac{x^{R-1}}{(R-1)!} (\tfrac{1}{2})^{r-R+1}.$$

Thus

$$S_n = \sum_{r=0}^{R-1} \frac{x^r}{r!} + \sum_{r=R}^{n-1} \frac{x^r}{r!} < S_R + \frac{x^{R-1}}{(R-1)!} \sum_{r=R}^{n-1} (\tfrac{1}{2})^{r-R+1}$$

$$= S_R + \frac{x^{R-1}}{(R-1)!} \cdot \frac{1}{2} \cdot \left(\frac{1 - (\frac{1}{2})^{n-R}}{1 - \frac{1}{2}} \right) < S_R + \frac{x^{R-1}}{(R-1)!},$$

which shows that $\{S_n\}$ is bounded. Hence by the postulate of Section 3·2 it follows that $\lim_{n \to \infty} S_n$ exists, and we now define the sum of the infinite series (6·14) to be equal to the value of this limit. The infinite series (6·14) is thus defined for all positive x.

As we have agreed to write $E(1) = e$, it follows from Eqn (6·14), by setting $x = 1$, that

$$e = 1 + 1 + \frac{1}{2!} + \frac{1}{3!} + \cdots + \frac{1}{n!} + \cdots, \tag{6·16}$$

which, to 15 decimal places, has the numerical value

$$e = 2 \cdot 718281828459045.$$

A modified argument shows that $E(x)$ is also defined for all negative x, so that taking account of Eqn (6·9) we have proved the following result:

THEOREM 6·1 (exponential theorem)   For all x it is true that if

$$e = \sum_{n=0}^{\infty} \frac{1}{n!},$$

then

$$e^x = \sum_{n=0}^{\infty} \frac{x^n}{n!}.$$

Finally, it remains for us to establish the equivalence of the function $E(x)$ defined by Eqn (3·6) and the one denoted by the same symbols in Eqn (6·14). We shall only give the details for positive x. Our best method is first to expand Eqn (3·6), obtaining

$$\left(1 + \frac{x}{n} \right)^n = 1 + x + \frac{n(n-1)}{2!} \left(\frac{x}{n} \right)^2 + \frac{n(n-1)(n-2)}{3!} \left(\frac{x}{n} \right)^3 + \cdots + \left(\frac{x}{n} \right)^n.$$

Then, setting $E_{n+1} = [1 + (x/n)]^n$, we rewrite the result in the form

$$E_{n+1} = 1 + x + \frac{x^2}{2!} \left(1 - \frac{1}{n} \right) + \frac{x^3}{3!} \left(1 - \frac{1}{n} \right) \left(1 - \frac{2}{n} \right) + \cdots$$

$$+ \frac{x^n}{n!} \left(1 - \frac{1}{n} \right) \left(1 - \frac{2}{n} \right) \cdots \left(1 - \frac{n-1}{n} \right). \tag{6·17}$$

Defining the number $g(r, n)$ by

$$g(r, n) = \left(1 - \frac{1}{n} \right) \left(1 - \frac{2}{n} \right) \cdots \left(1 - \frac{r}{n} \right),$$

we next write Eqn (6·17) as

$$E_{n+1} = 1 + x + \frac{x^2}{2!} g(1, n) + \frac{x^3}{3!} g(2, n) + \cdots + \frac{x^n}{n!} g(n - 1, n). \quad (6·18)$$

Now the difference $S_{n+1} - E_{n+1}$ is

$$S_{n+1} - E_{n+1} = \frac{x^2}{2!} (1 - g(1, n)) + \frac{x^3}{3!} (1 - g(2, n)) + \cdots$$
$$+ \frac{x^n}{n!} [1 - g(n - 1, n)]$$

which is obviously positive since $0 < g(r, n) < 1$.

However, it is readily seen that for any given r

$$\lim_{n \to \infty} g(r, n) = 1,$$

showing that

$$\lim_{n \to \infty} (S_{n+1} - E_{n+1}) = 0.$$

From Theorem 3·1 (a) it then follows that

$$\lim_{n \to \infty} E_{n+1} = \lim_{n \to \infty} S_{n+1} = e^x,$$

thereby establishing the equivalence of our two alternative definitions when x is positive. A similar argument also establishes the equivalence when x is negative.

Having now achieved a working definition for $E(x)$ we shall henceforth always denote this function, known as the *exponential* function, either by e^x or by $\exp(x)$.

It is worth formally recording the differentiability properties of this function e^x. However, we first remark that if $f(x) = e^{g(x)}$, where $g(x)$ is a differentiable function of x, then, setting $g(x) = u$ so that $f(x) = e^u$ and using the chain rule in the form displayed in Eqn (5·6), we find that

$$\frac{df}{dx} = \frac{df}{du} \cdot \frac{du}{dx} = e^u g'(x) = g'(x)e^{g(x)}.$$

THEOREM 6.2   If $f(x) = e^{g(x)}$, where $g(x)$ is a differentiable function of x, then

$$\frac{d}{dx} \{e^{g(x)}\} = g'(x)e^{g(x)}.$$

In particular, if $g(x) = \alpha x$, where α is a constant, then,

$$\frac{d}{dx} (e^{\alpha x}) = \alpha e^{\alpha x}.$$

Let us now establish an important property of e^x. Consider the quotient e^x/x^p, where p is any positive integer. Then from Eqn (6·14) it follows that,

$$\frac{e^x}{x^p} = \frac{1 + x + \dfrac{x^2}{2!} + \cdots + \dfrac{x^p}{p!} + \dfrac{x^{p+1}}{(p+1)!} + \cdots}{x^p} > \frac{x}{(p+1)!}.$$

Hence we have shown that

$$\lim_{x \to \infty} \frac{e^x}{x^p} > \lim_{x \to \infty} \frac{x}{(p+1)!} \to \infty.$$

We have proved the following result:

THEOREM 6·3 The function e^x increases more quickly than any positive power of x as $x \to \infty$.

We have already noted that $\lim\limits_{x \to \infty} e^x \to \infty$, and as $e^x = 1/e^{-x}$ it follows that $\lim\limits_{x \to -\infty} e^x = 0$ or, equivalently, $\lim\limits_{x \to \infty} e^{-x} = 0$. From Theorem 6·1 it follows that the function e^x is everywhere positive and since, by virtue of its definition, its derivative is everywhere a strictly monotonic increasing function of x it must be a *convex* function. A graph of e^x is shown in Fig. 6·1.

These last properties are frequently of help when studying limiting problems involving the exponential function, as illustrated in the following examples.

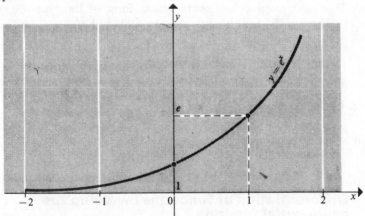

Fig. 6·1 The exponential function.

Example 6·1 Deduce the values of the following limits:

(a) $\lim\limits_{x \to \infty} \dfrac{3e^x + x^3 + 1}{2e^x + x^7}$;

(b) $\lim\limits_{x \to \infty} \dfrac{2e^{2x} + x^2 + 2}{3e^{3x} + 7}$;

(c) $\lim\limits_{x \to 0} \dfrac{e^{ax} - e^{bx}}{2x}$.

Solution (a) We have

$$\frac{3e^x + x^3 + 1}{2e^x + x^7} = \frac{3 + (x^3/e^x) + (1/e^x)}{2 + (x^7/e^x)},$$

and from Theorem 6·3 it then follows that all but the initial terms in numerator and denominator must vanish as $x \to \infty$, so that

$$\lim_{x \to \infty} \frac{3e^x + x^3 + 1}{2e^x + x^7} = \frac{3}{2}.$$

(b) In this case we have

$$\frac{2e^{2x} + x^2 + 2}{3e^{3x} + 7} = \frac{2e^{-x} + (x^2/e^{3x}) + (2/e^{3x})}{3 + (7/e^{3x})}.$$

However, this time as $x \to \infty$ so all the numerator tends to zero whilst the denominator approaches the value 3. Hence we have

$$\lim_{x \to \infty} \frac{2e^{2x} + x^2 + 2}{3e^{3x} + 7} = 0.$$

(c) This limit involves an indeterminate form of the type 0/0, so we appeal to Theorem 5·14. Writing $f(x) = e^{ax} - e^{bx}$ and $g(x) = 2x$ we see that $f(0) = g(0) = 0$, and

$$\lim_{x \to 0} \frac{f'(x)}{g'(x)} = \lim_{x \to 0} \frac{ae^{ax} - be^{bx}}{2} = \frac{a - b}{2}.$$

Hence, by the conditions of Theorem 5·14,

$$\lim_{x \to 0} \frac{e^{ax} - e^{bx}}{2x} = \lim_{x \to 0} \frac{ae^{ax} - be^{bx}}{2} = \frac{a - b}{2}.$$

6·2 Differentiation of functions involving the exponential function

The exponential function occurs frequently in mathematics, and all of its differentiability properties follow from Theorem 6·2 combined with the fundamental differentiation theorems of Chapter 5. These results are straightforward and are best illustrated by examples. The first example illustrates the ordinary differentiation of simple combinations of functions.

Example 6·2 Differentiate the following functions $f(x)$:

(a) $f(x) = x^2 e^{3x}$;   (b) $f(x) = 2 \exp (x^3 + 2x + 1)$.

Solution

(a) $f'(x) = e^{3x} \dfrac{d}{dx} (x^2) + x^2 \dfrac{d}{dx} (e^{3x})$

so that

$f'(x) = 2x e^{3x} + 3x^2 e^{3x}$.

(b) This is a more complicated example of a composite function or, more simply, of a function of a function. Set $u = x^3 + 2x + 1$ so that

$f(x) = 2e^u$.

Then, by the chain rule,

$f'(x) = \dfrac{df}{du} \cdot \dfrac{du}{dx}$

but

$\dfrac{df}{du} = \dfrac{d}{du} (2e^u) = 2e^u = 2 \exp (x^3 + 2x + 1)$ and $\dfrac{du}{dx} = 3x^2 + 2$

so that, finally,

$f'(x) = (6x^2 + 4) \exp (x^3 + 2x + 1)$.

Higher order derivatives are defined, as usual, by repeating the differentiation process the requisite number of times.

Example 6·3 Find $f''(x)$, given that:

$f(x) = (x - 1)e^x$.

Solution

$f'(x) = e^x + (x - 1)e^x = xe^x$

so that

$f''(x) = e^x + xe^x$.

Partial differentiation of functions involving the exponential function is also straightforward, as the following example indicates.

Example 6·4 Determine f_x, f_y, given that

$f(x, y) = (x^2 + y^2) \exp (x^2 - y^2)$.

Solution

$$\frac{\partial f}{\partial x} = 2x \exp(x^2 - y^2) + (x^2 + y^2) \frac{\partial}{\partial x} [\exp(x^2 - y^2)]$$

$$= 2x \exp(x^2 - y^2) + 2x(x^2 + y^2) \exp(x^2 - y^2).$$

Notice that $\partial f/\partial x$ comprises the sum of everywhere continuous functions and so is itself everywhere continuous.

$$\frac{\partial f}{\partial y} = 2y \exp(x^2 - y^2) + (x^2 + y^2) \frac{\partial}{\partial y} [\exp(x^2 - y^2)]$$

$$= 2y \exp(x^2 - y^2) - 2y(x^2 + y^2) \exp(x^2 - y^2).$$

6·3   The logarithmic function

Having introduced the exponential function there is now a need for an inverse function. The implicit function theorem (Theorem 5·22) tells us that such an inverse function exists and, furthermore, that it is differentiable whenever $(d/dx)(e^x) \neq 0$. However, this is always the case since we have already seen that $(d/dx)(e^x) = e^x$, which is never zero for x in the interval $-\infty < x < \infty$. Hence a differentiable function, inverse to the exponential function, exists for all x. We call it the *natural logarithmic* function and denote it by $\log_e$ whenever it is necessary to indicate that it has the *base* e.

DEFINITION 6·1   We define the natural logarithmic function $\log_e x$ by the requirement that

$$y = \log_e x \Leftrightarrow x = e^y.$$

We may use this definition, together with the Corollary to the implicit function theorem, to compute the derivative of $\log_e x$. As $dy/dx = 1/(dx/dy)$ and $x = e^y$, it follows that $dx/dy = e^y$, whence

$$\frac{dy}{dx} = \frac{1}{e^y} = \frac{1}{x}.$$

Now e^y is essentially positive, so that

$$\frac{d}{dx}(\log_e x) = \frac{1}{x} \qquad \text{for } x > 0. \tag{6·19}$$

It is obvious that $\log_e 1 = 0$ and, as x increases strictly monotonically with y, it also follows that $\log_e x \to +\infty$ as $x \to +\infty$, and $\log_e x \to -\infty$ as $x \to 0$.

Let us now prove that

$$\lim_{x \to \infty} \frac{\log_e x}{x^\alpha} = 0 \quad \text{for all } \alpha > 0.$$

As $x = e^y$ we have

$$\frac{\log_e x}{x^\alpha} = \frac{y}{e^{\alpha y}}$$

and so

$$\lim_{x \to \infty} \frac{\log_e x}{x^\alpha} = \lim_{y \to \infty} \frac{y}{e^{\alpha y}} = \frac{1}{\alpha} \lim_{y \to \infty} \frac{\alpha y}{e^{\alpha y}}.$$

Setting $u = \alpha y$ we arrive at

$$\lim_{x \to \infty} \frac{\log_e x}{x^\alpha} = \frac{1}{\alpha} \lim_{u \to \infty} \frac{u}{e^u} = 0,$$

by virtue of Theorem 6·3.

Collecting the previous results we arrive at the following theorem.

THEOREM 6·4   If $y = \log_e x$, then

(a) $\dfrac{dy}{dx} = \dfrac{1}{x}$ for $x > 0$;

(b) $\displaystyle\lim_{x \to \infty} \dfrac{\log_e x}{x^\alpha} = 0$ for all $\alpha > 0$.

Logarithms to other bases can be used if convenient. They are defined as follows.

DEFINITION 6·2   We define the logarithmic function to the base c, denoted by $\log_c x$ where c is a positive number, by the requirement that

$$y = \log_c x \Leftrightarrow x = c^y.$$

For reference purposes we record the following familiar properties of the logarithmic function, established in elementary courses.

Basic properties of the logarithmic function

Let $\log_e$ and $\log_c$ represent logarithms to the bases e and c respectively, and a, b, r be real numbers; then:

(a) $\log_e ab = \log_e a + \log_e b$;

(b) $\log_e a^r = r \log_e a$;

(c) $\log_c a = \dfrac{\log_e a}{\log_e c}$;

(d) $\log_c e = \dfrac{1}{\log_e c}$.

Results (c) and (d) quoted above are immediately useful if it is necessary to differentiate $\log_a x$. For we have

$$\log_a x = \frac{\log_e x}{\log_e a}$$

so that

$$\frac{d}{dx}(\log_a x) = \frac{1}{\log_e a} \cdot \frac{d}{dx}(\log_e x)$$

whence,

$$\frac{d}{dx}(\log_a x) = \frac{1}{x \log_e a} = \frac{\log_a e}{x}. \qquad (6.20)$$

Let us now find the derivative of the function a^x, where a is any positive number. Notice first that, by virtue of Definition 6·1,

$$a = e^{\log_e a}$$

so that

$$a^x = (e^{\log_e a})^x = e^{x \log_e a}.$$

Now $\log_e a$ is simply a constant, so we have

$$\frac{d}{dx}(a^x) = \frac{d}{dx}(e^{x \log_e a}) = \log_e a \, e^{x \log_e a} = a^x \log_e a.$$

We have thus established the useful result

$$\frac{d}{dx}(a^x) = a^x \log_e a. \qquad (6.21)$$

This result can also be obtained in another manner. We set

$$y = a^x,$$

so that taking the natural logarithm gives

$$\log_e y = x \log_e a.$$

Differentiating this result with respect to x we obtain

$$\frac{d}{dx}(\log_e y) = \frac{d}{dx}(x \log_e a)$$

or

$$\frac{1}{y} \cdot \frac{dy}{dx} = \log_e a,$$

and so

$$\frac{dy}{dx} = \frac{d}{dx}(a^x) = y \log_e a = a^x \log_e a.$$

For our final general result we consider the differentiation of the function $y = \log_e g(x)$, where $g(x)$ is a differentiable function. Setting $u = g(x)$ so that $y = \log_e u$ and using the chain rule gives

$$\frac{dy}{dx} = \frac{dy}{du} \cdot \frac{du}{dx} = \frac{1}{u} g'(x)$$

so that, finally,

$$\frac{d}{dx}[\log_e g(x)] = \frac{g'(x)}{g(x)}. \tag{6·22}$$

Henceforth, unless otherwise stated, the natural logarithm will always be used, so for simplicity of notation we shall write log in place of $\log_e$. Often, in other texts, the notation ln is used to denote the natural logarithmic function.

Let us now examine some representative cases of limits involving logarithms.

Example 6·5 Evaluate the following limits:

(a) $\lim\limits_{x \to \infty} \dfrac{\log x^3}{x}$;

(b) $\lim\limits_{x \to 0} \dfrac{\log (1 + 3x)}{2x}$.

Solution (a) We have

$$\frac{\log x^3}{x} = \frac{3 \log x}{x}$$

so that by Theorem 6·4 (b) it follows at once that

$$\lim_{x \to \infty} \frac{\log x^3}{x} = 0.$$

(b) This is an indeterminate form of the type 0/0. It is easily verified that Theorem 5·14 (L'Hospital's rule) is applicable so that

$$\lim_{x \to 0} \frac{f(x)}{g(x)} = \lim_{x \to 0} \frac{f'(x)}{g'(x)},$$

with $f(x) = \log(1 + 3x)$ and $g(x) = 2x$. As $f'(x) = 3/(1 + 3x)$ and $g'(x) = 2$ it thus follows that

$$\lim_{x \to 0} \frac{\log(1 + 3x)}{2x} = \lim_{x \to 0} \frac{3}{2(1 + 3x)} = \frac{3}{2}.$$

Example 6·6 Determine the derivative dy/dx for each of the following functions $y = f(x)$ where:

(a) $f(x) = 3^x x^2$;   (b) $f(x) = (\sin x)^x$.

Solution (a) We have

$$\frac{d}{dx}(3^x x^2) = 3^x \frac{d}{dx}(x^2) + x^2 \frac{d}{dx}(3^x)$$

which, by virtue of Eqn (6·25), becomes

$$\frac{d}{dx}(3^x x^2) = 2x . 3^x + x^2 3^x \log 3$$

giving

$$\frac{d}{dx}(3^x x^2) = (2x + x^2 \log 3)3^x.$$

(b) We set $y = (\sin x)^x$ and take logarithms to get

$$\log y = x \log \sin x.$$

Now, differentiating, we find that

$$\frac{1}{y} \cdot \frac{dy}{dx} = \log \sin x + x \frac{d}{dx}(\log \sin x)$$

or

$$\frac{dy}{dx} = (\sin x)^x (\log \sin x + x \cot x).$$

Partial differentiation involving the logarithmic function is equally straightforward. The final example illustrates a typical situation.

Example 6·7 If $u = x \log[1 + (x/y)] + y \log[1 + (y/x)]$, show that

$$x \frac{\partial u}{\partial x} + y \frac{\partial u}{\partial y} = u.$$

Solution We start by computing $\partial u/\partial x$. It is readily seen that

$$\frac{\partial u}{\partial x} = \log\left(1 + \frac{x}{y}\right) + x\frac{\partial}{\partial x}\log\left(1 + \frac{x}{y}\right) + y\frac{\partial}{\partial x}\log\left(1 + \frac{y}{x}\right)$$

$$= \log\left(1 + \frac{x}{y}\right) + x \cdot \frac{1}{1 + x/y} \cdot \frac{1}{y} + y \cdot \frac{1}{1 + y/x}\left(\frac{-y}{x^2}\right),$$

and so

$$\frac{\partial u}{\partial x} = \log\left(1 + \frac{x}{y}\right) + \frac{x}{x + y} - \frac{y^2}{x(x + y)}.$$

The symmetry of x and y in u then allows us to interchange x and y in the above partial derivative in order to derive $\partial u/\partial y$ without further calculation. We obtain

$$\frac{\partial u}{\partial y} = \log\left(1 + \frac{y}{x}\right) + \frac{y}{x + y} - \frac{x^2}{y(x + y)}.$$

Hereafter, direct substitution verifies that

$$x\frac{\partial u}{\partial x} + y\frac{\partial u}{\partial y} = u.$$

6·4 Hyperbolic functions

It is useful to define new functions called the *hyperbolic sine*, written sinh x, and the *hyperbolic cosine*, written cosh x, which are related to the exponential function. This is achieved as follows.

DEFINITION 6·3 (hyperbolic functions)   For all real x we define sinh x and cosh x by the requirement that

$$\sinh x = \frac{e^x - e^{-x}}{2}, \qquad \cosh x = \frac{e^x + e^{-x}}{2}.$$

It is an immediate consequence of the series for e^x and e^{-x} that

$$\sinh x = x + \frac{x^3}{3!} + \frac{x^5}{5!} + \frac{x^7}{7!} + \cdots + \frac{x^{2n+1}}{(2n + 1)!} + \cdots, \qquad (6·23)$$

and

$$\cosh x = 1 + \frac{x^2}{2!} + \frac{x^4}{4!} + \frac{x^6}{6!} + \cdots + \frac{x^{2n}}{(2n)!} + \cdots. \qquad (6·24)$$

Furthermore, it also follows from Definition 6·3 that sinh x is an odd function and cosh x is an even function.

We now define the *hyperbolic tangent, cotangent, cosecant,* and *secant,* denoted by tanh x, coth x, cosech x, and sech x, as follows.

DEFINITION 6·4

$$\tanh x = \frac{\sinh x}{\cosh x}; \qquad \coth x = \frac{\cosh x}{\sinh x};$$

$$\operatorname{cosech} x = \frac{1}{\sinh x}; \qquad \operatorname{sech} x = \frac{1}{\cosh x}.$$

We illustrate how useful identities may be established directly from Definition 6·3. Let us prove that

$$\sinh a \cosh b + \cosh a \sinh b = \sinh (a + b).$$

Substituting for $\sinh a$ and $\cosh b$ from Definition 6·3 we obtain

$$\frac{e^a - e^{-a}}{2} \cdot \frac{e^b + e^{-b}}{2} + \frac{e^a + e^{-a}}{2} \cdot \frac{e^b - e^{-b}}{2} = \frac{e^{(a+b)} - e^{-(a+b)}}{2},$$

which proves our result since $[e^{(a+b)} - e^{-(a+b)}]/2 = \sinh (a + b)$. Similar manipulation establishes the validity of all the identities listed below in Table 6·1.

Table 6·1   Identities for hyperbolic functions

$\sinh (x \pm y) = \sinh x \cosh y \pm \cosh x \sinh y;$	(6·25)
$\cosh (x \pm y) = \cosh x \cosh y \pm \sinh x \sinh y;$	(6·26)
$\cosh^2 x - \sinh^2 x = 1;$	(6·27)
$\tanh^2 x + \operatorname{sech}^2 x = 1;$	(6·28)
$1 + \operatorname{cosech}^2 x = \coth^2 x.$	(6·29)

Table 6·2   Derivatives of hyperbolic functions

$\dfrac{d}{dx} (\sinh x) = \cosh x;$	(6·30)
$\dfrac{d}{dx} (\cosh x) = \sinh x;$	(6·31)
$\dfrac{d}{dx} (\tanh x) = \operatorname{sech}^2 x;$	(6·32)
$\dfrac{d}{dx} (\coth x) = - \operatorname{cosech}^2 x;$	(6·33)
$\dfrac{d}{dx} (\operatorname{cosech} x) = - \operatorname{cosech} x \coth x;$	(6·34)
$\dfrac{d}{dx} (\operatorname{sech} x) = - \operatorname{sech} x \tanh x.$	(6·35)

Appeal to Definitions 6·3 and 6·4 together with the differentiability properties of the exponential function establishes Table 6·2, the table of derivatives.

The behaviour of the hyperbolic functions is indicated graphically in Fig. 6·2 and for comparison the graphs of $y = \frac{1}{2}e^x$ and $y = \frac{1}{2}e^{-x}$ have been added to Fig. 6·2 (a).

Functions inverse to the hyperbolic sine and cosine are introduced through the following definitions.

DEFINITION 6·5   The *inverse hyperbolic sine*, arcsinh x, and the *inverse hyperbolic cosine*, arccosh x, are defined by the relationships:

(a) $y = \operatorname{arcsinh} x \Leftrightarrow x = \sinh y$;

(b) $y = \operatorname{arccosh} x \Leftrightarrow x = \cosh y$.

Their derivatives are readily obtained by direct use of this definition and we illustrate the process by deriving d/dx arcsinh x.

If $y = \operatorname{arcsinh} x$, then $x = \sinh y$ and so, differentiating with respect to x, we obtain

$$1 = \cosh y \, \frac{dy}{dx},$$

and so

$$\frac{dy}{dx} = \frac{1}{\cosh y} = \frac{1}{\sqrt{(1 + \sinh^2 y)}},$$

by virtue of identity (6·31) and the fact that $\cosh y$ is essentially positive. Hence, using the fact that $x = \sinh y$, we find that

$$\frac{d}{dx}(\operatorname{arcsinh} x) = \frac{1}{\sqrt{(1 + x^2)}} \text{ for all } x.$$

In the case of $y = \operatorname{arccosh} x$ we must proceed with more care.

If $y = \operatorname{arccosh} x$, so that $x = \cosh y$, then, as before, differentiating with respect to x gives

$$1 = \sinh y \cdot \frac{dy}{dx}$$

or,

$$\frac{dy}{dx} = \frac{1}{\sinh y}.$$

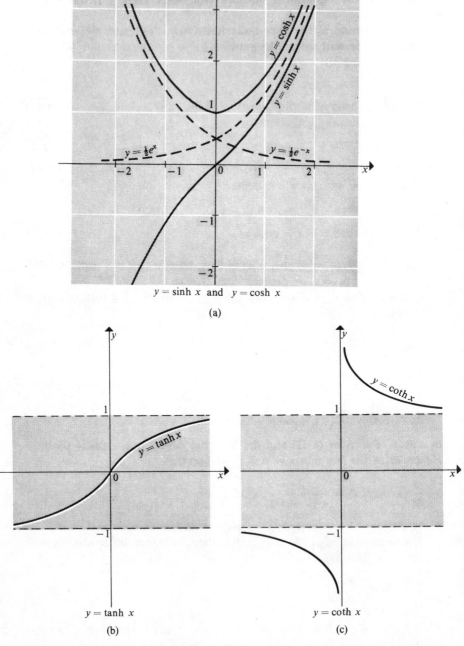

$y = \sinh x$ and $y = \cosh x$

(a)

$y = \tanh x$

(b)

$y = \coth x$

(c)

Fig. 6·2 Hyperbolic functions: (a) $y = \sinh x$ and $y = \cosh x$; (b) $y = \tanh x$; (c) $y = \coth x$;

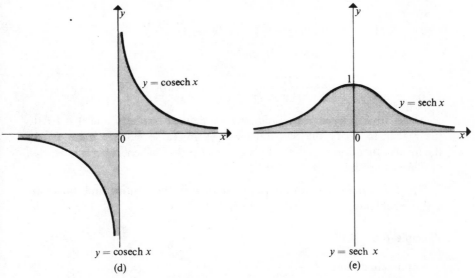

Fig. 6·2 (*continued*)   (d) $y = \operatorname{cosech} x$; (e) $y = \operatorname{sech} x$.

Now from the graph in Fig. 6·2 (a) we see that $\sinh y$ is positive if its argument arccosh $x > 0$ and negative if arccosh $x < 0$. Thus two different inverse functions must be defined.

If arccosh $x > 0$, then

Table 6·3   Derivatives of inverse hyperbolic functions

$$\frac{d}{dx}\left(\operatorname{arcsinh} \frac{x}{a}\right) = \frac{1}{\sqrt{(x^2 + a^2)}}, \qquad \text{for all } x; \tag{6·36}$$

$$\frac{d}{dx}\left(\operatorname{arccosh} \frac{x}{a}\right) = \frac{1}{\sqrt{(x^2 - a^2)}}, \qquad \text{for arccosh } \frac{x}{a} > 0 \text{ and } \frac{x}{a} > 1; \tag{6·37}$$

$$\frac{d}{dx}\left(\operatorname{arccosh} \frac{x}{a}\right) = \frac{-1}{\sqrt{(x^2 - a^2)}}, \qquad \text{for arccosh } \frac{x}{a} < 0 \text{ and } \frac{x}{a} > 1; \tag{6·38}$$

$$\frac{d}{dx}\left(\operatorname{arctanh} \frac{x}{a}\right) = \frac{a}{a^2 - x^2}, \qquad \text{for } x^2 < a^2; \tag{6·39}$$

$$\frac{d}{dx}\left(\operatorname{arccoth} \frac{x}{a}\right) = \frac{a}{a^2 - x^2}, \qquad \text{for } x^2 > a^2; \tag{6·40}$$

$$\frac{d}{dx}\left(\operatorname{arccosech} \frac{x}{a}\right) = \frac{-a}{x\sqrt{(x^2 + a^2)}}, \qquad \text{for all } x; \tag{6·41}$$

$$\frac{d}{dx}\left(\operatorname{arcsech} \frac{x}{a}\right) = \frac{-a}{x\sqrt{(a^2 - x^2)}}, \qquad \text{for arcsech } \frac{x}{a} > 0 \text{ and } 0 < \frac{x}{a} < 1; \tag{6·42}$$

$$\frac{d}{dx}\left(\operatorname{arcsech} \frac{x}{a}\right) = \frac{a}{x\sqrt{(a^2 - x^2)}}, \qquad \text{for arcsech } \frac{x}{a} < 0 \text{ and } 0 < \frac{x}{a} < 1. \tag{6·43}$$

$$\frac{dy}{dx} = \frac{1}{\sinh y} = \frac{1}{\sqrt{(\cosh^2 y - 1)}} = \frac{1}{\sqrt{(x^2 - 1)}} \text{ for } x > 1.$$

Conversely, if arccosh $x < 0$, then

$$\frac{dy}{dx} = \frac{1}{\sinh y} = \frac{-1}{\sqrt{(\cosh^2 y - 1)}} = \frac{-1}{\sqrt{(x^2 - 1)}} \text{ for } x > 1.$$

Other inverse hyperbolic functions are defined similarly and it is left to the reader to verify the remaining entries in Table 6·3. (In many books the inverse function is denoted by a superscript -1, when $\sinh^{-1} x$ is written in place of arcsinh x, etc.)

The following examples are representative of the limiting and differentiability problems encountered with hyperbolic functions.

Example 6·8

(a) Evaluate $\lim\limits_{x \to \infty} \dfrac{5 \sinh 3x + xe^x}{4e^{3x}}$;

(b) Find $f'(x)$ if $f(x) = \sinh (x^2 + 3x + 1)^{1/2}$;

(c) Find $f'(x)$ given that $f(x) < 0$ is given by $f(x) = \text{arccosh } (\sin^2 x)$;

(d) Determine f_x and f_y given that $f(x, y) = xy \cosh (x^2 + y^2)$.

Solution (a) From Definition 6·3 it is easily seen that for large x

$\sinh 3x \doteq \frac{1}{2}e^{3x}$.

Hence, applying the usual arguments, it follows at once that

$$\lim\limits_{x \to \infty} \frac{5 \sinh 3x + xe^x}{4e^{3x}} = \lim\limits_{x \to \infty} \frac{(5e^{3x}/2) + xe^x}{4e^{3x}} = \frac{5}{8}.$$

(b) $f'(x) = [\cosh (x^2 + 3x + 1)^{1/2}] \cdot \dfrac{1}{2} \cdot \dfrac{(2x + 3)}{(x^2 + 3x + 1)^{1/2}}$

so that

$$f'(x) = \frac{(2x + 3)}{2(x^2 + 3x + 1)^{1/2}} \cosh (x^2 + 3x + 1)^{1/2}.$$

(c) Set $y = \text{arccosh } (\sin^2 x)$ so that

$\sin^2 x = \cosh y$.

Differentiation with respect to x then gives

$$2 \sin x \cdot \cos x = \sinh y \cdot \frac{dy}{dx}.$$

or

$$\frac{dy}{dx} = \frac{2\sin x \cdot \cos x}{\sinh y}.$$

As we are told that $y = f(x) < 0$ it then follows that

$$\frac{dy}{dx} = \frac{-2\sin x \cdot \cos x}{\sqrt{(\cosh^2 y - 1)}} = \frac{-2\sin x \cdot \cos x}{\sqrt{(\sin^4 x - 1)}}$$

provided $\sin x \neq 1$.

(d) $\dfrac{\partial f}{\partial x} = y\cosh(x^2 + y^2) + xy\,\partial/\partial x \cosh(x^2 + y^2)$

$\qquad = y\cosh(x^2 + y^2) + 2x^2 y \sinh(x^2 + y^2).$

Similarly,

$$\frac{\partial f}{\partial y} = x\cosh(x^2 + y^2) + 2xy^2 \sinh(x^2 + y^2).$$

6·5   Exponential function with a complex argument

If we formally replace x by ix in the series expansion of e^x in Theorem 6·1 we obtain

$$e^{ix} = 1 + ix - \frac{x^2}{2!} - i\frac{x^3}{3!} + \frac{x^4}{4!} + i\frac{x^5}{5!} - \frac{x^6}{6!} + \cdots + i^n\frac{x^n}{n!} + \cdots.$$

Clearly e^{ix} is a complex number for any fixed real number x and, writing it in the form $e^{ix} = C(x) + iS(x)$, it follows by equating real and imaginary parts that

$$C(x) = 1 - \frac{x^2}{2!} + \frac{x^4}{4!} - \frac{x^6}{6!} + \cdots + (-1)^n\frac{x^{2n}}{(2n)!} + \cdots$$

and

$$S(x) = x - \frac{x^3}{3!} + \frac{x^5}{5!} - \frac{x^7}{7!} + \cdots + (-1)^n\frac{x^{2n+1}}{(2n + 1)!} + \cdots.$$

Thus, in fact, if x is regarded as a variable, $S(x)$ and $C(x)$ are functions of x and e^{ix} is, in some sense yet to be properly defined, a function of a complex variable.

Assuming that the series for $C(x)$ may be differentiated term by term it is easily verified that

$$C'(x) = -x + \frac{x^3}{3!} - \frac{x^5}{5!} + \frac{x^7}{7!} + \cdots + (-1)^{n+1}\frac{x^{2n+1}}{(2n + 1)!} + \cdots.$$

Next, differentiating $C'(x)$ again with respect to x yields

$$C''(x) = -1 + \frac{x^2}{2!} - \frac{x^4}{4!} + \frac{x^6}{6!} + \cdots + (-1)^{n+1}\frac{x^{2n}}{(2n)!} + \cdots,$$

showing that in fact

$$C''(x) = -C(x).$$

Now, setting $x = 0$ in the series for $C(x)$ and $C'(x)$, we find that

$$C(0) = 1 \quad \text{and} \quad C'(0) = 0.$$

Hence the function $C(x)$ is seen to be the solution of the special differential equation

$$\frac{d^2y}{dx^2} + y = 0$$

with $y(0) = 1$ and $y'(0) = 0$.

This same differential equation with the conditions on y was encountered in Example 5·12 (a), where it was derived as the equation satisfied by $y = \cos x$ and its derivatives. Thus the function $C(x)$ is, in reality, the function $\cos x$. An analogous argument establishes that $S(x) \equiv \sin x$. On account of this identification of $C(x)$ and $S(x)$ we may write

$$e^{ix} = \cos x + i \sin x. \tag{6·44}$$

As a direct consequence of replacing x by $-x$ in Eqn (6·48) and using the fact that $\cos x$ is even, but $\sin x$ is odd, we find that

$$e^{-ix} = \cos x - i \sin x. \tag{6·45}$$

Combination of Eqns (6·44) and (6·45) leads to the following definitions of the sine and cosine functions.

DEFINITION 6·6

$$\sin x = \frac{e^{ix} - e^{-ix}}{2i} \quad \text{and} \quad \cos x = \frac{e^{ix} + e^{-ix}}{2}.$$

Comparison of Eqns (4·15) and (6·44) shows that e^{ix} represents a complex number of unit modulus lying on the unit circle drawn about the origin. The argument of e^{ix} is x.

Slightly more general than Eqn (6·44) is the complex number $e^{(x+iy)}$ for, by the property of indices together with Eqn (6·44), we have

$$e^{(x+iy)} = e^x \cdot e^{iy} = e^x(\cos y + i \sin y), \tag{6·46}$$

showing that

$$|e^{(x+iy)}| = e^x \quad \text{and} \quad \arg e^{(x+iy)} = y. \tag{6·47}$$

Thus the *modulus-argument* form of a general non-zero complex number z may be written

$$z = re^{i\theta},$$

where

$$r = |z| \quad \text{and} \quad \theta = \arg z. \tag{6·48}$$

This is, of course, an alternative form of Eqn (4·15).

As it is true for any exponent α that $(a^x)^\alpha = a^{\alpha x}$, it follows that $(e^{ix})^\alpha = e^{i\alpha x}$, so that from Eqn (6·44) we arrive at the result

$$(\cos x + i \sin x)^\alpha = \cos \alpha x + i \sin \alpha x. \tag{6·49}$$

This is simply de Moivre's theorem (Theorem 4·2) for *any* exponent α and not just for the integral values used in the first proof of this important theorem.

To close, let us apply these results to give an alternative derivation of the results of Example 4·10, and also to express $\sin^n \theta$ and $\cos^n \theta$ in terms of sums involving $\sin r\theta$ and $\cos r\theta$, as promised in that example. As in Chapter 4, the argument is best presented by example.

Example 6·9

(a) Express $\sin n\theta$ and $\cos n\theta$ in terms of $\cos \theta$ and $\sin \theta$. Deduce the form taken by the result when $n = 4$.

(b) Express $\cos^7 \theta$ in terms of $\cos r\theta$.

(c) Express $\sin^5 \theta$ in terms of $\sin r\theta$.

Solution

(a)

$$\cos n\theta = \text{Re}(e^{in\theta}) = \text{Re}[(e^{i\theta})^n] = \text{Re}[(\cos \theta + i \sin \theta)^n].$$

$$\sin n\theta = \text{Im}(e^{in\theta}) = \text{Im}[(e^{i\theta})^n] = \text{Im}[(\cos \theta + i \sin \theta)^n].$$

When $n = 4$ we have

$$(\cos \theta + i \sin \theta)^4 = \cos^4 \theta + 4i \cos^3 \theta \sin \theta - 6 \cos^2 \theta \sin^2 \theta$$
$$- 4i \cos \theta \sin^3 \theta + \sin^4 \theta.$$

Hence

$$\cos 4\theta = \text{Re}[(\cos \theta + i \sin \theta)^4] = \cos^4 \theta - 6 \cos^2 \theta \sin^2 \theta + \sin^4 \theta$$

and

$$\sin 4\theta = \text{Im}[(\cos \theta + i \sin \theta)^4] = 4(\cos^3 \theta \sin \theta - \cos \theta \sin^3 \theta).$$

(b) From Definition 6·6 we may write

$$\cos^7 \theta = \left(\frac{e^{i\theta} + e^{-i\theta}}{2} \right)^7.$$

Expanding the right-hand side by the Binomial theorem, simplifying and grouping terms, we obtain

$$\cos^7 \theta = \frac{1}{2^6} \left(\frac{e^{7i\theta} + e^{-7i\theta}}{2} + 7 \frac{e^{5i\theta} + e^{-5i\theta}}{2} + 21 \frac{e^{3i\theta} + e^{-3i\theta}}{2} \right.$$
$$\left. + 35 \frac{e^{i\theta} + e^{-i\theta}}{2} \right).$$

Again using Definition 6·6, we see that this immediately simplifies to

$$\cos^7 \theta = \frac{1}{64} (\cos 7\theta + 7 \cos 5\theta + 21 \cos 3\theta + 35 \cos \theta).$$

(c) From Definition 6·6 we may write

$$\sin^5 \theta = \left(\frac{e^{i\theta} - e^{-i\theta}}{2i} \right)^5.$$

Expanding the right-hand side, simplifying and grouping terms gives

$$\sin^5 \theta = \frac{1}{2^4} \left(\frac{e^{5i\theta} - e^{-5i\theta}}{2i} - 5 \frac{e^{3i\theta} - e^{-3i\theta}}{2i} + 10 \frac{e^{i\theta} - e^{-i\theta}}{2i} \right).$$

Again appealing to Definition 6·6, we see that this immediately reduces to

$$\sin^5 \theta = \frac{1}{16} (\sin 5\theta - 5 \sin 3\theta + 10 \sin \theta).$$

A variant of the method used here and in example (b) above is to be found outlined in Problems 6·23 and 6·24.

PROBLEMS

Section 6·1

6·1 Solve the differential equation $dy/dx = y$, with $y(0) = c$, as in Section 6·1, by substituting

$$y = \sum_{r=0}^{\infty} a_r x^r.$$

Hence deduce that, provided $c \neq 0$, the differential equation has the non-trivial solution $y = ce^x$.

6·2 Evaluate the following limits:

(a) $\lim\limits_{x \to \infty} \dfrac{4e^{2x} + xe^x + 3}{5xe^{3x} + e^x + 1}$;

(c) $\lim\limits_{x \to -\infty} \dfrac{(2 - x^2)e^x + 3}{1 + (2 + x)e^{2x}}$;

(b) $\lim\limits_{x \to \infty} \dfrac{(x^2 + 1)e^{3x} + e^x + 1}{(2x^2 - 3x + 1)e^{3x}}$;

(d) $\lim\limits_{x \to 0} \dfrac{3(2e^{-3x} + x^2 + 1)}{4e^{2x} + 2x + 1}$.

6·3 Make use of the series expansion of e^x to evaluate the following limits and verify your result by using Theorem 5·14:

(a) $\lim\limits_{x \to 0} \dfrac{e^{2x} - 1}{3x}$;

(c) $\lim\limits_{x \to 0} \dfrac{e^x - 1 - x}{3x^2}$.

(b) $\lim\limits_{x \to 0} \dfrac{1 - e^{-x}}{\sin 4x}$;

6·4 Differentiate the following functions:

(a) $f(x) = 2e^x \cos x$;

(c) $f(x) = e^x/x^2$;

(b) $f(x) = e^{3x} \arcsin x$;

(d) $f(x) = e^{x \sin x}$.

6·5 Differentiate the following functions:

(a) $f(x) = \arcsin e^{2x}$;

(c) $f(x) = \sin (xe^x + 2)$;

(b) $f(x) = \sqrt{(xe^x + x)}$;

(d) $f(x) = (e^x - 1)/(e^x + 1)$.

Section 6·2

6·6 Differentiate the following functions:

(a) $f(x) = 3 \exp [-(x^2 + x + 1)]$;

(c) $f(x) = \cos [\exp (x \sin x + 2)]$.

(b) $f(x) = e^{\sin^2 x}$;

6·7 Find $\partial f/\partial x$ and $\partial f/\partial y$, given that

$$f(x, y) = e^{\sin (y/x)}.$$

6·8 Show that $u = xy + xe^{y/x}$ satisfies the equation

$$x \frac{\partial u}{\partial x} + y \frac{\partial u}{\partial y} = xy + u.$$

Section 6·3

6·9 Evaluate the following limits:

(a) $\lim\limits_{x \to \infty} \dfrac{(x - 1) \log x^2}{x^2}$;

(b) $\lim\limits_{x \to \infty} \dfrac{\log (3 + 2^x)}{4x + 1}$;

(c) $\lim\limits_{x \to 0} \dfrac{\log (3 \sin x) - \log [(1 + x) \sin x]}{2e^x - 1}$;

(d) $\lim\limits_{x \to \infty} [\log (3x + 1) - \log (2x + 5)]$;

(e) $\lim\limits_{x \to \infty} \dfrac{\log (1 + 2e^x)}{x}$.

6·10 Differentiate the following functions:

(a) $f(x) = \log (x^3 + 7x^2 + 2)$;

(b) $f(x) = \log \sin 2x$;

(c) $f(x) = \log \cos \left(\dfrac{x - 1}{x} \right)$.

6·11 If $y = [f(x)]^{g(x)}$ then, taking the natural logarithm,

$$\log y = g(x) \log f(x).$$

Hence, differentiating with respect to x, it follows that

$$\frac{dy}{dx} = \left[g'(x) \log f(x) + \frac{g(x)}{f(x)} f'(x) \right] [f(x)]^{g(x)}$$

Use this result to differentiate the following functions:

(a) $y = x^x$;

(b) $y = (\sin 2x)^x$;

(c) $y = x^{\sin x}$;

(d) $y = 10^{\log \sin x}$.

6·12 If $u = x \log (1 + x/y) + y \log (1 + y/x)$

show that $x^2 \dfrac{\partial^2 u}{\partial x^2} = y^2 \dfrac{\partial^2 u}{\partial y^2}$.

6·13 By taking logarithms deduce $\partial u/\partial x$, $\partial u/\partial y$, and $\partial u/\partial z$ if $u = (xy)^z$.

Section 6·4

6·14 Prove by means of the definition that

$$(\cosh x + \sinh x)^n = \cosh nx + \sinh nx.$$

6·15 Prove by means of the definitions that:

(a) $2 \sinh x \cosh y = \sinh (x + y) + \sinh (x - y)$;

(b) $2 \cosh x \cosh y = \cosh (x + y) + \cosh (x - y)$;

(c) $2 \sinh x \sinh y = \cosh (x + y) - \cosh (x - y)$.

6·16 Evaluate the following limits, using the series (6·23) and (6·24) where necessary:

(a) $\lim\limits_{x \to \infty} \dfrac{x^3 \cosh 2x + e^x}{(2x^3 + x + 1)e^{2x} + x^3 e^{-2x}}$;

(b) $\lim\limits_{x \to -\infty} \dfrac{x^3 \cosh 2x + e^x}{(2x^3 + x + 1)e^{2x} + x^3 e^{-2x}}$;

(c) $\lim\limits_{x \to 0} \dfrac{\sinh \alpha x}{x}$;

(d) $\lim\limits_{x \to 0} \dfrac{1 - \cosh 2x}{3x^2}$.

6·17 Differentiate the following functions:

(a) $f(x) = \sinh 2x \cosh^2 x$;

(b) $f(x) = \exp (1 + \cosh 3x)$;

(c) $f(x) = \log (\tanh x)$;

(d) $f(x) = \text{arcsech} (x^2 + \frac{1}{2})$ if $f(x) > 0$;

(e) $f(x) = \cosh (\sin 2x)$.

6·18 Evaluate $\partial u/\partial x$ and $\partial u/\partial y$ given that:

(a) $u(x, y) = \sin x \cosh xy$;

(b) $u(x, y) = \sinh (x^2 + x \sin y + 3y^2)$;

(c) $u(x, y) = x^{\cosh(x^2 + 2y^2)}$.

Section 6·5

6·19 Establish by means of the definitions that:

(a) $\sin (iz) = i \sinh z$;

(b) $\cos (iz) = \cosh z$;

(c) $\sinh (iz) = i \sin z$;

(d) $\cosh (iz) = \cos z$.

6·20 Given that a, b are positive real numbers, deduce four trigonometric identities by equating real and imaginary parts in each of the following results

$$e^{ia} \cdot e^{ib} = e^{i(a+b)} \quad \text{and} \quad e^{ia} \cdot e^{-ib} = e^{i(a-b)}.$$

6·21 Express the following complex numbers in the form $re^{i\theta}$:

(a) $1 + i$; (b) $1 - i$; (c) $-8(i\sqrt{3} - 1)$;
(d) $(-1 + i)^6$; (e) $(5 + 14i)/(4 + i)$.

6·22 Show by means of de Moivre's theorem that:

(a) $32\cos^6\theta = 10 + 15\cos 2\theta + 6\cos 4\theta + \cos 6\theta$;
(b) $\sin 7\theta = 7\sin\theta - 56\sin^3\theta + 112\sin^5\theta - 64\sin^7\theta$.

6·23 Verify that if $z = e^{i\theta}$, then

$$\cos\theta = \frac{1}{2}\left(z + \frac{1}{z}\right) \quad \text{and} \quad \sin\theta = -\frac{i}{2}\left(z - \frac{1}{z}\right)$$

and, more generally,

$$\cos r\theta = \frac{1}{2}\left(z^r + \frac{1}{z^r}\right) \quad \text{and} \quad \sin r\theta = -\frac{i}{2}\left(z^r - \frac{1}{z^r}\right).$$

By replacing $\cos\theta$ and $\sin\theta$ by their equivalent expressions involving z, make use of these results to express $\cos^2\theta \sin^3\theta$ in terms of $\sin n\theta$.

6·24 Use the method of Problem 6·23 to express $\sin^8\theta$ in terms of $\cos n\theta$.

6·25 Consider the function $\cosh z$, where $z = x + iy$. Then, using Definition 6·3, deduce that $\cosh z = 0$ when $z = (2n + 1)\pi i/2$, with $n = 0, \pm 1, \pm 2, \ldots \ldots$. Use the results of Problem 6·19 to deduce the zeros of $\cos z$.

6·26 Consider the function $\sin z$, where $z = x + iy$. Then, using Definition 6·6, deduce that $\sin z = 0$ when $z = n\pi$, with $n = 0, \pm 1, \pm 2, \ldots \ldots$ Use the results of Problem 6·19 to deduce the zeros of $\sinh z$.

7 Fundamentals of integration

7·1 Definite integrals and areas

The work of this chapter is concerned with the theory of the operation known as *integration*, which occupies a central position in the calculus. The connection between differentiation and integration is basic to the whole of the calculus and is contained in a result we shall prove later known as the fundamental theorem of calculus. Once again, limiting operations will play an essential part in the development of our argument. In fact we will show not only how they enable a satisfactory general theory of integration to be established, but also how they provide a tool, albeit a clumsy one, for the actual integration of functions. However, aside from a number of simple but important examples, the practical details of the evaluation of integrals of specific classes of function will be deferred until Chapter 8.

We begin by seeking to determine the shaded area I of Fig. 7·1 which is interior to the region bounded above and below by the curve $y = f(x)$ and the x-axis, respectively, and to the left and right by the lines $x = a$, $x = b$.

This approach will lead naturally to what is called the *definite integral* of $f(x)$ over the interval $a \leq x \leq b$, and it illustrates a valuable geometrical interpretation of the process of integration. Although we use the definite integral to give precise meaning to the notion of the area contained within a closed curve, this appeal to geometry is not actually necessary when defining an integral. Indeed, we shall also show how a purely analytical definition of

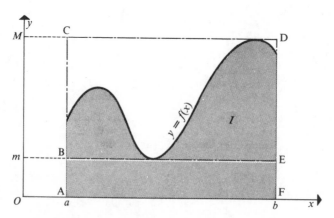

Fig. 7·1   Area I defined by $y = f(x)$.

a definite integral, quite independent of any geometrical arguments, may be formulated.

Let $f(x)$ be a non-negative continuous function defined in the closed interval $[a, b]$ and consider, for a moment, the conceptual problem that arises when trying to determine the area I defined by it in Fig. 7·1. The only simple plane geometrical figure for which the concept of area is defined in an elementary and unambiguous manner is the rectangle, so that we shall seek to define the area I in terms of the limit of a sum of rectangular areas. It should perhaps be remarked at this point that the derivation of the formula πr^2 for the area of a circle of radius r involves the concept of integration, although this is invariably avoided in any first encounter by the employment of arguments that are at best only plausible.

We shall start our discussion from the postulates that (a) the area of a rectangle is given by the product length $\times$ breadth, (b) the area of the union of two non-overlapping rectangles is the sum of their separate areas, and (c) if a rectangle is divided into two parts by a curve, then the sum of the separate non-rectangular areas comprising these two parts is equal to the area of the rectangle.

On the basis of postulate (c), we at once see that the area I in Fig. 7·1 exceeds the rectangular area ABEF, but is less than the rectangular area ACDF. Letting m, M denote, respectively, the minimum and maximum values attained by $f(x)$ in $[a, b]$, this result becomes

$$m(b - a) \leq I \leq M(b - a). \tag{7·1}$$

This inequality, although interesting, must obviously be refined if it is ever to lead to the actual value of I. In principle, our approach will be simple, for we shall begin by dividing $[a, b]$ into n adjacent sub-intervals in each of which an inequality of type (7·1) will apply, after which we shall use postulate (b) to find better upper and lower bounds for I.

Specifically, we start by choosing any sequence of $n + 1$ numbers x_0, $x_1, \ldots, x_n$ subject only to the requirements that $x_0 = a$, $x_n = b$, and

$$x_0 < x_1 < \cdots < x_{n-1} < x_n.$$

The sequence $\{x_r\}_{r=0}^n$ so defined is called a *partition* P of the interval $[a, b]$, and for any given value of n it is obviously not unique. Next, on each sub-interval $[x_{i-1}, x_i]$, let the function $f(x)$ attain a minimum value m_i and a maximum value M_i and denote the length of the ith sub-interval by Δ_i, so that

$$\Delta_i = x_i - x_{i-1}.$$

We now define numbers $\underline{S}_P$ and $\bar{S}_P$ called, respectively, the *lower* and *upper* sums taken over the partition P, by the expressions

$$\underline{S}_P = m_1\Delta_1 + m_2\Delta_2 + \cdots + m_n\Delta_n = \sum_{r=1}^{n} m_r\Delta_r \tag{7·2}$$

and

$$\bar{S}_P = M_1\Delta_1 + M_2\Delta_2 + \cdots + M_n\Delta_n = \sum_{r=1}^{n} M_r\Delta_r. \qquad (7\cdot3)$$

Clearly, as Figs. 7·2 (a), (b) illustrate, $\underline{S}_P$ and $\bar{S}_P$ are, respectively, under- and over-estimates of the area I.

The fact that $\underline{S}_P \leq \bar{S}_P$ is apparent on geometrical grounds, but it also follows without appeal to geometry by considering the difference

$$\bar{S}_P - \underline{S}_P = (M_1 - m_1)\Delta_1 + (M_2 - m_2)\Delta_2 + \cdots + (M_n - m_n)\Delta_n. \qquad (7\cdot4)$$

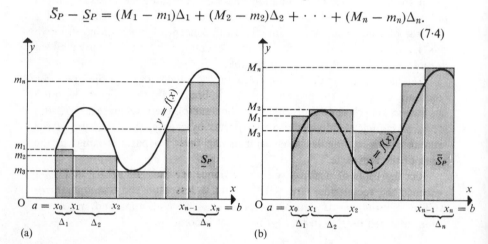

Fig. 7·2 (a) Shaded area represents lower sum S_p; (b) shaded area represents upper sum S_p.

In this equation we have, by definition, $\Delta_r > 0$ and $M_r \geq m_r$ for $r = 1$, $2, \ldots, n$, so that

$$\bar{S}_P - \underline{S}_P \geq 0 \qquad \text{or,} \qquad \underline{S}_P \leq \bar{S}_P,$$

and thus by postulate (c),

$$\underline{S}_P \leq I \leq \bar{S}_P. \qquad (7\cdot5)$$

It would seem reasonable to suppose that as the number n of points in a partition increases, provided the lengths of all intervals shrink to zero, the limit of both the lower and upper sums must be I, the desired area. We prove this in two stages, first considering the effect on the lower and upper sums of the refinement of the partition P by the inclusion of extra points.

It will suffice here to consider only the effect of the inclusion of one extra point x_r' between x_{r-1} and x_r in the partition P. The resulting partition P' is called a *refinement* of P, in the sense that although P' has more points than P, all points of P are also points of P'.

Suppose that in the intervals $[x_{r-1}, x_r']$ and $[x_r', x_r]$ the function $f(x)$ attains the minimum values m_r' and m_r'', respectively. Then the effect of the extra point is to replace the term $m_r(x_r - x_{r-1})$ in the lower sum S_P by the sum $m_r'(x_r' - x_{r-1}) + m_r''(x_r - x_r')$ thereby generating the sum S_P' appropriate to the refinement P' of the partition P. As it must be true that $m_r \leq m_r'$ and $m_r \leq m_r''$, it thus follows that

$$m_r'(x_r' - x_{r-1}) + m_r''(x_r - x_r') \geq m_r(x_r - x_{r-1}),$$

whence

$$\underline{S}_P \leq \underline{S}_{P'}. \tag{7·6}$$

Identical reasoning involving the maxima M_r' and M_r'' attained by $f(x)$ in the intervals $[x_{r-1}, x_r']$ and $[x_r', x_r]$ establishes that

$$\bar{S}_{P'} \leq \bar{S}_P. \tag{7·7}$$

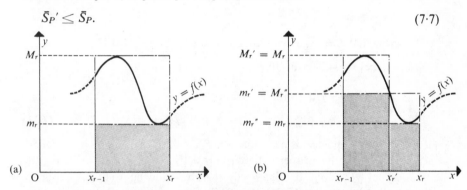

Fig. 7·3 Effect of refinement of a partition: (a) area inequality on interval $[x_{r-1}, x_r]$ of P; (b) area inequality on interval $[x_{r-1}, x_r]$ of P'.

The inequalities leading to results (7·6) and (7·7) are illustrated geometrically in Figs. 7·3. Thus in Fig. 7·3 (a) the area inequalities associated with the interval $[x_{r-1}, x_r]$ of P are displayed, whilst in Fig. 7·3 (b) the corresponding situation is displayed for the refinement P' produced by inserting an additional point x_r' in $[x_{r-1}, x_r]$.

The further refinement of the partition P' by the inclusion of additional points only serves to reinforce results (7·6) and (7·7). We have thus established that if the partitions $P_1, P_2, \ldots, P_m$ are successive refinements of the partition P, then

$$m(b - a) \leq \underline{S}_{P_1} \leq \underline{S}_{P_2} \leq \cdots \leq \underline{S}_{P_m} \leq I \leq \bar{S}_{P_m} \leq \bar{S}_{P_{m-1}} \leq \cdots$$
$$\cdots \leq \bar{S}_{P_1} \leq M(b - a). \tag{7·8}$$

Expressed in words, the effect of refinement of a partition is to *increase* the corresponding lower sum and to *decrease* the corresponding upper sum, so that $\{\underline{S}_{P_r}\}$ is a monotonic increasing sequence of numbers, and $\{\bar{S}_{P_r}\}$ is a monotonic decreasing sequence of numbers.

For the second and final stage of our argument we introduce the *norm* $\| \Delta \|_P$ of a partition P by means of the definition

$$|| \Delta ||_P = \max_i (x_i - x_{i-1}).$$ (7·9)

That is to say, for any partition P of the interval $[a, b]$, the norm $|| \Delta ||_P$ is the length of the longest sub-interval.of $[a, b]$ produced by the partition.

Let us consider a sequence of partitions which are successive refinements of P and are such that

$$\lim_{m \to \infty} || \Delta ||_{P_m} = 0.$$

Then by the postulate of Section 3·2, as $\{\underline{S}_{P_r}\}$ is monotonic increasing and bounded above it must tend to a limit $\underline{S}$ and, similarly, as $\{\bar{S}_{P_r}\}$ is monotonic decreasing and bounded below it must tend to a limit $\bar{S}$, where

$$\underline{S} \leq I \leq \bar{S}.$$ (7·10)

To show that $\underline{S} = \bar{S}$, as would be expected, observe that if

$$\delta_P = \max_i (M_i - m_i) \text{ for all } i,$$

then Eqn (7·4) gives rise to the inequality

$$\bar{S}_P - \underline{S}_P \leq \delta_P(\Delta_1 + \Delta_2 + \cdots + \Delta_n) = \delta_P(b - a).$$ (7·11)

Hence, for any sequence of partitions $P_1, P_2, \ldots, P_m, \ldots$ which are refinements of P with the property that $\lim || \Delta ||_{P_m} \to 0$, it follows from the *continuity* of $f(x)$ that $\lim \delta_{p}P \to 0$, thereby showing that the sequence $\{\bar{S}_{Pm} - \underline{S}_{Pm}\}$ has the limit zero. Thus $\{\underline{S}_{Pm}\}$ and $\{\bar{S}_{Pm}\}$ both have the same limit.

Taken in conjunction with Eqn (7·10), we have proved that because of the continuity of $f(x)$, the limit of the lower sums is equal to the limit of the upper sums, and each is equal to the limit I which has been interpreted as the shaded area in Fig. 7·1.

This result suggests the following form of definition for the definite integral.

DEFINITION 7·1 (definite integral of a continuous non-negative function) Let $f(x)$ be a continuous non-negative function on the closed interval $a \leq x \leq b$, and let $P_1, P_2, \ldots, P_m, \ldots$ be a sequence of successive refinements of some partition P of $[a, b]$ with the property that $\lim ||\Delta||P_m = 0$. Then, if ξ_i is any point in the ith sub-interval of length Δ_i generated by the partition P_m, the *definite integral* of $f(x)$ integrated over the interval $[a, b]$, and written symbolically

$$\int_a^b f(x)\mathrm{d}x,$$

is defined to be

$$\int_a^b f(x)\mathrm{d}x = \lim_{|| \Delta ||P_m \to 0} \sum_{i=1}^n f(\xi_i)\Delta_i.$$

In the context of a definite integral, the function $f(x)$ is called the *integrand*, the numbers a, b are called the *lower and upper limits of integration*, respectively, and the sign $\int$ itself is called the *integral sign*.

In summary then, a *definite integral* of a positive continuous function $f(x)$ integrated over the interval $[a, b]$ is a positive number defined by means of a limiting process. It may be interpreted geometrically as the shaded area I below the curve $y = f(x)$ as shown in Fig. 7·1.

To show that this is a working definition, in the sense that it can be used to yield a useful answer, let us now apply it to a simple function.

Example 7·1   Evaluate the definite integral

$$\int_a^b x^2 \, dx, \quad \text{where} \quad a < b.$$

Solution   As x^2 is everywhere continuous and is non-negative on the stated interval Definition 7·1 applies. Thus we start by considering a convenient partition P_n in which $[a, b]$ is divided into n equal sub-intervals, each of length $\Delta = (b - a)/n$. Then, if for convenience we identify ξ_i with the right-hand end-point of the ith sub-interval, we have

$$\xi_1 = a + \Delta, \xi_2 = a + 2\Delta, \xi_3 = a + 3\Delta, \ldots, \xi_n = a + n\Delta.$$

Hence, from Definition 7·1,

$$I = \lim_{n \to \infty} \sum_{i=1}^{n} (a + i\Delta)^2 \Delta.$$

Expanding and grouping the terms of the summation then gives

$$I = \lim_{n \to \infty} [na^2\Delta + 2a\Delta^2(1 + 2 + 3 + \cdots + n)$$
$$+ \Delta^3(1^2 + 2^2 + 3^2 + \cdots + n^2)].$$

Using the fact that $\Delta = (b - a)/n$ together with the well-known results

$$1 + 2 + 3 + \cdots + n = \frac{n}{2}(n + 1)$$

and

$$1^2 + 2^2 + 3^2 + \cdots + n^2 = \frac{n(n + 1)(2n + 1)}{6},$$

it follows that

$$I = \lim_{n \to \infty} \left\{ a^2(b - a) + a(b - a)^2 \left[\frac{n(n + 1)}{n^2}\right] \right.$$
$$\left. + (b - a)^3 \left[\frac{(n + 1)(2n + 1)}{6n^2}\right] \right\}.$$

Thus, taking the limit, we find

$$I = \tfrac{1}{3}(b^3 - a^3),$$

and so

$$\int_a^b x^2 \, dx = \tfrac{1}{3}(b^3 - a^3).$$

In terms of numbers, if $a = 1$, $b = 2$, then

$$\int_1^2 x^2 \, dx = \tfrac{1}{3}(2^3 - 1^3) = \frac{7}{3}.$$

When the behaviour of $f(x)$ is monotonic over the interval $a \le x \le b$, then Theorem 7·1 coupled with Definition 7·1 can often be used to derive interesting and useful series approximations to the definite integral as the following example illustrates.

Example 7·2   Show that

$$\sum_{r=1}^{n} \left(\frac{1}{n + r - 1} \right) \ge \int_1^2 \frac{dx}{x} \ge \sum_{r=1}^{n} \left(\frac{1}{n + r} \right).$$

Solution   In this case $f(x) = 1/x$, which is continuous, positive, and monotonic decreasing on the interval $[1, 2]$ so that Theorem 7·1 and Definition 7·1 apply. We again choose a partition P_n which divides the interval $[1, 2]$ into n equal sub-intervals of length $\Delta = 1/n$. The general point x_r in the partition P_n is, of course, $x_r = 1 + r/n$ so that

$$f(x_r) = \frac{n}{n + r}.$$

Thus as $f(x)$ is monotonic decreasing, it follows that on the interval $[x_{r-1}, x_r]$, $f(x)$ attains its maximum value M_r at x_{r-1} and its minimum value m_r at x_r, where

$$M_r = \frac{n}{n + r - 1} \quad \text{and} \quad m_r = \frac{n}{n + r}.$$

Hence

$$\underline{S}_{P_n} = \sum_{r=1}^{n} \left(\frac{n}{n + r} \right) \frac{1}{n} \quad \text{and} \quad \bar{S}_{P_n} = \sum_{r=1}^{n} \left(\frac{n}{n + r - 1} \right) \frac{1}{n},$$

so that from Theorem 7·1 and Definition 7·1, we deduce that

$$\sum_{r=1}^{n} \left(\frac{1}{n + r - 1} \right) \ge \int_1^2 \frac{dx}{x} \ge \sum_{r=1}^{n} \left(\frac{1}{n + r} \right).$$

A few numbers might help here, so we show in the table below the behaviour of the upper and lower sums $\bar{S}_{P_n}$ and $\underline{S}_{P_n}$ as a function of n.

n	$\bar{S}_{P_n}$	$\underline{S}_{P_n}$
5	0·7456	0·6456
10	0·7188	0·6688
15	0·7101	0·6768
∞	0·6931	0·6931

We shall discover later that the exact result, which is shown in this table against the entry $n = \infty$, is in fact $\log_e 2$.

7·2 Integration of arbitrary continuous functions

As most functions assume both positive and negative values in their domain of definition, our notion of a definite integral as formulated so far is rather restrictive, for it requires that the integrand be non-negative. A brief examination of the introductory arguments used in the previous section shows that this restriction stems from our idea of area as being an essentially positive quantity, although this was not stated explicitly at any stage in our argument.

Nothing in the limiting arguments that we used requires either the upper and lower sums themselves, or any of the terms comprising them to be non-negative. Since a term in either of these sums will be negative when m_r or M_r is negative, that is, when $f(x)$ is negative, it follows that the interpretation of a definite integral as an area may be extended to continuous functions $f(x)$ which assume negative values provided that areas *below* the x-axis are regarded as negative. This is illustrated in Fig. 7·4 in which the positive and negative area contributions to the definite integral of $f(x)$ integrated over the interval $[a, b]$ are marked accordingly.

Thus using this convention when interpreting a definite integral as an area, we may remove the condition that the integrand $f(x)$ be non-negative throughout all of Section 7·1. Because it simply amounts to the deletion of the word 'non-negative', we shall not trouble to reformulate our earlier definitions and theorems to take account of this result. It is interesting to observe that had we introduced the definite integral via the upper and lower sums, without any appeal to graphs and areas, this artificial restriction would never have arisen.

The definition of a definite integral of a function $f(x)$ integrated over the interval $[a, b]$ immediately implies a number of important general results which we now state in the form of a theorem. No proofs will be offered since the results are virtually self-evident.

THEOREM 7·1 (properties of definite integrals)   Let $f(x)$, $g(x)$ be continuous functions defined on the closed interval $a \le x \le b$, and let c be a constant and k be such that $a < k < b$. Then

(a) $\displaystyle\int_a^b f(x)\mathrm{d}x = \int_a^k f(x)\mathrm{d}x + \int_k^b f(x)\mathrm{d}x$ (Additivity with respect to interval of integration),

(b) $\displaystyle\int_a^b cf(x)\mathrm{d}x = c\int_a^b f(x)\mathrm{d}x$ (Homogeneity),

(c) $\displaystyle\int_a^b (f(x) + g(x))\mathrm{d}x = \int_a^b f(x)\mathrm{d}x + \int_a^b g(x)\mathrm{d}x$ (Linearity).

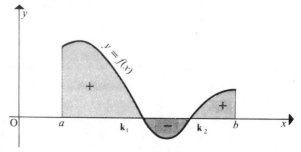

Fig. 7·4   Positive and negative areas defined by $y = f(x)$.

By virtue of these results, the definite integral of the function $f(x)$ appropriate to Fig. 7·4 could, if desired, be written in terms of the sum of three integrals involving non-negative integrands. To achieve this, notice that $f(x)$ is negative for $k_1 \le x \le k_2$, so that for all x in this interval, $-f(x)$ is positive. Then, first expressing our integral as the sum of three separate integrals over adjacent intervals

$$\int_a^b f(x)\mathrm{d}x = \int_a^{k_1} f(x)\mathrm{d}x + \int_{k_1}^{k_2} f(x)\mathrm{d}x + \int_{k_2}^b f(x)\mathrm{d}x, \qquad (7\cdot12)$$

we can replace $-f(x)$ by $|f(x)|$ in the second of these integrals to obtain

$$\int_p^b f(x)\mathrm{d}x = \int_a^{k_1} f(x)\mathrm{d}x - \int_{k_1}^{k_2} |f(x)|\,\mathrm{d}x + \int_{k_2}^b f(x)\mathrm{d}x. \qquad (7\cdot13)$$

Each of these integrands is now the definite integral of a non-negative function as required.

We must now take account of the fact that so far it has been implicit in our definition of a definite integral that x increases positively from a to b, where $b > a$. This *sense*, or *direction*, of integration is indicated in the definite integral by writing a at the bottom of the integral sign $\int$ to signify the *lower limit* of integration and by writing b at the top to signify the *upper limit* of integration. If, despite the fact that $b > a$, their positions as upper and lower limits of integration are reversed, this implies that integration is to be carried out in the direction in which x increases negatively. Because we are now allowing areas to have both magnitude and sign, to be consistent we must compensate for a reversal of the limits of integration by changing the sign of the integral. Hence we arrive at our next definition.

DEFINITION 7·2 (reversal of limits of integration)   If $a < b$, then we define the definite integral

$$\int_b^a f(x)\mathrm{d}x$$

of a continuous function $f(x)$ by the equation

$$\int_b^a f(x)\mathrm{d}x = -\int_a^b f(x)\mathrm{d}x.$$

Example 7·3   Evaluate the definite integral

$$\int_3^1 2x^2 \,\mathrm{d}x.$$

Solution   From Definition 7·2 we have

$$\int_3^1 2x^2 \,\mathrm{d}x = -\int_1^3 2x^2 \,\mathrm{d}x.$$

Hence an application of Theorem 7·1 (b) together with the result of Example 7·1 shows that

$$\int_3^1 2x^2 \,\mathrm{d}x = -2\int_1^3 x^2 \,\mathrm{d}x = -2(\tfrac{1}{3})(3^3 - 1^3) = -\frac{52}{3}.$$

Since a definite integral is simply a number, the choice of symbol used to denote the argument of the function f forming the integrand is arbitrary, and often it is convenient to replace x by some other variable, say t. Thus

$$\int_a^b f(x)\mathrm{d}x \qquad \text{and} \qquad \int_a^b f(t)\mathrm{d}t$$

are identical in meaning, so that

$$\int_a^b f(x)\mathrm{d}x = \int_a^b f(t)\mathrm{d}t. \tag{7·14}$$

On account of this fact, the variable in the integrand of a definite integral is often called a *dummy* variable, and it is sometimes said to be 'integrated out' when the integral is evaluated. This fact is usually recognized in modern accounts of the theory of the definite integral by simply writing

$$\int_a^b f$$

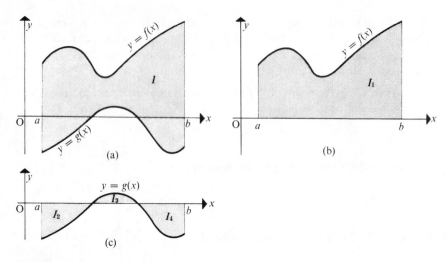

Fig. 7·5 (a) Area I bounded by curves $y = f(x)$ and $y = g(x)$; (b) area below $y = f(x)$; (c) positive and negative areas defined by $y = g(x)$.

in place of either of the expressions in Eqn (7·14). The full significance of the symbol dx, which is suggestive of a differential, comes when changes of variable of the form $x = g(u)$ are made in Eqn (7·14) and it is for this reason that we choose to retain it. This matter will be taken up in detail in the next chapter, where it is shown that because of the chain rule for differentiation, dx can indeed be interpreted as a differential.

Now that the definite integral has been extended to arbitrary continuous integrands we are in a position to determine quite general areas. Consider, for example, the situation illustrated in Fig. 7·5 (a) in which it is desired to determine the area I of the shaded region. Then obviously, referring to Figs. 7·5 (b), (c) we have

$$I = I_1 + I_2 - I_3 + I_4,$$

where I_1 to I_4 represent the *positive* areas identified by these symbols. However, we know that

$$I_1 = \int_a^b f(x)dx,$$

and from the form of argument leading to Eqn (7·13) we also know that

$$-I_2 = \int_a^{k_1} g(x)dx, \quad I_3 = \int_{k_1}^{k_2} g(x)dx, \quad -I_4 = \int_{k_2}^b g(x)dx,$$

where k_1 and k_2 are the first and second points of intersection of $y = g(x)$ with the x-axis as x increases from a to b.

However, by Theorem 7·1 (a) we have

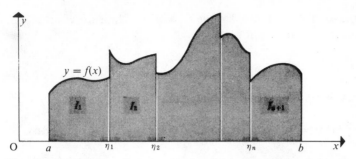

Fig. 7·6 Piecewise continuous function $y = f(x)$ defining a sequence of areas $I_1, I_2, \ldots, I_{n-1}$.

$$-I_2 + I_3 - I_4 = \int_a^b g(x)\mathrm{d}x,$$

so that combining these results we obtain

$$I = \int_a^b f(x)\mathrm{d}x - \int_a^b g(x)\mathrm{d}x.$$

From Theorem 7·1 (b) it then finally follows that

$$I = \int_a^b (f(x) - g(x))\mathrm{d}x. \tag{7·15}$$

Example 7·4 Find the area I between the two curves $y = \mathrm{e}^{2x}$ and $y = -x^2$, which is bounded to the left by the line $x = 1$ and to the right by the line $x = 3$.

Solution We start by making the obvious identifications $f(x) = \mathrm{e}^{2x}$, $g(x) = -x^2$, $a = 1$ and $b = 3$. Then from Eqn (7·15) it follows that

$$I = \int_1^3 (\mathrm{e}^{2x} + x^2)\mathrm{d}x$$

whence, using the results of Example 7·1 and Problem 7·3, we find

$$I = \tfrac{1}{2}(\mathrm{e}^6 - \mathrm{e}^2) + \frac{26}{3}.$$

The fact that a definite integral is additive with respect to its interval of integration enables a function to be integrated even when it has discontinuities, provided only that they are finite in number and that elsewhere the function is continuous and bounded. This result is perhaps best seen diagrammatically, though an analytical justification can easily be given without appeal to geometry. By way of example, consider the function $y = f(x)$ illustrated in Fig. 7·6 which is bounded and continuous everywhere except at the discrete number of points $\eta_1, \eta_2, \ldots, \eta_n$. Such a function is said to be *piecewise continuous*, for obvious reasons.

Using the valid interpretation of a definite integral in terms of area we see that the total shaded area I is the sum of the sequence of areas $I_1, I_2, \ldots, I_{n+1}$, so that we may still write

$$I = \int_a^b f(x)\mathrm{d}x, \tag{7.16}$$

but this time with the understanding that

$$\int_a^b f(x)\mathrm{d}x = \int_a^{\eta_1-} f(x)\mathrm{d}x + \int_{\eta_1+}^{\eta_2-} f(x)\mathrm{d}x + \cdots + \int_{\eta_n+}^b f(x)\mathrm{d}x. \tag{7.17}$$

Here, as before, we have used η_i- to signify the limiting process of approaching the point $x = \eta_i$ from the *left*, and η_i+ to signify the limiting process of approaching the point $x = \eta_i$ from the right.

Example 7·5   Evaluate the definite integral

$$I = \int_0^2 f(x)\mathrm{d}x$$

when

$$f(x) = \begin{cases} x^2 & \text{for } 0 \le x < 1 \\ e^{5x} & \text{for } 1 \le x \le 2. \end{cases}$$

Solution   From Eqn (7·17) we have

$$I = \int_0^{1-} x^2\,\mathrm{d}x + \int_{1+}^2 e^{5x}\,\mathrm{d}x,$$

so that evaluating the integrals and then taking the appropriate limits gives

$$I = \tfrac{1}{3} + \tfrac{1}{5}(e^{10} - e^5).$$

Sometimes a more difficult situation than this arises in which either the integrand tends to infinity at some point in the interval of integration or, perhaps, the interval of integration itself is infinite in length. Such definite integrals are called *improper* integrals, and the way in which to attribute a value to any such integral is suggested by Eqn (7·17).

Let us illustrate something of the difficulty that can arise if ideas are not made precise. Consider the integral

$$\int_{-1}^1 \frac{\mathrm{d}x}{x^2}.$$

Then since $y = 1/x^2$ is essentially positive, the area under the curve must also be positive. Now if we apply the result of Problem 7·4 we have

$$\int_{-1}^1 \frac{\mathrm{d}x}{x^2} = -\frac{1}{1} + \frac{1}{-1} = -2$$

which, since it is negative, contradicts our previous conclusion. What has gone wrong? The trouble is that $1/x^2$ tends to infinity as $x \to 0$, so that the arguments of Problem 7·4 are not applicable, for its was pre-supposed there that the interval of integration excluded the origin. When dealing with improper integrals of this type in which the integrand has an infinity within the interval of integration we shall assign a value to the integral according to the following definition.

DEFINITION 7·3 (improper integral due to infinity of integrand)   Let the function $f(x)$ be continuous throughout the intervals $a \leq x < c$ and $c < x \leq b$, and suppose that $f(x)$ has a *singularity* at $x = c$ in the sense that $f(x)$ tends to infinity as $x \to c$. Then the integral of $f(x)$ over the interval of integration $[a, b]$ is said to be *improper*, and it is defined to have the value

$$I = \lim_{\varepsilon \to 0} \int_a^{c-\varepsilon} f(x)dx + \lim_{\delta \to 0} \int_{c+\delta}^b f(x)dx,$$

whenever both limits involved exist. Under these circumstances the improper integral will be said to *converge* to the value I. When either of the limits does not exist, the integral will be said to be *divergent*. If the point c coincides with an end-point of the interval $[a, b]$, then I is defined to be equal to the limit of the single integral for which the interval of integration lies within $[a, b]$.

On the basis of this definition we are now able to determine the value to be attributed to the improper integral used as an illustration above. Let us do this in the form of an example.

Example 7·6   Evaluate the improper integrals:

(a) $I_1 = \int_{-1}^1 \frac{dx}{x^2}$ and   (b) $I_2 = \int_{-1}^0 \left(\frac{x^2 + 1}{x^2} \right) dx.$

Solution   The integrand $1/x^2$ tends to infinity as $x \to 0$, so that for case (a), when appealing to Definition 7·3, we need to make the identifications $a = -1$, $b = 1$, $c = 0$ and $f(x) = 1/x^2$. Thus,

$$I_1 = \lim_{\varepsilon \to 0} \int_{-1}^{-\varepsilon} \frac{dx}{x^2} + \lim_{\delta \to 0} \int_{\delta}^1 \frac{dx}{x^2}.$$

Using the result of Problem 7·4 we find that

$$I_1 = \lim_{\varepsilon \to 0} \left(\frac{1}{\varepsilon} - 1 \right) + \lim_{\delta \to 0} \left(-1 + \frac{1}{\delta} \right) \to \infty.$$

Thus the improper integral (a) is divergent.

In case (b) the integrand is $(x^2 + 1)/x^2$, which again tends to infinity as $x \to 0$. However, in this case we must make the identifications $a = -1$, $b = 0$, $c = 0$, and $f(x) = 1 + 1/x^2$, so that this time the singularity in the integrand occurs at the right-hand end-point of the interval of integration

[−1, 0] (that is, at the upper limit of integration).

It then follows from Definition 7·3 that

$$I_2 = \lim_{\varepsilon \to 0} \int_{-1}^{-\varepsilon} \left(1 + \frac{1}{x^2}\right) dx,$$

which, from the results of Problems 7·2 (b) and 7·4, becomes

$$I_2 = \lim_{\varepsilon \to 0} \left\{(-\varepsilon + 1) + \left(\frac{1}{\varepsilon} - 1\right)\right\} \to \infty.$$

Hence the improper integral (b) is also divergent.

The one remaining form of improper integral requiring consideration occurs when the interval of integration is infinite. In these circumstances we shall assign a value to the integral according to the following definition.

DEFINITION 7·4 (improper integral due to infinite interval of integration) Let the function $f(x)$ be continuous on the interval $[a, \infty)$, then the integral of $f(x)$ over the interval of integration $[a, \infty)$ is said to be *improper*, and it is defined to have the value

$$I_1 = \lim_{k \to \infty} \int_a^k f(x)dx,$$

whenever this limit exists. Under these circumstances the improper integral will be said to converge to the value I_1. When the limit does not exist, the integral will be said to be *divergent*. Similarly, if the interval of integration is $(-\infty, a]$, then when the limit exists, the improper integral of $f(x)$ over the interval of integration $(-\infty, b]$ is defined to have the value

$$I_2 = \lim_{k \to \infty} \int_{-k}^{b} f(x)dx.$$

Symbolically, these improper integrals will be denoted, respectively, by

$$I_1 = \int_a^\infty f(x)dx \quad \text{and} \quad I_2 = \int_{-\infty}^b f(x)dx.$$

Example 7·7   Evaluate the improper integral

$$I = \int_3^\infty \frac{dx}{x^2}.$$

Solution   It follows at once from Definition 7·4 that

$$I = \lim_{k \to \infty} \int_3^k \frac{dx}{x^2},$$

so that by virtue of the result of Problem 7·4,

$$I = \lim_{k \to \infty} \left[\frac{-1}{k} + \frac{1}{3} \right] = \frac{1}{3}.$$

Hence this improper integral converges to the value $1/3$.

7·3 Integral inequalities

A number of useful inequalities may be deduced concerning definite integrals, the simplest of which has already been stated in Eqn (7·1). Let us now derive our first result of this type, of which Eqn (7·1) represents a special case.

Suppose that the definite integrals of $f(x)$ and $g(x)$ taken over the interval $[a, b]$ both exist. In brief, let us agree to say that $f(x)$ and $g(x)$ are *integrable* over the interval $[a, b]$. Now suppose that $f(x) \leq g(x)$ for $a \leq x \leq b$. Then if P_m is a partition of $[a, b]$, we have from Definition 7·1 that

$$\int_a^b g(x)\mathrm{d}x - \int_a^b f(x)\mathrm{d}x = \int_a^b (g(x) - f(x))\mathrm{d}x$$

$$= \lim_{||\Delta||_{P_m} \to 0} \sum_{i=1}^n (g(\xi_i) - f(\xi_i))\Delta_i, \qquad (7·18)$$

where ξ_i is some point in the ith sub-interval of length Δ_i generated by the partition P_m. Now since by hypothesis $f(x) \leq g(x)$, it follows that $f(\xi_i) \leq g(\xi_i)$, so that the right-hand side of Eqn (7·18) must be non-negative. Thus we have proved the following theorem.

THEOREM 7·2 (inequality between two definite integrals) Let $f(x) \leq g(x)$ be two integrable functions over the interval $[a, b]$. Then,

$$\int_a^b f(x)\mathrm{d}x \leq \int_a^b g(x)\mathrm{d}x.$$

Equation (7·1) follows as a trivial consequence of this result, for the theorem implies that if $\phi(x) \leq f(x) \leq \psi(x)$ are three integrable functions over the interval $[a, b]$, then

$$\int_a^b \phi(x)\mathrm{d}x \leq \int_a^b f(x)\mathrm{d}x \leq \int_a^b \psi(x)\mathrm{d}x.$$

Hence, if m, M are, respectively, the minimum and maximum values of $f(x)$ on $[a, b]$, our required result follows by setting $\phi(x) = m$, $\psi(x) = M$, when we obtain

$$m(b - a) \leq \int_a^b f(x)\mathrm{d}x \leq M(b - a). \qquad (7·19)$$

This last simple result implies a more important result which we now derive by appeal to the intermediate value theorem of Chapter 5. Writing inequality (7·19) in the form

$$m \le \frac{1}{b-a} \int_a^b f(x)\mathrm{d}x \le M$$

shows that the number

$$\frac{1}{b-a} \int_a^b f(x)\mathrm{d}x$$

is intermediate between m and M which are extreme values of the function $f(x)$ itself. Hence, provided $f(x)$ is continuous, it then follows from the intermediate value theorem that some number ξ exists, strictly between a and b, such that

$$f(\xi) = \frac{1}{b-a} \int_a^b f(x)\mathrm{d}x. \tag{7·20}$$

This result is called the *first mean value theorem for integrals*, and it constitutes our next theorem.

THEOREM 7·3 (first mean value theorem for integrals) Let $f(x)$ be continuous on the interval $[a, b]$, then there exists a number ξ, strictly between a and b, for which

$$\int_a^b f(x)\mathrm{d}x = (b-a)f(\xi).$$

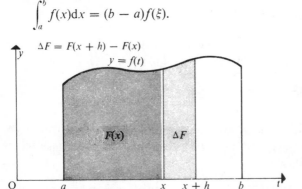

Fig. 7·7 Area below $y = f(t)$ as a function of the upper limit of integration x.

7·4 The definite integral as a function of its upper limit-indefinite integral

If the lower limit of a definite integral is held constant, but the upper limit is replaced by the variable x, then the numerical value of the integral will clearly depend on x. Another way of describing this situation is if we say that a definite integral with a variable upper limit x defines a *function* of x. In Fig. 7·7 this idea is illustrated in terms of areas, with the shaded region marked

$F(x)$ denoting the area below the curve $y = f(t)$ which is bounded on the left by the line $t = a$, and on the right by the line $t = x$.

In terms of the definite integral we have

$$F(x) = \int_a^x f(t)dt. \tag{7·21}$$

Now let us suppose that $f(t)$ is continuous in some interval $[a, b]$, with $a \leq x \leq b$. Notice here that for the first time it is necessary to use the dummy variable t, because x and t are fulfilling two different roles in Eqn (7·21). To be precise, x represents the upper limit of integration, whilst the dummy variable t represents the general variable in the interval of integration $a \leq t \leq x$.

Consider the difference

$$F(x + h) - F(x) = \int_a^{x+h} f(t)dt - \int_a^x f(t)dt$$

$$= \int_x^{x+h} f(t)dt. \tag{7·22}$$

Then the first mean value theorem for integrals allows us to rewrite Eqn (7·22) in the form

$$F(x + h) - F(x) = hf(\xi), \tag{7·23}$$

where $x < \xi < x + h$.

Now, forming the difference quotient $\{F(x + h) - F(x)\}/h$, we find

$$\frac{F(x + h) - F(x)}{h} = f(\xi),$$

so that taking the limit as $h \to 0$ gives,

$$F'(x) = \lim_{h \to 0} \left\{ \frac{F(x + h) - F(x)}{h} \right\} = f(x). \tag{7·24}$$

This important result shows that the integrand of integral (7·21) at the upper limit of integration $t = x$ is equal to the derivative of $F(x)$ with respect to x.

Suppose now that $G(x)$ is any function for which $G'(x) = f(x)$. Then,

$$G'(x) - F'(x) = \frac{d}{dx} \{G(x) - F(x)\} = 0,$$

and so from Corollary 5·12

$$G(x) = F(x) + \text{constant}. \tag{7·25}$$

Combining Eqns (7·21) and (7·25) shows that the most general function $G(x)$ whose derivative is equal to $f(x)$ must be of the form

$$G(x) = \int_a^x f(t)dt + C, \tag{7·26}$$

where C is a constant.

The first term on the right-hand side of Eqn (7·26) is called an *indefinite integral*. The function $G(x)$ itself is called either a *primitive* of f or an *antiderivative* of f. We shall usually use the name antiderivative, since this offers an accurate description of the process by which it is to be found. Namely, an antiderivative arises from the process of reversing the operation of differentiation, and the most frequent method of finding antiderivatives utilizes this idea by employing tables of derivatives in reverse. That is to say, by matching an integrand with an entry in a table of derivatives and thereby finding the functional form of $G(x)$ apart from the additive arbitrary constant.

Usually the antiderivative $G(x)$ defined in either Eqn (7·25) or Eqn (7·26) is written symbolically in the form

$$\int f(x)\mathrm{d}x = F(x) + C. \tag{7·27}$$

In this notation, the fact that an antiderivative is a *function* related to the operation of integration, and not just a number as in an ordinary definite integral, is indicated by again employing the integral sign, but this time without limits. On occasions the reader will find books in which an antiderivative is signified by the notation

$$\int^x f(x)\mathrm{d}x,$$

rather than the notation used in Eqn (7·27).

The following short table lists a few of the antiderivatives which are of most frequent occurrence in mathematics.

Table 7.1

$\int f(x)\mathrm{d}x = F(x) + C$		
	$f(x)$	$F(x)$
1	a (const)	ax
2	x^n	$\dfrac{x^{n+1}}{n+1}$
3	$e^{\lambda x}$	$\dfrac{1}{\lambda}e^{\lambda x}$
4	$\sin x$	$-\cos x$
5	$\cos x$	$\sin x$

Other useful elementary antiderivatives that should be memorized, together with an account of systematic methods for finding antiderivatives, are given in the next chapter.

Let us now return to Eqn (7·25) and notice that it follows from this that

$$G(b) - G(a) = F(b) - F(a) = F(b) = \int_a^b f(x)\mathrm{d}x. \qquad (7 \cdot 28)$$

Hence we have proved that

$$\int_a^b f(x)\mathrm{d}x = G(b) - G(a), \qquad (7 \cdot 29)$$

where $G'(x) = f(x)$. This provides a method for the evaluation of definite integrals, for expressed in words it asserts that the definite integral of $f(x)$ taken over an interval $[a, b]$ is the difference between the value of any antiderivative of $f(x)$ at $x = b$ and $x = a$.

It is now time to express results (7·24) and (7·29) in the form of two basic theorems known, respectively, and the *first* and *second fundamental theorems of calculus*.

THEOREM 7·4 (first fundamental theorem of calculus)   If $f(x)$ is continuous for $a \leq x \leq b$, and

$$F(x) = \int_a^x f(t)\mathrm{d}t,$$

then $F'(x) = f(x)$ for all points x in $[a, b]$.

Alternatively expressed, this result may also be written

$$\frac{\mathrm{d}}{\mathrm{d}x} \int_a^x f(t)\mathrm{d}t = f(x).$$

THEOREM 7·5 (second fundamental theorem of calculus)   If $f(x)$ is continuous for $a \leq x \leq b$ and $G(x)$ is any antiderivative of $f(x)$, then

$$\int_a^x f(t)\mathrm{d}t = G(x) - G(a).$$

The statement of Theorem 7·5 is often written in the form

$$\int_a^b f(x)\mathrm{d}x = G(x)\Big|_{x=a}^{x=b},$$

with the understanding that

$$G(x)\Big|_{x=a}^{x=b} = G(b) - G(a).$$

It follows from Theorem 7·5 that the definite integral calculated so laboriously in Example 7·1 may be evaluated directly by appeal to entry number 2 in Table 7·1. To see this set $n = 2$, so that $f(x) = x^2$, then $F(x) = x^3/3$, and by Theorem 7·5 we immediately deduce that

$$\int_a^b x^2 \, \mathrm{d}x = \tfrac{1}{3}(b^3 - a^3).$$

The systematic employment of the fundamental theorems of calculus will be taken up in detail in Chapter 8, since our concern here is primarily with the theory rather than the practice of integration.

Finally, to emphasize that the indefinite integral is a function, we now give an example of such an integral which defines an important mathematical function. Since we have the relationship

$$\frac{d}{dx} \log_e x = \frac{1}{x}, \quad \text{for} \quad x > 0,$$

it follows from Theorem 7·5 that, provided $a > 0$,

$$\int_a^x \frac{dt}{t} = \log_e x - \log_e a.$$

Hence, setting $a = 1$ gives the result

$$\log_e x = \int_1^x \frac{dt}{t}, \tag{7·30}$$

which is illustrated as the shaded area in Fig. 7·8.

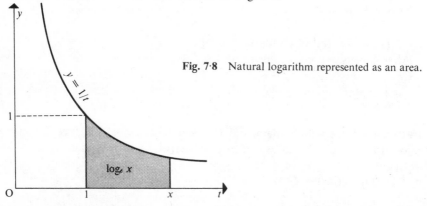

Fig. 7·8   Natural logarithm represented as an area.

7·5 Other geometrical applications of definite integrals

This section offers a brief discussion of the application of the definite integral to the determination of arc length for plane curves, the surface area of a surface of revolution, and the volume of a volume of revolution. Each result will be derived by appeal to the basic definition of a definite integral, since it will first be necessary to define the precise meaning of the concepts that are involved.

(a) Arc length of a plane curve

Consider the plane curve Γ with the equation $y = f(x)$ illustrated in Fig. 7·9 (a). Then our task here will be first to define the meaning of the length s of the arc MN, and then to deduce a method by which it may be

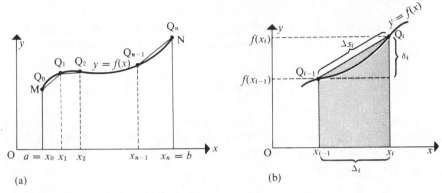

Fig. 7·9 (a) Arc length of curve; (b) element of arc length.

found once the equation of Γ has been given. Let $Q_0, Q_1, \ldots, Q_n$ represent any set of points on Γ, the first of which coincides with the left-hand end-point M, and the last of which coincides with the right-hand end-point N. Then if Δs_i denotes the length of the chord joining Q_{i-1} to Q_i, the length S_n of the polygonal line joining M to N is

$$S_n = \sum_{i=1}^{n} \Delta s_i.$$

Now the projection of the set of points $Q_0, Q_1, \ldots, Q_n$ onto the x-axis defines a set of points $a = x_0 < x_1 < \ldots < x_n = b$ which form a partition P_n of the interval $[a, b]$. Thus, denoting the norm of P_n by $|| \Delta ||_{P_n}$, we shall define the length s of the arc Γ from M to N to be

$$s = \lim_{||\Delta||_{P_n} \to 0} \sum_{i=1}^{n} \Delta s_i. \tag{7·31}$$

Now, setting $\Delta_i = x_i - x_{i-1}$ and $\delta_i = f(x_i) - f(x_{i-1})$, it follows directly by an application of Pythagoras' theorem (Fig. 7·9 (b)) that

$$\Delta s_i = \sqrt{(\Delta_i^2 + \delta_i^2)} = \sqrt{\left(1 + \left(\frac{\delta_i}{\Delta_i}\right)^2\right)}\, \Delta_i.$$

However, by virtue of the mean value theorem for derivatives we may write, provided that $f(x)$ is differentiable on $[a, b]$,

$$\frac{\delta_i}{\Delta_i} = \frac{f(x_i) - f(x_{i-1})}{x_i - x_{i-1}} = f'(\xi_i),$$

where $x_{i-1} < \xi_i < x_i$, and so

$$\Delta s_i = \sqrt{(1 + [f'(\xi_i)]^2)}\, \Delta_i. \tag{7·32}$$

Thus the desired arc length s will be determined by evaluating

$$s = \lim_{||\Delta||_{P_n} \to 0} \sum_{i=1}^{n} \sqrt{(1 + [f'(\xi_i)]^2)}\, \Delta_i. \tag{7·33}$$

We see from Definition 7·1 that this is simply the definite integral of the function $\sqrt{(1 + [f'(x)]^2)}$ integrated from $x = a$ to $x = b$, and hence

$$s = \int_a^b \sqrt{(1 + [f'(x)]^2)}dx = \int_a^b \sqrt{\left(1 + \left(\frac{dy}{dx}\right)^2\right)}dx. \qquad (7\cdot34)$$

THEOREM 7·6 (arc length of plane curve) Let $y = f(x)$ be a differentiable function on the interval $[a, b]$. Then the length s of the plane curve Γ defined by the graph of this function in the (x, y)-plane between the points $(a, f(a))$, $(b, f(b))$ is given by

$$s = \int_a^b \sqrt{\left(1 + \left(\frac{dy}{dx}\right)^2\right)}dx.$$

Example 7·8 Determine the length of arc of the curve $y = \cosh x$ between the points $(1, \cosh 1)$ and $(3, \cosh 3)$.

Solution We have $a = 1$, $b = 3$, $y = \cosh x$, and so $dy/dx = \sinh x$, whence

$$s = \int_1^3 \sqrt{(1 + \sinh^2 x)}\, dx = \int_1^3 \cosh x\, dx.$$

Now since $d/dx (\sinh x) = \cosh x$, it follows that $\sinh x + C$ is an anti-derivative of $\cosh x$, so that by Theorem 7·5 we have

$$s = \int_1^3 \cosh x\, dx = (\sinh x + C)|_1^3 = \sinh 3 - \sinh 1.$$

(b) Area of surface of revolution

The name *surface of revolution* is given to any surface which is generated by rotating a plane curve $y = f(x)$ about either the x-axis or the y-axis. Since the determination of the area in either case is exactly similar, we shall discuss only the case of the revolution of the curve $y = f(x)$ about the x-axis, as shown in Fig. 7·10.

A problem arises here as to how to define the area of a non-cylindrical curved surface. We propose to approach the problem by sectioning the surface into annular strips of width Δ. as shown in Fig. 7·10, and then to approximate the area ΔS of each such annular strip by representing it by the conical area which is obtained by rotating the chord PQ of length Δs_i about the x-axis. Then if this element of area of cone between the planes $x = x_{i-1}$ and $x = x_i$ is ΔS_i, this will be given by

$$\Delta S_i = 2\pi\left(\frac{y_{i-1} + y_i}{2}\right)\Delta s_i.$$

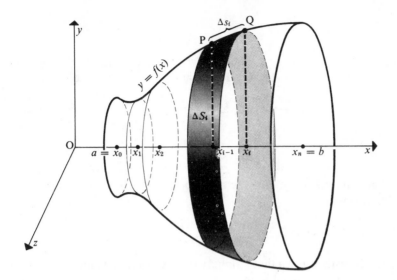

Fig. 7·10 Area of surface of revolution.

Similar elements of area may be defined for each of the other annular strips defined by some partition P_n of the interval $[a, b]$ by the set of points $a = x_0 < x_1 < \cdots < x_n = b$. Thus, denoting the norm of P_n by $|| \Delta ||_{P_n}$, we shall define the area S of the surface of revolution generated by rotating $y = f(x)$ about the x-axis, and contained between the planes $x = a$ and $x = b$, to be

$$S = \lim_{||\Delta||_{P_n} \to 0} \sum_{i=1}^{n} \Delta S_i = \lim_{||\Delta||_{P_n} \to 0} \pi \sum_{i=1}^{n} (y_{i-1} + y_i) \Delta s_i. \qquad (7·35)$$

Hence, if $f(x)$ is differentiable in $a \le x \le b$, by using result (7·32) we find

$$S = \lim_{||\Delta||_{P_n} \to 0} \pi \sum_{i=1}^{n} (y_{i-1} + y_i) \sqrt{(1+[f'(\xi_i)]^2)} \Delta_i, \qquad (7·36)$$

where $x_{i-1} < \xi_i < x_i$.

Once again our previous form of argument shows that this is just the definite integral of the function $2\pi f(x)\sqrt{(1 + [f'(x)]^2)}$ integrated from $x = a$ to $x = b$, and so

$$S = 2\pi \int_a^b f(x)\sqrt{(1 + [f'(x)]^2)}\, \mathrm{d}x. \qquad (7·37)$$

THEOREM 7·7 (area of surface of revolution) Let $f(x)$ be a differentiable function on $a \le x \le b$. Then the area S of the surface of revolution generated by rotating the graph of the function $y = f(x)$ about the x-axis, and contained between the planes $x = a$ and $x = b$ is given by

$$S = 2\pi \int_a^b f(x)\sqrt{(1 + [f'(x)]^2)}\, dx.$$

Example 7·9   Find the area contained between the planes $x = -1$ and $x = 2$ of the surface of revolution about the x-axis of the curve $y = \cosh x$.

Solution   We have $a = -1$, $b = 2$, and $f(x) = \cosh x$, and so $f'(x) = \sinh x$, whence

$$S = 2\pi \int_{-1}^2 \cosh x \sqrt{(1 + \sinh^2 x)}\, dx = 2\pi \int_{-1}^2 \cosh^2 x\, dx.$$

To evaluate this result we now use the hyperbolic identity $\cosh^2 x = \tfrac{1}{2}(1 + \cosh 2x)$ to obtain

$$S = \pi \int_{-1}^2 (1 + \cosh 2x)dx.$$

Then, as it is easily verified that $\tfrac{1}{2}\sinh 2x + C$ is an antiderivative of $\cosh 2x$, we have from Theorem 7·5 that

$$S = \pi \int_{-1}^2 (1 + \cosh 2x)dx = \pi(x + \tfrac{1}{2}\sinh 2x + C)\big|_{-1}^2$$

$$= \tfrac{1}{2}\pi(6 + \sinh 4 + \sinh 2).$$

(c) Volume of revolution

Finally, let us determine the *volume of revolution V* of the volume shown in Fig. 7·10. This time, to define the volume of such a figure, we consider cylindrical elements of volume of thickness Δ_i, and place upper and lower bounds on that element of volume by the obvious inequality:

$$\pi \times \text{(least radius of annulus)}^2 \times \Delta_i \le \text{element of volume} \le$$

$$\pi \times \text{(greatest radius of annulus)}^2 \times \Delta_i.$$

Then, if $x_{i-1} < \xi_i < x_i$, a volume element ΔV_i satisfying this inequality and bounded to the left by the plane $x = x_{i-1}$ and to the right by the plane $x = x_i$ is

$$\Delta V_i = \pi[f(\xi_i)]^2\Delta_i. \tag{7·38}$$

The volume of revolution generated by rotating $y = f(x)$ about the x-axis, and contained between the planes $x = a$ and $x = b$ will then be defined to be

$$V = \lim_{\|\Delta\|_{P_n}\to 0} \pi \sum_{i=1}^n [f(\xi_i)]^2\Delta_i. \tag{7·39}$$

A repetition of the previous form of argument then yields

$$V = \pi \int_a^b [f(x)]^2\, dx. \tag{7·40}$$

Notice that we have imposed no differentiability requirements on $f(x)$, so that result (7·40) is applicable even if $f(x)$ is only piecewise continuous.

THEOREM 7·8 (volume of solid of revolution) Let $f(x)$ be a piecewise continuous function on $a \leq x \leq b$. Then the volume of the solid of revolution generated by rotating the curve $y = f(x)$ about the x-axis, and contained between the planes $x = a$ and $x = b$, is given by

$$V = \pi \int_a^b [f(x)]^2 \, dx.$$

Example 7·10 Determine the volume of revolution generated by rotating the parabola $y = 1 + x^2$ about the x-axis, and contained between the planes $x = 1$ and $x = 2$.

Solution Here we have $a = 1$, $b = 2$, and $f(x) = 1 + x^2$, so that

$$V = \pi \int_1^2 (1 + x^2)^2 \, dx = \pi \int_1^2 (1 + 2x^2 + x^4) dx$$

$$= \pi \left(x + \frac{2x^3}{3} + \frac{x^5}{5} \right) \Bigg|_1^2 = \frac{178\pi}{15}.$$

7·6 Numerical integration

From the second fundamental theorem of calculus we have seen that the successful analytical evaluation of a definite integral involves the determination of an antiderivative of the integrand. Although in many practical cases of importance an antiderivative can be found, the fact remains that in general this is not possible and Theorem 7·5 is therefore of no avail. Such, for example, is the case with an integral as simple as

$$\int_1^3 e^{-x^2} \, dx,$$

for although an antiderivative of e^{-x^2} certainly exists on theoretical grounds, it is not expressible in terms of elementary functions.

Of the many possible methods whereby a numerical estimate of the value of a definite integral may be made, we choose to mention only the very simplest ones here. The general process of evaluating a definite integral by numerical means will be referred to as *numerical integration*, though the old fashioned term *numerical quadrature* is still often employed for such a process.

(a) Trapezoidal rule

Although a strictly analytical derivation of the so called *trapezoidal rule* for integration may be given we shall not use this approach, and instead

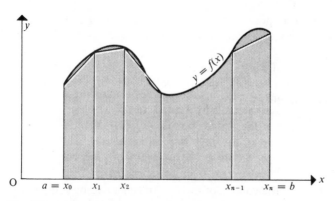

Fig. 7.11 Trapezoidal approximation of area.

make appeal to the area representation of a definite integral. Consider Fig. 7·11, and let us estimate the shaded area below the curve $y = f(x)$ which we know has the value

$$\int_a^b f(x)\mathrm{d}x.$$

Let us begin by taking any set of $n + 1$ points $a = x_0 < x_1 < \cdots < x_n = b$, and on each interval $[x_{i-1}, x_i]$, approximate the true area above it by the trapezium obtained by replacing the arc of the curve through the points $(x_{i-1}, f(x_{i-1}))$, $(x_i, f(x_i))$ by the chord joining these two points.

Then the area of the trapezium on the interval $[x_{i-1}, x_i]$ is

$$\tfrac{1}{2}(f(x_{i-1}) + f(x_i))\Delta x_i,$$

where $\Delta x_i = x_i - x_{i-1}$.

Thus, adding the n contributions of this type, we arrive at the *general trapezoidal rule*

$$\int_a^b f(x)\mathrm{d}x \approx \tfrac{1}{2}(f(x_0) + f(x_1))\Delta x_1 + \tfrac{1}{2}(f(x_1) + f(x_2))\Delta x_2 + \cdots$$

$$+ \tfrac{1}{2}(f(x_{n-1}) + f(x_n))\Delta x_n. \qquad (7\cdot41)$$

If the interval $[a, b]$ is divided into n equal parts of length $h = (b - a)/n$, then (7·41) becomes the *trapezoidal rule for equal intervals*

$$\int_a^b f(x)\mathrm{d}x = h[\tfrac{1}{2}f(x_0) + f(x_1) + f(x_2) + \cdots + f(x_{n-1})$$

$$+ \tfrac{1}{2}f(x_n)] + \varepsilon(h), \qquad (7\cdot42)$$

where an equality sign has now been used because we have included the *error term* $\varepsilon(h)$, which recognizes that the error is, in part, dependent on the magnitude of h.

(b) Simpson's rule

A different approach involves dividing $[a, b]$ into an *even* number n of sub-intervals of equal length $h = (b - a)/n$, and then approximating the function over consecutive pairs of sub-intervals by a quadratic polynomial. That is to say fitting a parabola to the three points $(a, f(a))$, $(a + h, f(a + h))$, $(a + 2h, f(a + 2h))$ comprising the first two sub-intervals, and thereafter repeating the process until the whole of the interval $[a, b]$ has been covered. The value of the definite integral can then be estimated by integrating the successive quadratic approximations over their respective intervals of length $2h$ and adding the results. This simple idea leads to *Simpson's rule* for numerical integration which we now formulate in analytical terms.

Consider the first interval $[a, a + 2h]$, and represent the function $y = f(x)$ in this interval by the quadratic

$$y = c_0 + c_1 x + c_2 x^2. \tag{7·43}$$

Then the approximation to the desired integral taken over this interval is

$$\int_a^{a+2h} f(x)\mathrm{d}x \approx \int_a^{a+2h} (c_0 + c_1 x + c_2 x^2)\mathrm{d}x$$

$$= \left(c_0 x + \frac{c_1 x^2}{2} + \frac{c_2 x^3}{3} \right)\Bigg|_a^{a+2h} \tag{7·44}$$

To determine the coefficients c_0, c_1, and c_2 in order that the quadratic should pass through the three points $(a, f(a))$, $(a + h, f(a + h))$, $(a + 2h, f(a + 2h))$ we must solve the three simultaneous equations

$$f(a) = c_0 + c_1 a + c_2 a^2,$$
$$f(a + h) = c_0 + c_1(a + h) + c_2(a + h)^2,$$
$$f(a + 2h) = c_0 + c_1(a + 2h) + c_2(a + 2h)^2. \tag{7·45}$$

When this is done and the results are substituted into Eqn (7·44) we arrive at the desired result

$$\int_a^{a+2h} f(x)\mathrm{d}x = \frac{h}{3}(f(a) + 4f(a + h) + f(a + 2h)) + \varepsilon(h), \tag{7·46}$$

where again we have included the error term by $\varepsilon(h)$. In its simplest form Eqn (7·46), together with its error term, is called Simpson's rule. An explicit form for $\varepsilon(h)$ in both the trapezium rule and Simpson's rule will be given later.

If, now, result (7·46) is applied to the intervals $[a, a + 2h]$, $[a + 2h, a + 4h]$, . . ., $[a + (n - 2)h, b]$ and the results are added, we arrive at Simpson's rule for an even number n of intervals

$$\int_a^b f(x)\mathrm{d}x = \frac{h}{3}[f(a) + 4f(a + h) + 2f(a + 2h) + 4f(a + 3h) + \cdots$$

$$+ 4f(a + (n - 1)h) + f(b)] + \varepsilon(h), \tag{7·47}$$

where $h = (b - a)/n$.

It can be shown that the error term $\varepsilon(h)$ in Simpson's rule (7·46) satisfies the inequality $\varepsilon(h) \leq (2h)^5 M/90$, where $M = \max |f^{(4)}(x)|$, for $a \leq x \leq a + 2h$. The corresponding result for the trapezoidal rule (7·42) is $\varepsilon(h) \leq (2h)^3 M/12$, where $M = \max |f^{(2)}(x)|$.

Example 7·11 Calculate the definite integral

$$I = \int_1^2 \frac{dx}{x}$$

by the trapezoidal rule and by Simpson's rule, taking ten integration steps of length $h = 0·1$.

Solution We start by tabulating the functional values of the integrand $1/x$ at intervals of 0·1.

x	$f(x) = \dfrac{1}{x}$
1·0	1·0000
1·1	0·9091
1·2	0·8333
1·3	0·7692
1·4	0·7143
1·5	0·6667
1·6	0·6250
1·7	0·5882
1·8	0·5556
1·9	0·5263
2·0	0·5000

Then, using the trapezoidal rule (7·42), we find

$$I \approx 0·1 \times [0·5000 + 0·9091 + 0·8333 + 0·7692 + 0·7143 + 0·6667$$
$$+ 0·6250 + 0·5882 + 0·5556 + 0·5263 + 0·25],$$

whence $I \approx 0·6938$.

The same calculation using Simpson's rule, (7·47), gives

$$I \approx \frac{0-1}{3} \times [1·0000 + 4 \times (0·9091) + 2 \times (0·8333) + 4 \times (0·7692)$$
$$+ 2 \times (0·7143) + 4 \times (0·6667) + 2 \times (0·6250) + 4 \times (0·5882)$$
$$+ 2 \times (0·5556) + 4 \times (0·5263) + 0·5000],$$

whence $I \approx 0·6932$.

In actual fact the exact result of this definite integral is $\log_e 2 = 0·69315$.

As would have been expected on intuitive grounds, Simpson's rule is more accurate than the trapezoidal rule.

PROBLEMS

Section 7·1

7·1 Let $f(x) = \lambda x$ on some closed interval $a \leq x \leq b$ lying in the positive part of the x-axis, where $\lambda > 0$ is a constant. Then, if P_n is a partition of $[a, b]$ into n sub-intervals of equal length, determine the form of the lower and upper sums $\underline{S}_{P_n}$, $\bar{S}_{P_n}$ for $f(x)$ taken over this partition and prove directly by taking the limit that

$$\lim_{n \to \infty} \underline{S}_{P_n} = \lim_{n \to \infty} \bar{S}_{P_n}.$$

Hence deduce that

$$\int_a^b \lambda x \, dx = \frac{\lambda}{2} (b^2 - a^2).$$

7·2 Let λ, $\mu > 0$ be constants, and set $f(x) = \mu + \lambda x$ on some closed interval $a \leq x \leq b$ lying in the positive part of the x-axis. Show, using the method of Problem 7·1, that

$$\int_a^b (\mu + \lambda x) \, dx = \mu(b - a) + \frac{\lambda}{2} (b^2 - a^2). \tag{A}$$

Show also by this method that

$$\int_a^b \mu \, dx = \mu(b - a), \tag{B}$$

and deduce from (A), (B) and the result of Problem 7·1 that

$$\int_a^b (\mu + \lambda x) \, dx = \int_a^b \mu \, dx + \int_a^b \lambda x \, dx.$$

This provides a direct proof of the linearity of the operation of integration in the special case that $f(x) = \mu + \lambda x$.

7·3 Let $f(x) = e^{\lambda x}$, and take P_n to be a partition of the closed interval $[a, b]$ into n sub-intervals of equal length. By taking the numbers ξ_i of Definition 7·1 to be at the left-hand end points of the sub-intervals, compute the approximating sum S_{P_n} corresponding to $f(x) = e^{\lambda x}$, and by finding its limit prove that

$$\int_a^b e^{\lambda x} \, dx = \frac{1}{\lambda} (e^{\lambda b} - e^{\lambda a}).$$

7·4 Let $[a, b]$ be any closed interval not containing the origin, and denote by P_m the partition of this interval into m equal sub-intervals each of length $(b - a)/m$. Denote by x_r the point $x_r = a + (r/m)(b - a)$ lying at the right-hand end point of the rth interval. Then, by setting $\xi_r = \sqrt{(x_{r-1}x_r)}$ show, by considering $x_{r-1} - \xi_r$ and $x_r - \xi_r$, that $x_{r-1} < \xi_r < x_{r+1}$. By writing $f(x) = 1/x^2$ in Definition 7·1; and taking P_m and the points ξ_r in that definition to be as defined above, prove that

$$\int_a^b \frac{dx}{x^2} = \left(\frac{1}{a} - \frac{1}{b}\right).$$

$$\left[\text{Hint: Use the fact that } \sum_{r=1}^n \frac{1}{x_{r-1}x_r} = \sum_{r=1}^n \left(\frac{1}{x_r - x_{r-1}}\right)\left(\frac{1}{x_{r-1}} - \frac{1}{x_r}\right).\right]$$

7·5 Determine the lower bounds m_r and the upper bounds M_r of the function $f(x) = 1/(1 + x^2)$ in each of the n adjacent sub-intervals of length $1/n$ comprising a partition P_n of the closed interval $[0, 1]$. Use these results to deduce the form taken by the upper and lower sums $\bar{S}_{P_n}$, $\underline{S}_{P_n}$ and show that

$$\lim_{n \to \infty} (\bar{S}_{P_n} - \underline{S}_{P_n}) = 0.$$

Deduce from this that

$$\int_0^1 \frac{dx}{1 + x^2} = \lim_{n \to \infty} n \left\{\frac{1}{n^2 + 1^2} + \frac{1}{n^2 + 2^2} + \frac{1}{n^2 + 3^2} + \cdots + \frac{1}{n^2 + n^2}\right\}$$

or, equivalently,

$$= \lim_{n \to \infty} n \left\{\frac{1}{n^2} + \frac{1}{n^2 + 1^2} + \frac{1}{n^2 + 2^2} + \cdots + \frac{1}{n^2 + (n-1)^2}\right\}.$$

We shall see later that this integral has the value $\frac{1}{4}\pi$, and so each of these different expressions has this same interesting limit.

Section 7·2

7·6 Outline the proofs of the results of Theorem 7·1.

7·7 Use the result of Problem 7·3 to evaluate the definite integral

$$\int_4^2 e^{-3x} \, dx.$$

7·8 Find the area I between the curves $y = x^2 + 2$ and $y = -x + 1$, which is bounded to the left by the line $x = -1$ and to the right by the line $x = 2$.

7·9 Evaluate the integral

$$I = \int_0^3 f(x) \, dx,$$

given that

$$f(x) = \begin{cases} x & \text{for } 0 \le x < 1; \\ 2 + 2x & \text{for } 1 \le x < 2; \\ x - 1 & \text{for } 2 \le x \le 3. \end{cases}$$

7·10 On the assumption that the definite integral

$$\int_a^b \frac{dx}{\sqrt{(1 - x^2)}} = \arcsin b - \arcsin a,$$

prove that the improper integral

$$I = \int_0^1 \frac{dx}{\sqrt{(1 - x^2)}}$$

is convergent, and determine its value.

7·11 Sketch the area bounded below by the positive x-axis, and above by the line $y = x$ on the interval $0 \le x \le 1$, and by the curve $y = 1/x^2$ on the interval $1 \le x < \infty$. Determine this area I by the use of an improper integral combined with elementary geometrical arguments.

Section 7·3

7·12 Use Theorem 7·2 to place bounds on the value of the definite integral

$$I = \int_{\frac{1}{4}\pi}^{\frac{1}{3}\pi} e^{-x^2} \cos^3 x \, dx.$$

7·13 Evaluate the definite integral

$$\int_{-1}^3 x^2 \, dx,$$

and use the result to determine the number ξ in Theorem 7·3 when it is applied to this definite integral. Is the number ξ unique? Repeat the argument, but this time applying it to the definite integral

$$\int_{-2}^2 x^2 \, dx.$$

Is there a unique number ξ in this case?

Section 7·4

7·14 Use Theorem 7·5 to evaluate the following definite integrals:

(a) $\int_a^b (x^{5/2} + 3e^x)dx$, (b) $\int_0^\pi \sin x \, dx$, (c) $\int_0^{2\pi} \sin x \, dx$,

(d) $\int_0^{2\pi} |\sin x| \, dx$.

7·15 Use Theorem 7·5 to determine the area contained between the x-axis and the curve $y = 1 + x^3 + 2 \sin x$, which is bounded to the left by the line $x = 0$ and to the right by the line $x = \pi$.

Section 7·5

7·16 Express in terms of a definite integral the arc length of the curve $y = 1 + x^2 + \sin 2x$, that lies between the points on the curve corresponding to $x = 1$ and $x = 4$.

7·17 Prove that the circumference of a circle of radius a is $2\pi a$ by using the parametric equations of a circle $x = a \cos t$, $y = a \sin t$ with $0 \le t \le 2\pi$.

7·18 Find the area contained between the planes $x = -2$ and $x = 3$ of the surface of revolution about the x-axis generated by the curve $y = 2 + \cosh x$. [Hint: An antiderivative of $\cosh x$ is $\sinh x + C$.]

7·19 Determine the volume contained between the parabola $y = 2 + x + x^2$ and the cubic $y = 5 + 2x + x^3$, which lies between the planes $x = 1$ and $x = 2$.

Section 7.6

7·20 Evaluate the definite integral

$$\int_1^3 (x^3 + 2x + 1)\mathrm{d}x$$

by the trapezoidal rule using four intervals of équal length and then by Simpson's rule for the same intervals. Compare the result with that obtained by direct integration. Infer from your result that Simpson's rule is exact for cubic equations despite the fact that it is based on a parabolic fitting of the function.

Systematic integration

8·1 Integration of elementary functions

The main objective of this chapter is to explore some of the systematic methods for determining an *antiderivative*, that is, a function $F(x)$ whose derivative is equal to some given function $f(x)$. As described in the previous chapter, we shall denote the antiderivative of the function f by $\int f(x)\mathrm{d}x$ with the understanding that

$$\int f(x)\mathrm{d}x = F(x) + C \tag{8·1}$$

with C an arbitrary constant.

Alternatively, as any *indefinite integral* of f must also be an antiderivative of f, we may identify $F(x)$ in Eqn (8·1) with $\displaystyle\int_a^t f(t)\mathrm{d}t$ where a is arbitrary, to obtain the equivalent expression

$$\int f(x)\mathrm{d}x = \int_a^t f(t)\mathrm{d}t + C. \tag{8·2}$$

Remember that the symbol $\int f(x)\mathrm{d}x$ for the antiderivative of f derives from differentiation and denotes the most general *function* whose derivative is f. The allied symbol $\displaystyle\int_a^b f(x)\mathrm{d}x$, denoting a definite integral of f, derives from integration and is simply a real *number*. Considering the definition of an antiderivative, we shall say that two antiderivatives are equal if they only differ by a constant.

It should be recalled that the connection between the concepts of an antiderivative and a definite integral is provided by the fundamental theorem of calculus, which asserts that

$$\int_a^b f(x)\mathrm{d}x = \left\{\int f(x)\mathrm{d}x\right\}\Bigg|_{x=b} - \left\{\int f(x)\mathrm{d}x\right\}\Bigg|_{x=a}.$$

In view of Eqn (8.1) this may be written

$$\int_a^b f(x)\mathrm{d}x = F(b) - F(a). \tag{8·3}$$

Very often in texts the term *indefinite integral* is loosely ascribed to the entire right-hand side of Eqn (8·2) instead of, as here, only to its first term. This is usually justified by the fact that a is arbitrary though, of course, it does not necessarily follow that all possible constants C can be absorbed into the integral by a suitable choice of a. For example, we have the antiderivative

$$\int \cos x \, dx = \sin x + C,$$

though if for some particular problem it was appropriate to set $C = 3$, say, then no choice of the arbitrary constant a would enable us to equate $\int_a^x \cos x \, dx$ and $\sin x + 3$, for this would imply that $\sin a = -3$.

Unfortunately, the theorems for the differentiation of wide classes of functions seldom have any counterpart for determining antiderivatives. Ultimately, success in finding an antiderivative depends on whether or not the function f can be so simplified that one may be recognized by using tables of derivatives in reverse: that is, matching the desired derivative f with one in the table, and reading backwards to deduce an antiderivative. Thus, to find the antiderivative of $3 \sec x \tan x$, we first glean from Table 5·1 that

$$\frac{d}{dx} (\sec x) = \sec x \tan x$$

or, equivalently,

$$\frac{d}{dx} (3 \sec x) = 3 \sec x \tan x$$

showing that the antiderivative is

$$\int 3 \sec x \tan x \, dx = 3 \sec x + C.$$

In colloquial terms, the process of finding the most general antiderivative of the function $f(x)$ is called the 'integration of $f(x)$'.

Table 8·1 gives a preliminary working list of important integrals which has been compiled from the tables of derivatives in Chapters 5 and 6.

The two separate results shown against number 3 are usually contracted to

$$\int \frac{dx}{x} = \log |x| + C,$$

with the tacit understanding that the arbitrary constant C differs according as x is positive or negative. With obvious modifications, this convention will be extended to include all integrals involving the logarithmic function. Specific examples involving this convention are to be found in Problems 8·1–8·3.

The following statement is equivalent to both Eqn (8·1) and Eqn (8·2), and it arises as a direct consequence of the definition of an antiderivative. We formulate it as a general theorem.

Table 8·1   Basic table for integrals

---

1. $\int x^n \, dx = \dfrac{x^{n+1}}{n+1} + C$ $(n \neq -1)$;

2. $\int a^x \, dx = \dfrac{a^x}{\log a} + C$ $(a > 0)$;

3. $\int \dfrac{dx}{x} = \begin{cases} \log x + C \\ \log(-x) + C \end{cases}$ for $x > 0$
 for $x < 0$;

4. $\int e^{ax} \, dx = \dfrac{1}{a} e^{ax} + C$ $(a \neq 0)$;

5. $\int \cos ax \, dx = \dfrac{1}{a} \sin ax + C$ $(a \neq 0)$;

6. $\int \sin ax \, dx = -\dfrac{1}{a} \cos ax + C$ $(a \neq 0)$;

7. $\int \dfrac{dx}{\sqrt{(a^2 - x^2)}} = \arcsin \dfrac{x}{a} + C$ for $|x| < |a|$;

8. $\int \dfrac{dx}{a^2 + x^2} = \dfrac{1}{a} \arctan \dfrac{x}{a} + C$ $(a \neq 0)$;

9. $\int \dfrac{dx}{\sqrt{(a^2 + x^2)}} = \operatorname{arcsinh} \dfrac{x}{a} + C$ $(a \neq 0)$;

10. $\int \dfrac{dx}{\sqrt{(x^2 - a^2)}} = \begin{cases} \operatorname{arccosh} \dfrac{x}{a} + C \\ -\operatorname{arccosh}\left(\dfrac{-x}{a}\right) + C \end{cases}$ for $x > a$, <br> for $x < -a$;

11. $\int \dfrac{dx}{a^2 - x^2} = \dfrac{1}{a} \operatorname{arctanh} \dfrac{x}{a} + C$ for $|x| < |a|$;

12. $\int \dfrac{dx}{x^2 - a^2} = -\dfrac{1}{a} \operatorname{arccoth} \dfrac{x}{a} + C$ for $|x| > |a|$.

THEOREM 8·1

$$\frac{d}{dx} \int f(x) \, dx = f(x).$$

In words, this general result merely asserts the obvious fact that the derivative of the antiderivative of a function $f(x)$ is the function $f(x)$ itself. Its most frequent application is probably to the verification of antiderivatives. For example, let us use the theorem to verify the antiderivative

$$\int \frac{g'\,dx}{\sqrt{(a^2 - g^2)}} = \arcsin\left(\frac{g}{a}\right) + C, \tag{A}$$

where $g = g(x)$ is some differentiable function of x and $|g| < a$.

By Theorem 8·1 we must have

$$\frac{d}{dx}\int \frac{g'\,dx}{\sqrt{(a^2 - g^2)}} = \frac{g'}{\sqrt{(a^2 - g^2)}}. \tag{B}$$

Now, differentiating the right-hand side of (A) we find

$$\frac{d}{dx}\left[\arcsin\left(\frac{g}{a}\right) + C\right] = \frac{1}{\sqrt{(1 - (g/a)^2)}} \cdot \frac{g'}{a}$$

$$= \frac{g'}{\sqrt{(a^2 - g^2)}},$$

which is identical with (B). Thus, (A) is verified.

A final general result of great value is the fact that the derivative of a linear combination of functions is equal to the same linear combination of their derivatives (Theorem 5·4). Expressed in terms of antiderivatives this implies the following general theorem.

THEOREM 8·2

$$\int (k_1 f + k_2 g)\,dx = k_1 \int f\,dx + k_2 \int g\,dx.$$

It is, of course, this theorem that permits us to simplify many expressions to the point at which antiderivatives may be deduced from tables of standard integrals (antiderivatives) such as Table 8·1. Hence we have

$$\int (5x^2 - 2\cos x)\,dx = 5\int x^2\,dx - 2\int \cos x\,dx$$

$$= \frac{5x^3}{3} - 2\sin x + C.$$

The separate arbitrary constants associated with each of the antiderivatives on the right-hand side have, of course, been combined into the single arbitrary constant C.

The remaining sections of this chapter are concerned with outlining the details of the main techniques available for finding antiderivatives.

8·2 Integration by substitution

Possibly the most frequently used technique of integration is that in which the variable under the integral sign is changed in a manner which simplifies the task of finding the antiderivative. This process is known as *integration by substitution* or *integration by change of variable*. It is in this technique that the full significance of the symbol dx in Eqn (8·1) is first realized. Indeed, by making a straightforward application of the chain rule for differentiation (Theorem 5·7) we shall arrive at a simple mechanical rule for effecting a variable change by using differentials.

Let us begin with an antiderivative of the form

$$I = \int f(x)\, dx, \tag{8·4}$$

and suppose that we wish to change the variable x to u by using the substitution

$$x = h(u), \tag{8·5}$$

where h is a differentiable function of u. Then by Theorem 8·1 and the chain rule for differentiation in Eqn (5·6), we have

$$\frac{dI}{du} = \frac{dI}{dx} \cdot \frac{dx}{du} = f(x)\frac{dx}{du} \tag{8·6}$$

or, from Eqn. (8·5),

$$\frac{dI}{du} = f[h(u)]h'(u). \tag{8·7}$$

Integration then gives

$$I = \int f[h(u)]h'(u)du, \tag{8·8}$$

so that by comparing Eqns (8·4) and (8·8) we arrive at the next result which we formulate as a theorem.

THEOREM 8·3 (integration by substitution)   If $x = h(u)$ is a differentiable function of u, then

$$\int f(x)dx = \int f[h(u)]h'(u)du.$$

Now the substitution method depends for its success on the fact that although the antiderivative of the function f on the left-hand side of our theorem might not be known, the corresponding antiderivative of the function $f[h(u)]h'(u)$ on the right-hand side might be simpler, and so be recognizable. There is no straightforward rule by which a substitution may be found which will automatically simplify the determination of an anti-derivative of a function. The choice of a substitution is usually dictated by experience coupled with an attempt to simplify some awkward feature associated with the function, such as a square root.

Usually the substitution involved is not as simple as in Eqn (8·5), but is of the form

$$g(x) = h(u), \tag{8·9}$$

where g, h are both differentiable functions. As we now show, this generalization does not involve a major modification of our arguments that led to Theorem 8·3. All that is necessary is to differentiate Eqn (8·9) with respect to x to obtain

$$g'(x) = h'(u)\frac{du}{dx} \tag{8·10}$$

and then, if $g'(x) \neq 0$, to write this in the differential form

$$dx = \frac{h'(u)}{g'(x)}du. \tag{8·11}$$

If dx in Eqn (8·4) is now replaced by the expression in Eqn (8·11) then, apart from the variable x occurring in $f(x)$ and $g'(x)$, the antiderivative now involves the new variable u. This x can be replaced by a function of u if Eqn (8·9) is solved for x in terms of u to obtain

$$x = g^{-1}[h(u)], \tag{8·12}$$

and the result then substituted into $f(x)$ and $g'(x)$.

All of these results can be conveniently summarized in the form of a simple mechanical rule for changing the variable in an antiderivative. To give the most useful form of this rule we choose to express it in terms of a slightly more general integrand than was used in Eqn (8·4).

Rule 1   (Integration by substitution)

We suppose that in the antiderivative

$$I = \int k(x) \cdot f[g(x)]dx$$

it is required to change from the variable x to the variable u by means of the relationship $g(x) = h(u)$, where g and h are differentiable functions, with $g'(x) \neq 0$. The result may be deduced from I above by:

(a) replacing $g(x)$ in $f[g(x)]$ by $h(u)$;
(b) solving $g(x) = h(u)$ for x in the form $x = g^{-1}[h(u)]$ and then replacing x in $k(x)$ by this result;
(c) replacing dx by du, where du is obtained from the differential relationship $g'(x)dx = h'(u)du$;
(d) replacing x in $g'(x)$ by $x = g^{-1}[h(u)]$.

We now illustrate the application of this rule in a series of examples. Unfortunately, although the rule tells us how to change the variable, it offers us no information on the type of variable change that should be made. That is to say it does not tell us the functional form of f and g. Only experience can help here.

Example 8·1   Evaluate the antiderivative

$$I = \int x^3\sqrt{(1 + x^2)}dx.$$

Solution   First we make the obvious identification $k(x) = x^3$ and then, to remove the square root function which is difficult to manipulate, we shall try setting

$$1 + x^2 = u^2.$$

That is to say, in the hope that it will lead to a simpler expression, we make the further identifications

$$g(x) = 1 + x^2 \quad \text{and} \quad h(u) = u^2.$$

The function f in Rule 1 then becomes the square root function, with $\sqrt{(1 + x^2)} = u$. Rather than solving for x, for the moment we shall use the result $x^3 = x \cdot x^2 = x(u^2 - 1)$, when we find $x^3\sqrt{(1 + x^2)} = xu(u^2 - 1)$.

Now $g'(x) = 2x$ and $h'(u) = 2u$, so that the differential relation $g'(x)\mathrm{d}x = h'(u)\mathrm{d}u$ gives rise to $x\mathrm{d}x = u\mathrm{d}u$. Hence, in differential form,

$$x^3\sqrt{(1 + x^2)}\mathrm{d}x = u(u^2 - 1)x\mathrm{d}x = u^2(u^2 - 1)\mathrm{d}u,$$

and so,

$$I = \int x^3\sqrt{(1 + x^2)}\mathrm{d}x = \int u^2(u^2 - 1)\mathrm{d}u.$$

The antiderivative on the right-hand side is now straightforward and may be integrated on sight to give

$$I = \frac{u^5}{5} - \frac{u^3}{3} + C$$

or,

$$I = \frac{(1 + x^2)^{5/2}}{5} - \frac{(1 + x^2)^{3/2}}{3} + C.$$

Example 8·2   Evaluate the antiderivative

$$I = \int \sqrt{(1 + x^2)}\mathrm{d}x.$$

Solution   In this antiderivative $k(x) \equiv 1$, but it is not immediately clear how best to change the variable. It is left to the reader to see why neither of the possible substitutions $u^2 = 1 + x^2$ or $u = 1 + x^2$ bring about any effective simplification. Instead, let us seek to remove the square root by making the substitution $x = \sinh u$. Then $1 + x^2 = 1 + \sinh^2 u = \cosh^2 u$, so that $\sqrt{(1 + x^2)} = \cosh u$.

Next, as $g(x) = x$ and $h(u) = \sinh u$, $g'(x) = 1$, $h'(u) = \cosh u$ and so $\mathrm{d}x = \cosh u\mathrm{d}u$. Applying the rule then gives

$$\sqrt{(1 + x^2)}\mathrm{d}x = \cosh u \cdot \cosh u\mathrm{d}u = \cosh^2 u\mathrm{d}u,$$

whence

$$I = \int \cosh^2 u\mathrm{d}u.$$

Now use the identity $\cosh^2 u = \frac{1}{2}(\cosh 2u + 1)$ to give

$I = \frac{1}{2}\int(\cosh 2u + 1)du$

$= \frac{1}{4}\sinh u + \dfrac{u}{2} + C.$

To return to the variable x it is necessary to use the results $u = \operatorname{arcsinh} x$, $\cosh u = \sqrt{(1 + x^2)}$ together with the identity $\sinh 2u = 2 \sinh u \cdot \cosh u$ to obtain

$I = \frac{1}{2}[x\sqrt{(1 + x^2)} + \operatorname{arcsinh} x] + C.$

Example 8·3 Evaluate the antiderivative

$I = \int \cos(1 + 3x)dx.$

Solution   This antiderivative has $k(x) \equiv 1$, and by setting $1 + 3x = u$ so that $g(x) = 1 + 3x$, $h(u) = u$ we find that $\cos(1 + 3x) = \cos u$ and $3dx = du$, whence

$I = \int \frac{1}{3}\cos u\, du$

$= \frac{1}{3}\sin u + C,$

and thus

$I = \frac{1}{3}\sin(1 + 3x) + C.$

Example 8·4 Evaluate the antiderivative

$I = \int 2x\sqrt{(1 + x^2)}dx.$

Solution   Setting $u = 1 + x^2$ it follows that $du = 2xdx$, so that

$2x\sqrt{(1 + x^2)}dx = \sqrt{u}\,du,$

whence

$I = \int \sqrt{u}\,du = \frac{2}{3}u^{3/2} + C$

$= \frac{2}{3}(1 + x^2)^{3/2} + C.$

It is an immediate consequence of Eqn (8·3) that Theorem 8·3 also applies to definite integrals provided the limits are also transformed by the same substitution rule. The restatement of Theorem 8·3 in terms of definite integrals is as follows:

THEOREM 8·4 (integration of definite integrals by substitution)   If $x = h(u)$ is a differentiable function of u, and $u = h^{-1}(x)$, then

$$\int_a^b f(x)dx = \int_{h^{-1}(a)}^{h^{-1}(b)} f[h(u)]h'(u)du.$$

As with Theorem 8·3, this result can be generalized and stated in terms of a mechanical rule which is as easy to apply as was our previous rule.

Rule 2   (Integrating definite integrals by substitution)

We suppose that in the definite integral

$$I = \int_a^b k(x) \cdot f[g(x)]dx$$

it is required to change from the variable x to the variable u by means of the relationship $g(x) = h(u)$, where g and h are differentiable functions, with $g'(x) \neq 0$. The result may be deduced from I above by:

(a) transforming the differential expression $k(x) \cdot f[g(x)]dx$ as indicated in Rule 1;
(b) solving $g(x) = h(u)$ for u in the form $u = h^{-1}[g(x)]$ and replacing the upper limit b by $h^{-1}[g(b)]$ and the lower limit a by $h^{-1}[g(a)]$.

Example 8·5   Evaluate the definite integral

$$I = \int_0^1 x^2 \sqrt{(1 - x^2)}dx.$$

Solution   Let us make the substitution $x = \sin u$, so that $dx = \cos u du$, when

$$x^2 \sqrt{(1 - x^2)}dx = \sin^2 u \cdot \cos u \cdot \cos u du$$
$$= \sin^2 u \cdot \cos^2 u du.$$

Then, as $u = \arcsin x$, using the principal branch of the sine function, we find from Rule 2 that

$$I = \int_0^1 x^2 \sqrt{(1 - x^2)}dx = \int_{\arcsin 0}^{\arcsin 1} \sin^2 u \cdot \cos^2 u du$$
$$= \int_0^{\frac{1}{2}\pi} \sin^2 u \cdot \cos^2 u du.$$

To evaluate this last definite integral we use a technique from Chapter 6 which is often helpful. From Definition 6·6 we may write

$$\sin^2 u \cdot \cos^2 u = \left(\frac{e^{iu} - e^{-iu}}{2i}\right)^2 \left(\frac{e^{iu} + e^{-iu}}{2}\right)^2$$
$$= \left(\frac{e^{2iu} - 2 + e^{-2iu}}{-4}\right)\left(\frac{e^{2iu} + 2 + e^{-2iu}}{4}\right)$$
$$= \frac{e^{4iu} + e^{-4iu} - 2}{-16},$$

and thus

$$\sin^2 u \,.\, \cos^2 u = \tfrac{1}{8}(1 - \cos 4u).$$

Using this result in the definite integral, which may then be evaluated on sight, we finally obtain

$$I = \tfrac{1}{8} \int_0^{\frac{1}{2}\pi} (1 - \cos 4u)\mathrm{d}u$$

$$= \tfrac{1}{8} \left[u - (\tfrac{1}{4} \sin 4u) \right]\Big|_0^{\frac{1}{2}\pi} = \tfrac{1}{16}\pi,$$

and so

$$\int_0^1 x^2 \sqrt{(1 - x^2)}\mathrm{d}x = \tfrac{1}{16}\pi.$$

Example 8·6 Evaluate the definite integral

$$I = \int_0^1 (2x + 5) \cosh (x^2 + 5x + 1)\mathrm{d}x.$$

Solution Inspection shows that this example is of the form $f \equiv \cosh$ and $g(x) = x^2 + 5x + 1$.
    As $g(0) = 1, g(1) = 7$, by setting $u = g(x)$ we at once obtain

$$I = \int_1^7 \cosh u\mathrm{d}u = (\sinh 7 - \sinh 1).$$

8·3   Integration by parts

This most valuable technique is based on Theorem 5·5, concerning the derivative of the product of two functions. That theorem asserts that if f, g are two differentiable functions of x, then

$$\frac{\mathrm{d}}{\mathrm{d}x} [f(x) g(x)] = [f(x) g'(x)] + [f'(x) g(x)].$$

Taking the antiderivative of this result gives

$$f(x) g(x) = \int f(x) g'(x)\mathrm{d}x + \int g(x) f'(x)\mathrm{d}x$$

which, on rearrangement, becomes

$$\int f(x) g'(x)\mathrm{d}x = f(x) g(x) - \int g(x) f'(x)\mathrm{d}x. \tag{8.13}$$

This is one form of the required result. Using the differential notation $\mathrm{d}f = f'(x)\mathrm{d}x$, $\mathrm{d}g = g'(x)\mathrm{d}x$ enables this to be contracted to the equivalent and easily remembered alternative form

$$\int f \,\mathrm{d}g = fg - \int g \,\mathrm{d}f. \tag{8·14}$$

These results are now formulated as our next theorem:

THEOREM 8·5 (integration by parts)　If f, g are differentiable functions of x, then

$$\int f(x)\, g'(x)\mathrm{d}x = f(x)\, g(x) - \int g(x) f'(x)\mathrm{d}x$$

or, expressed in differential notation,

$$\int f\, \mathrm{d}g = fg - \int g\, \mathrm{d}f.$$

This useful theorem is the nearest possible approach to a general theorem for finding the antiderivative of the product of two functions. It depends on the fact that often the antiderivative $\int g\, \mathrm{d}f$ is easier to determine than the antiderivative $\int f\, \mathrm{d}g$. Naturally, the technique of integration by substitution can also be employed when evaluating $\int g\, \mathrm{d}f$.

When definite integrals are involved it is not difficult to see that the result is still valid provided the limits are also applied to the product fg. The general result is as follows:

THEOREM 8·6 (integration by parts: definite integral)　If f, g are differentiable functions of x in $[a, b]$, then

$$\int_a^b f(x)\, g'(x)\mathrm{d}x = f(x)\, g(x)\Big|_a^b - \int_a^b g(x) f'(x)\mathrm{d}x$$

$$= [f(b)\, g(b)] - [f(a)\, g(a)] - \int_a^b g(x) f'(x)\mathrm{d}x.$$

As before, we illustrate both of these theorems by means of a series of examples. These have been carefully chosen to demonstrate a variety of situations in which integration by parts is useful.

Example 8·7　Evaluate the antiderivative

$$I = \int x^k \log x\, \mathrm{d}x \qquad \text{for } x > 0,\ k \neq -1.$$

Solution　The problem here, as with all applications of the technique of integration by parts, is to decide upon the functions f and g. A little experimentation will soon convince the reader that I will only simplify if we set $f(x) = \log x$ and $g(x) = x^{k+1}/(k + 1)$, for then $g'(x) = x^k$ and $f'(x) = 1/x$. Accordingly we write I in the form

$$I = \int \log x\, \mathrm{d}\left[\frac{x^{k+1}}{k + 1}\right].$$

Applying Theorem 8·5 gives

$$I = \frac{x^{k+1} \log x}{k+1} - \int \frac{x^{k+1}}{k+1} \cdot \frac{1}{x} \, dx$$

$$= \frac{x^{k+1} \log x}{k+1} - \frac{x^{k+1}}{(k+1)^2} + C.$$

Example 8·8 Evaluate the definite integral

$$\int_0^{1/2} \arcsin x \, dx.$$

Solution This time we make the identifications $f(x) = \arcsin x$ and $g(x) = x$ and write

$$\int_0^{1/2} \arcsin x \, d[x] = x \arcsin x \Big|_0^{1/2} - \int_0^{1/2} \frac{x \, dx}{\sqrt{(1 - x^2)}}. \tag{A}$$

We have

$$x \arcsin x \Big|_0^{1/2} = \pi/12 - 0 = \pi/12$$

but the definite integral on the right-hand side is still not recognizable. To simplify it let us now set $u = 1 - x^2$ so that $x \, dx = -\frac{1}{2} \, du$; using Theorem 8·4 we obtain

$$\int_0^{1/2} \frac{x \, dx}{\sqrt{(1 - x^2)}} = -\frac{1}{2} \int_1^{3/4} \frac{du}{\sqrt{u}} = -\frac{1}{2} \cdot 2u^{1/2} \Big|_1^{3/4} = 1 - \frac{\sqrt{3}}{2}.$$

Combining this result with (A) gives

$$\int_0^{1/2} \arcsin x \, dx = \pi/12 + \frac{\sqrt{3}}{2} - 1.$$

Example 8·9 Evaluate the antiderivative

$$I = \int e^{ax} \sin bx \, dx.$$

Solution This time we choose to make the identification $f(x) = \sin bx$, $g(x) = (1/a)e^{ax}$ and to write I in the form

$$I = \int \sin bx \, d\left(\frac{1}{a} e^{ax}\right).$$

Integrating by parts we find

$$\int e^{ax} \sin bx \, dx = \frac{1}{a} e^{ax} \sin bx - \frac{b}{a} \int e^{ax} \cos bx \, dx.$$

Now let us use this same device on the second term above to obtain

$$\int e^{ax} \sin bx \, dx = \frac{1}{a} e^{ax} \sin bx - \frac{b}{a} \int \cos bx \, d\left(\frac{1}{a} e^{ax}\right)$$

$$= \frac{1}{a} e^{ax} \sin bx - \frac{b}{a^2} e^{ax} \cos bx - \frac{b^2}{a^2} \int e^{ax} \sin bx \, dx + C.$$

Combining terms gives

$$\left(1 + \frac{b^2}{a^2}\right) \int e^{ax} \sin bx \, dx = \frac{e^{ax} (a \sin bx - b \cos bx)}{a^2} + C,$$

and so

$$\int e^{ax} \sin bx \, dx = \frac{e^{ax} (a \sin bx - b \cos bx)}{a^2 + b^2} + C^*$$

where C^* is related to C by $C^* = a^2 C/(a^2 + b^2)$. In fact there is no necessity to distinguish between C and C^*, since as C was an arbitrary constant of integration, C^* is also an arbitrary constant. For this reason it is not customary to redefine arbitrary constants when, as above, they are simply multiplied by a constant factor.

8·4 Reduction formulae

It not infrequently happens that an antiderivative I involving a parameter m may be reduced by means of the technique of integration by parts to an expression in which the parameter has a value differing by an integer k from its original value. If we denote such an antiderivative by I_m, then a typical situation is the one in which we arrive at an expression of the form

$$I_m = A(m) + I_{m-1}, \tag{8·15}$$

where $A(m)$ is some known function.

Expressions of this form provide an algorithm for the computation of any antiderivative of the given type once one of them is known, for the I_m are then defined recursively by this relation in terms of I_1, say. It is customary to refer to expressions of the general form of Eqn (8·15) as *reduction formulae*. The same idea is equally applicable, without essential modification, to definite integrals.

Example 8·10 Determine the reduction formula for

$$I_m = \int \cos^m \theta \, d\theta.$$

Use the result to determine I_7.

Solution We rewrite I_m as follows and use integration by parts.

$$I_m = \int \cos^{m-1} \theta \, d(\sin \theta)$$
$$= \cos^{m-1} \theta \cdot \sin \theta - \int \sin \theta \cdot (m-1) \cos^{m-2} \theta (-\sin \theta) d\theta$$
$$= \cos^{m-1} \theta \cdot \sin \theta + (m-1) \int \cos^{m-2} \theta \cdot \sin^2 \theta \, d\theta$$
$$= \cos^{m-1} \theta \cdot \sin \theta + (m-1) \int \cos^{m-2} \theta (1 - \cos^2 \theta) d\theta$$
$$= \cos^{m-1} \theta \cdot \sin \theta + (m-1) \int \cos^{m-2} \theta \, d\theta - (m-1) \int \cos^m \theta \, d\theta.$$

Recalling the definition of I_m we discover that this may be re-expressed in terms of I_m and I_{m-2} as

$$I_m = \cos^{m-1}\theta \,.\, \sin\theta + (m-1)I_{m-2} - (m-1)I_m,$$

whence we arrive at the required reduction formula

$$I_m = \frac{\cos^{m-1}\theta \,.\, \sin\theta}{m} + \left(\frac{m-1}{m}\right) I_{m-2}.$$

Setting $m = 7$ gives

$$I_7 = \frac{\cos^6\theta \,.\, \sin\theta}{7} + \frac{6}{7} I_5$$

$$= \frac{\cos^6\theta \,.\, \sin\theta}{7} + \frac{6}{7}\left(\frac{\cos^4\theta \,.\, \sin\theta}{5} + \frac{4}{5} I_3\right)$$

$$= \frac{\cos^6\theta \,.\, \sin\theta}{7} + \frac{6}{35}\cos^4\theta \,.\, \sin\theta + \frac{24}{35}\left(\frac{\cos^2\theta \,.\, \sin\theta}{3} + \frac{2}{3} I_1\right).$$

As $I_1 = \int \cos\theta \, d\theta = \sin\theta + C$ this gives the result

$$\int \cos^7\theta \, d\theta = \frac{1}{7}\cos^6\theta \,.\, \sin\theta + \frac{6}{35}\cos^4\theta \,.\, \sin\theta + \frac{8}{35}\cos^2\theta \,.\, \sin\theta$$

$$+ \frac{16}{35}\sin\theta + C.$$

Example 8·11   Evaluate the definite integral

$$J_m = \int_0^{\frac{1}{2}\pi} \cos^m\theta \, d\theta.$$

Solution   We can make use of the reduction formula determined in the previous example. It follows from

$$I_m = \frac{\cos^{m-1}\theta \,.\, \sin\theta}{m} + \left(\frac{m-1}{m}\right) I_{m-2}$$

that the definite integral J_m obeys the reduction formula

$$J_m = \frac{\cos^{m-1}\theta \,.\, \sin\theta}{m}\bigg|_0^{\frac{1}{2}\pi} + \left(\frac{m-1}{m}\right) J_{m-2} = \left(\frac{m-1}{m}\right) J_{m-2}.$$

We must now consider separately even and odd values of m. Firstly, if m is *even*, so that we may write $m = 2n$, then

$$J_{2n} = \frac{2n-1}{2n} \cdot \frac{2n-3}{2n-2} \cdots \frac{1}{2} J_0.$$

Secondly, if m is *odd*, so that we may write $m = 2n + 1$, then

$$J_{2n+1} = \frac{2n}{2n + 1} \cdot \frac{2n - 2}{2n - 1} \cdots \frac{2}{3} J_1.$$

So, using the fact that $J_0 = \int_0^{\frac{1}{2}\pi} 1 \, d\theta = \frac{1}{2}\pi$ and $J_1 = \int_0^{\frac{1}{2}\pi} \cos \theta \, d\theta = 1$, we obtain:

$$J_{2n} = \frac{1 \cdot 3 \cdot 5 \ldots (2n - 1)}{2 \cdot 4 \cdot 6 \ldots 2n} \cdot \tfrac{1}{2}\pi;$$

$$J_{2n+1} = \frac{2 \cdot 4 \cdot 6 \ldots 2n}{3 \cdot 5 \cdot 7 \ldots (2n + 1)}.$$

Reduction formulae may involve more than one parameter, as the final example illustrates.

Example 8·12   Show that

$$I_{m,n} = \int \sin^m x \cos^n x \, dx$$

satisfies the reduction formula

$$(m + n)I_{m,n} = -\sin^{n-1} x \cdot \cos^{n+1} x + (m - 1)I_{m-2,n}.$$

Solution   Write $I_{m,n}$ in the form shown below and integrate by parts.

$$I_{m,n} = \int \sin^{m-1} x \cdot \cos^n x \, d(-\cos x)$$

$$= -\sin^{m-1} x \cdot \cos^{n+1} x - \int (-\cos x)[(m - 1) \sin^{m-2} x \cdot \cos^{n+1} x$$

$$-n \sin^m x \cdot \cos^{n-1} x] dx$$

$$= -\sin^{m-1} x \cdot \cos^{n+1} x + (m - 1)I_{m-2,n+2} - nI_{m,n}.$$

Next reduce $I_{m-2,n+2}$ to a simpler form by writing

$$I_{m-2,n+2} = \int \sin^{m-2} x \cdot \cos^{n+2} x \, dx = \int \sin^{m-2} x \cdot \cos^n x(1 - \sin^2 x) dx$$

which shows that

$$I_{m-2,n+2} = I_{m-2,n} - I_{m,n}.$$

Using this to eliminate $I_{m-2,n+2}$ from the previous result gives

$$I_{m,n} = -\sin^{m-1} x \cdot \cos^{n+1} x + (m - 1)I_{m-2,n} - (m - 1)I_{m,n} - nI_{m,n}$$

or,

$$(m + n)I_{m,n} = -\sin^{m-1} x \cdot \cos^{n+1} x + (m - 1)I_{m-2,n}.$$

8·5 Integration of rational functions—partial fractions

It will be recalled from Chapter 2 that a rational fraction is a quotient $N(x)/D(x)$, in which $N(x)$ and $D(x)$ are polynomials. Antiderivatives of rational fractions are often required and in this section we indicate ways of expressing the fractions as the sum of simpler expressions, the antiderivatives of which are either known or may be found by standard methods. Our approach to the general problem of finding the antiderivative

$$I = \int \frac{N(x)}{D(x)}\,dx$$

will be to first consider some important special cases.

Case (a)   Suppose that $N(x)$ is of degree 0 and $D(x)$ is a polynomial of degree 1 and write

$$\frac{N(x)}{D(x)} = \frac{1}{cx + d}.$$

Then, making the substitution $u = cx + d$, we find

$$\int \frac{dx}{cx + d} = \frac{1}{c} \int \frac{du}{u} = \frac{1}{c} \log |u| + C$$

and so

$$\int \frac{dx}{cx + d} = \frac{1}{c} \log |cx + d| + C.$$

A similar argument establishes that

$$\int \frac{dx}{(cx + d)^n} = \frac{-1}{c(n - 1)} \cdot \frac{1}{(cx + d)^{n-1}} + C.$$

Case (b)   Suppose $N(x)$ is of degree 0 and $D(x)$ is of degree 2 and write

$$\frac{N(x)}{D(x)} = \frac{1}{ax^2 + bx + c}.$$

Then completing the square in the denominator $D(x)$ gives

$$ax^2 + bx + c = a\left[\left(x + \frac{b}{2a}\right)^2 + \left(\frac{c}{a} - \frac{b^2}{4a^2}\right)\right] = a\left[\left(x + \frac{b}{2a}\right)^2 + \alpha\right],$$

where $\alpha = (c/a) - (b^2/4a^2)$ may be positive, negative, or zero. Making the variable change $u = x + (b/2a)$ then shows that

$$I = \int \frac{dx}{ax^2 + bx + c} = \frac{1}{a} \int \frac{du}{u^2 + \alpha}.$$

This is a standard integral which may be identified from Table 8·1 once the sign of α has been determined. It will involve either the function arctan or the function arctanh.

Case (c)   Suppose $N(x)$ is of degree 1 and $D(x)$ is of degree 2 and write

$$\frac{N(x)}{D(x)} = \frac{px + q}{ax^2 + bx + c}.$$

Then we can write

$$I = \int \frac{px + q}{ax^2 + bx + c} \, dx = \int \frac{(p/2a)(2ax + b) + [q - (pb/2a)]}{ax^2 + bx + c} \, dx,$$

from which we find

$$I = \frac{p}{2a} \int \frac{2ax + b}{ax^2 + bx + c} \, dx + \left(\frac{2aq - pb}{2a}\right) \int \frac{dx}{ax^2 + bx + c}.$$

The second antiderivative is the one discussed in (b) above, and by setting $u = ax^2 + bx + c$, the first antiderivative reduces to

$$\int \frac{2ax + b}{ax^2 + bx + c} \, dx = \int \frac{du}{u} = \log |u| + C = \log |ax^2 + bx + c| + C.$$

Combining this result with that of Case (b) then leads to the desired anti-derivative I.

Case (d)   Suppose $N(x)$ is of degree 1 and $D(x)$ is a quadratic raised to the power $n > 1$ and write

$$\frac{N(x)}{D(x)} = \frac{px + q}{(ax^2 + bx + c)^n}.$$

Then, using the identity

$$px + q = \left(\frac{p}{2a}\right)(2ax + b) + \left(q - \frac{pb}{2a}\right)$$

enables us to write

$$\int \frac{px + q}{(ax^2 + bx + c)^n} \, dx = \left(\frac{p}{2a}\right) \int \frac{2ax + b}{(ax^2 + bx + c)^n} \, dx$$

$$+ \left(q - \frac{pb}{2a}\right) \int \frac{dx}{(ax^2 + bx + c)^n}.$$

Setting $u = ax^2 + bx + c$ in the first antiderivative on the right-hand side then leads to

$$\int \frac{2ax + b}{(ax^2 + bx + c)^n}\, dx = \int \frac{du}{u^n} = \left(\frac{-1}{n - 1}\right)\frac{1}{u^{n-1}} + C$$

$$= \left(\frac{-1}{n - 1}\right)\frac{1}{(ax^2 + bx + c)^{n-1}} + C.$$

The second antiderivative on the right-hand side must be evaluated by means of a reduction formula.

In the case $n = 1$ we have the obvious result

$$\int \frac{2ax + b}{ax^2 + bx + c}\, dx = \log | ax^2 + bx + c | + C.$$

Having considered a number of special cases we must now examine how we should proceed when $D(x)$ is any polynomial with real coefficients, and the degree of the polynomial $N(x)$ is less than that of $D(x)$. The coefficient a_0 of the highest power of x in $D(x)$ will be assumed to be unity, since if this is not the case it can always be made so by division of $N(x)$ and $D(x)$ by a_0. Now we know from Corollary 4·1 (b) that $D(x)$ may be factorized into real factors of the form

$$D(x) = (x - a)^k(x - b)^l \ldots (x^2 + px + q)^m, \tag{8·16}$$

where $x = a, b, \ldots$, are real roots with multiplicities $k, l, \ldots$, and $(x^2 + px + q)^m$ represents an m-fold repeated pair of complex conjugate roots.

Then from elementary algebraic considerations it may be shown that when the degree of $N(x)$ is less than that of $D(x)$ we may *always* set

$$\frac{N(x)}{D(x)} = \frac{A_1}{(x - a)} + \frac{A_2}{(x - a)^2} + \cdots + \frac{A_k}{(x - a)^k} + \frac{B_1}{(x - b)} +$$

$$\frac{B_2}{(x - b)^2} + \cdots + \frac{B_l}{(x - b)^l} + \cdots + \frac{P_1x + Q_1}{(x^2 + px + q)} +$$

$$\frac{P_2x + Q_2}{(x^2 + px + q)^2} + \cdots + \frac{P_mx + Q_m}{(x^2 + px + q)^m}. \tag{8·17}$$

That is to say, *every* rational fraction may be expressed as a sum of simple fractions of the types whose antiderivatives were obtained in Cases (a) to (d).

The expression on the right-hand side of Eqn (8·17) is called a *partial fraction* expansion of the rational fraction $N(x)/D(x)$ and the coefficients $A_1, A_2, \ldots, P_m, Q_m$ are called *undetermined coefficients*. The undetermined coefficients may be found by cross-multiplication of this expression, followed by equating the coefficients of equal powers of x. Antiderivatives of rational fractions $N(x)/D(x)$ may thus be found by a combination of the method of partial fractions and the results of Cases (a) to (d).

If the degree of $N(x)$ exceeds that of $D(x)$ by n, then the situation may be reduced to the one just described by simply adding to the partial fraction expansion (8·17) the extra terms

$$R_0 + R_1 x + R_2 x^2 + \cdots + R_n x^n.$$

This result can also be achieved by first dividing $N(x)$ by $D(x)$. The circumstances usually dictate which approach is the easier.

Example 8·13   Evaluate

$$I = \int \left(\frac{x^3 + 5x^2 + 9x + 5}{x^2 + 3x + 1} \right) dx.$$

Solution   Here, as the degree of $N(x)$ only exceeds that of $D(x)$ by one, we shall start by dividing the integrand to get

$$\frac{x^3 + 5x^2 + 9x + 5}{x^2 + 3x + 1} = x + 2 + \frac{2x + 3}{x^2 + 3x + 1},$$

when

$$I = \int (x + 2)dx + \int \frac{2x + 3}{x^2 + 3x + 1} dx.$$

The first antiderivative is trivial, whilst the second is of the form discussed in Case (c), so that

$$I = \frac{x^2}{2} + 2x + \log | x^2 + 3x + 1 | + C.$$

Example 8·14   Evaluate

$$I = \int \frac{x \, dx}{(x + 2)^2(x - 1)}.$$

Solution   In this case we must adopt the partial fraction expansion

$$\frac{x}{(x + 2)^2(x - 1)} = \frac{A}{x + 2} + \frac{B}{(x + 2)^2} + \frac{C}{x - 1}.$$

Cross-multiplication gives

$$x = A(x + 2)(x - 1) + B(x - 1) + C(x + 2)^2$$

or

$$x = A(x^2 + x - 2) + B(x - 1) + C(x^2 + 4x + 4).$$

Equating coefficients of equal powers of x gives:

Coefficient of x^2: $0 = A + C$

Coefficient of x: $1 = A + B + 4C$

Coefficient of x^0: $0 = -2A - B + 4C,$

showing that $A = -1/9$, $B = 2/3$, and $C = 1/9$. We may thus write

$$\int \frac{x \, dx}{(x + 2)^2(x - 1)} = -\frac{1}{9} \int \frac{dx}{x + 2} + \frac{2}{3} \int \frac{dx}{(x + 2)^2} + \frac{1}{9} \int \frac{dx}{x - 1}.$$

These antiderivatives were all discussed in Case (a), so that using those results we obtain

$$I = -\frac{1}{9} \log |x + 2| - \frac{2}{3} \frac{1}{(x + 2)} + \frac{1}{9} \log |x - 1| + C.$$

Example 8·15   Find the antiderivative

$$I = \int \frac{x^4 - x^3 + 5x^2 + x + 3}{(x + 1)(x^2 - x + 1)^2} \, dx.$$

Solution   Here $N(x) = x^4 - x^3 + 5x^2 + x + 3$ and $D(x) = (x + 1)(x^2 - x + 1)^2$, so that the degree of $N(x)$ is 4 and the degree of $D(x)$ is 5. Following on from our earlier reasoning we must set

$$\frac{x^4 - x^3 + 5x^2 + x + 3}{(x + 1)(x^2 - x + 1)^2} = \frac{A}{x + 1} + \frac{Bx + C}{x^2 - x + 1} + \frac{Dx + E}{(x^2 - x + 1)^2}.$$

Cross-multiplication gives the identity

$$x^4 - x^3 + 5x^2 + x + 3 = A(x^2 - x + 1)^2$$
$$+ (Bx + C)(x + 1)(x^2 - x + 1) + (Dx + E)(x + 1).$$

Instead of expanding the right-hand side and then equating coefficients of equal powers of x as in the previous example, we shall use the fact that $(x + 1)$ is a factor of $D(x)$ to simplify this expression. Setting $x = -1$ we find that $9 = 9A$, or $A = 1$ and so

$$x^4 - x^3 + 5x^2 + x + 3 = (x^2 - x + 1)^2 + (Bx + C)(x^3 + 1)$$
$$+ (Dx + E)(x + 1),$$

whence

$$x^3 + 2x^2 + 3x + 2 = (Bx + C)(x^3 + 1) + (Dx + E)(x + 1).$$

Having eliminated A we now proceed as before and equate coefficients of equal powers of x to find B, C, D, and E:

Coefficient of x^4: $0 = B$

Coefficient of x^3: $1 = C$

Coefficient of x^2: $2 = D$

Coefficient of x: $3 = B + E + D$

Coefficient of x^0: $2 = C + E$.

Thus, $B = 0$, $C = 1$, $D = 2$, $E = 1$ and so

$$I = \int \frac{dx}{x + 1} + \int \frac{dx}{x^2 - x + 1} + \int \frac{2x + 1}{(x^2 - x + 1)^2} \, dx = I_1 + I_2 + I_3.$$

Now

$$I_1 = \int \frac{dx}{x + 1} = \log | x + 1 | + C_1$$

and

$$I_2 = \int \frac{dx}{(x - \frac{1}{2})^2 + (\sqrt{3}/2)^2} = \frac{2}{\sqrt{3}} \arctan \left(\frac{2x - 1}{\sqrt{3}} \right) + C_2.$$

To evaluate I_3 write

$$I_3 = \int \frac{2x - 1}{(x^2 - x + 1)^2} \, dx + \int \frac{2 \, dx}{(x^2 - x + 1)^2}$$

$$= \frac{-1}{(x^2 - x + 1)} + \int \frac{2 \, dx}{[(x - \frac{1}{2})^2 + (\sqrt{3}/2)^2]^2}.$$

Next, setting $x - \frac{1}{2} = (\sqrt{3}/2) \tan \theta$, so that $dx = (\sqrt{3}/2) \sec^2 \theta \, d\theta$, gives

$$\int \frac{2 \, dx}{[(x - \frac{1}{2})^2 + (\sqrt{3}/2)^2]^2} = \int \frac{\sqrt{3} \sec^2 \theta \, d\theta}{(\frac{3}{4} \sec^2 \theta)^2} = \frac{16\sqrt{3}}{9} \int \cos^2 \theta \, d\theta.$$

Using the identity $\cos^2 \theta = \frac{1}{2}(1 + \cos 2\theta)$ this may be evaluated to give

$$\int \frac{2 \, dx}{[(x - \frac{1}{2})^2 + (\sqrt{3}/2)^2]^2} = \frac{8\sqrt{3}}{9} [\theta + \frac{1}{2} \sin 2\theta] + C_3$$

$$= \frac{8\sqrt{3}}{9} \left\{ \arctan \left(\frac{2x - 1}{\sqrt{3}} \right) + \frac{\sqrt{3}}{4} \cdot \frac{2x - 1}{(x^2 - x + 1)} \right\} + C_3.$$

Hence we have shown that

$$I_3 = \frac{-1}{(x^2 - x + 1)} + \frac{8\sqrt{3}}{9} \left\{ \arctan \left(\frac{2x - 1}{\sqrt{3}} \right) + \frac{\sqrt{3}}{4} \cdot \frac{2x - 1}{(x^2 - x + 1)} \right\} + C_3.$$

Adding I_1, I_2, and I_3 to find I finally gives

$$I = \log | x + 1 | + \frac{14\sqrt{3}}{9} \arctan \frac{2x - 1}{\sqrt{3}} + \frac{4x - 5}{3(x^2 - x + 1)} + C.$$

A factor $(x^2 - x + 1)^3$ in the denominator would have led to $\int \cos^4 \theta \, d\theta$ and so, in general, we would obtain antiderivatives of the form $\int \cos^{2n} \theta \, d\theta$.

8·6   Other special techniques of integration

A great variety of different methods exist for evaluating particular types of antiderivative, and in this final section we illustrate only a few specially useful ones with the help of some examples. Extensive tables of integrals are readily available and, where possible, should be used to minimise tedious manipulation.

8·6 (a)   Substitution $t = \tan x/2$

If we write $t = \tan x/2$ it is easily proved by means of trigonometric identities that

$$\sin x = \frac{2t}{1 + t^2} \quad \text{and} \quad \cos x = \frac{1 - t^2}{1 + t^2}. \tag{8·18}$$

Using these results we can also establish the differential relation

$$dx = \frac{2dt}{1 + t^2}. \tag{8·19}$$

Consequently, in principle, any rational fraction $R(\sin x, \cos x)$ that involves only the sine and cosine functions may be transformed by means of (8·18) into a rational fraction involving t. On account of this result and (8·19), it then follows that

$$I = \int R(\sin x, \cos x)dx = \int R\left[\frac{2t}{1 + t^2}, \frac{1 - t^2}{1 + t^2}\right]\frac{2\,dt}{1 + t^2}.$$

Thus I has been transformed into an antiderivative of a rational function involving t.

Example 8·16   Evaluate

$$I = \int \frac{\cos x\,dx}{1 + \sin x}.$$

Solution   Transforming to the variable t as indicated above gives

$$I = \int \frac{2(1 - t)}{(1 + t^2)(1 + t)}\,dt.$$

It is readily established that

$$\frac{2(1 - t)}{1 + t^2(1 + t)} = \frac{2}{1 + t} - \frac{2t}{1 + t^2},$$

showing that

$$I = \int \frac{2\,dt}{1 + t} - \int \frac{2t}{1 + t^2}\,dt$$

$$= \log(1 + t)^2 - \log(1 + t^2) + C.$$

Thus

$$I = \log\left[\frac{(1 + t)^2}{1 + t^2}\right] + C = \log\left[1 + \frac{2t}{1 + t^2}\right] + C$$

whence from (8·18),

$$I = \log(1 + \sin x) + C.$$

8·6 (b) · Integration of $R[x, \sqrt{(ax^2 + bx + c)}]$

We define $R[x, \sqrt{(ax^2 + bx + c)}]$ to be a rational fraction involving x and $\sqrt{(ax^2 + bx + c)}$. Special cases of this general type in which $b = 0$ have been encountered in Examples 8·2 and 8·5 where it was shown that the substitutions $x = \sin u$ and $x = \sinh u$ can be used to reduce the integrand to one involving only trigonometric or hyperbolic functions. If it is of trigonometric type then the technique of (a) above may be used to reduce the integrand further to a rational function. If the integrand is of hyperbolic type then the substitution

$$t = \tanh x/2,$$

together with

$$\sinh x = \frac{2t}{1 - t^2} \quad \text{and} \quad \cosh x = \frac{1 + t^2}{1 - t^2} \tag{8·20}$$

and the differential relation

$$dx = \frac{2\,dt}{1 - t^2}, \tag{8·21}$$

will again reduce the integrand to a rational function.

If $b \neq 0$, then completing the square under the square root sign gives

$$\sqrt{(ax^2 + bx + c)} = \sqrt{a\left[\left(x + \frac{b}{2a}\right)^2 + \left(\frac{c}{a} - \frac{b^2}{4a^2}\right)\right]}.$$

The substitution $u = x + (b/2a)$ will then reduce the problem to one of the two special cases just discussed, according to the signs of a and $[(c/a) - (b^2/4a^2)]$.

Example 8·17   Evaluate

$$I = \int \frac{dx}{\sqrt{(2 - 3x - 4x^2)}}.$$

Solution   First we complete the square under the square root sign to obtain

$$I = \int \frac{dx}{\sqrt{\{4[41/64 - (x + 3/8)^2]\}}}.$$

Then, setting $u = x + \frac{3}{8}$ this becomes

$$I = \frac{1}{2} \int \frac{du}{\sqrt{[(41/64)^2 - u^2]}} = \frac{1}{2} \arcsin \frac{8u}{\sqrt{41}} + C$$

and thus

$$I = \frac{1}{2} \arcsin \left(\frac{8x + 3}{\sqrt{41}} \right) + C.$$

8·6 (c) Integration of trigonometric functions involving multiple angles

Antiderivatives of products of trigonometric functions involving multiple angles are of considerable importance and the most frequently occurring ones are:

$$I_1 = \int \sin mx \cos nx \, dx, \tag{8·22}$$

$$I_2 = \int \sin mx \sin nx \, dx, \tag{8·23}$$

$$I_3 = \int \cos mx \cos nx \, dx. \tag{8·24}$$

These are easily evaluated by appeal to the trigonometric identities:

$$\sin mx \cos nx = \frac{1}{2}[\sin (m + n)x + \sin (m - n)x], \tag{8·25}$$

$$\sin mx \sin nx = \frac{1}{2}[\cos (m - n)x - \cos (m + n)x], \tag{8·26}$$

$$\cos mx \cos nx = \frac{1}{2}[\cos (m + n)x + \cos (m - n)x]. \tag{8·27}$$

Substitution of these identities into the above antiderivatives produces:

$$I_1 = \begin{cases} -\dfrac{1}{2} \left[\dfrac{\cos (m - n)x}{(m - n)} + \dfrac{\cos (m + n)x}{(m + n)} \right] + C \quad \text{for} \quad m^2 \neq n^2 \\[4mm] -\dfrac{1}{4m} \cos 2mx + C \quad \text{for} \quad m = n, \end{cases} \tag{8·28}$$

$$I_2 = \begin{cases} \dfrac{1}{2} \left[\dfrac{\sin (m - n)x}{(m - n)} - \dfrac{\sin (m + n)x}{(m + n)} \right] + C \quad \text{for} \quad m^2 \neq n^2 \\[4mm] \dfrac{1}{2m} (mx - \sin mx \cos mx) + C \quad \text{for} \quad m = n, \end{cases} \tag{8·29}$$

$$I_3 = \begin{cases} \dfrac{1}{2} \left[\dfrac{\sin (m - n)x}{(m - n)} + \dfrac{\sin (m + n)x}{m + n} \right] + C \quad \text{for} \quad m^2 \neq n^2 \\[4mm] \dfrac{1}{2m} (mx + \sin mx \cos mx) + C \quad \text{for} \quad m = n. \end{cases} \tag{8·30}$$

Example 8·18 Evaluate the following two antiderivatives:

$$I_1 = \int \sin 3x \cos 5x \, dx, \qquad I_2 = \int \sin^2 3x \, dx.$$

Solution The antiderivatives follow immediately by substitution in (8·28) and (8·29):

$$I_1 = \frac{\cos 2x}{4} - \frac{\cos 8x}{16} + C, \qquad I_2 = \frac{x}{2} - \frac{\sin 3x \cos 3x}{6} + C.$$

PROBLEMS

Section 8·1

8·1 Find the following antiderivatives:

(a) $\int \dfrac{3 \, dx}{4x^2 - 16}$; (b) $\int \sin 3x \, dx$; (c) $\int \dfrac{dx}{9 - x^2}$;

(d) $\int \dfrac{dx}{4 + x^2}$; (e) $\int \tfrac{1}{3} \cos 4x \, dx$; (f) $\int 3^x \, dx$.

8·2 Verify by means of differentiation that

$$\int \frac{dx}{\sqrt{(x^2 - a^2)}} = \log | x + \sqrt{(x^2 - a^2)} | + C.$$

Compare this form of result with that shown against entry 10 of Table 8.1.

8·3 Verify by means of differentiation that

$$\int \frac{dx}{a^2 - b^2 x^2} = \frac{1}{2ab} \log \left| \frac{a + bx}{a - bx} \right| + C.$$

Compare this more general result with those shown against entries 11 and 12 of Table 8.1.

8·4 Verify by means of differentiation that

$$\int \frac{dx}{\sqrt{(a^2 + x^2)}} = \log [x + \sqrt{(a^2 + x^2)}] + C.$$

Compare this form of result with that shown against entry 9 of Table 8·1.

8·5 Evaluate the following antiderivatives:

(a) $\int (x^2 + 3 \sin x + 1) dx$;   (b) $\int (4^x + 2 \cos 2x) dx$;

(c) $\int (4 \sinh x + \sin x) dx$;   (d) $\int (e^{ax} + 3) dx$.

8·6 Use the following identities to evaluate the four antiderivatives listed below:

$$\sinh mx \cosh mx = \tfrac{1}{2}[\sinh (m + n)x + \sinh (m - n)x]$$
$$\sinh mx \sinh nx = \tfrac{1}{2}[\cosh (m + n)x - \cosh (m - n)x]$$
$$\cosh mx \cosh nx = \tfrac{1}{2}[\cosh (m + n)x + \cosh (m - n)x]$$

(a) $\int \sinh 4x \cosh 2x \, dx$;   (b) $\int \sinh x \sinh 3x \, dx$;

(c) $\int \cosh 4x \cosh 2x \, dx$;   (d) $\int \cosh^2 2x \, dx$.

Section 8·2

Use the indicated substitutions to evaluate the following antiderivatives.

8·7 $\displaystyle\int \frac{dx}{x\sqrt{(x^2 - 4)}}, \quad x = 1/u.$

8·8 $\int \sqrt{(1 - x^2)}dx, \quad x = \sin u.$

8·9 $\displaystyle\int \frac{\tanh x\,dx}{2\sqrt{(\cosh x - 1)}}, \quad \cosh x = 1 + u^2.$

8·10 $\int \cos x\sqrt{\sin x}\,dx, \quad \sin x = u.$

8·11 $\int x(3x^2 + 1)^5\,dx, \quad 3x^2 + 1 = u.$

Evaluate the following antiderivatives by means of a suitable trigonometric substitution.

8·12 $\displaystyle\int \frac{\sqrt{(x^2 + 1)}}{x}\,dx.$

8·13 $\displaystyle\int \frac{\sqrt{(x^2 - 1)}}{x}\,dx.$

Evaluate the following definite integrals.

8·14 $\displaystyle\int_0^1 (3x + 1)\sinh(x^3 + x + 3)dx.$

8·15 $\displaystyle\int_0^1 x^5\sqrt{(1 + x^2)}dx.$

8·16 $\displaystyle\int_2^6 \sqrt{(x - 2)}dx.$

8·17 $\displaystyle\int_1^2 \left(\frac{4x + 6}{x^2 + 3x + 1}\right)dx.$

Section 8·3

Evaluate the following antiderivatives using the technique of integration by parts.

8·18 $\int e^{ax}\sin x\,dx.$

8·19 $\int xe^{ax}\,dx.$

8·20 $\int 7^x\cos x\,dx.$

8·21 $\int \log^2 x\,dx.$

8·22 $\int x\arcsin x\,dx.$

Section 8·4

8·23 Given that $I_n = \int (1 - x^3)^n\,dx$, where n is an integer, show that

$$(3n + 1)I_n = x(1 - x^3)^n + 3n\,I_{n-1}.$$

Hence prove that

$$\int_0^1 (1 - x^3)^5 = 3^6/2^4 . 7 . 13.$$

8·24 The integral I_m is defined by

$$I_m = \int_0^\infty \frac{x^{2m-1}}{(x^2 + 1)^{m+3}} \, dx \text{ for integral } m \geq 0.$$

Show that

$$I_{m-1} = \frac{m + 2}{m - 1} I_m,$$

and by using the substitution $x = \tan \theta$ prove that

$$\int_0^{\frac{1}{2}\pi} \sin^7 \theta \cos^5 \theta \, d\theta = \frac{1}{120}.$$

8·25 If

$$T_n = \int \tan^n \theta \, d\theta,$$

where $n \neq 1$ is a positive integer, show that

$$T_n = \frac{\tan^{n-1} \theta}{n - 1} - T_{n-2}.$$

Use this result to evaluate

$$\int_0^{\frac{1}{4}\pi} \tan^6 \theta \, d\theta.$$

8·26 The function $I_{m,n}$ is defined by

$$I_{m,n} = \int x^m(a + bx)^n \, dx,$$

in which m,n are positive integers. Prove that

$$b(m + n + 1)I_{m,n} + ma \, I_{m-1,n} = x^m(a + bx)^{n+1}.$$

Section 8·5

Evaluate the following antiderivatives by means of partial fractions.

8·27 $\int \dfrac{x^2 - 5x + 9}{x^2 - 5x + 6} \, dx.$

8·28 $\int \dfrac{dx}{x^3 - 2x^2 + x}.$

8·29 $\int \dfrac{x^2 - 8x + 7}{(x^2 - 3x - 10)^2} \, dx.$

8·30 $\int \dfrac{x^2 + 2}{(x + 1)^3(x - 2)} \, dx.$

Section 8·6

Evaluate the following antiderivatives by means of the substitution $t = \tan x/2$.

8·31 $\displaystyle \int \frac{dx}{8 - 4\sin x + 7\cos x}$.

8·32 $\displaystyle \int \frac{\sin x}{(1 - \cos x)^3}\,dx$.

Evaluate the following antiderivatives by means of one or more suitable substitutions.

8·33 $\displaystyle \int \frac{dx}{\sqrt{(2 + 3x - 2x^2)}}$.

8·34 $\displaystyle \int \frac{3x - 6}{\sqrt{(x^2 - 4x + 5)}}\,dx$.

8·35 $\displaystyle \int \frac{dx}{x\sqrt{(1 - x^2)}}$.

Evaluate the following trigonometric antiderivatives.

8·36 $\int \sin ax \sin (ax + \varepsilon)dx$, a, ε non-zero constants.

8·37 $\int \cos x \cos^2 3x\,dx$.

8·38 $\int \sin x \sin 2x \sin 3x\,dx$.

Use the results of this chapter together with Definitions 7·3 and 7·4 of Chapter 7 to classify the following improper integrals as convergent or divergent. Determine the value of all improper integrals that are convergent stating any conditions that must be imposed to ensure this.

8·39 $\displaystyle \int_0^1 \frac{dx}{x^\lambda}$.

8·40 $\displaystyle \int_1^\infty \frac{dx}{1 + x^2}$.

8·41 $\displaystyle \int_0^\infty \frac{dx}{(1 + x)\sqrt{x}}$.

8·42 $\displaystyle \int_0^\infty \cos x\,dx$.

9 Matrices

9·1 Introductory ideas

This chapter is concerned with the branch of mathematics known as *linear algebra*. One aspect of this subject has already been encountered, namely vectors, and it is now necessary to develop in a more general context various of the ideas that were first introduced there. Central to the entire subject is the fundamental idea that the algebraic operations of addition, subtraction, and multiplication can be made meaningful when applied to an *array* of numbers or functions considered as a single entity.

An example will help here to indicate one of the many different ways in which such an array may arise, and at the same time to show something of the type of algebra it is reasonable to want to perform on an array. Three chemical plants numbered 1 to 3 each have separate sources of raw material from which each one produces the same four products numbered 1 to 4. Let plant number m produce product number n at a cost a_{mn} units per ton, then the production costs of the complex of chemical plants is conveniently characterized by the following table of the twelve quantities a_{mn}.

Table 9·1

		Product			
		1	2	3	4
Plant	1	a_{11}	a_{12}	a_{13}	a_{14}
	2	a_{21}	a_{22}	a_{23}	a_{24}
	3	a_{31}	a_{32}	a_{33}	a_{34}

In writing this table or array of quantities a_{mn} we have used the convention that the first of the two suffixes attached to the quantity a_{mn} refers to the row number in which a_{mn} appears, and the second to the column number. Thus the entry a_{23} occurs in row 2, column 3, whilst the entry a_{32} occurs in row 3, column 2. The important use of suffixes in this way is strictly analogous to a map reference in which the first entry is a latitude and the second a longitude. Thus the double suffix notation used here serves to identify the position in the array to which the associated quantity is assigned.

On account of the use to which the suffixes have been put, we can now dispense with the extreme left-hand column and the top row of Table 9·1, which only serve for identification purposes, and write instead

$$A = \begin{bmatrix} a_{11} & a_{12} & a_{13} & a_{14} \\ a_{21} & a_{22} & a_{23} & a_{24} \\ a_{31} & a_{32} & a_{33} & a_{34} \end{bmatrix},$$ (9·1)

with the understanding that the symbol **A** represents the array of quantities originally contained in Table 9·1.

Returning now to the physical situation from which the array (9·1) was derived, let us suppose that at some time the quality of the raw materials changes, so that a revised Table 9·1 then applies in which entry a_{mn} is replaced by the new entry b_{mn}. Then, in terms of our concise notation, we can characterize this new situation by defining an array **B** as follows:

$$B = \begin{bmatrix} b_{11} & b_{12} & b_{13} & b_{14} \\ b_{21} & b_{22} & b_{23} & b_{24} \\ b_{31} & b_{32} & b_{33} & b_{34} \end{bmatrix}.$$ (9·2)

In terms of the information at our disposal, we know that the change in the cost of product n from chemical plant m is $a_{mn} - b_{mn}$, whilst the average cost of product n from plant m is $\frac{1}{2}(a_{mn} + b_{mn})$. Hence, if **C** is the array of change in costs of products and **D** is the array of the average costs of products, in our new notation we may write:

$$C = \begin{bmatrix} a_{11} - b_{11} & a_{12} - b_{12} & a_{13} - b_{13} & a_{14} - b_{14} \\ a_{21} - b_{21} & a_{22} - b_{22} & a_{23} - b_{23} & a_{24} - b_{24} \\ a_{31} - b_{31} & a_{32} - b_{32} & a_{33} - b_{33} & a_{34} - b_{34} \end{bmatrix}$$ (9·3)

and

$$D = \begin{bmatrix} \frac{1}{2}(a_{11} + b_{11}) & \frac{1}{2}(a_{12} + b_{12}) & \frac{1}{2}(a_{13} + b_{13}) & \frac{1}{2}(a_{14} + b_{14}) \\ \frac{1}{2}(a_{21} + b_{21}) & \frac{1}{2}(a_{22} + b_{22}) & \frac{1}{2}(a_{23} + b_{23}) & \frac{1}{2}(a_{24} + b_{24}) \\ \frac{1}{2}(a_{31} + b_{31}) & \frac{1}{2}(a_{32} + b_{32}) & \frac{1}{2}(a_{33} + b_{33}) & \frac{1}{2}(a_{34} + b_{34}) \end{bmatrix}.$$ (9·4)

The form of these results is suggestive, for it would seem that by defining subtraction of two similar arrays to mean the array formed by the subtraction of corresponding elements, we may write

$$C = A - B.$$ (9·5)

Similarly, if addition of two similar arrays is taken to mean the array formed by the addition of corresponding entries, and the multiplication of an array by a factor is taken to mean the array formed by the multiplication of each entry by that factor, we may write

$$D = \frac{1}{2}(A + B).$$ (9·6)

Hence, in a natural manner, we are starting to perform what appears to be conventional algebraic operations on an entire array of numbers, rather than on the individual entries in the arrays themselves. In mathematical terms an array of the form shown on the right-hand side of Eqn (9·1) is called a *matrix* of *order* (3 × 4). Here, analogous to the double suffix notation already introduced, the first number is taken to refer to the total number of rows in the matrix and the second number to refer to the total number of columns in the matrix.

In terms of the simple physical situation used to introduce the notion of a matrix and its associated algebra we have so far given no indication of the interpretation to be placed upon multiplication. To elucidate the form taken by this operation when applied to matrices, we again return to our physical situation and consider the cost of buying c_1, c_2, c_3, and c_4 tons, respectively, of products 1, 2, 3, and 4 from each of the three chemical plants in turn. If the product costs are as shown in Table 9·1, and the costs of the orders are denoted by d_1, d_2, and d_3, it is readily seen that

$$d_1 = a_{11}c_1 + a_{12}c_2 + a_{13}c_3 + a_{14}c_4$$
$$d_2 = a_{21}c_1 + a_{22}c_2 + a_{23}c_3 + a_{24}c_4 \qquad (9·7)$$
$$d_3 = a_{31}c_1 + a_{32}c_2 + a_{33}c_3 + a_{34}c_4.$$

In terms of the matrix $\mathbf{A}$ in Eqn (9·1), the right-hand side of the first equation in (9·7) is obtained by multiplying successive entries in the first row of $\mathbf{A}$ by c_1, c_2, c_3, and c_4, respectively, and then adding the four products. The same process will generate the right-hand side of both the second and third equation in (9·7), provided that the entries in the second and third rows of matrix $\mathbf{A}$ are used in place of those in the first row. If the four numbers c_1, c_2, c_3, and c_4 are arranged in a column which is then regarded as a (4 × 1) matrix, the basic operation of matrix multiplication is seen to be the multiplication of a row of the first matrix into the column of the second to yield a single number. Thus, in terms of the first row of $\mathbf{A}$ expressed as a (1 × 4) matrix, we have the definition

$$a_{11}c_1 + a_{12}c_2 + a_{13}c_3 + a_{14}c_4 = [a_{11} \ a_{12} \ a_{13} \ a_{14}] \begin{bmatrix} c_1 \\ c_2 \\ c_3 \\ c_4 \end{bmatrix},$$

where juxtaposition is used to imply multiplication of the row and column matrices on the right-hand side.

Similarly, in terms of the second row of $\mathbf{A}$ expressed as a (1 × 4) matrix, our definition yields

$$a_{21}c_1 + a_{22}c_2 + a_{23}c_3 + a_{24}c_4 = [a_{21} \ a_{22} \ a_{23} \ a_{24}] \begin{bmatrix} c_1 \\ c_2 \\ c_3 \\ c_4 \end{bmatrix},$$

and a corresponding result is also true for the third row of **A** when expressed as a (1 × 4) matrix. This special form of product is called either the *inner* product or the *scalar* product of a row matrix and a column matrix.

Collectively these results suggest that we should write Eqns (9·7) in the matrix form

$$\begin{bmatrix} d_1 \\ d_2 \\ d_3 \end{bmatrix} = \begin{bmatrix} a_{11} & a_{12} & a_{13} & a_{14} \\ a_{21} & a_{22} & a_{23} & a_{24} \\ a_{31} & a_{32} & a_{33} & a_{34} \end{bmatrix} \begin{bmatrix} c_1 \\ c_2 \\ c_3 \\ c_4 \end{bmatrix}, \tag{9·8}$$

with the understanding, as before, that multiplication is implied by juxtaposition and means the inner product of rows of the first matrix with the column of the second matrix. To be consistent, equality of two matrices must then be taken to mean the equality of corresponding entries in two matrices of similar order. Using this convention our suffix notation works for us in the sense that the row number and the column number, taken in that order, which are involved in an inner product are the row and column numbers of the location into which that product is to be put. Thus in matrix equation (9·8), the number d_2 is in row 2, column 1 of the left-hand column matrix, and it is the result of forming the inner product of row 2 of the first matrix on the right-hand side with column 1 of the second matrix. (The second matrix here only has one column.)

If the column matrix with entries d_1, d_2, d_3 is denoted by **D**, and the column matrix with entries c_1, c_2, c_3, and c_4 is denoted by **C**, then Eqn (9·8) can be reduced to the deceptively simple equation

D = AC. (9·9)

It should be noticed that the resemblance to the algebra of real numbers ends here, because although multiplication is a commutative operation for real numbers, it is an easy task for the reader to verify that the matrix product **CA** is not even defined for the matrices involved here. Later we shall see that the non-commutative character of matrix multiplication is not the only difference between the field of real numbers and matrices. The result of matrix multiplication using numbers is illustrated in the following example:

$$\begin{bmatrix} 1 & 2 & 1 & 0 \\ 0 & 1 & 1 & 3 \\ 1 & 2 & 1 & 4 \end{bmatrix} \begin{bmatrix} 2 \\ 1 \\ 0 \\ -1 \end{bmatrix} = \begin{bmatrix} 4 \\ -2 \\ 0 \end{bmatrix}.$$

We remark in passing that the name scalar product of a row matrix and a column matrix derives from a comparison with the scalar product of two vectors. Namely, if $\boldsymbol{\alpha} = \alpha_1\mathbf{i} + \alpha_2\mathbf{j} + \alpha_3\mathbf{k}$, $\boldsymbol{\beta} = \beta_1\mathbf{i} + \beta_2\mathbf{j} + \beta_3\mathbf{k}$ are two vectors, then $\boldsymbol{\alpha} \cdot \boldsymbol{\beta} = \alpha_1\beta_1 + \alpha_2\beta_2 + \alpha_3\beta_3$, which is just the result of forming the inner product of a row matrix with entries α_1, α_2, α_3 and a column matrix with entries β_1, β_2, β_3. Because of this similarity it is customary to refer to matrices comprising only one row or one column as *row vectors* or *column vectors*, respectively. Thus a general $(1 \times n)$ row vector may be considered as a matrix representation of an ordinary form of vector having n components, and which belongs to an n-dimensional space.

This simple idea proves to be very fruitful in more advanced accounts of linear algebra where it leads to the study of what are called n-dimensional vector spaces. These spaces have properties very similar to those discussed in Chapter 4 and, as in three dimensions, the scalar product is related to the geometrical operation of projection in the space. In an n-dimensional vector space a fundamental set of row or column vectors called a *basis* takes the place of the unit vectors $\mathbf{i}$, $\mathbf{j}$, and $\mathbf{k}$ and lead to the important idea of linear independence which will be examined later.

Because of the shape of the array, a general $(m \times n)$ matrix is called a *rectangular* matrix. The rule just devised for the product of a (3×4) matrix and a (4×1) column vector also applies to the product $\mathbf{AB}$ of two rectangular matrices $\mathbf{A}$ and $\mathbf{B}$, provided only that the number of columns in $\mathbf{A}$ is equal to the number of rows of $\mathbf{B}$. This last requirement follows directly from the concept of an inner product which is only defined when the number of entries in a row of $\mathbf{A}$ is equal to the number of entries in a column of $\mathbf{B}$. Once again the suffix notation works for us, because the inner product of row p of matrix $\mathbf{A}$ and column q of matrix $\mathbf{B}$ is the number c_{pq}, which is found in row p and column q of the product matrix $\mathbf{C} = \mathbf{AB}$. Consider the following example which illustrates the application of this rule:

$$\begin{bmatrix} 1 & 2 & 1 & 0 \\ 0 & 1 & 1 & 3 \\ 1 & 2 & 1 & 4 \end{bmatrix} \begin{bmatrix} 2 & 1 \\ 1 & 2 \\ 0 & 2 \\ -1 & 1 \end{bmatrix} = \begin{bmatrix} 4 & 7 \\ -2 & 7 \\ 0 & 11 \end{bmatrix}.$$

Then, for example, the entry in row 3, column 2 of the product matrix is the number 11, which is the inner product of row 3 of the first matrix involved in the product and column 2 of the second matrix involved in the product.

Notice that the rule for forming an inner product also determines the shape of the product matrix $C = AB$, for C must have as many rows as A and as many columns as B. (Think about this and check it.) In fact these arguments may be formulated into a useful short-hand rule for checking that two matrices are conformable for multiplication, and at the same time displaying the shape of the product matrix.

Rule 1   (Multiplication conformability rule)

If A is an $(m \times n)$ matrix and B is a $(p \times q)$ matrix, then the matrix product AB may be formed provided $n = p$. The resultant product matrix then has the form $(m \times q)$. Symbolically we write this

$$(m \times n)(p \times q) = (m \times q) \quad \text{only if} \quad n = p.$$

Thus matrix products of the form $(3 \times 7)(7 \times 2)$ are conformable for multiplication and yield a (3×2) matrix. Matrix products of the form $(7 \times 3)(5 \times 4)$ are *not* defined and certainly do *not* yield a (7×4) matrix.

This rule has various important implications, and at this stage in our argument we would draw attention to the fact that even when for two matrices A and B, both the matrix products AB and BA are defined, they are not usually equal. Indeed, the order of the two product matrices may be different, as the following example shows. If

$$A = \begin{bmatrix} 1 & 2 \\ 0 & -1 \\ 4 & 1 \end{bmatrix}, \quad B = \begin{bmatrix} 1 & 2 & 1 \\ -1 & 1 & 0 \end{bmatrix}$$

then

$$AB = \begin{bmatrix} -1 & 4 & 1 \\ 1 & -1 & 0 \\ 3 & 9 & 4 \end{bmatrix} \quad \text{and} \quad BA = \begin{bmatrix} 5 & 1 \\ -1 & -3 \end{bmatrix}.$$

A different but most important way in which matrices can arise is in dealing with sets of simultaneous equations. Consider the following set of simultaneous equations:

$$x + y + 2z = 4$$
$$2x - y + 3z = 9$$
$$3x - y - z = 2.$$

These equations may be written in matrix form by introducing a column vector with entries x, y, z and then using the rule of matrix multiplication to write

$$\begin{bmatrix} 1 & 1 & 2 \\ 2 & -1 & 3 \\ 3 & -1 & -1 \end{bmatrix} \begin{bmatrix} x \\ y \\ z \end{bmatrix} = \begin{bmatrix} 4 \\ 9 \\ 2 \end{bmatrix}.$$

With only a little practice, the reader will quickly learn to transcribe systems of equations into matrix form, for the patterns of numbers involved in the two numerical matrices are identical to the patterns of numbers in the equations themselves.

For obvious reasons the (3 × 3) matrix is called the *coefficient* matrix of the simultaneous equations. As in this case there are three equations and three unknowns, the coefficient matrix is square in shape. In general the name *square matrix* will be given to any (n × n) matrix. If the coefficient matrix above is denoted by **A**, and the column vectors with entries x, y, z and 4, 9, 2 are denoted, respectively, by **X** and **K**, we arrive at the matrix equation

AX = K.

There is a great temptation to attempt to solve this for **X** by dividing by **A**, but as it is meaningless to divide two arrays of numbers this approach must be abandoned. Later we will return to this matter and resolve the difficulty by introducing the concept of the inverse of a square matrix via the operation of multiplication.

9·2   Matrix algebra

In this section we return to the fundamental ideas connected with matrices and their algebra which were outlined on an intuitive basis in Section 9·1. This time, however, our discussion will be more formal and, relying on our introductory account to provide motivation, we shall proceed quickly through the basic definitions and theorems, which will be illustrated by example. The problem of the solution of systems of linear equations and a discussion of linear transformations and some of their applications will be presented in subsequent sections.

DEFINITION 9·1 (matrix and its order)   A *matrix* is a rectangular array of *elements* a_{ij} involving m *rows* and n *columns*. The first suffix i in element a_{ij} is called the *row index* of the element and the second suffix j is called the *column index* of the element. These indices specify the row number and column number in which the element is located, with row 1 occurring at the top of the array and column 1 at the extreme left. A matrix with m rows and n columns is said to be of *order m* by n and this is written (m × n). The order describes the shape of the matrix.

Special names are given to certain types of matrix and we now describe and give examples of some of the more frequently used terms.

(a) A *row matrix* or *row vector* is any matrix of order $(1 \times n)$. The following is an example of a row vector of order (1×4):

[3 0 7 2].

(b) A *column matrix* or *column vector* is any matrix of order $(n \times 1)$. The following is an example of a column vector of order (3×1):

$$\begin{bmatrix} 11 \\ 2 \\ 5 \end{bmatrix}.$$

(c) A *square matrix* is any matrix of order $(n \times n)$. The following is an example of a square matrix of order (3×3):

$$\begin{bmatrix} 1 & 2 & 4 \\ 3 & 0 & 2 \\ 5 & 1 & 3 \end{bmatrix}.$$

Three particular cases of square matrices that are worthy of note are the diagonal matrix, the symmetric matrix and the skew-symmetric matrix. Of these, the *diagonal matrix* has non-zero elements only on what is called the *principal diagonal*, which runs from the top left of the matrix to the bottom right. The principal diagonal is also often referred to as the *leading diagonal*. The following is an example of a diagonal matrix of order (4×4):

$$\begin{bmatrix} 3 & 0 & 0 & 0 \\ 0 & 0 & 0 & 0 \\ 0 & 0 & 2 & 0 \\ 0 & 0 & 0 & 5 \end{bmatrix}.$$

The diagonal matrix in which every element of the principal diagonal is a unity is called either the *unit matrix* or the *identity matrix*, and it is usually denoted by **I**. The unit matrix of order (3×3) thus has the form

$$\mathbf{I} = \begin{bmatrix} 1 & 0 & 0 \\ 0 & 1 & 0 \\ 0 & 0 & 1 \end{bmatrix}.$$

A *symmetric matrix* is one in which the elements obey the rule $a_{ij} = a_{ji}$, so that the pattern of numbers has a reflection symmetry about the principal diagonal. A typical symmetric matrix of order (3×3) is:

$$\begin{bmatrix} 5 & 1 & 3 \\ 1 & 2 & -2 \\ 3 & -2 & 7 \end{bmatrix}.$$

A *skew-symmetric matrix* is one in which the elements obey the rule $a_{ij} = -a_{ji}$, so that the principal diagonal must contain zeros, whilst the pattern of numbers has a reflection symmetry about the principal diagonal but with a reversal of sign. A typical skew-symmetric matrix of order (3 × 3) is:

$$\begin{bmatrix} 0 & 1 & 5 \\ -1 & 0 & -3 \\ -5 & 3 & 0 \end{bmatrix}.$$

(d) A *null matrix* is the name given to a matrix of any order which contains only zero elements. It is usually denoted by the symbol **0**. The null matrix of order (2 × 3) has the form

$$\mathbf{0} = \begin{bmatrix} 0 & 0 & 0 \\ 0 & 0 & 0 \end{bmatrix}.$$

DEFINITION 9·2 (equality of matrices)   Two matrices **A** and **B** with general elements a_{ij} and b_{ij}, respectively, are equal only when they are both of the same order and $a_{ij} = b_{ij}$ for all possible pairs of indices (i, j).

Example 9·1   Is it possible for the following pair of matrices to be equal and, if so, for what value of a does equality occur:

$$\begin{bmatrix} 5 & a^3 \\ a^2 & 1 \end{bmatrix} \quad \text{and} \quad \begin{bmatrix} 5 & -27 \\ 9 & 1 \end{bmatrix}.$$

Solution   The matrices are both of the same order and hence they will be equal when their corresponding elements are equal. As corresponding elements on the principal diagonal are indeed equal, we need only confine attention to the off-diagonal elements. Thus the matrices will be equal if there is a common solution to the two equations $a^2 = 9$ and $a^3 = -27$. Obviously, equality will occur if $a = -3$.

DEFINITION 9·3 (addition of matrices)   Two matrices **A** and **B** with general elements a_{ij} and b_{ij}, respectively, will be said to be *conformable for addition* only if they are both of the same order. Their sum $\mathbf{C} = \mathbf{A} + \mathbf{B}$ is the matrix **C** with elements $c_{ij} = a_{ij} + b_{ij}$.

As addition of real numbers is commutative we have $a_{ij} + b_{ij} = b_{ij} + a_{ij}$. This shows that addition of conformable matrices must also be commutative, whence

$$\mathbf{A} + \mathbf{B} = \mathbf{B} + \mathbf{A}. \tag{9·10}$$

Now addition of real numbers is also associative so that $(a_{ij} + b_{ij}) +$

$c_{ij} = a_{ij} + (b_{ij} + c_{ij})$. Hence if a_{ij}, b_{ij}, and c_{ij} are general elements of matrices **A**, **B**, and **C** which are conformable for addition, then this also implies that addition of matrices is associative, whence

$$(A + B) + C = A + (B + C). \qquad (9·11)$$

Results (9·10) and (9·11) comprise our first theorem.

THEOREM 9·1 (matrix addition is both commutative and associative) If **A**, **B**, and **C** are matrices which are conformable for addition, then

(a) $A + B = B + A$ (Matrix Addition is Commutative);

(b) $(A + B) + C = A + (B + C)$ (Matrix Addition is Associative).

Example 9·2 Determine the constants a, b, c, and d in order that the following matrix equation should be valid:

$$\begin{bmatrix} 0 & a & 3 \\ b & 2 & 2 \end{bmatrix} + \begin{bmatrix} c & 1 & 2 \\ 1 & 1 & d \end{bmatrix} = \begin{bmatrix} 4 & 3 & 5 \\ 7 & 3 & 5 \end{bmatrix}.$$

Solution Adding the two matrices on the left-hand side we arrive at the matrix equation

$$\begin{bmatrix} c & (a + 1) & 5 \\ (b + 1) & 3 & (d + 2) \end{bmatrix} = \begin{bmatrix} 4 & 3 & 5 \\ 7 & 3 & 5 \end{bmatrix}.$$

Equating corresponding elements shows that $a = 2$, $b = 6$, $c = 4$, and $d = 3$.

DEFINITION 9·4 (multiplication by scalar) If k is a scalar and the matrix **A** has elements a_{ij}, then the matrix $B = kA$ is the same order as **A** and has elements ka_{ij}.

Example 9·3 Determine $2A + 5B$, given that:

$$A = \begin{bmatrix} 1 & 2 \\ 3 & 4 \end{bmatrix} \quad \text{and} \quad B = \begin{bmatrix} -1 & 3 \\ 4 & 2 \end{bmatrix}.$$

Solution $2A + 5B = 2\begin{bmatrix} 1 & 2 \\ 3 & 4 \end{bmatrix} + 5\begin{bmatrix} -1 & 3 \\ 4 & 2 \end{bmatrix}$

or,

$$2A + 5B = \begin{bmatrix} 2 & 4 \\ 6 & 8 \end{bmatrix} + \begin{bmatrix} -5 & 15 \\ 20 & 10 \end{bmatrix},$$

whence

$$2A + 5B = \begin{bmatrix} -3 & 19 \\ 26 & 18 \end{bmatrix}.$$

DEFINITION 9·5 (difference of two matrices)   If the matrices A and B are both of the same order, then their *difference* $A - B$ is defined by the relation

$$A - B = A + (-1)B.$$

Example 9·4   Determine $A - B$, given that:

$$A = \begin{bmatrix} 1 & 3 \\ 4 & -2 \\ 1 & 6 \end{bmatrix} \quad \text{and} \quad B = \begin{bmatrix} 4 & 2 \\ 3 & 1 \\ 0 & -2 \end{bmatrix}.$$

Solution

$$A - B = \begin{bmatrix} 1 & 3 \\ 4 & -2 \\ 1 & 6 \end{bmatrix} + (-1) \begin{bmatrix} 4 & 2 \\ 3 & 1 \\ 0 & -2 \end{bmatrix},$$

and so

$$A - B = \begin{bmatrix} 1 & 3 \\ 4 & -2 \\ 1 & 6 \end{bmatrix} + \begin{bmatrix} -4 & -2 \\ -3 & -1 \\ 0 & 2 \end{bmatrix} = \begin{bmatrix} -3 & 1 \\ 1 & -3 \\ 1 & 8 \end{bmatrix}.$$

DEFINITION 9·6 (matrix multiplication)   The two matrices A and B with general elements a_{ij} and b_{ij} are said to be *conformable for matrix multiplication* provided that the number of columns in A equals the number of rows in B. If A is of order $(m \times n)$ and B is of order $(n \times r)$, then the matrix product AB is the matrix C of order $(m \times r)$ with elements c_{ij}, where

$$c_{ij} = a_{i1}b_{1j} + a_{i2}b_{2j} + \cdots + a_{in}b_{nj}.$$

The number c_{ij} is called the *inner product* of the ith row of A with the jth column of B.

Example 9·5   Determine $A + BC$, given that:

$$A = \begin{bmatrix} 1 & 4 \\ 2 & 3 \end{bmatrix}, \quad B = \begin{bmatrix} 1 & 4 & 2 \\ 2 & 1 & 1 \end{bmatrix}, \quad \text{and} \quad C = \begin{bmatrix} 3 & 4 \\ 1 & 0 \\ 0 & 2 \end{bmatrix}.$$

Solution   Matrix B is of order (2×3) and matrix C is of order (3×2), showing that BC are conformable for multiplication. We have

$$BC = \begin{bmatrix} 1 & 4 & 2 \\ 2 & 1 & 1 \end{bmatrix} \begin{bmatrix} 3 & 4 \\ 1 & 0 \\ 0 & 2 \end{bmatrix} = \begin{bmatrix} 7 & 8 \\ 7 & 10 \end{bmatrix},$$

and so

$$A + BC = \begin{bmatrix} 1 & 4 \\ 2 & 3 \end{bmatrix} + \begin{bmatrix} 7 & 8 \\ 7 & 10 \end{bmatrix} = \begin{bmatrix} 8 & 12 \\ 9 & 13 \end{bmatrix}.$$

On account of the fact that matrix multiplication is not normally commutative, it is important to use a terminology that distinguishes between matrix multipliers that appear on the left or the right in a matrix product. This is achieved by adopting the convention that when matrix **B** is multiplied by matrix **A** from the *left* to form the product **AB**, we shall say that **B** is *pre-multiplied* by **A**. Conversely, when the matrix **B** is multiplied by **A** from the *right* to form the product **BA**, we shall say that **B** is *post-multiplied* by **A**.

The most important results concerning matrix multiplication are contained in the following theorem, which asserts that matrix multiplication is distributive with respect to addition and that it is also associative.

THEOREM 9·2 (matrix multiplication is distributive and associative)   If matrices **A**, **B**, and **C** are conformable for multiplication, then:

(a) matrix multiplication is *distributive with respect to addition*, so that

A(B + C) = AB + AC;

(b) matrix multiplication is *associative*, so that

A(BC) = (AB)C.

Proof   To establish result (a) let **B** and **C** be of order $(m \times n)$, and denote their general elements by b_{ij} and c_{ij}, respectively, so that the general element of **B** + **C** is $b_{ij} + c_{ij}$. Then if **A** is of order $(r \times m)$ with general element a_{ij}, and d_{ij}, is the general element of **D** = A(B + C) which is of order $(r \times n)$, we have from Definition 9·6 that

$$d_{ij} = a_{i1}(b_{1j} + c_{1j}) + a_{i2}(b_{2j} + c_{2j}) + \cdots + a_{im}(b_{mj} + c_{mj}).$$

Performing the indicated multiplications and re-grouping we have

$$d_{ij} = (a_{i1}b_{1j} + a_{i2}b_{2j} + \cdots + a_{im}b_{mj}) + (a_{i1}c_{1j} + a_{i2}c_{2j} + \\ \cdots + a_{im}c_{mj}).$$

However, from Definition 9·6 this is seen to be equivalent to

D = AB + AC,

which was to be proved.

Result (b) may be established in similar fashion, and to achieve this we assume **A**, **B**, and **C** to be respectively of order $(p \times q)$, $(q \times m)$, and $(m \times n)$ with general elements a_{ij}, b_{ij}, and c_{ij}.

From Definition 9·6 we know that the general element occurring in row i, column j of the product **BC** has the form

$$b_{i1}c_{1j} + b_{i2}c_{2j} + \cdots + b_{im}c_{mj},$$

so that the general element d_{ij} occurring in row i column j of the product **D** = **A(BC)** which is of order $(p \times n)$ must have the form

$$
\begin{aligned}
d_{ij} = \; & a_{i1}(b_{11}c_{1j} + b_{12}c_{2j} + \cdots + b_{1m}c_{mj}) \\
+ \; & a_{i2}(b_{21}c_{1j} + b_{22}c_{2j} + \cdots + b_{2m}c_{mj}) \\
+ \; & . \quad . \quad . \quad . \quad . \quad . \quad . \\
+ \; & a_{iq}(b_{q1}c_{1j} + b_{q2}c_{2j} + \cdots + b_{qm}c_{mj}).
\end{aligned}
$$

Re-grouping of the terms then gives

$$
\begin{aligned}
d_{ij} = \; & (a_{i1}b_{11} + a_{i2}b_{21} + \cdots + a_{iq}b_{q1})c_{1j} \\
+ \; & (a_{i1}b_{12} + a_{i2}b_{22} + \cdots + a_{iq}b_{q2})c_{2j} \\
+ \; & . \quad . \quad . \quad . \quad . \quad . \quad . \\
+ \; & (a_{i1}b_{1m} + a_{i2}b_{2m} + \cdots + a_{iq}b_{qm})c_{mj}.
\end{aligned}
$$

Appealing once more to Definition 9·6 we find that this is equivalent to

$$\mathbf{D} = (\mathbf{AB})\mathbf{C},$$

which was to be proved.

Example 9·6 If

$$A = [1 \quad 2], \quad B = \begin{bmatrix} 1 & 3 \\ -1 & 2 \end{bmatrix}, \quad C = \begin{bmatrix} 2 & 1 \\ 3 & 1 \end{bmatrix},$$

verify that

(a) $\mathbf{A(B + C)} = \mathbf{AB} + \mathbf{AC}$,

(b) $\mathbf{A(BC)} = \mathbf{(AB)C}$.

Solution

(a) We have

$$B + C = \begin{bmatrix} 3 & 4 \\ 2 & 3 \end{bmatrix},$$

so that

$$A(B + C) = \begin{bmatrix} 1 & 2 \end{bmatrix} \begin{bmatrix} 3 & 4 \\ 2 & 3 \end{bmatrix} = \begin{bmatrix} 7 & 10 \end{bmatrix};$$

whereas

$$AB = \begin{bmatrix} -1 & 7 \end{bmatrix} \quad \text{and} \quad AC = \begin{bmatrix} 8 & 3 \end{bmatrix},$$

so that

$$AB + AC = \begin{bmatrix} 7 & 10 \end{bmatrix}.$$

(b) We have

$$BC = \begin{bmatrix} 1 & 3 \\ -1 & 2 \end{bmatrix} \begin{bmatrix} 2 & 1 \\ 3 & 1 \end{bmatrix} = \begin{bmatrix} 11 & 4 \\ 4 & 1 \end{bmatrix}$$

so that

$$A(BC) = \begin{bmatrix} 1 & 2 \end{bmatrix} \begin{bmatrix} 11 & 4 \\ 4 & 1 \end{bmatrix} = \begin{bmatrix} 19 & 6 \end{bmatrix};$$

whereas

$$AB = \begin{bmatrix} 1 & 2 \end{bmatrix} \begin{bmatrix} 1 & 3 \\ -1 & 2 \end{bmatrix} = \begin{bmatrix} -1 & 7 \end{bmatrix},$$

whence

$$(AB)C = \begin{bmatrix} -1 & 7 \end{bmatrix} \begin{bmatrix} 2 & 1 \\ 3 & 1 \end{bmatrix} = \begin{bmatrix} 19 & 6 \end{bmatrix}.$$

An important matrix operation involves the interchange of rows and columns of a matrix, thereby changing a matrix of order ($m \times n$) into one of order ($n \times m$). Thus a row vector is changed into a column vector and a matrix of order (3×2) is changed into a matrix of order (2×3). This operation is called the operation of *transposition* and is denoted by the addition of a prime to the matrix in question.

DEFINITION 9·7 (transposition operation)   If **A** is a matrix of order ($m \times n$), then its transpose **A**′ is the matrix of order ($n \times m$) which is derived from **A** by the interchange of rows and columns. Symbolically, if a_{ij} is the element in the ith row and jth column of **A**, then a_{ji} is the element in the corresponding position in **A**′.

Example 9·7   Find **A**′ and (**A**′)′, given that:

$$A = \begin{bmatrix} 1 & 4 & 7 & 3 \\ 2 & -1 & 4 & -1 \end{bmatrix}.$$

Solution   Writing the first row in place of the first column and the second row in place of the second column, as is required by Definition 9·7, we find that

$$\mathbf{A}' = \begin{bmatrix} 1 & 2 \\ 4 & -1 \\ 7 & 4 \\ 3 & -1 \end{bmatrix}.$$

The same argument shows that

$$(\mathbf{A}')' = \begin{bmatrix} 1 & 4 & 7 & 3 \\ 2 & -1 & 4 & -1 \end{bmatrix}.$$

It is obvious from the definition of the transpose operation that $(\mathbf{A}')' = \mathbf{A}$, as was indeed illustrated in the last example. It is also obvious from Definitions 9·3 and 9·5 that if $\mathbf{A}$ and $\mathbf{B}$ are conformable for addition, then

$$(\mathbf{A} \pm \mathbf{B})' = \mathbf{A}' \pm \mathbf{B}'. \tag{9·12}$$

Now if $\mathbf{A}$ is of order $(m \times n)$ and $\mathbf{B}$ is of order $(n \times r)$, and the general matrix elements are a_{ij} and b_{ij}, respectively, the element c_{ij} in the ith row and jth column of the matrix product $\mathbf{C} = \mathbf{AB}$ is

$$c_{ij} = a_{i1}b_{1j} + a_{i2}b_{2j} + \cdots + a_{in}b_{nj}.$$

By definition, this is the element that will appear in the jth row and ith column of $(\mathbf{AB})'$.

Applying the transpose operation separately to $\mathbf{A}$ and $\mathbf{B}$ we find that $\mathbf{A}'$ is of order $(n \times m)$ and $\mathbf{B}'$ is of order $(r \times n)$, so that only the matrix product $\mathbf{B}'\mathbf{A}'$ is conformable.

Now the elements of the jth row of $\mathbf{B}'$ are the elements of the jth column of $\mathbf{B}$, and the elements of the ith column of $\mathbf{A}'$ are the elements of the ith row of $\mathbf{A}$, so that the element d_{ji} in the jth row and ith column of the product $\mathbf{D} = \mathbf{B}'\mathbf{A}'$ must be

$$d_{ji} = b_{1j}a_{i1} + b_{2j}a_{i2} + \cdots + b_{nj}a_{in}$$

or, equivalently,

$$d_{ji} = a_{i1}b_{1j} + a_{i2}b_{2j} + \cdots + a_{in}b_{nj}.$$

However, equating elements in the jth row and ith column of $(\mathbf{AB})'$ and $\mathbf{B}'\mathbf{A}'$ we find that $c_{ij} = d_{ji}$, and so

$$(\mathbf{AB})' = \mathbf{B}'\mathbf{A}'. \tag{9·13}$$

We summarize these results into a final theorem.

THEOREM 9·3 (properties of transposition operation)   If $\mathbf{A}$ and $\mathbf{B}$ are conformable for addition or multiplication, as required, then:

(a) $(\mathbf{A}')' = \mathbf{A}$ (Transposition is *Reflexive*);

(b) $(\mathbf{A} + \mathbf{B})' = \mathbf{A}' + \mathbf{B}'$;

(c) $(\mathbf{A} - \mathbf{B})' = \mathbf{A}' - \mathbf{B}'$;

(d) $(\mathbf{AB})' = \mathbf{B}'\mathbf{A}'$.

Example 9·8   Verify that $(\mathbf{AB})' = \mathbf{B}'\mathbf{A}'$, given that:

$$\mathbf{A} = \begin{bmatrix} 1 & 3 \\ 2 & 4 \end{bmatrix} \quad \text{and} \quad \mathbf{B} = \begin{bmatrix} 2 & -1 \\ 3 & 1 \end{bmatrix}.$$

Solution   We have

$$\mathbf{AB} = \begin{bmatrix} 1 & 3 \\ 2 & 4 \end{bmatrix}\begin{bmatrix} 2 & -1 \\ 3 & 1 \end{bmatrix} = \begin{bmatrix} 11 & 2 \\ 16 & 2 \end{bmatrix},$$

so that

$$(\mathbf{AB})' = \begin{bmatrix} 11 & 16 \\ 2 & 2 \end{bmatrix}.$$

However,

$$\mathbf{B}'\mathbf{A}' = \begin{bmatrix} 2 & 3 \\ -1 & 1 \end{bmatrix}\begin{bmatrix} 1 & 2 \\ 3 & 4 \end{bmatrix} = \begin{bmatrix} 11 & 16 \\ 2 & 2 \end{bmatrix},$$

which is equal to $(\mathbf{AB})'$.

9·3   Determinants

The notion of a determinant, when first introduced in Chapter 4, was that of a single *number* associated with a square array of numbers. In its subsequent application in that chapter it was used in a subsidiary role to simplify the manipulation of the vector product, and in that capacity it gave rise to a *vector*.

These are but two of the situations in which determinants occur in different branches of mathematics, and it is the object of this section to examine some of the most important algebraic properties of determinants. Our results will only be proved for determinants of order 3 but they are, in fact, all true for determinants of any order.

We begin by rewriting Definition 4·16 using the matrix element notation as follows:

DEFINITION 9·8 (third order determinant)   Let **A** be the square matrix of order (3 × 3)

$$A = \begin{bmatrix} a_{11} & a_{12} & a_{13} \\ a_{21} & a_{22} & a_{23} \\ a_{31} & a_{32} & a_{33} \end{bmatrix}.$$

Then the expression

$$|A| = \begin{vmatrix} a_{11} & a_{12} & a_{13} \\ a_{21} & a_{22} & a_{23} \\ a_{31} & a_{32} & a_{33} \end{vmatrix}$$

is called the third order *determinant* associated with the square matrix A, and it is defined to be the number

$$|A| = a_{11} \begin{vmatrix} a_{22} & a_{23} \\ a_{32} & a_{33} \end{vmatrix} - a_{12} \begin{vmatrix} a_{21} & a_{23} \\ a_{31} & a_{33} \end{vmatrix} + a_{13} \begin{vmatrix} a_{21} & a_{22} \\ a_{31} & a_{32} \end{vmatrix};$$

where for any numbers a, b, c, and d,

$$\begin{vmatrix} a & b \\ c & d \end{vmatrix} \equiv ad - bc.$$

The notation det A is also frequently used in place of $|A|$ to signify the determinant of A.

This definition has a number of consequences of considerable value in simplifying the manipulation of determinants. Let us confine attention to the third order determinant which is typical of all orders of determinant, and expand the last line of Definition 9·8. We have

$$\begin{vmatrix} a_{11} & a_{12} & a_{13} \\ a_{21} & a_{22} & a_{23} \\ a_{31} & a_{32} & a_{33} \end{vmatrix} = \begin{aligned} & a_{11}a_{22}a_{33} - a_{11}a_{23}a_{32} + a_{12}a_{23}a_{31} \\ & - a_{12}a_{21}a_{33} + a_{13}a_{21}a_{32} - a_{13}a_{22}a_{31}, \end{aligned} \quad (9 \cdot 14)$$

showing that one, and only one, element of each row and each column of the determinant appears in each of the products on the right-hand side defining $|A|$. Hence, if any row or column of a determinant is multiplied by a factor λ, then the value of the determinant is multiplied by λ, since a factor λ will appear in each product on the right-hand side of Eqn (9·14). Conversely, if any row or column of a determinant is divided by a factor λ, then the value of the determinant is divided by λ. It is also obvious from Eqn (9·14) that $|A| = 0$ if all the elements of a row or column of $|A|$ are zero, or if all the corresponding elements of two rows or columns of $|A|$ are equal.

Suppose, for example, that $\lambda = 3$ and

$$|\mathbf{A}| = \begin{vmatrix} 1 & 2 & 3 \\ 2 & 1 & 1 \\ 4 & 1 & 2 \end{vmatrix}$$

Then it is easily shown that $|\mathbf{A}| = -5$, so that $3|\mathbf{A}| = -15$. Now this result could have been obtained equally well by using the above argument and multiplying any row or any column of $|\mathbf{A}|$ by 3. If the first row of $|\mathbf{A}|$ is multiplied by 3 we have

$$3|\mathbf{A}| = \begin{vmatrix} 3 & 6 & 9 \\ 2 & 1 & 1 \\ 4 & 1 & 2 \end{vmatrix} = -15$$

or, alternatively, if the third column is multiplied by 3 we have

$$3|\mathbf{A}| = \begin{vmatrix} 1 & 2 & 9 \\ 2 & 1 & 3 \\ 4 & 1 & 6 \end{vmatrix} = -15.$$

It is readily verified from Eqn (9·14) that interchanging any two rows or columns of $|\mathbf{A}|$ changes its sign. Thus we have

$$\begin{vmatrix} 1 & 4 & 3 \\ 2 & 1 & 4 \\ 9 & 4 & -6 \end{vmatrix} = -\begin{vmatrix} 1 & 3 & 4 \\ 2 & 4 & 1 \\ 9 & -6 & 4 \end{vmatrix},$$

in which the determinant on the left has been obtained from the one on the right by interchanging the second and third columns.

A particularly simple case arises when $|\mathbf{A}|$ is the determinant associated with a diagonal matrix $\mathbf{A}$, for then all off-diagonal elements are automatically zero. This implies that Eqn (9·14) reduces to $|\mathbf{A}| = a_{11}a_{22}a_{33}$, which is just the product of the elements of the principal diagonal. Thus if

$$|\mathbf{A}| = \begin{vmatrix} 3 & 0 & 0 \\ 0 & -2 & 0 \\ 0 & 0 & 4 \end{vmatrix},$$

then $|\mathbf{A}| = (3)(-2)(4) = -24$.

Another useful result is that the value of a determinant is unchanged when elements of a row (or column) have added to them some multiple of the corresponding elements of some other row (or column). We prove this result by direct expansion in the following typical case. Consider the determinant $|\mathbf{D}|$ obtained from $|\mathbf{A}|$ by adding to the elements of column 3 of $|\mathbf{A}|$, λ times the corresponding elements in column 2 of $|\mathbf{A}|$ to obtain:

$$|\mathbf{D}| = \begin{vmatrix} a_{11} & a_{12} & a_{13} + \lambda a_{12} \\ a_{21} & a_{22} & a_{23} + \lambda a_{22} \\ a_{31} & a_{32} & a_{33} + \lambda a_{32} \end{vmatrix}.$$

Then at once Definition 9·8 asserts that

$$|\mathbf{D}| = a_{11} \begin{vmatrix} a_{22} & a_{23} \\ a_{32} & a_{33} \end{vmatrix} + a_{11} \begin{vmatrix} a_{22} & \lambda a_{22} \\ a_{32} & \lambda a_{32} \end{vmatrix} - a_{12} \begin{vmatrix} a_{21} & a_{23} \\ a_{31} & a_{33} \end{vmatrix}$$

$$- a_{12} \begin{vmatrix} a_{21} & \lambda a_{22} \\ a_{31} & \lambda a_{32} \end{vmatrix} + a_{13} \begin{vmatrix} a_{21} & a_{22} \\ a_{31} & a_{32} \end{vmatrix} + \lambda a_{12} \begin{vmatrix} a_{21} & a_{22} \\ a_{31} & a_{32} \end{vmatrix}.$$

Now the second term on the right-hand side is zero, whilst the fourth and last terms cancel leaving only three remaining terms. These are seen to comprise the definition of $|\mathbf{A}|$, so that we have proved that $|\mathbf{D}| = |\mathbf{A}|$ or, in symbols, that

$$\begin{vmatrix} a_{11} & a_{12} & a_{13} + \lambda a_{12} \\ a_{21} & a_{22} & a_{23} + \lambda a_{22} \\ a_{31} & a_{32} & a_{33} + \lambda a_{32} \end{vmatrix} = \begin{vmatrix} a_{11} & a_{12} & a_{13} \\ a_{21} & a_{22} & a_{23} \\ a_{31} & a_{32} & a_{33} \end{vmatrix}.$$

A similar result would have been obtained had different columns been used or, indeed, had rows been used instead of columns.

An obvious implication of this result is that if a row (or column) of a determinant is expressible as the sum of multiples of other rows (or columns) of the determinant, then the value of the determinant must be zero. This is so because by subtraction of this sum of multiples of other rows (or columns) from the row (or column) in question, it is possible to produce a row (or column) containing only zero elements.

Let us illustrate how a determinant may be simplified by means of this result. Consider the determinant

$$|\mathbf{A}| = \begin{vmatrix} 7 & 18 & 8 \\ 1 & 5 & 7 \\ 3 & 9 & 4 \end{vmatrix}.$$

Subtracting twice the third row from the first row we find

$$|A| = \begin{vmatrix} 1 & 0 & 0 \\ 1 & 5 & 7 \\ 3 & 9 & 4 \end{vmatrix},$$

whence $|A| = -43$.

Let us summarize our findings in the form of a theorem.

THEOREM 9·4 (properties of determinants)

(a) A determinant in which all the elements of a row or column are zero, itself has the value zero;

(b) A determinant in which all corresponding elements in two rows (or columns) are equal has the value zero;

(c) If the elements of a row (or column) of a determinant are multiplied by a factor λ, then the value of the determinant is multiplied by λ;

(d) The value of a determinant associated with a diagonal matrix is equal to the product of the elements on the principal diagonal;

(e) The value of a determinant is unaltered by adding to the elements of any row (or column), a constant multiple of the corresponding elements of any other row (or column);

(f) If a row (or column) of a determinant is expressible as the sum of multiples of other rows (or columns) of the determinant, then its value is zero.

Higher order determinants can be defined with exactly similar properties to those enumerated in the theorem above. Thus the determinant $|A|$ of order n associated with the square matrix A of order $(n \times n)$ has $n!$ terms in its expansion, each of which contains one, and only one, element from each row and column of A.

DEFINITION 9·9 (fourth order determinant) If A is the square matrix of order (4×4)

$$A = \begin{bmatrix} a_{11} & a_{12} & a_{13} & a_{14} \\ a_{21} & a_{22} & a_{23} & a_{24} \\ a_{31} & a_{32} & a_{33} & a_{34} \\ a_{41} & a_{42} & a_{43} & a_{44} \end{bmatrix},$$

then the expression

$$|A| = \begin{vmatrix} a_{11} & a_{12} & a_{13} & a_{14} \\ a_{21} & a_{22} & a_{23} & a_{24} \\ a_{31} & a_{32} & a_{33} & a_{34} \\ a_{41} & a_{42} & a_{43} & a_{44} \end{vmatrix}$$

is called the *fourth order determinant* associated with the square matrix **A**, and it is defined to be the number

$$|\mathbf{A}| = a_{11} \begin{vmatrix} a_{22} & a_{23} & a_{24} \\ a_{32} & a_{33} & a_{34} \\ a_{42} & a_{43} & a_{44} \end{vmatrix} - a_{12} \begin{vmatrix} a_{21} & a_{23} & a_{24} \\ a_{31} & a_{33} & a_{34} \\ a_{41} & a_{43} & a_{44} \end{vmatrix}$$

$$+ a_{13} \begin{vmatrix} a_{21} & a_{22} & a_{24} \\ a_{31} & a_{32} & a_{34} \\ a_{41} & a_{42} & a_{44} \end{vmatrix} - a_{14} \begin{vmatrix} a_{21} & a_{22} & a_{23} \\ a_{31} & a_{32} & a_{33} \\ a_{41} & a_{42} & a_{43} \end{vmatrix}.$$

An inductive argument applied to Definitions 9·8 and 9·9 shows one way in which higher order determinants may be defined, but clearly our notation needs some simplification to avoid unwieldy expressions of the type given above. This is achieved by the introduction of the *minor* and the *cofactor* of an element of a square matrix.

DEFINITION 9·10 (minors and cofactors)   Let **A** be a square matrix of order $(n \times n)$ with general element a_{ij}, and let $|\,\mathbf{A}\,|$ be the determinant of order n associated with **A**. Denote by M_{ij} the determinant of order $(n - 1)$ associated with the matrix of order $(n - 1, n - 1)$ derived from **A** by the deletion of row i and column j. Then M_{ij} is called the *minor* of the element a_{ij} of **A**, and $A_{ij} = (-1)^{i+j} M_{ij}$ is called the *cofactor* of the element a_{ij} of **A**.

Example 9·9   Find the minors and cofactors of the matrix

$$\mathbf{A} = \begin{bmatrix} 1 & 0 & 3 \\ 2 & 1 & 4 \\ 1 & 2 & 1 \end{bmatrix}.$$

Solution   The minor M_{11} is derived from **A** by deleting row 1 and column 1 and equating M_{11} to the determinant formed by the remaining elements. That is,

$$M_{11} = \begin{vmatrix} 1 & 4 \\ 2 & 1 \end{vmatrix} = -7.$$

Similarly, minor M_{12} is derived from **A** by deleting row 1 and column 2 and equating M_{12} to the determinant formed by the remaining elements. That is,

$$M_{12} = \begin{vmatrix} 2 & 4 \\ 1 & 1 \end{vmatrix} = -2.$$

Identical reasoning then shows that $M_{13} = 3$, $M_{21} = -6$, $M_{22} = -2$, $M_{23} = 2$, $M_{31} = -3$, $M_{32} = -2$, and $M_{33} = 1$. As the cofactors $A_{ij} = (-1)^{i+j} M_{ij}$, it follows that $A_{11} = -7$, $A_{12} = 2$, $A_{13} = 3$, $A_{21} = 6$, $A_{22} = -2$, $A_{23} = -2$, $A_{31} = -3$, $A_{32} = 2$, and $A_{33} = 1$.

If **A** is a square matrix with general element a_{ij} and corresponding cofactor A_{ij}, it is easily seen that:

(a) if **A** is of order (2×2), then $| \, \mathbf{A} \, | = a_{11}A_{11} + a_{12}A_{12}$,

(b) if **A** is of order (3×3), then $| \, \mathbf{A} \, | = a_{11}A_{11} + a_{12}A_{12} + a_{13}A_{13}$,

(c) if **A** is of order (4×4), then $| \, \mathbf{A} \, | = a_{11}A_{11} + a_{12}A_{12} + a_{13}A_{13} + a_{14}A_{14}$.

This suggests that if **A** is of order $(n \times n)$, then for $| \, \mathbf{A} \, |$ we could adopt the definition

$$| \, \mathbf{A} \, | = a_{11}A_{11} + a_{12}A_{12} + \cdots + a_{1n}A_{1n}. \tag{9.15}$$

This is a true statement and could be accepted as a definition, but it is not the most general one which may be adopted. To see this we return to Eqn (9·14) and re-arrange the terms on the right-hand side to give

$$| \, \mathbf{A} \, | = a_{31}(a_{12}a_{23} - a_{13}a_{22}) - a_{32}(a_{11}a_{23} - a_{13}a_{21})$$
$$+ a_{33}(a_{11}a_{22} - a_{12}a_{21}).$$

Hence, working backwards, we have

$$| \, \mathbf{A} \, | = a_{31} \begin{vmatrix} a_{12} & a_{13} \\ a_{22} & a_{23} \end{vmatrix} - a_{32} \begin{vmatrix} a_{11} & a_{13} \\ a_{21} & a_{23} \end{vmatrix} + a_{33} \begin{vmatrix} a_{11} & a_{12} \\ a_{21} & a_{22} \end{vmatrix},$$

thereby showing that it is also true that

$$| \, \mathbf{A} \, | = a_{31}A_{31} + a_{32}A_{32} + a_{33}A_{33}. \tag{9.16}$$

We now have two equivalent but different looking expressions for $| \, \mathbf{A} \, |$ either of which could be taken as the definition of $| \, \mathbf{A} \, |$. The expression in (b) above involves the elements and cofactors of the first row of **A** and the expression in Eqn (9·16) involves the elements and cofactors of the third row of **A**. A repetition of this argument involving other rearrangements of the terms of Eqn (9·14) shows that $| \, \mathbf{A} \, |$ may be evaluated as the sum of the products of the elements and their cofactors of *any* row or column of **A**. This very valuable and general result is known as the *Laplace expansion theorem*, and it is true for determinants of any order though we have only proved it for a third order determinant. Let us state this result formally as it would apply to a determinant of order n.

THEOREM 9·5 (Laplace expansion theorem)   The determinant $| \, \mathbf{A} \, |$ associated with any $(n \times n)$ square matrix **A** is obtained by summing the products of the elements and their cofactors in any row or column of **A**. If **A** has the general element a_{ij} and the corresponding cofactor is A_{ij}, then this result is equivalent to:

Expansion by elements of a row

$$|\,\mathbf{A}\,| = \sum_{j=1}^{n} a_{ij}A_{ij}$$

for $i = 1, 2, \ldots, n$;

Expansion by elements of a column

$$|\,\mathbf{A}\,| = \sum_{i=1}^{n} a_{ij}A_{ij}$$

for $j = 1, 2, \ldots, n$.

Example 9·10   Evaluate the determinant

$$|\,\mathbf{A}\,| = \begin{vmatrix} 1 & 4 & 2 \\ 3 & -2 & 1 \\ 1 & 5 & 2 \end{vmatrix}$$

by expanding it (a) in terms of the elements of row 2, and (b) in terms of the elements of column 3.

Solution

(a) $|\,\mathbf{A}\,| = -3 \begin{vmatrix} 4 & 2 \\ 5 & 2 \end{vmatrix} - 2 \begin{vmatrix} 1 & 2 \\ 1 & 2 \end{vmatrix} - 1 \begin{vmatrix} 1 & 4 \\ 1 & 5 \end{vmatrix} = 5$

(b) $|\,\mathbf{A}\,| = 2 \begin{vmatrix} 3 & -2 \\ 1 & 5 \end{vmatrix} - 1 \begin{vmatrix} 1 & 4 \\ 1 & 5 \end{vmatrix} + 2 \begin{vmatrix} 1 & 4 \\ 3 & -2 \end{vmatrix} = 5.$

An important extension of Theorem 9·5 asserts that the sum of the products of the elements of any row (or column) of a square matrix **A** with the cofactors corresponding to the elements of a *different* row (or column) is zero. This is easily proved as follows.

Let **A** be a matrix of order $(n \times n)$, and let **B** be obtained from **A** by replacing row q of **A** by row p. Then **B** has the elements of rows p and q equal, so that by Theorem 9·4 (b) it follows that $|\,\mathbf{B}\,| = 0$. Expanding $|\,\mathbf{B}\,|$ in terms of elements of row q by Theorem 9·5 we then find

$$|\,\mathbf{B}\,| = a_{p1}A_{q1} + a_{p2}A_{q2} + \cdots + a_{pn}A_{qn} = 0,$$

which was to be proved. A similar argument establishes the corresponding result for columns and so we have proved our assertion.

THEOREM 9·6 The sum of the products of the elements of any row (or column) of a square matrix **A** with the cofactors corresponding to the elements of a different row (or column) is zero. Symbolically, if a_{ij} is the general element of **A** and A_{ij} is its cofactor, then:

Expansion by elements of a row

$$\sum_{i=1}^{n} a_{pi} A_{qi} = 0$$

if $p \neq q$; and

Expansion by elements of a column

$$\sum_{i=1}^{n} a_{ip} A_{iq} = 0$$

if $p \neq q$.

Example 9·11 Verify that the sum of the products of the elements of column 1 and the corresponding cofactors of column 2 of the following matrix is zero:

$$\mathbf{A} = \begin{bmatrix} 1 & 3 & 2 \\ 4 & 1 & 2 \\ 3 & 1 & 3 \end{bmatrix}.$$

Solution The elements of column 1 are $a_{11} = 1$, $a_{21} = 4$, $a_{31} = 3$. The cofactors corresponding to the elements of the second column are $A_{12} = -6$, $A_{22} = -3$, $A_{32} = 6$. Hence

$$\dot{a}_{11} A_{12} + a_{21} A_{22} + a_{31} A_{32} = (1)(-6) + (4)(-3) + (3)(6) = 0.$$

9·4 Linear dependence and linear independence

We are now in a position to discuss the important idea of linear independence. This concept has already been used implicitly in Chapter 4 when the three mutually orthogonal unit vectors **i**, **j**, and **k** were introduced comprising what in linear algebra is called a *basis* for the vector space. By this we mean that all other vectors are expressible in terms of the vectors comprising the basis through the operations of scaling and vector addition, but that no member of the basis itself is expressible in terms of the other members of the basis.

Thus *no* choice of the scalars λ, μ can ever make the vectors **i** and $\lambda\mathbf{j} + \mu\mathbf{k}$ equal. It is in this sense that the unit vectors **i**, **j**, **k** comprising the basis for ordinary vector analysis are *linearly independent*, and obviously any other set of unit vectors **a**, **b**, **c** which are not co-planar, and no two of which are parallel, would serve equally well as a basis for this space.

The same idea carries across to matrices when the term vector is interpreted to mean either a matrix row vector or a matrix column vector. Thus the three column vectors

$$\mathbf{C}_1 = \begin{bmatrix} 1 \\ 3 \\ -2 \end{bmatrix}, \quad \mathbf{C}_2 = \begin{bmatrix} 2 \\ 1 \\ 4 \end{bmatrix}, \quad \text{and} \quad \mathbf{C}_3 = \begin{bmatrix} 5 \\ 5 \\ 6 \end{bmatrix}$$

are *not* linearly independent because $\mathbf{C}_3 = \mathbf{C}_1 + 2\mathbf{C}_2$, whereas the three row vectors

$$\mathbf{R}_1 = [1 \quad 0 \quad 0], \quad \mathbf{R}_2 = [0 \quad 1 \quad 0], \quad \text{and} \quad \mathbf{R}_3 = [0 \quad 0 \quad 1]$$

are obviously linearly independent, because no choice of the scalars λ, μ can ever make the vectors $\mathbf{R}_1$ and $\lambda\mathbf{R}_2 + \mu\mathbf{R}_3$ equal. It is these ideas that underlie the formulation of the following definition.

DEFINITION 9·11 (linear dependence and linear independence)   The set of n matrix row or column vectors $\mathbf{V}_1$, $\mathbf{V}_2$, . . ., $\mathbf{V}_n$ which are conformable for addition will be said to be *linearly dependent* if there exist n scalars α_1, α_2, . . ., α_n, not all zero, such that

$$\alpha_1\mathbf{V}_1 + \alpha_2\mathbf{V}_2 + \cdots + \alpha_n\mathbf{V}_n = \mathbf{0}.$$

When no such set of scalars exists, so that this relationship is only true when $\alpha_1 = \alpha_2 = \cdots = \alpha_n = 0$, then the n matrix vectors $\mathbf{V}_1$, $\mathbf{V}_2$, . . ., $\mathbf{V}_n$ will be said to be *linearly independent*.

In the event that the n matrix vectors in Definition 9·11 represent the rows or columns of a rectangular matrix **A**, the linear dependence or independence of the vectors $\mathbf{V}_1$, $\mathbf{V}_2$, . . ., $\mathbf{V}_n$ becomes a statement about the linear dependence or independence of the rows or columns of **A**. In particular, if **A** is a square matrix, and linear dependence exists between its rows (or columns), then by definition it is possible to express at least one row (or column) of **A** as the sum of multiples of the other rows (or columns). Thus from Theorem 9·4 (f), we see that linear dependence amongst the rows or columns of a square matrix **A** implies the condition $|\mathbf{A}| = 0$. Similarly, if $|\mathbf{A}| \neq 0$ then the rows and columns of **A** cannot be linearly dependent.

THEOREM 9·7 (test for linear independence)   The rows and columns of a square matrix **A** are linearly independent if, and only if, $|\mathbf{A}| \neq 0$. Conversely, linear dependence is implied between rows or columns of a square matrix **A** if $|\mathbf{A}| = 0$.

Example 9·12   Test the following matrices for linear independence between rows or columns:

$$A = \begin{bmatrix} 1 & 4 & 3 \\ -2 & 18 & 7 \\ 4 & -6 & 1 \end{bmatrix} \quad \text{and} \quad B = \begin{bmatrix} 1 & 1 & 0 \\ 3 & 2 & 1 \\ 1 & 1 & 3 \end{bmatrix}.$$

Solution   We shall apply Theorem 9·7 by examining $|\,A\,|$ and $|\,B\,|$. A simple calculation shows that $|\,A\,| = 0$, so that linear dependence exists between either the rows or the columns of **A**. In fact, denoting the columns of **A** by C_1, C_2, and C_3, we have $C_2 = 2(C_3 - C_1)$. As $|\,B\,| = -3$ the rows and columns of **B** are linearly independent.

Let us now give consideration to any linear independence that may exist between the rows or columns of a rectangular matrix **A** of order ($m \times n$). If r rows (or columns) of **A** are linearly independent, where $r \leq \min(m, n)$, then Theorem 9·7 implies that there is at least one determinant of order r that may be formed by taking these r rows (or columns) which is *non-zero*, but that all determinants of order greater than r must of necessity vanish. This number r is called the *rank* of the matrix **A**, and it represents the greatest number of linearly independent rows or columns existing in **A**. If, for example, **A** is a square matrix of order ($n \times n$) and $|\,A\,| \neq 0$, this implies that the rank of **A** must be n.

DEFINITION 9·12 (rank of a matrix)   The *rank r* of a matrix **A** is the greatest number of linearly independent rows or columns that exist in the matrix **A**. Numerically, r is equal to the *order* of the largest order non-vanishing determinant $|\,B\,|$ associated with any square matrix **B** which can be constructed from **A** by combination of r rows and r columns.

Example 9·13   Find the rank of the following matrix:

$$A = \begin{bmatrix} 1 & 0 & 0 & 1 & 0 \\ -1 & 1 & 1 & -1 & 1 \\ -3 & 0 & 1 & -1 & 0 \end{bmatrix}.$$

Solution   The largest order of determinant that can be constructed in this case from the rows and columns of **A** is 3. As there is certainly one such determinant that is non-vanishing, namely the one associated with the first three columns of **A**, the rank of **A** must be 3. The fact that other non-vanishing determinants of order three may be constructed from **A** is immaterial (e.g., take the last three columns).

9·5 Inverse and adjoint matrix

The operation of division is not defined for matrices, but a multiplicative inverse matrix denoted by $\mathbf{A}^{-1}$ can be defined for any *square* matrix $\mathbf{A}$ for which $|\mathbf{A}| \neq 0$. This multiplicative inverse $\mathbf{A}^{-1}$ is unique and has the property that

$$\mathbf{A}^{-1}\mathbf{A} = \mathbf{A}\mathbf{A}^{-1} = \mathbf{I}$$

where $\mathbf{I}$ is the unit matrix, and it is defined in terms of what is called the matrix *adjoint* to $\mathbf{A}$. The uniqueness follows from the fact that if $\mathbf{B}$ and $\mathbf{C}$ are each inverse to $\mathbf{A}$, then $\mathbf{B}(\mathbf{A}\mathbf{C}) = (\mathbf{B}\mathbf{A})\mathbf{C}$, so that $\mathbf{B}\mathbf{I} = \mathbf{I}\mathbf{C}$, or $\mathbf{B} = \mathbf{C}$.

DEFINITION 9·13 (adjoint matrix) Let $\mathbf{A}$ be a square matrix, then the transpose of the matrix of cofactors of $\mathbf{A}$ is called the matrix *adjoint* to $\mathbf{A}$, and it is denoted by adj $\mathbf{A}$. A square matrix and its adjoint are both of the same order.

Example 9·14 Find the matrix adjoint to:

$$\mathbf{A} = \begin{bmatrix} 1 & 2 & 1 \\ 3 & 1 & 0 \\ 2 & 1 & 2 \end{bmatrix}.$$

Solution The cofactors A_{ij} of $\mathbf{A}$ are: $A_{11} = 2$, $A_{12} = -6$, $A_{13} = 1$, $A_{21} = -3$, $A_{22} = 0$, $A_{23} = 3$, $A_{31} = -1$, $A_{32} = 3$, and $A_{33} = -5$. Hence the matrix of cofactors has the form

$$\begin{bmatrix} 2 & -6 & 1 \\ -3 & 0 & 3 \\ -1 & 3 & -5 \end{bmatrix},$$

so that its transpose, which by definition is adj $\mathbf{A}$, is

$$\text{adj } \mathbf{A} = \begin{bmatrix} 2 & -3 & -1 \\ -6 & 0 & 3 \\ 1 & 3 & -5 \end{bmatrix}.$$

Now from Theorems 9·5 and 9·6, we see that the effect of forming either the product (adj $\mathbf{A}$)$\mathbf{A}$ or the product $\mathbf{A}$(adj $\mathbf{A}$) is to produce a diagonal matrix in which each element of the leading diagonal is $|\mathbf{A}|$. That is, we have shown that

$$(\text{adj } A)A = A(\text{adj } A) = \begin{bmatrix} |A| & 0 & 0 & \ldots & 0 \\ 0 & |A| & 0 & \ldots & 0 \\ \cdot & \cdot & \cdot & \cdot & \cdot \\ 0 & 0 & 0 & \ldots & |A| \end{bmatrix},$$

whence

$$(\text{adj } A)A = A(\text{adj } A) = |A| I. \tag{9.17}$$

Thus, provided $|A| \neq 0$, by writing

$$A^{-1} = \frac{\text{adj } A}{|A|}, \tag{9.18}$$

we arrive at the result

$$A^{-1}A = AA^{-1} = I. \tag{9.19}$$

The matrix A^{-1} is called the matrix *inverse* to A and it is only defined for square matrices A for which $|A| \neq 0$. A square matrix whose associated determinant is non-vanishing is called a *non-singular* matrix. Although the inverse matrix is only defined for non-singular square matrices, the adjoint matrix is defined for any square matrix, irrespective of whether or not it is non-singular.

DEFINITION 9.14 (inverse matrix) If A is a square matrix for which $|A| \neq 0$, the matrix *inverse* to A which is denoted by A^{-1} is defined by the relationship

$$A^{-1} = \frac{\text{adj } A}{|A|}.$$

Example 9.15 Find the matrix inverse to the matrix A of Example 9.14 above.

Solution It is easily found from the cofactors already computed that $|A| = -9$. This follows, for example, by expanding $|A|$ in terms of elements of the first row to obtain $|A| = (1)(2) + (2)(-6) + (1)(1) = -9$. Hence from Definition 9.14, we have

$$A^{-1} = \frac{\text{adj } A}{|A|} = (-1/9) \begin{bmatrix} 2 & -3 & -1 \\ -6 & 0 & 3 \\ 1 & 3 & -5 \end{bmatrix} = \begin{bmatrix} -2/9 & 1/3 & 1/9 \\ 2/3 & 0 & -1/3 \\ -1/9 & -1/3 & 5/9 \end{bmatrix}.$$

The steps in the determination of an inverse matrix are perhaps best remembered in the form of a rule.

Rule 2 (Determination of inverse matrix)

To determine the matrix A^{-1} which is inverse to the square matrix A proceed as follows:

   (a) Construct the matrix of cofactors of A;
   (b) Transpose the matrix of cofactors of A to obtain adj A;
   (c) Calculate $|A|$ and, if it is not zero, divide adj A by $|A|$ to obtain A^{-1};
   (d) If $|A| = 0$, then A^{-1} is *not* defined.

It is a trivial consequence of Definition 9·14 and the fact that for any square matrix A, $|A| = |A'|$, that

$$(A^{-1})' = (A')^{-1}. \tag{9·20}$$

Also, if A and B are non-singular matrices of the same order, then

$$(B^{-1}A^{-1})AB = I = AB(B^{-1}A^{-1}),$$

showing that

$$(AB)^{-1} = B^{-1}A^{-1}. \tag{9·21}$$

Although we shall not prove it, let us accept as being valid for square matrices A, B of arbitrary order $(n \times n)$, that $|AB| = |A||B|$. We are then able to prove another useful result concerning the inverse matrix. If $|A| \neq 0$, then $AA^{-1} = I$ showing that $|AA^{-1}| = 1$, or $|A||A^{-1}| = 1$. It follows from this that:

$$|A| = 1/|A^{-1}|. \tag{9·22}$$

One final result follows directly from the obvious fact that $(A^{-1})^{-1}A^{-1} = I$, which is always true provided $|A^{-1}| \neq 0$. If we post-multiply this result by A we find

$$(A^{-1})^{-1}A^{-1}A = IA$$

giving

$$(A^{-1})^{-1}I = A,$$

whence

$$(A^{-1})^{-1} = A. \tag{9·23}$$

THEOREM 9·8 (properties of inverse matrix)   If A and B are non-singular square matrices of the same order, then:

  (a) $\mathbf{AA}^{-1} = \mathbf{A}^{-1}\mathbf{A} = \mathbf{I}$;
  (b) $(\mathbf{AB})^{-1} = \mathbf{B}^{-1}\mathbf{A}^{-1}$;
  (c) $(\mathbf{A}^{-1})' = (\mathbf{A}')^{-1}$;
  (d) $(\mathbf{A}^{-1})^{-1} = \mathbf{A}$;
  (e) $|\mathbf{A}| = 1/|\mathbf{A}^{-1}|$.

Example 9·16   Verify that $(\mathbf{A}^{-1})' = (\mathbf{A}')^{-1}$, given that

$$\mathbf{A} = \begin{bmatrix} 1 & 3 \\ 2 & 4 \end{bmatrix}.$$

Solution   We have

$$\mathbf{A}^{-1} = \begin{bmatrix} -2 & 3/2 \\ 1 & -1/2 \end{bmatrix},$$

so that

$$(\mathbf{A}^{-1})' = \begin{bmatrix} -2 & 1 \\ 3/2 & -1/2 \end{bmatrix}.$$

However,

$$\mathbf{A}' = \begin{bmatrix} 1 & 2 \\ 3 & 4 \end{bmatrix},$$

so that

$$(\mathbf{A}')^{-1} = \begin{bmatrix} -2 & 1 \\ 3/2 & -1/2 \end{bmatrix},$$

confirming that $(\mathbf{A}^{-1})' = (\mathbf{A}')^{-1}$.

9·6   Solution of systems of linear equations

A system of m linear inhomogeneous equations in the n variables x_1, x_2, . . ., x_n has the general form

$$a_{11}x_1 + a_{12}x_2 + \cdots + a_{1n}x_n = k_1$$
$$a_{21}x_1 + a_{22}x_2 + \cdots + a_{2n}x_n = k_2 \qquad (9·24)$$
$$\cdot \quad \cdot \quad \cdot \quad \cdot \quad \cdot \quad \cdot \quad \cdot \quad \cdot$$
$$a_{m1}x_1 + a_{m2}x_2 + \cdots + a_{mn}x_n = k_m,$$

where the term *inhomogeneous* refers to the fact that not all of the numbers $k_1, k_2, \ldots, k_m$ are zero. Defining the matrices

$$
\mathbf{A} = \begin{bmatrix} a_{11} & a_{12} & \cdots & a_{1n} \\ a_{21} & a_{22} & \cdots & a_{2n} \\ \cdot & \cdot & \cdot & \cdot \\ a_{m1} & a_{m2} & \cdots & a_{mn} \end{bmatrix}, \quad \mathbf{X} = \begin{bmatrix} x_1 \\ x_2 \\ \cdot \\ \cdot \\ x_n \end{bmatrix}, \quad \text{and} \quad \mathbf{K} = \begin{bmatrix} k_1 \\ k_2 \\ \cdot \\ \cdot \\ k_m \end{bmatrix},
$$

this system can be written

$$\mathbf{AX} = \mathbf{K}. \tag{9·25}$$

Here $\mathbf{A}$ is called the *coefficient matrix*, $\mathbf{X}$ the *solution vector*, and $\mathbf{K}$ the *inhomogeneous vector*.

In the event that $m = n$ and $|\mathbf{A}| \neq 0$ it follows that $\mathbf{A}^{-1}$ exists, so that pre-multiplication of Eqn (9·25) by $\mathbf{A}^{-1}$ gives for the solution vector,

$$\mathbf{X} = \mathbf{A}^{-1}\mathbf{K}. \tag{9·26}$$

This method of solution is of more theoretical than practical interest because the task of computing $\mathbf{A}^{-1}$ becomes prohibitive when n is much greater than three. However, one useful method of solution for small systems of such equations ($n \leq 4$) known as *Cramer's rule* may be deduced from Eqn (9·26).

Consideration of Eqn (9·26) and Definitions 9·14 shows that x_i, the ith element in the solution vector $\mathbf{X}$, is given by

$$x_i = \frac{1}{|\mathbf{A}|}(k_1 A_{1i} + k_2 A_{2i} + \cdots + k_n A_{ni}) \tag{9·27}$$

for $i = 1, 2, \ldots, n$, where A_{ij} is the cofactor of $\mathbf{A}$ corresponding to element a_{ij}. Using Laplace's expansion theorem we then see that the numerator of Eqn (9·27) is simply the expansion of $|\mathbf{A}_i|$, where $\mathbf{A}_i$ denotes the matrix derived from $\mathbf{A}$ by replacing the ith column of $\mathbf{A}$ by the column vector $\mathbf{K}$. Thus we have derived the simple result

$$x_i = \frac{|\mathbf{A}_i|}{|\mathbf{A}|}$$

for $i = 1, 2, \ldots, n,$ (9·28)

which expresses the elements of the solution vector K of Eqn (9·24) in terms of determinants.

Rule 3 (Cramer's rule)

To solve n linear inhomogeneous equations in n variables proceed as follows:

(a) Compute $|\mathbf{A}|$ the determinant of the coefficient matrix and, if

$| \mathbf{A} | \neq 0$, proceed to the next step;
(b) Compute the modified coefficient determinants $| \mathbf{A}_i |$, $i = 1, 2, \ldots,$ n where $\mathbf{A}_i$ is derived from $\mathbf{A}$ by replacing the ith column of $\mathbf{A}$ by the inhomogeneous vector $\mathbf{K}$;
(c) Then the solutions $x_1, x_2, \ldots, x_n$ are given by

$$x_i = \frac{| \mathbf{A}_i |}{| \mathbf{A} |}$$

for $i = 1, 2, \ldots, n$;
(d) If $| \mathbf{A} | = 0$ the method fails.

Example 9·17 Use Cramer's rule to solve the equations:

$$x_1 + 3x_2 + x_3 = 8$$
$$2x_1 + x_2 + 3x_3 = 7$$
$$x_1 + x_2 - x_3 = 2.$$

Solution The coefficient matrix $\mathbf{A}$ and the modified coefficient matrices $\mathbf{A}_1$, $\mathbf{A}_2$, and $\mathbf{A}_3$ are obviously:

$$\mathbf{A} = \begin{bmatrix} 1 & 3 & 1 \\ 2 & 1 & 3 \\ 1 & 1 & -1 \end{bmatrix}, \quad \mathbf{A}_1 = \begin{bmatrix} 8 & 3 & 1 \\ 7 & 1 & 3 \\ 2 & 1 & -1 \end{bmatrix}, \quad \mathbf{A}_2 = \begin{bmatrix} 1 & 8 & 1 \\ 2 & 7 & 3 \\ 1 & 2 & -1 \end{bmatrix}, \quad \text{and}$$

$$\mathbf{A}_3 = \begin{bmatrix} 1 & 3 & 8 \\ 2 & 1 & 7 \\ 1 & 1 & 2 \end{bmatrix}.$$

Hence $| \mathbf{A} | = 12$, $| \mathbf{A}_1 | = 12$, $| \mathbf{A}_2 | = 24$, and $| \mathbf{A}_3 | = 12$, so that

$$x_1 = \frac{| \mathbf{A}_1 |}{| \mathbf{A} |} = 1, \quad x_2 = \frac{| \mathbf{A}_2 |}{| \mathbf{A} |} = 2, \quad x_3 = \frac{| \mathbf{A}_3 |}{| \mathbf{A} |} = 1.$$

In the more general case in which $m = n$, but $| \mathbf{A} | = 0$, the inverse matrix does not exist and so any method using $\mathbf{A}^{-1}$ must fail. In these circumstances we must consider more carefully what is meant by a solution. In general, when a solution vector $\mathbf{X}$ exists whose elements simultaneously satisfy all the equations in the system, the equations will be said to be *consistent*. If no solution vector exists having this property then the equations will be said to be *inconsistent*. Consider the following equations:

$$x_1 + x_2 + 2x_3 = 9$$
$$4x_1 - 2x_2 + x_3 = 4$$
$$5x_1 - x_2 + 3x_3 = 1.$$

These equations are obviously inconsistent, because the left-hand side of the third equation is just the sum of the left-hand sides of the first two equations, whereas the right-hand sides are not so related (that is, $1 \neq 9 + 4$). In effect, what we are saying is that there is a linear dependence between the rows of the left-hand side of the equations which is not shared by the inhomogeneous terms. The row linear dependence in the coefficient matrix $\mathbf{A}$ is obviously dependent upon the rank of $\mathbf{A}$ and we now offer a brief discussion of one way in which the general problem of consistency may be approached.

Obviously, when working conventionally with the individual equations comprising (9·24) we know that: (a) equations may be scaled, (b) equations may be interchanged, and (c) multiples of one equation may be added to another. This implies that if we consider the coefficient matrix $\mathbf{A}$ of the system and supplement it on the right by the elements of the inhomogeneous vector $\mathbf{K}$ to form what is called the *augmented matrix*, then these same operations are valid for the rows of the augmented matrix. Clearly, the rank will not be affected by these operations. If the ranks of $\mathbf{A}$ and of the augmented matrix denoted by $(\mathbf{A}, \mathbf{K})$ are the same, then the equations must be consistent; otherwise they must be inconsistent.

DEFINITION 9·15 (augmented matrix and elementary row operations) Suppose that $\mathbf{AX} = \mathbf{K}$, where

$$
\mathbf{A} = \begin{bmatrix} a_{11} & a_{12} & \cdots & a_{1n} \\ a_{21} & a_{22} & \cdots & a_{2n} \\ \cdot & \cdot & \cdot & \cdot \\ a_{n1} & a_{n2} & \cdots & a_{nn} \end{bmatrix}, \quad \mathbf{X} = \begin{bmatrix} x_1 \\ x_2 \\ \cdot \\ x_n \end{bmatrix}, \quad \text{and} \quad \mathbf{K} = \begin{bmatrix} k_1 \\ k_2 \\ \cdot \\ k_n \end{bmatrix}.
$$

Then the *augmented matrix*, written $(\mathbf{A}, \mathbf{K})$, is defined to be the matrix

$$
(\mathbf{A}, \mathbf{K}) = \begin{bmatrix} a_{11} & a_{12} & \cdots & a_{1n} & k_1 \\ a_{21} & a_{22} & \cdots & a_{2n} & k_2 \\ \cdot & \cdot & \cdot & \cdot \\ a_{n1} & a_{n2} & \cdots & a_{nn} & k_n \end{bmatrix}.
$$

An *elementary row operation* performed on an augmented matrix is any one of the following:

(a) scaling of all elements in a row by a factor λ;
(b) interchange of any two rows;
(c) addition of a multiple of one row to another row.

An augmented matrix will be said to have been reduced to *echelon* form by elementary row operations when the first non-zero element in any row is a unity, and it lies to the right of the unity in the row above.

Example 9·18 Perform elementary row operations on the augmented matrix corresponding to the inconsistent equations above to reduce them to echelon form. Find the ranks of $\mathbf{A}$ and $(\mathbf{A}, \mathbf{K})$.

Solution The augmented matrix

$$(\mathbf{A}, \mathbf{K}) = \begin{bmatrix} 1 & 1 & 2 & 9 \\ 4 & -2 & 1 & 4 \\ 5 & -1 & 3 & 1 \end{bmatrix}.$$

Subtract from the elements of row 3 the sum of the corresponding elements in rows 1 and 2 to obtain

$$\begin{bmatrix} 1 & 1 & 2 & 9 \\ 4 & -2 & 1 & 4 \\ 0 & 0 & 0 & -12 \end{bmatrix}.$$

Subtract from the elements of row 2, four times the corresponding elements in row 1 to obtain

$$\begin{bmatrix} 1 & 1 & 2 & 9 \\ 0 & -6 & -7 & -32 \\ 0 & 0 & 0 & -12 \end{bmatrix}.$$

Divide row 2 by -6 and row 3 by -12 to obtain

$$\begin{bmatrix} 1 & 1 & 2 & 9 \\ 0 & 1 & 7/6 & 16/3 \\ 0 & 0 & 0 & 1 \end{bmatrix}.$$

This is now in echelon form and the rank of the matrix comprising the first three columns is 2, which must be the same as the rank of the coefficient matrix $\mathbf{A}$. The rank of $(\mathbf{A}, \mathbf{K})$ must be the same as the rank of the echelon equivalent of the augmented matrix which is clearly 3.

The general conclusion that may be reached from the echelon form of an augmented matrix $(\mathbf{A}, \mathbf{K})$, is that equations are consistent only when the ranks of $\mathbf{A}$ and $(\mathbf{A}, \mathbf{K})$ are the same. If the equations are consistent, and $\mathbf{A}$ is of order $(n \times n)$ and the rank $r < n$, we shall have fewer equations than variables. In these circumstances we may solve for any r of the variables x_i in terms of the $n - r$ remaining ones which can then be assigned arbitrary values.

THEOREM 9·9 (solution of inhomogeneous systems) The inhomogeneous system of equations

$$AX = K,$$

where A is of order $(n \times n)$ and X, K are of order $(n \times 1)$ has a unique solution if $|A| \neq 0$. If $|A| = 0$, then the equations are only consistent when the ranks of A and (A, K) are equal. In this case, if the rank $r < n$, it is possible to solve for r variables in terms of the $n - r$ remaining variables which may then be assigned arbitrary values.

Example 9·19 Solve the following equations by reducing the augmented matrix to echelon form:

$$x_1 + 3x_2 - x_3 = 6$$
$$8x_1 + 9x_2 + 4x_3 = 21$$
$$2x_1 + x_2 + 2x_3 = 3.$$

Solution The augmented matrix

$$(A, K) = \begin{bmatrix} 1 & 3 & -1 & 6 \\ 8 & 9 & 4 & 21 \\ 2 & 1 & 2 & 3 \end{bmatrix}.$$

Subtract from the elements of row 2, the sum of three times the corresponding element in row 3 and twice the corresponding element in row 1 to obtain

$$\begin{bmatrix} 1 & 3 & -1 & 6 \\ 0 & 0 & 0 & 0 \\ 2 & 1 & 2 & 3 \end{bmatrix}.$$

Interchange rows two and three to obtain

$$\begin{bmatrix} 1 & 3 & -1 & 6 \\ 2 & 1 & 2 & 3 \\ 0 & 0 & 0 & 0 \end{bmatrix}.$$

Subtract twice row 1 from row 2 and divide the resulting row 2 by -5 to obtain

$$\begin{bmatrix} 1 & 3 & -1 & 6 \\ 0 & 1 & -4/5 & 9/5 \\ 0 & 0 & 0 & 0 \end{bmatrix}.$$

This is now in echelon form and clearly the ranks of A and (A, K) are both 2 showing that the equations are consistent. However, only two equations exist between the three variables x_1, x_2, and x_3, for the echelon form of the augmented matrix may be seen to be equivalent to the two scalar equations

$$x_1 + 3x_2 - x_3 = 6 \quad \text{and} \quad x_2 - \frac{4}{5}x_3 = \frac{9}{5}.$$

Hence, assigning x_3 arbitrarily, we find that

$$x_1 = \frac{3}{5} - \frac{7}{5}x_3 \quad \text{and} \quad x_2 = \frac{9}{5} + \frac{4}{5}x_3.$$

When the inhomogeneous vector $\mathbf{K} = \mathbf{0}$, the resulting system of equations $\mathbf{AX} = \mathbf{0}$ is said to be *homogeneous*. Consider the case of a homogeneous system of n equations involving the n variables x_1, x_2, . . ., x_n. Then it is obvious that a *trivial* solution $x_1 = x_2 = \cdots = x_n = 0$ corresponding to $\mathbf{X} = \mathbf{0}$ always exists, but a *non-trivial* solution, in the sense that not all x_1, x_2, . . ., x_n are zero, can only occur if $|\mathbf{A}| = 0$. To see this notice that if $|\mathbf{A}| \neq 0$ then $\mathbf{A}^{-1}$ exists, so that premultiplication of $\mathbf{AX} = \mathbf{0}$ by $\mathbf{A}^{-1}$ gives at once the trivial solution $\mathbf{X} = \mathbf{0}$ as being the only possible solution. Conversely, if $|\mathbf{A}| = 0$, then certainly at least one row of $\mathbf{A}$ is linearly dependent upon the other rows, showing that not all of the variables x_1, x_2, . . ., x_n can be zero.

When a non-trivial solution exists to a homogeneous system of n equations involving n variables it cannot be unique, for if $\mathbf{X}$ is a solution vector, then so also is $\lambda\mathbf{X}$, where λ is a scalar. As in our previous discussion, if the rank of $\mathbf{A}$ which is of order $(n \times n)$ is r, then we may solve for r of the variables x_1, x_2, . . ., x_n in terms of the $n - r$ remaining ones which can then be assigned arbitrary values.

THEOREM 9·10 (solution of homogeneous systems) The homogeneous system of equations

$$\mathbf{AX} = \mathbf{0},$$

where $\mathbf{A}$ is of order $(n \times n)$ and $\mathbf{X}$, $\mathbf{0}$ are of order $(n \times 1)$ always has the trivial solution $\mathbf{X} = \mathbf{0}$. It has a non-trivial solution only when $|\mathbf{A}| = 0$. If $\mathbf{A}$ is of rank $r < n$, it is possible to solve for r variables in terms of the $n - r$ remaining variables which may then be assigned arbitrary values. If $\mathbf{X}$ is a non-trivial solution, so also is $\lambda\mathbf{X}$, where λ is an arbitrary scalar.

Example 9·20 Solve the equations

$$\begin{aligned}
x_1 - x_2 + x_3 &= 0 \\
2x_1 + x_2 - x_3 &= 0 \\
x_1 + 5x_2 - 5x_3 &= 0.
\end{aligned}$$

Solution   There is the trivial solution $x_1 = x_2 = x_3 = 0$ and, since the determinant associated with the coefficient matrix vanishes, there are also non-trivial solutions. The augmented matrix is now

$$(\mathbf{A}, \mathbf{0}) = \begin{bmatrix} 1 & -1 & 1 & 0 \\ 2 & 1 & -1 & 0 \\ 1 & 5 & -5 & 0 \end{bmatrix},$$

which is easily reduced by elementary row transformations to the echelon form

$$\begin{bmatrix} 1 & -1 & 1 & 0 \\ 0 & 1 & -1 & 0 \\ 0 & 0 & 0 & 0 \end{bmatrix}.$$

This shows that there are only two equations between the three variables x_1, x_2, and x_3, for the echelon form of the augmented matrix is seen to be equivalent to the two scalar equations

$$x_1 - x_2 + x_3 = 0 \qquad \text{and} \qquad x_2 - x_3 = 0.$$

Hence, assigning x_3 arbitrarily, we have for our solution $x_1 = 0$ and $x_2 = x_3 = k$ (say).

A practical numerical method of solution called *Gaussian elimination* is usually used when dealing with inhomogeneous systems of n equations involving n variables. This is essentially the same method as the one described above for the reduction of an augmented matrix to echelon form. The only difference is that it is not necessary to make the first non-zero element appearing in any row in the position corresponding to the leading diagonal equal to unity. We illustrate the method by example.

Example 9·21   Solve the following equations by Gaussian elimination:

$$x_1 - x_2 - x_3 = 0$$
$$3x_1 + x_2 + 2x_3 = 6$$
$$2x_1 + 2x_2 + x_3 = 2.$$

Solution   The augmented matrix

$$(\mathbf{A}, \mathbf{K}) = \begin{bmatrix} 1 & -1 & -1 & 0 \\ 3 & 1 & 2 & 6 \\ 2 & 2 & 1 & 2 \end{bmatrix}.$$

Subtracting three times row 1 from row 2 and twice row 1 from row 3 gives

$$\begin{bmatrix} 1 & -1 & -1 & 0 \\ 0 & 4 & 5 & 6 \\ 0 & 4 & 3 & 2 \end{bmatrix}.$$

Subtraction of row 2 from row 3 gives

$$\begin{bmatrix} 1 & -1 & -1 & 0 \\ 0 & 4 & 5 & 6 \\ 0 & 0 & -2 & -4 \end{bmatrix}.$$

The solution is now found by the process of 'back-substitution' using the scalar equations corresponding to this modified augmented matrix. That is, the equations

$$x_1 - x_2 - x_3 = 0$$
$$4x_2 + 5x_3 = 6$$
$$- 2x_3 = -4.$$

The last equation gives $x_3 = 2$ and, using this result in the second then gives $x_2 = -1$. Combination of these results in the first equation then gives $x_1 = 1$.

It is not proposed to offer more than a few general remarks about the solutions of m equations involving n variables. If the equations are consistent, but there are more equations than variables so that $m > n$, it is clear that there must be linear dependence between the equations. In the case that the rank of the coefficient matrix is equal to n there will obviously be a unique solution for, despite appearances, there will be only n linearly independent equations involving n variables. If, however, the rank is less than n we are in the situation of solving for r variables $x_1, x_2, \ldots$, in terms of the remaining $n - r$ variables whose values may be assigned arbitrarily. In the remaining case where there are fewer equations than variables we have $m < n$. When this system is consistent it follows that at least $n - m$ variables must be assigned arbitrary values.

When the number of variables involved becomes large enough to make Gaussian elimination impracticable, then approximate methods of solution must be used. The idea underlying the most important of these, which are known as *iterative methods*, is simply that given an approximation to the solution, a numerical rule or *algorithm* is developed which when starting from this approximation gives rise to a better one. If such an algorithm is applied repetitively, then each successive calculation to determine the next better approximation to the solution is called an *iteration*. The successive approximations themselves are called *iterates*. Of the many possible methods we mention only the *Jacobi* and the *Gauss–Seidel* iterative schemes.

These apply to systems of equations having coefficient matrices $\mathbf{A} = [a_{ij}]$

with the property that for each i the coefficient a_{ii} in the ith row of matrix **A** has magnitude $|a_{ii}|$ greater than the sum of the magnitudes of the other coefficients in that same row. Such matrices are called *diagonally dominant* matrices because in iterative schemes the elements a_{ii} in the leading diagonal of **A** dominate the calculations. This type of matrix frequently arises in connection with certain numerical methods for the solution of differential equations where the matrices **A** that occur are both large and sparse (they contain numerous zero elements).

Both of the methods we now describe have essentially the same starting point. We shall work with a system of linear simultaneous equations of the form shown in (9·24), but it will be assumed here that $m = n$ so that there are as many equations as there are unknowns. On the assumption that the system is diagonally dominant we begin by rewriting it in the form

$$x_1 = (k_1 - a_{12}x_2 - a_{13}x_3 - \cdots - a_{1n}x_n)/a_{11},$$
$$x_2 = (k_2 - a_{21}x_1 - a_{23}x_3 - \cdots - a_{2n}x_n)/a_{22}, \qquad (9\cdot24')$$
$$\qquad \cdot \qquad \cdot$$
$$x_n = (k_n - a_{n1}x_1 - a_{n2}x_2 - \cdots - a_{nn-1}x_{n-1})a_{nn}.$$

Now suppose that the $(r - 1)$th iteration gives rise to the approximations $x_1^{(r-1)}, x_2^{(r-1)}, \ldots, x_n^{(r-1)}$ to the required solutions $x_1, x_2, \ldots, x_n$. Then in the *Jacobi iterative* method the rth iterates are determined from the $(r - 1)$th iterates by means of the algorithm

$$x_1^{(r)} = (k_1 - a_{12}x_2^{(r-1)} - a_{13}x_3^{(r-1)} - \cdots - a_{1n}x_n^{(r-1)})/a_{11},$$
$$x_2^{(r)} = (k_2 - a_{21}x_1^{(r-1)} - a_{23}x_3^{(r-1)} - \cdots - a_{2n}x_n^{(r-1)})/a_{22},$$
$$x_3^{(r)} = (k_3 - a_{31}x_1^{(r-1)} - a_{32}x_2^{(r-1)} - \cdots - a_{3n}x_n^{(r-1)})/a_{33},$$
$$\qquad \cdot \qquad \cdot \qquad \cdot$$
$$x_n^{(r)} = (k_n - a_{n1}x_1^{(r-1)} - a_{n2}x_2^{(r-1)} - \cdots - a_{nn-1}x_{n-1}^{(r-1)})/a_{nn}.$$

This has been obtained from the rewritten system (9·24′) by simply employing the $(r - 1)$th iterates in the right-hand side to determine the rth iterates. To start the iteration procedure any values may be assumed for $x_1^{(0)}, x_2^{(0)}, \ldots, x_n^{(0)}$. Unless some approximate solution is known it is customary to start either by setting each $x_i^{(0)} = 1$ or, alternatively, each $x_i^{(0)} = 0$. The iteration process is terminated once the desired accuracy has been attained. This is achieved by stopping the calculation when the n numbers $|x_i^{(r)} - x_i^{(r-1)}|$, for $i = 1, 2, \ldots, n$, are all less than some predetermined small number $\varepsilon > 0$.

The *Gauss–Seidel* iterative method differs from the Jacobi method in that it uses each approximation $x_i^{(r)}$ in the rth iteration as soon as it is available. The algorithm for the Gauss–Seidel method is thus as follows

$$x_1^{(r)} = (k_1 - a_{12}x_2^{(r-1)} - a_{13}x_3^{(r-1)} - \cdots - a_{1n}x_n^{(r-1)})/a_{11},$$
$$x_2^{(r)} = (k_2 - a_{21}x_1^{(r)} - a_{23}x_3^{(r-1)} - \cdots - a_{2n}x_n^{(r-1)})/a_{22},$$
$$x_3^{(r)} = (k_3 - a_{31}x_1^{(r)} - a_{32}x_2^{(r)} - \cdots - x_{3n}x_n^{(r-1)})/a_{33},$$
$$\cdots \cdots$$
$$x_n^{(r)} = (k_n - a_{n1}x_1^{(r)} - a_{n2}x_2^{(r)} - \cdots - a_{nn-1}x_{n-1}^{(r)})/a_{nn}.$$

The Gauss–Seidel iterative method is started and terminated in exactly the same fashion as the Jacobi method, though its rate of convergence is roughly double that of the Jacobi method. The condition of diagonal dominance imposed on each of these methods is, in fact, just a sufficient condition to ensure their convergence. If this condition is not satisfied then successive iterations may diverge.

Example 9·22 Solve by the Jacobi and the Gauss–Seidel iterative methods the system of equations

$$5x_1 + x_2 - x_3 = 4$$
$$x_1 + 4x_2 + 2x_3 = 15$$
$$x_1 - 2x_2 + 5x_3 = 12,$$

which has the exact solution $x_1 = 1$, $x_2 = 2$ and $x_3 = 3$.

Solution As $|5| > |1| + |-1|$, $|4| > |1| + |2|$ and $|5| > |1| + |-2|$ the system is diagonally dominant and so the two iterative methods may be used. Both calculations start from the equations derived by rewriting the system in the form

$$x_1 = 0 \cdot 8 - 0 \cdot 2x_2 + 0 \cdot 2x_3$$
$$x_2 = 3 \cdot 75 - 0 \cdot 25x_1 - 0 \cdot 5x_3$$
$$x_3 = 2 \cdot 4 - 0 \cdot 2x_1 + 0 \cdot 4x_2.$$

Starting by setting $x_i^{(0)} = 1$, for $i = 1, 2, 3$, the result of iteration by each method is as follows and clearly shows the faster convergence of the Gauss–Seidel method.

Iteration number	Jacobi:			Gauss–Seidel:		
r	$x_1^{(r)}$	$x_2^{(r)}$	$x_3^{(r)}$	$x_1^{(r)}$	$x_2^{(r)}$	$x_3^{(r)}$
0	1·0	1·0	1·0	1·0	1·0	1·0
1	0·8	3·0	2·6	0·8	3·05	3·46
2	0·72	2·25	3·44	0·882	1·7995	2·934
3	1·038	1·85	3·632	1·0288	2·0211	3·0026
4	1·1564	1·6745	2·9324	0·9963	1·9996	3·0005
5	1·0516	1·9947	2·8385	1·0002	1·9997	2·9999
6	0·9688	2·0681	2·9876	1·0	2·0	3·0
7	0·9839	2·0140	3·0334	—	—	—
8	1·0039	1·9873	3·0088	—	—	—

9·7   Eigenvalues and eigenvectors

Let us examine the consequence of requiring that in the system

$$AX = K, \tag{9.29}$$

where A is of order ($n \times n$) and X, K are of order ($n \times 1$), the vector K is proportional to the vector X itself. That is, we are requiring that $K = \lambda X$, where λ is some scalar multiplier as yet unknown. This requires us to solve the system

$$AX = \lambda X, \tag{9.30}$$

which is equivalent to the homogeneous system

$$(A - \lambda I)X = 0, \tag{9.31}$$

where I is the unit matrix.

Now we know from Theorem 9·10 that Eqn (9·31) can only have a nontrivial solution when the determinant associated with the coefficient matrix vanishes, so that we must have

$$| A - \lambda I | = 0. \tag{9.32}$$

When expanded, this determinant gives rise to an algebraic equation of degree n in λ of the form

$$\lambda^n + \alpha_1 \lambda^{n-1} + \alpha_2 \lambda^{n-2} + \cdots + \alpha_n = 0. \tag{9.33}$$

The determinant (9·32) is called the *characteristic determinant* associated with A and Eqn (9·33) is called the *characteristic equation*. It has n roots λ_1, λ_2, . . ., λ_n, each of which is called either an *eigenvalue*, a *characteristic root*, or, in some texts, a *latent root* of A.

Example 9·23   Find the characteristic equation and the eigenvalues corresponding to

$$A = \begin{bmatrix} 1 & 2 \\ 3 & 0 \end{bmatrix}.$$

Solution   We have

$$A - \lambda I = \begin{bmatrix} 1 & 2 \\ 3 & 0 \end{bmatrix} - \lambda \begin{bmatrix} 1 & 0 \\ 0 & 1 \end{bmatrix} = \begin{bmatrix} 1 - \lambda & 2 \\ 3 & -\lambda \end{bmatrix},$$

so that

$$| A - \lambda I | = \begin{vmatrix} 1 - \lambda & 2 \\ 3 & -\lambda \end{vmatrix} = \lambda^2 - \lambda - 6.$$

Thus the characteristic equation is

$$\lambda^2 - \lambda - 6 = 0,$$

and its roots, the eigenvalues of $\mathbf{A}$, are $\lambda = 3$ and $\lambda = -2$.

No consideration will be given here to the interpretation that is to be placed on the appearance of repeated roots of the characteristic equation, and henceforth we shall always assume that all the eigenvalues (roots) are distinct.

Returning to Eqn (9·31) and setting $\lambda = \lambda_i$, where λ_i is any one of the eigenvalues, we can then find a corresponding solution vector $\mathbf{X}_i$ which because of Theorem 9·10, will only be determined to within an arbitrary scalar multiplier. This vector $\mathbf{X}_i$ is called either an *eigenvector*, a *characteristic vector* or, a *latent vector* of $\mathbf{A}$ corresponding to λ_i. The eigenvectors of a square matrix $\mathbf{A}$ are of fundamental importance in both the theory of matrices and in their application,

Example 9·24   Find the eigenvectors of the matrix $\mathbf{A}$ in Example 9·23.

Solution   Use the fact that the eigenvalues have been determined as being $\lambda = 3$ and $\lambda = -2$ and make the identifications $\lambda_1 = 3$ and $\lambda_2 = -2$. Now let the eigenvectors $\mathbf{X}_1$ and $\mathbf{X}_2$, corresponding to λ_1 and λ_2, be denoted by

$$\mathbf{X}_1 = \begin{bmatrix} x_1^{(1)} \\ x_2^{(1)} \end{bmatrix} \quad \text{and} \quad \mathbf{X}_2 = \begin{bmatrix} x_1^{(2)} \\ x_2^{(2)} \end{bmatrix}.$$

Then for the case $\lambda = \lambda_1$, Eqn (9·31) becomes

$$\begin{bmatrix} (1-3) & 2 \\ 3 & (0-3) \end{bmatrix} \begin{bmatrix} x_1^{(1)} \\ x_2^{(1)} \end{bmatrix} = \mathbf{0},$$

whence

$$-2x_1^{(1)} + 2x_2^{(1)} = 0 \quad \text{and} \quad 3x_1^{(1)} - 3x_2^{(1)} = 0.$$

These are automatically consistent by virtue of their manner of definition, so that we find that $x_1^{(1)} = x_2^{(2)}$. So, arbitrarily assigning to $x_1^{(1)}$ the value $x_1^{(1)} = 1$, we find that the eigenvector $\mathbf{X}_1$ corresponding to $\lambda_1 = 3$ is

$$\mathbf{X}_1 = \begin{bmatrix} 1 \\ 1 \end{bmatrix}.$$

A similar argument for $\lambda = \lambda_2$ gives

$$\begin{bmatrix} (1+2) & 2 \\ 3 & (0+2) \end{bmatrix} \begin{bmatrix} x_1^{(2)} \\ x_2^{(2)} \end{bmatrix} = \mathbf{0},$$

whence

$$3x_1^{(2)} + 2x_2^{(2)} = 0.$$

Again, arbitrarily assigning to $x_1^{(2)}$ the value $x_1^{(2)} = 1$, we find that $x_2^{(2)} = -3/2$. Thus the eigenvector $\mathbf{X}_2$ corresponding to $\lambda_2 = -2$ is

$$\mathbf{X}_2 = \begin{bmatrix} 1 \\ -\dfrac{3}{2} \end{bmatrix}.$$

Obviously $\mu\mathbf{X}_1$ and $\mu\mathbf{X}_2$ are also eigenvectors for any arbitrary scalar μ.

PROBLEMS

Section 9·1

9·1 Suggest two physical situations in which the outcomes may be displayed in the form of a matrix.

9·2 Find the sum $\mathbf{A} + \mathbf{B}$ and difference $\mathbf{A} - \mathbf{B}$ of the matrices

$$\mathbf{A} = \begin{bmatrix} 1 & 2 & 3 & 4 \\ 2 & 1 & 2 & 2 \\ 1 & 2 & 0 & 0 \end{bmatrix}, \quad \mathbf{B} = \begin{bmatrix} 2 & 3 & 1 & 2 \\ 0 & 2 & 2 & 0 \\ 1 & -2 & 1 & 1 \end{bmatrix}.$$

9·3 Evaluate the following inner products:

(a) $[2 \ \ 1 \ \ 1 \ \ 3] \begin{bmatrix} 1 \\ 2 \\ 2 \\ 1 \end{bmatrix}$; (b) $[1 \ \ -2 \ \ 7 \ \ 4] \begin{bmatrix} 2 \\ 3 \\ 0 \\ 1 \end{bmatrix}$; (c) $[2 \ \ -1 \ \ 3 \ \ 1] \begin{bmatrix} 2 \\ -1 \\ 3 \\ 1 \end{bmatrix}$.

9·4 Evaluate the following matrix products:

(a) $\begin{bmatrix} 0 & 3 & 1 & 2 \\ 1 & 2 & 2 & 2 \\ 1 & 1 & 1 & 0 \end{bmatrix} \begin{bmatrix} 1 & 2 \\ 1 & -1 \\ 0 & 0 \\ 1 & 1 \end{bmatrix}$; (b) $\begin{bmatrix} -1 & 2 & 1 & 3 \\ 1 & -1 & 1 & -1 \\ 1 & 0 & 0 & 1 \end{bmatrix} \begin{bmatrix} -1 & 1 \\ 2 & 2 \\ 1 & 1 \\ 3 & 2 \end{bmatrix}$.

9·5 State which of the following forms of matrix product are defined and, where appropriate, give the shape of the resulting product matrix:

(a) $(7 \times 3)(3 \times 9)$;   (b) $(5 \times 3)(2 \times 3)$;

(c) $(1 \times 9)(9 \times 1)$;   (d) $(3 \times 1)(1 \times 4)$.

9·6 Display each of the following sets of simultaneous equations in matrix form:

(a) $2x + 4y + z = 9$
$x - 3y + 2z = -4$
$x + y - z = 1,$

(c) $3w + x - 2y + 4z = 1$
$w - 3x + y - 3z = 4$
$w + 7x + 2y + 5z = 2,$

(b) $w + 2x - y = 4$
$x - 3y + 2z = -1$
$2w + 5x - 3z = 0$
$4w - y + 4z = 2,$

Section 9·2

9·7 State which of the following pairs of matrices can be made equal by assigning suitable values to the constants a, b, and c. Where appropriate, determine what these values must be.

(a) $\begin{bmatrix} 1 & 2 & 1 & 0 \\ 3 & a & b & 2 \\ 1 & 2 & c & 1 \end{bmatrix}$ and $\begin{bmatrix} 1 & 2 & 1 & 0 \\ 3 & 1 & 2 & 2 \\ 1 & 2 & 4 & 1 \end{bmatrix}$,

(b) $\begin{bmatrix} 1 & 5 & a & 2 \\ 2 & a^2 & 3 & b \\ 4 & 3 & 2 & c \end{bmatrix}$ and $\begin{bmatrix} 1 & 5 & 1 & 2 \\ 2 & 4 & 3 & 4 \\ 4 & 3 & 2 & 1 \end{bmatrix}$,

(c) $\begin{bmatrix} 1 & (a+b) & 3 \\ (a+c) & 2 & 4 \\ 1 & 2 & (b+c) \end{bmatrix}$ and $\begin{bmatrix} 1 & 4 & 3 \\ 0 & 2 & 4 \\ 1 & 2 & 2 \end{bmatrix}$.

9·8 Determine $3A + 2B$ and $2A - 6B$ given that

$$A = \begin{bmatrix} 1 & 3 & 7 \\ 2 & -1 & 6 \end{bmatrix} \text{ and } B = \begin{bmatrix} 2 & -1 & 4 \\ 3 & -3 & 2 \end{bmatrix}.$$

9·9 If

$$A = \begin{bmatrix} 2 & 1 & 0 \\ 3 & 2 & 0 \\ 1 & 0 & 1 \end{bmatrix}, \quad B = \begin{bmatrix} 1 & 1 & 1 & 0 \\ 2 & 1 & 1 & 0 \\ 2 & 3 & 1 & 2 \end{bmatrix}, \quad C = \begin{bmatrix} 2 & 3 & 4 \\ 1 & 5 & 6 \end{bmatrix},$$

and

$$D = \begin{bmatrix} 1 \\ 2 \\ 3 \end{bmatrix},$$

find the matrix products A B and C D.

9·10 This example shows that the matrix product $\mathbf{A}\,\mathbf{B} = \mathbf{0}$ does not necessarily imply either that $\mathbf{A} = \mathbf{0}$ or that $\mathbf{B} = \mathbf{0}$. If,

$$\mathbf{A} = \begin{bmatrix} 1 & -1 & 1 \\ -3 & 2 & -1 \\ -2 & 1 & 0 \end{bmatrix} \quad \text{and} \quad \mathbf{B} = \begin{bmatrix} 1 & 2 & 3 \\ 2 & 4 & 6 \\ 1 & 2 & 3 \end{bmatrix},$$

find $\mathbf{A}\,\mathbf{B}$ and $\mathbf{B}\,\mathbf{A}$ and show that $\mathbf{A}\,\mathbf{B} \neq \mathbf{B}\,\mathbf{A}$.

9·11 Show that the matrix equation

$$\mathbf{A}\,\mathbf{X} = \mathbf{K},$$

where

$$\mathbf{A} = \begin{bmatrix} 1 & 3 & 1 \\ 1 & 1 & 2 \\ 2 & 2 & 0 \end{bmatrix}, \quad \mathbf{X} = \begin{bmatrix} x_1 \\ x_2 \\ x_3 \end{bmatrix}, \quad \text{and} \quad \mathbf{K} = \begin{bmatrix} 1 \\ 2 \\ 3 \end{bmatrix},$$

may be solved for x_1, x_2, and x_3 by pre-multiplication by $\mathbf{B}$, where

$$\mathbf{B} = \begin{bmatrix} -\tfrac{1}{2} & \tfrac{1}{4} & \tfrac{5}{8} \\ \tfrac{1}{2} & -\tfrac{1}{4} & -\tfrac{1}{8} \\ 0 & \tfrac{1}{2} & -\tfrac{1}{4} \end{bmatrix}.$$

9·12 Use matrix multiplication to verify the results of Theorem 9·2 when $\mathbf{A}$, $\mathbf{B}$, and $\mathbf{C}$ are of the form

$$\mathbf{A} = \begin{bmatrix} 1 & 3 & 2 \\ 0 & 1 & 4 \\ 2 & 3 & 1 \end{bmatrix}, \quad \mathbf{B} = \begin{bmatrix} -1 & 2 & 1 \\ 3 & -2 & -1 \\ 1 & 4 & 2 \end{bmatrix}, \quad \text{and} \quad \mathbf{C} = \begin{bmatrix} -2 & 2 & 1 \\ 0 & 2 & 4 \\ 1 & 3 & 1 \end{bmatrix}.$$

9·13 If $\mathbf{A}$ is a square matrix, then the associative property of matrices allows us to write $\mathbf{A}^n$ without ambiguity because, for example, $\mathbf{A}^3 = \mathbf{A}(\mathbf{A}\,\mathbf{A}) = (\mathbf{A}\,\mathbf{A})\mathbf{A}$. If

$$\mathbf{A} = \begin{bmatrix} \cosh x & \sinh x \\ \sinh x & \cosh x \end{bmatrix},$$

use the hyperbolic identities to express $\mathbf{A}^2$ and $\mathbf{A}^3$ in their simplest form and use induction to deduce the form of $\mathbf{A}^n$.

9·14 Transpose the following matrices:

(a) $[1 \quad 4 \quad 17 \quad 3]$; (b) $\begin{bmatrix} 2 & 5 & 7 & 9 \\ 4 & 3 & 0 & 1 \end{bmatrix}$; (c) $\begin{bmatrix} 1 & 4 & 19 \\ 4 & 0 & 2 \\ 19 & 2 & 4 \end{bmatrix}$;

(d) $\begin{bmatrix} 0 & 3 & -2 \\ -3 & 0 & 1 \\ 2 & -1 & 0 \end{bmatrix}$; (e) $\begin{bmatrix} 4 \\ 3 \\ 1 \\ 0 \end{bmatrix}$.

9·15 Use Definition 9·7 and Theorem 9·3 to prove that:
    (a) the sum of a square matrix and its transpose is a *symmetric* matrix;
    (b) the difference of a square matrix and its transpose is a *skew-symmetric* matrix.

Illustrate each of these results by an example.

9·16 Verify that $(\mathbf{A}\,\mathbf{B})' = \mathbf{B}'\,\mathbf{A}'$, given that

$$\mathbf{A} = \begin{bmatrix} 1 & 4 & 7 \\ 9 & -3 & 1 \end{bmatrix} \quad \text{and} \quad \mathbf{B} = \begin{bmatrix} -4 & 2 \\ 3 & 1 \\ -5 & 6 \end{bmatrix}.$$

Section 9·3

9·17 Evaluate the determinants

$$\text{(a)} \begin{vmatrix} 1 & 2 \\ 4 & 7 \end{vmatrix}; \quad \text{(b)} \begin{vmatrix} 1 & 0 & 3 \\ 2 & 0 & 5 \\ 1 & 3 & 7 \end{vmatrix}; \quad \text{(c)} \begin{vmatrix} 1 & 2 & 5 \\ 3 & 1 & 5 \\ -5 & 0 & -5 \end{vmatrix}.$$

9·18 Without expanding the determinant, prove that

$$\begin{vmatrix} 1 + a_1 & a_1 & a_1 \\ a_2 & 1 + a_2 & a_2 \\ a_3 & a_3 & 1 + a_3 \end{vmatrix} = (1 + a_1 + a_2 + a_3).$$

9·19 Use Theorem 9·4 to simplify the following determinants before expansion:

$$\text{(a)} \quad |\mathbf{A}| = \begin{vmatrix} 42 & 61 & 50 \\ 3 & 0 & 2 \\ 4 & 6 & 5 \end{vmatrix}; \quad \text{(b)} \quad |\mathbf{A}| = \begin{vmatrix} 0 & 9 & 3 \\ 2 & 16 & 4 \\ 1 & 2 & 1 \end{vmatrix};$$

$$\text{(c)} \quad |\mathbf{A}| = \begin{vmatrix} 2 & 1 & 5 \\ 5 & 17 & 56 \\ 4 & 1 & 7 \end{vmatrix}.$$

9·20 Show without expansion that

$$\begin{vmatrix} a^2 & b^2 & c^2 \\ a & b & c \\ 1 & 1 & 1 \end{vmatrix} = (a - b)(a - c)(b - c).$$

This determinant is called an *alternant* determinant. Illustrate the result by means of a numerical example and verify it by direct expansion.

9·21 Find the minors M_{ij} and cofactors A_{ij} of each element a_{ij} in the matrix

$$\begin{bmatrix} -\tfrac{1}{3} & -\tfrac{2}{3} & -\tfrac{2}{3} \\ \tfrac{2}{3} & \tfrac{1}{3} & -\tfrac{2}{3} \\ \tfrac{2}{3} & -\tfrac{2}{3} & \tfrac{1}{3} \end{bmatrix}.$$

Section 9·4

9·22 Which of the following sets of vectors are linearly independent, and where linear dependence exists determine its form:

(a) $C_1 = \begin{bmatrix} 3 \\ 0 \\ 0 \end{bmatrix}$, $C_2 = \begin{bmatrix} 0 \\ -7 \\ 0 \end{bmatrix}$, $C_3 = \begin{bmatrix} 0 \\ 0 \\ 15 \end{bmatrix}$;

(b) $R_1 = [1 \quad 9 \quad -2 \quad 14]$, $R_2 = [-2 \quad -18 \quad 4 \quad -28]$;

(c) $C_1 = \begin{bmatrix} 2 \\ 1 \\ 0 \end{bmatrix}$, $C_2 = \begin{bmatrix} 1 \\ 1 \\ 2 \end{bmatrix}$, $C_3 = \begin{bmatrix} 1 \\ 2 \\ 1 \end{bmatrix}$, $C_4 = \begin{bmatrix} 5 \\ 6 \\ 4 \end{bmatrix}$.

9·23 Test the following matrices for linear independence between their rows or columns:

(a) $\begin{bmatrix} 1 & 2 & -1 & 0 \\ 2 & 3 & 1 & 1 \\ -1 & 1 & 0 & 2 \\ 0 & 1 & 2 & 3 \end{bmatrix}$; (b) $\begin{bmatrix} 0 & 2 & 3 & 1 \\ -2 & 0 & -1 & 2 \\ -3 & 1 & 0 & -2 \\ -1 & -2 & 2 & 0 \end{bmatrix}$; (c) $\begin{bmatrix} 1 & 2 & 1 & 5 \\ 2 & 1 & 2 & 0 \\ 1 & 0 & 2 & 1 \\ 5 & 3 & 7 & 7 \end{bmatrix}$.

Section 9·5

9·24 Show that adj $A = A$ when

$$A = \begin{bmatrix} -4 & -3 & -3 \\ 1 & 0 & 1 \\ 4 & 4 & 3 \end{bmatrix}.$$

9·25 Find the matrix adjoint to each of the following matrices:

(a) $\begin{bmatrix} 1 & 2 & 3 \\ 2 & 3 & 2 \\ 3 & 3 & 4 \end{bmatrix}$; (b) $\begin{bmatrix} 1 & 2 & 3 \\ 1 & 3 & 4 \\ 1 & 4 & 3 \end{bmatrix}$; (c) $\begin{bmatrix} a & b \\ c & d \end{bmatrix}$.

9·26 Set

$$\begin{bmatrix} 1 & 2 \\ 3 & 4 \end{bmatrix}\begin{bmatrix} a & b \\ c & d \end{bmatrix} = \begin{bmatrix} 1 & 0 \\ 0 & 1 \end{bmatrix}$$

and equate corresponding elements to determine the inverse of

$$\begin{bmatrix} 1 & 2 \\ 3 & 4 \end{bmatrix}.$$

9·27 Find the inverse of

$$A = \begin{bmatrix} 3 & -2 & -1 \\ -4 & 1 & -1 \\ 2 & 0 & 1 \end{bmatrix}.$$

Verify that:
(a) $A^{-1} A = A A^{-1} = I$;
(b) $(A^{-1})^{-1} = A$.

9·28 Given that A and B are

$$A = \begin{bmatrix} 1 & 2 & 1 \\ 1 & 4 & 2 \\ 0 & 3 & 2 \end{bmatrix} \quad \text{and} \quad B = \begin{bmatrix} 1 & -1 & 2 \\ 0 & 2 & 4 \\ 1 & 0 & 3 \end{bmatrix},$$

verify that $(A B)^{-1} = B^{-1} A^{-1}$.

Section 9·6

9·29 Solve the following equations using Cramer's rule:

$$\begin{aligned} x_1 + x_2 + x_3 &= 7 \\ 2x_1 - x_2 + 2x_3 &= 8 \\ 3x_1 + 2x_2 - x_3 &= 11. \end{aligned}$$

9·30 Solve the equations of the previous example using the inverse matrix method and compare the task with the previous method.

9·31 Solve the following equations using Cramer's rule:

$$\begin{aligned} x_1 - x_2 + x_3 - x_4 &= 1 \\ 2x_1 - x_2 + 3x_3 + x_4 &= 2 \\ x_1 + x_2 + 2x_3 + 2x_4 &= 3 \\ x_1 + x_2 + x_3 + x_4 &= 3. \end{aligned}$$

9·32 Write down the augmented matrix corresponding to the equations:

$$\begin{aligned} 2x_1 - x_2 + 3x_3 &= 1 \\ 3x_1 + 2x_2 - x_3 &= 4 \\ x_1 - 4x_2 + 7x_3 &= 3. \end{aligned}$$

Show, by reducing this matrix to its echelon equivalent, that these equations are inconsistent.

9·33 Write down the augmented matrix corresponding to the equations:

$$\begin{aligned} 3x_1 + 2x_2 - x_3 &= 4 \\ 2x_1 - 5x_2 + 2x_3 &= 1 \\ 5x_1 + 16x_2 - 7x_3 &= 10. \end{aligned}$$

Show, by reducing this matrix to its echelon equivalent, that these equations are consistent and solve them.

9·34 Solve the following equations, in which α is an arbitrary constant, by reducing the augmented matrix to echelon form:

$$x_1 + \alpha x_2 + \alpha x_3 = 1$$
$$\alpha x_1 + x_2 + 2\alpha x_3 = -4$$
$$\alpha x_1 - \alpha x_2 + 4x_3 = 2.$$

Consider the effect of α on the solution.

9·35 Discuss briefly, but do not solve, the following sets of equations:

(a) $$x_1 + x_2 = 1$$
$$2x_1 - x_2 = 5;$$

(b) $$x_1 + x_2 = 1$$
$$2x_1 - x_2 = 5$$
$$x_1 - x_2 = 0;$$

(c) $$x_1 + x_2 = 1$$
$$2x_1 - x_2 = 5$$
$$-x_1 - 2x_2 = 0;$$

(d) $$x_1 + x_2 - x_3 = 0$$
$$2x_1 - x_2 - 5x_3 = 0.$$

9·36 Solve the following equations using Gaussian elimination:

$$1{\cdot}202x_1 - 4{\cdot}371x_2 + 0{\cdot}651x_3 = 19{\cdot}447$$
$$-3{\cdot}141x_1 + 2{\cdot}243x_2 - 1{\cdot}626x_3 = -13{\cdot}702$$
$$0{\cdot}268x_1 - 0{\cdot}876x_2 + 1{\cdot}341x_3 = 6{\cdot}849.$$

9·37 Set $x_i^{(0)} = 1$, for $i = 1, 2, 3$, and complete four iterations of the Jacobi and Gauss–Seidel methods for the system of equations

$$7x_1 - x_2 + x_3 = 7{\cdot}3$$
$$2x_1 - 8x_2 - x_3 = -6{\cdot}4$$
$$x_1 + 2x_2 + 9x_3 = 13{\cdot}6.$$

Compare the results of the fourth iteration with the exact solution $x_1 = 1$, $x_2 = 0{\cdot}9$ and $x_3 = 1{\cdot}2$.

Section 9·7

9·38 Write down the characteristic equations for the following matrices:

(a) $\mathbf{A} = \begin{bmatrix} 1 & 4 \\ 3 & 7 \end{bmatrix}$; (b) $\mathbf{A} = \begin{bmatrix} 1 & 0 & 2 \\ 2 & 1 & 1 \\ 0 & 2 & 1 \end{bmatrix}$.

9·39 Find the eigenvalues and eigenvectors of

$$\mathbf{A} = \begin{bmatrix} 1 & -1 \\ -2 & 0 \end{bmatrix}.$$

9·40 Prove that the eigenvalues of a diagonal matrix of any order are given by the elements on the leading diagonal. What form do the eigenvectors take.

10 Functions of a complex variable

10·1 Curves and regions

The notions of a curve and a region in the real plane may be immediately extended to the complex plane. As a closed and not necessarily smooth curve is a connected set of points which serves to de-limit two areas of the plane, which we shall call the *interior* and *exterior* regions relative to that curve, we ought first to define a curve C in the complex plane. It is frequently convenient to give a parametric representation by expressing C as the set of points

$$z = x(s) + iy(s) \qquad \text{for } a \leq s \leq b, \tag{10·1}$$

where $x(s)$ and $y(s)$ are continuous real functions of the parameter s. It should be apparent from Section 2·5 and subsequent work that the requirement of continuity for the real functions $x(s)$, $y(s)$ will ensure that C is a continuous curve (that is, unbroken), but that it does not necessarily possess a tangent at every point. As a simple illustration C might be a rectangle, for then tangents would not be defined at the corners though the curve would be continuous everywhere. We shall return to these general matters later when a continuous function of a complex variable has been defined. For conciseness let us henceforth call such curves C, *continuous* curves.

For a less trivial example, suppose that the curve C in the complex plane is defined by $z = x(s) + iy(s)$, where

$$x(s) = \sin s \quad \text{for } -\tfrac{1}{2}\pi \leq s \leq \frac{3\pi}{2},$$

$$y(s) = \begin{cases} \sin^2 s & \text{for } -\tfrac{1}{2}\pi \leq s \leq \tfrac{1}{2}\pi \\[2mm] 1 & \text{for } \tfrac{1}{2}\pi < s \leq \dfrac{3\pi}{2}. \end{cases}$$

Then it is readily seen that C is the continuous closed curve comprising the parabola $y = x^2$ in the interval $-1 \leq x \leq 1$, together with the points of the line $y = 1$ common to that same interval. The curve C is shown in Fig. 10·1 and it is continuous everywhere, though it is not smooth everywhere for no tangent can be defined at points P and Q. The darkly shaded area in that Figure comprises points which are interior relative to C and form the interior region, whilst the lightly shaded area comprises points which are exterior relative to C and form the exterior region. When speaking in terms of regions, the points comprising the curve C itself are usually called the *boundary* points and they may, or may not, belong to a region.

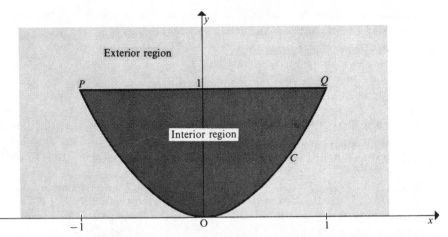

Fig. 10·1   Continuous curve C having no tangent defined at P and Q.

A parametric representation of a curve C is not always the most convenient method for its description in the complex plane and, on occasions, it is better to identify the points z comprising a curve directly in terms of z itself. When necessary, regions are usually defined in the complex plane by means of a combination of curves and inequalities, as was done in the real plane.

Example 10·1   Describe the curve C defined by the equation

$$|z - 2| = \tfrac{3}{2}$$

and use the result to define the region exterior to C.

Solution   This expression defines a connected set of points that all have a modulus 3/2 relative to the point $z = 2$ as origin, that is to say, the set of points which are all distant 3/2 from the point $z \doteq 2$. Hence the equation $|z - 2| = 3/2$ describes a circle C of radius 3/2 centred on the point $z = 2$. Algebraically, the same result is obtained by writing $z = x + iy$, when $|z - 2| = |(x - 2) + iy|$, so that from the definition of the modulus of a complex number, $|z - 2| = 3/2$ is seen to be equivalent to the algebraic equation $(x - 2)^2 + y^2 = 9/4$. This is a circle of radius 3/2 centred on the point $(2, 0)$. The region exterior to C is the entire complex plane less the points lying in and on this circle.

Example 10·2   Describe the region interior to and including the curve C defined by

$$\arg(z - 1) - \arg(z - i) = \tfrac{1}{2}\pi,$$

and also satisfying the inequalities

$$\tfrac{1}{4} < \operatorname{Re} z \leq \tfrac{3}{4} \qquad \text{and} \qquad \operatorname{Im} z \geq 0.$$

Solution   Consider the construction in Fig. 10·2 (a) in which P is the point $z = 1$, Q is the point $z = i$ and R is a general point z.

Simple geometrical arguments then establish that the angle γ is related to the angles α and β by the equation

$$\gamma = \pi + \alpha - \beta.$$

However, the line PR is the vector $z - 1$, whilst the line QR is the vector $z - i$, so that $\arg(z - i) = \alpha$ and $\arg(z - 1) = \beta$. Since by the conditions of the problem we must have $\beta - \alpha = \frac{1}{2}\pi$, it follows that $\gamma = \frac{1}{2}\pi$. The angle QRP is thus a right angle and hence the curve C must be a semi-circle drawn

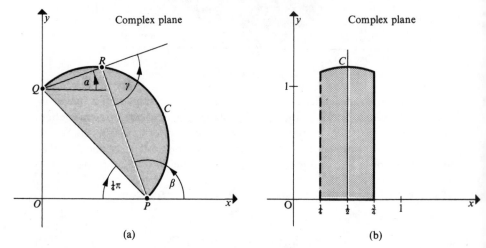

(a)   (b)

Fig. 10·2   Region in complex plane: (a) boundary curve; (b) region interior to C and satisfying stated inequalities.

from P to Q with PQ as its diameter. The semi-circle must lie *above* the diameter PQ, since were.the general point R to be taken below that line the equation relating the arguments would no longer be satisfied. To define the lower semi-circle the following condition would be needed:

$$\arg(z - 1) - \arg(z - i) = -\tfrac{1}{2}\pi.$$

To complete the solution to the problem it is now necessary to interpret the inequalities. The inequality $\frac{1}{4} < \operatorname{Re} z \le \frac{3}{4}$ describes the narrow strip bounded by the lines $x = \frac{1}{4}$ and $x = \frac{3}{4}$, with the points of the line $x = \frac{1}{4}$ excluded from consideration. The inequality $\operatorname{Im} z \ge 0$ is the half plane above and including the x-axis itself. Figure 10·2 (b) presents a composite diagram with the shaded area representing the region satisfying all the conditions of the problem. Boundary points belonging to the region are indicated by a heavy line and those excluded by a dotted line.

Notice from this and the previous example that there is more than one way of specifying a given curve and region. The condition

$\arg (z - 1) - \arg (z - i) = \tfrac{1}{2}\pi$

is an alternative expression of the condition

$$| z - \tfrac{1}{2} - \tfrac{1}{2}i | = \frac{\sqrt{2}}{2}$$

with $\operatorname{Re} z \geq 0$, $\operatorname{Im} z \geq 0$,

which, in turn, is an alternative expression of the algebraic condition

$$(x - \tfrac{1}{2})^2 + (y - \tfrac{1}{2})^2 = \tfrac{1}{2}$$

with $x \geq 0$, $y \geq 0$.

10·2   Function of a complex variable, limits, continuity and differentiability

In Chapter 2 we used the term 'a real valued function of a real variable' to mean any rule that associates with each real number from the domain of definition of the function a unique real number from the range of that function. Symbolically, if D denotes the set of points in the domain of a function f, and R denotes the set of points in the range of f, this relationship or mapping is given by

$$R = f(D).$$

These ideas still hold good when the domain D and the range R include complex numbers. Thus if z is any point in D, and w is the unique number assigned to z by the function f, we write

$$w = f(z). \tag{10·2}$$

The number $z = x + iy$ is allowed to assume any value in D and so, if desired, could be called a complex independent variable, when w could then properly be called a complex dependent variable. Usually we shall simply refer to z and w as *complex variables*. It must be appreciated that, like z, the variable w has a real part and an imaginary part, both of which are in general dependent on x and y through the variable $z = x + iy$. We summarize these ideas formally as follows.

DEFINITION 10·1 (function of a complex variable)   We shall say that f is a *function of the complex variable* $z = x + iy$, and write

$$w = f(z),$$

if f associates a unique complex number $w = u + iv$ with each complex number z belonging to some region D of the complex plane.

Specific examples of functions of a complex variable are:

(a) $w = iz + 1$; (b) $w = z\bar{z}$; (c) $w = z^2 + 2z + 1$; (d) $w = 1/(z - 2)$;
(e) $w = \sin z$.

With the exception of (d), which is not defined for $z = 2$, these functions are defined for all z.

The difference between a function of a complex variable and a real valued function of a real variable is made clear by expressing these examples in real and imaginary form. Thus writing $z = x + iy$ and $w = u + iv$ we find:

(a) $w = i(x + iy) + 1 = (1 - y) + ix$, showing that $u = 1 - y$, $v = x$;

(b) $w = (x + iy)(x - iy) = x^2 + y^2$, showing that $u = x^2 + y^2$, $v \equiv 0$. This is an example of a function that always maps a complex variable into a real variable.

(c) $w = (x + iy)^2 + 2(x + iy) + 1 = (x^2 + 2x - y^2 + 1) + i(2y + 2xy)$, showing that $u = x^2 + 2x - y^2 + 1$, $v = 2y(1 + x)$;

(d) $w = 1/(x + iy - 2) = [(x - 2) - iy]/(x^2 + y^2 - 4x + 4)$, showing that $u = (x - 2)/(x^2 + y^2 - 4x + 4)$, $v = -y/(x^2 + y^2 - 4x + 4)$, provided only that $x \neq 2$ and $y \neq 0$;

(e) $w = \sin z = \sin (x + iy) = \sin x \cos iy + \cos x \sin iy$, and so using the results of Problem 6·19, that $\cos iy = \cosh y$, $\sin iy = i \sinh y$, we arrive at $w = \sin x \cosh y + i \cos x \sinh y$. Thus in this case $u = \sin x \cosh y$, $v = \cos x \sinh y$.

Any function of x, y and complex constants that gives rise to a unique complex number when x and y are specified defines a function of the complex variable z by virtue of the relationship $z = x + iy$. For suppose that

$$(x + y + 1) + i(x - 2y) = f(z),$$

then to determine $f(z)$ when $z = 1 + 2i$ we simply write $x + iy = 1 + 2i$, showing that $x = 1$, $y = 2$, after which it follows from the form of $f(z)$ that $f(1 + 2i) = 4 - 3i$.

As when z is purely real the forms assumed by $f(z)$ and $f(x)$ are identical, we may deduce the form of f as a function of z from the following simple rule.

Rule 1 (Expression of f in terms of z)

The complex function $w = u(x, y) + iv(x, y)$ may be expressed in terms of z by formally setting $y = 0$ in the right-hand side and then replacing x by z.

The application of this rule is well illustrated by applying it to the examples (a) to (e) discussed above.

The concept of a limit can also be extended to complex functions $f(z) = u(x, y) + iv(x, y)$. We shall say that $w = f(z)$ has the *limit* w_0 as $z \to z_0$ if the two real functions $u(x, y)$ and $v(x, y)$ of the two real variables x, y each has a limit in the sense of Definition 3·7 as $(x, y) \to (x_0, y_0)$, where $z = x + iy$

and $z_0 = x_0 + iy_0$. That is to say, the complex limit w_0 of $w = f(z)$ must be independent of the path in the complex plane by which $z \to z_0$. By analogy with real functions of two real variables, we shall say that $w = f(z)$ is *continuous* at $z = z_0$ if

$$\lim_{z \to z_0} f(z) = w_0 \quad \text{and} \quad f(z_0) = w_0. \tag{10·3}$$

Using these ideas we now define the derivative dw/dz, or $f'(z)$, of $w = f(z)$ as the limit of the usual difference quotient $[f(z+h) - f(z)]/h$ as $h \to 0$, whenever this limit exists. Complex functions with this differentiability property throughout some region D are called *analytic* functions in D. Let us now examine the condition that this definition imposes on $u(x, y)$ and $v(x, y)$ on account of the required independence of the limit of the difference quotient of the manner in which $h \to 0$. We start by writing out the difference quotient in terms of u, v, setting $z = x + iy$ and $h = \lambda + i\mu$, where λ, μ are real numbers:

$$\frac{f(z+h) - f(z)}{h} = \frac{u(x+\lambda, y+\mu) + iv(x+\lambda, y+\mu) - u(x, y) - iv(x, y)}{\lambda + i\mu} \tag{10·4}$$

First, we choose to let $h \to 0$ through purely real values, so that $\mu \equiv 0$, when we find that

$$\begin{aligned} \frac{dw}{dz} &= \lim_{\lambda \to 0} \left[\frac{u(x+\lambda, y) - u(x, y)}{\lambda} + i \frac{v(x+\lambda, y) - v(x, y)}{\lambda} \right] \\ &= \frac{\partial u}{\partial x} + i \frac{\partial v}{\partial x}. \end{aligned} \tag{10·5}$$

Next, we deduce the form of dw/dz by choosing to let $h \to 0$ through purely imaginary values, so that $\lambda \equiv 0$, when the same form of argument then gives

$$\frac{dw}{dz} = \frac{1}{i} \left[\frac{\partial u}{\partial y} + i \frac{\partial v}{\partial y} \right]. \tag{10·6}$$

If the limit of difference quotient (10·4) exists, then it is unique, so that Eqns (10·5) and (10·6) must be alternative forms of the same result. Thus, equating real and imaginary parts, we obtain the two equations that must be satisfied simultaneously by the real and imaginary parts of $f(z)$:

$$\frac{\partial u}{\partial x} = \frac{\partial v}{\partial y} \quad \text{and} \quad \frac{\partial u}{\partial y} = -\frac{\partial v}{\partial x}. \tag{10·7}$$

These are called the *Cauchy–Riemann* equations.

Our argument has shown that a differentiable or *analytic* function $w = u + iv$ must satisfy the Cauchy–Riemann equations (10·7), and a similar argument establishes the converse result; namely, that if the Cauchy–Riemann equations are satisfied by a complex function, then it must have a unique derivative.

It is a direct consequence of our definitions of limit, continuity and differentiability that all the limit and continuity theorems (Theorems 3·3 and 3·4) and differentiation theorems (Theorems 5·4 to 5·8) for real functions apply also to analytic functions. Points at which $w = f(z)$ is not analytic are called *singularities* of $f(z)$. Thus the function $f(z) = 1/(z + 1)$ is easily seen to be analytic everywhere except at the point $z = -1$, which is a singularity.

Results (10·5) and (10·6) may be used to deduce the form of $f'(z)$ by using the simple observation that when z is purely real, so that $z = x$, the forms assumed by $f'(z)$ and $f'(x)$ are identical. Similarly, when z is purely imaginary, so that $z = iy$, the forms of $f'(z)$ and $f'(iy)$ are identical. This gives the following straightforward rule for determining the derivative $f'(z)$ of the function $f(z)$ which is sometimes helpful.

Rule 2 (Determination of the derivative of a complex function)

If $f(z) = u + iv$ satisfies the Cauchy–Riemann equations, then the derivative $f'(z)$ expressed in terms of z may either be deduced

(a) from the result

$$f'(z) = \frac{\partial u}{\partial x} + i \frac{\partial v}{\partial x}$$

by formally setting $y = 0$, and then replacing x by z; or

(b) from the result

$$f'(z) = \frac{\partial v}{\partial y} - i \frac{\partial u}{\partial y}$$

by formally setting $x = 0$, and then replacing iy by z.

Example 10·3 Determine which of the following functions satisfy the Cauchy–Riemann equations and thus possess uniquely defined derivatives. Give the form of this derivative when it is defined.

(a) $w = z^2$;

(b) $w = \cos z$;

(c) $w = |z|$.

Solution

(a) If $w = z^2$, then $w = (x + iy)^2 = x^2 - y^2 + i2xy$ and so $u = x^2 - y^2$, $v = 2xy$. So $u_x = 2x$, $u_y = -2y$, $v_x = 2y$, and $v_y = 2x$. It is readily seen that these expressions satisfy the Cauchy–Riemann equations and so we may conclude that $w = z^2$ possesses a unique derivative. It follows from Eqn (10·5) that

$$f'(z) = 2x + i2y = 2z.$$

This result was so simple that appeal to Rule 2 was not necessary.

(b) If $w = \cos z$, then $w = \cos(x + iy) = \cos x \cos iy - \sin x \sin iy$, when $w = \cos x \cosh y - i \sin x \sinh y$, and so $u = \cos x \cosh y$, $v = - \sin x \sinh y$. Hence, $u_x = - \sin x \cosh y$, $u_y = \cos x \cosh y$, $v_x = - \cos x \sinh y$ and $v_y = - \sin x \cosh y$. Here also it is immediately apparent that the expressions satisfy the Cauchy–Riemann equations, showing that $w = \cos z$ possesses a unique derivative.

Let us choose to work with Rule 2 (a) to determine $f'(z)$ in terms of z. We must therefore start with the equation

$$f'(z) = \frac{\partial u}{\partial x} + i \frac{\partial v}{\partial x}.$$

In this case we find

$$f'(z) = -\sin x \cosh y - i \cos x \sinh y.$$

Then, setting $y = 0$ and replacing x by z gives

$$f'(z) = - \sin z.$$

It is instructive to compare this rapid method with the direct approach we now indicate.

$$f'(z) = - \sin x \cosh y - i \cos x \sinh y$$
$$= - \sin x \cos iy - \cos x \sin iy$$
$$= - \sin(x + iy) = - \sin z.$$

(c) If $w = |z|$, then $w = (x^2 + y^2)^{1/2}$, showing that $u = (x^2 + y^2)^{1/2}$, $v = 0$. Then, as $u_x = x/(x^2 + y^2)^{1/2}$, $u_y = y/(x^2 + y^2)^{1/2}$, $v_x = v_y = 0$, it is clear that $w = |z|$ cannot satisfy the Cauchy–Riemann equations anywhere in the complex plane. We conclude that $w = |z|$ has no derivative at any point in the complex plane.

Example 10·4   Determine the constants a and b in order that

$$w = x^2 + ay^2 - 2xy + i(bx^2 - y^2 + 2xy)$$

should satisfy the Cauchy–Riemann equations. Deduce the derivative of w.

Solution   Here we have $u = x^2 + ay^2 - 2xy$, $v = bx^2 - y^2 + 2xy$ so that $u_x = 2x - 2y$, $u_y = 2ay - 2x$, $v_x = 2bx + 2y$, and $v_y = -2y + 2x$. It is certainly true that $u_x = v_y$, so that the first of the Cauchy–Riemann equations is automatically satisfied. For the second equation to be satisfied we must require that $u_y = -v_x$, or $2ay - 2x = -(2bx + 2y)$. This is only possible if $a = -1, b = 1$.

Now as $f'(z) = u_x + iv_x$, we have

$$f'(z) = 2x - 2y + i(2x + 2y).$$

Again, working with Rule 2 (a) gives

$$f'(z) = 2(1 + i)z.$$

Supposing that u_{xy}, v_{xy} exist and are continuous, it follows directly by partial differentiation of the Cauchy–Riemann equations $u_x = v_y$, $u_y = -v_x$ that

$$\frac{\partial^2 u}{\partial x^2} + \frac{\partial^2 u}{\partial y^2} = 0 \quad \text{and} \quad \frac{\partial^2 v}{\partial x^2} + \frac{\partial^2 v}{\partial y^2} = 0. \tag{10·8}$$

These equations are identical in form and are examples of an important partial differential equation called *Laplace's equation*, any solution of which is called a *harmonic* function. The harmonic functions u and v associated with an analytic function $f(z) = u + iv$ are called *conjugate* harmonic functions. For example,

$$\cos z = \cos x \cosh y - i \sin x \sinh y$$

is an analytic function with $u = \cos x \cosh y$, $v = -\sin x \sinh y$. Now both u and v are such that u_{xy}, v_{xy} are continuous, so it follows immediately that u and v satisfy Eqns (10·8). Hence $u = \cos x \cosh y$, $vv = -\sin x \sinh y$ are conjugate harmonic functions. The term conjugate is, of course, used here in a different sense from when discussing complex conjugates.

10·3 Conformal mapping

Thus far we have examined some of the analytical consequences of requiring that a function $w = f(z)$ be differentiable. Let us now pursue this matter further by studying some of the geometrical implications of differentiability.

Take two complex planes, which we shall refer to as the z-plane and the w-plane, the connection between their respective points being through the differentiable function $w = f(z)$. Because each value of z gives rise to a unique value of w, it follows that any curve γ in the z-plane must correspond to some other curve Γ in the w-plane. In this sense the w-plane can correctly be described as a mapping of the z-plane.

For a specific illustration, let us determine how the straight line $y = \alpha x$ in the z-plane is mapped by the function $w = iz + (1 + i)$ onto the w-plane. We begin by setting $w = u + iv$, $z = x + iy$, after which a simple calculation yields $u = 1 - y$, $v = x + 1$. Hence to find the line in the w-plane that corresponds to $y = \alpha x$ in the z-plane it is now only necessary to set $y = \alpha x$ in these expressions for u, v and then to eliminate x between them. Performing

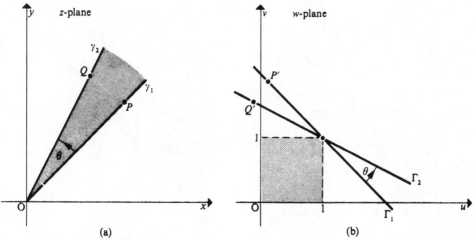

Fig. 10·3 Mapping by the function $w = iz + (1 + i)$.

these operations we find $u = 1 - \alpha x$, $v = x + 1$, whence

$$v = \left(\frac{1 + \alpha}{\alpha}\right) - \frac{1}{\alpha} u.$$

This is again an equation of a straight line but this time in the w-plane. The line passes through the point $(0, (1 + \alpha)/\alpha)$ and has the gradient $-1/\alpha$. Representative lines γ_1, γ_2 are shown in the z-plane of Fig. 10·3 and their respective maps or images are shown as the lines Γ_1, Γ_2 in the associated w-plane. The lines γ_1, γ_2 correspond, respectively, to $\alpha = 1$, $\alpha = 2$.

It is not difficult to see that the map in the w-plane has been obtained from the map in the z-plane by first rotating the original pair of lines anti-clockwise through an angle $\frac{1}{2}\pi$ and then translating the resulting picture to the point $1 + i$ as a new origin. More important than this, however, is the fact that the angle θ between the lines γ_1, γ_2 is equal to the angle between the lines Γ_1, Γ_2 and, moreover, the *sense of rotation* is preserved. That is to say if γ_2 is inclined to γ_1 at an angle θ, measured anti-clockwise, then Γ_2 is also inclined to Γ_1 an an angle θ, measured anti-clockwise.

This is no chance result and, indeed, we now prove that if a function $f(z)$ is analytic (that is, satisfies the Cauchy–Riemann equations and so has a uniquely defined derivative) then, except for points z_0 at which $f'(z_0) = 0$, the function $w = f(z)$ will preserve both the angle and the sense of rotation when mapping intersecting curves γ_1, γ_2 in the z-plane onto corresponding intersecting curves Γ_1, Γ_2 in the w-plane. These properties of a mapping or transformation are recognized by saying that the transformation is *conformal*.

To prove this general result we now consider a function $w = f(z)$ that is analytic in some region of the z-plane and take a point z_0 in that region at which $f'(z_0) \neq 0$. Let γ_1, γ_2 be two curves drawn in the z-plane that intersect

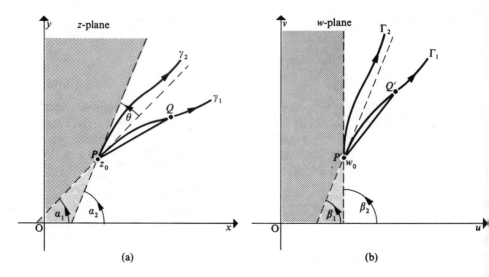

Fig. 10·4   Conformal mapping $w = f(z)$.

at z_0 and let z_1 denote a point Q on the curve γ_1 as indicated in Fig. 10·4. We shall suppose that as Q moves away from P along γ_1 in the direction indicated by an arrow in the Figure, so the point $w_1 = f(z_1)$, which we denote by Q', moves away from point P' in the direction indicated. This process thus associates a sense of direction with each of the corresponding curves γ_1 and Γ_1. A similar argument defines directions along γ_2 and Γ_2.

Now as Q approaches P, so the secant PQ will assume its limiting position in which, when it is inclined at an angle α_1 to the x-axis, it is tangent to γ_1 at z_0. As PQ $= z_1 - z_0$ we have

$$\alpha_1 = \lim_{z_1 \to z_0} \arg(z_1 - z_0).$$

Identical reasoning shows that

$$\beta_1 = \lim_{z_1 \to z_0} \arg(w_1 - w_0),$$

where β_1 is the angle of the tangent to Γ_1 at P' measured from the u-axis. Hence we have

$$\beta_1 - \alpha_1 = \lim_{z_1 \to z_0} \arg(w_1 - w_0) - \lim_{z_1 \to z_0} \arg(z_1 - z_0)$$

and, as $\arg a - \arg b = \arg a/b$, this may be written

$$\beta_1 - \alpha_1 = \lim_{z_1 \to z_0} \arg\left(\frac{w_1 - w_0}{z_1 - z_0}\right).$$

However, as we are assuming $f(z)$ is differentiable

$$f'(z_0) = \lim_{z_1 \to z_0} \left(\frac{w_1 - w_0}{z_1 - z_0} \right),$$

and provided $f'(z_0) \neq 0$ it then follows that

$$\beta_1 - \alpha_1 = \arg f'(z_0). \tag{10·9}$$

In the case that $f'(z_0) = 0$, the amplitude of $f'(z_0)$ is indeterminate. Such points are called *critical* points of $f(z)$, by analogy with the real variable case.

We have seen that $f'(z_0)$ is unique, so that the expression on the right-hand side of Eqn (10·9) is a constant. The result must, then, also be true for any other curve γ_2, say, and its map Γ_2. Hence we have

$$\beta_1 - \alpha_1 = \beta_2 - \alpha_2$$

or

$$\alpha_2 - \alpha_1 = \beta_2 - \beta_1.$$

The curves γ_1, γ_2 were any two curves which intersected at z_0, so we have proved the following result.

THEOREM 10·1 (conformal mapping) If $f(z)$ is analytic in some region, then apart from those points z_0 in that region for which $f'(z_0) = 0$, the mapping $w = f(z)$ preserves both the angle and the sense of rotation when mapping intersecting directed pairs of curves in the z-plane into corresponding intersecting directed pairs of curves in the w-plane. Such a mapping is said to be conformal.

To close this chapter we now examine some important special conformal mappings. Rather than emphasize the algebraic details of the transformations or mappings, we shall aim primarily at interpretation in terms of basic geometrical operations such as translation, rotation, and change of scale (dilatation).

10·3 (a)   The general linear transformation

The *general linear transformation* is the name given to the mapping described by the equation

$$w = az + b, \tag{10·10}$$

where a, b are arbitrary constants with $a \neq 0$. Our introductory example was of this form with $a = i$, $b = 1 + i$. The mapping (10·10) obviously satisfies the Cauchy–Riemann equations and, as $dw/dz = a \neq 0$, it has no critical points and so provides a conformal mapping of the entire z-plane. To appreciate the geometrical effect of this mapping consider first the case in which $a = 1$ so that $w = z + b$.

This has the effect of generating the w-plane by simply adding a constant complex number b to every point in the z-plane. Using the vectorial representation of complex numbers this is seen to be equivalent to generating the w-plane by shifting the entire z-plane through a distance $|b|$ parallel to the vector b. Such a mapping is accordingly called a *translation*. Another way of expressing this result is by saying that if the w- and z-planes were to be superimposed, then the $O\{u, v\}$ axes would be obtained by translating the $O\{x, y\}$ axes, without rotation, such that in their new position the origin coincided with the point $z = -b$. To see this, remember that b is a vector and that the position vector of the origin of $O\{u, v\}$ is b relative to $O\{x, y\}$, but that the position vector of the origin of $O\{x, y\}$ relative to $O\{u, v\}$ is $-b$. Consequently, we may conclude that the mapping $w = z + b$ leaves invariant the shape and size of any curve in the z-plane.

Next we consider the consequences of setting $b = 0$ so that $w = az$. If we write $a = \rho e^{i\alpha}$ and $z = re^{i\theta}$, we have $w = \rho r e^{i(\alpha + \theta)}$. This shows that the effect on the z-plane of the mapping $w = az$ is to multiply the modulus of z by a constant factor ρ and to increase the argument of z by a constant angle α. Hence $w = az$ corresponds to a *magnification*, or *dilatation*, of every z by a constant factor $|a|$, and a rotation about the origin of every z by a constant angle α. Thus we may deduce that the general linear transformation

$$w = az + b$$

of the z-plane may be described geometrically as the combination of a dilatation, a rotation, and a translation. In the trivial case $a = 1$, $b = 0$ the mapping reduces to an identity.

10.3 (b) The mapping $w = z^n$

A typical example of this form is provided by the function $w = z^2$. As it is interesting to interpret mappings in terms of both polar coordinates and cartesian coordinates, let us first study the polar representation. To do this we set $z = re^{i\theta}$, $w = \rho e^{i\phi}$, when we find

$$\rho(\cos\phi + i\sin\phi) = r^2(\cos 2\theta + i\sin 2\theta),$$

showing that $\rho = r^2$ and $\phi = 2\theta + 2n\pi$, where $n = 0, 1, 2, \ldots$. However, for our purposes we shall disregard this ambiguity of the angle ϕ with respect to multiples of 2π, since all angles in polar coordinates are indeterminate in this manner.

In words, the effect of the mapping $w = z^2$ is to square the modulus of every number z and to double its argument. This is very easily illustrated by appeal to Fig. 10·5 depicting the mapping of a shaded portion of an annular region in the z-plane into another, larger, annular region in the w-plane. The conformal nature of the mapping is reflected by the fact that at the corresponding corners of the figures the angles between the boundary lines together with their senses have been preserved. They are of course equal to $\frac{1}{2}\pi$ in this instance.

Because of the properties just outlined it is readily seen that the function

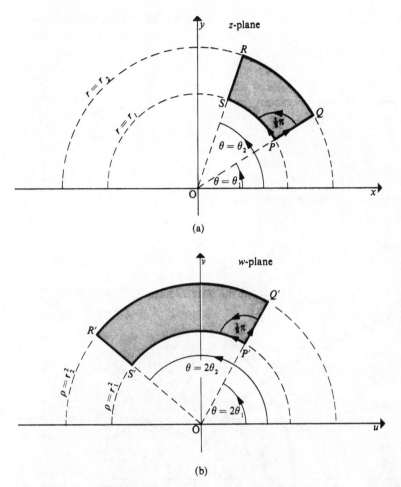

Fig. 10·5   The polar mapping $w = z^2$.

$w = z^2$ maps the upper half z-plane onto the entire w-plane. When this is done it is necessary to exclude the origin in the w-plane together with all the points on the positive u-axis, since these are mapped twice. In fact they correspond to points on both the positive and negative parts of the real axis in the z-plane. The origin in the w-plane is in fact a critical point, for $w' = 2z$ vanishes at $z = 0$. This exclusion of a line of points in the w-plane is often described by saying that the w-plane has been *cut* along the real axis.

The effect of the mapping is more striking if it is displayed in terms of x and y by again setting $w = u + iv$, but this time writing $z = x + iy$ to obtain $u = x^2 - y^2$, $v = 2xy$. These equations show, for example, that the straight line $x = \alpha$ maps into the curve $u = \alpha^2 - y^2$, $v = 2\alpha y$ in the w-plane which, after elimination of y, is seen to be equivalent to $v^2 = 4\alpha^2(\alpha^2 - u)$. Similarly, the straight line $y = \beta$ may be seen to map into the curve $v^2 =$

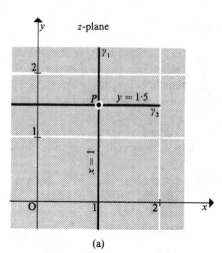

(a)

Fig. 10·6   The Cartesian mapping $w = z^2$.

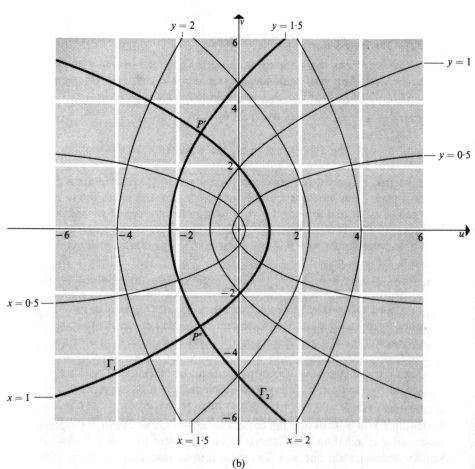

(b)

$4\beta^2(\beta^2 + u)$ in the w-plane. These equations describe two parabolas that are symmetrical about the u-axis, as shown in Fig. 10·6.

The lines $x = 1, y = 3/2$ denoted by γ_1 and γ_2, respectively, in the z-plane map into the parabolas Γ_1 and Γ_2 in the w-plane. This shows that the single point $z = 1 + 3i/2$ denoted by P in the z-plane (that is, the point $(1, 3/2)$) maps into the pair of points P$'$ and P$''$ in the w-plane determined by the two points of intersection of parabolas Γ_1 and Γ_2. Again the conformal nature of the transformation is reflected in the easily checked geometrical fact that the two families of parabolas are mutually orthogonal, as are the lines $x = \text{const}, y = \text{const}$ in the z-plane.

The more general mapping $w = z^n$ may be analysed in similar fashion, though the algebraic complexity is naturally greater. When n is integral the mapping may be seen to transform the segment $0 \le \arg z < 2\pi/n$ into the complete w-plane with a suitable cut along the u-axis. (Care must be exercised when n is fractional for then the mapping is many valued. We shall not pursue this matter further.)

10·3 (c)   The inversion $w = 1/z$

For obvious reasons the mapping $w = 1/z$ is called the *inversion* mapping. Its geometrical effect may be deduced by setting $w = \rho e^{i\phi}$, $z = re^{i\theta}$ to find

$$\rho(\cos \phi + i \sin \phi) = \frac{1}{r}(\cos \theta - i \sin \theta).$$

Arguing as with the function $w = z^2$, we then see that this implies that $\rho = 1/r, \phi = -\theta$.

Expressed in words, the inversion mapping $w = 1/z$ transforms a point in the z-plane with modulus r and argument θ into a point in the w-plane with modulus $1/r$ and argument $-\theta$. This may be interpreted geometrically by appeal to Fig. 10·7 in which the w- and z-planes are shown superimposed with a common origin, and P is any point in the z-plane with P$'$ denoting its image in the w-plane.

The circle shown in Fig. 10·7 is the unit circle $|z| = 1$, and point Q on the radius vector drawn from O to P is such that OP . OQ $= 1$. Hence if OP $= r$, then OQ $= 1/r$. In geometrical terms point Q is said to have been obtained by inverting point P with respect to the unit circle. Point P$'$, which is the image in the w-plane of the point P in the z-plane, is then obtained by reflecting Q in the x-axis.

Thus the mapping $w = 1/z$ corresponds to the inversion of points z with respect to the unit circle, followed by their reflection in the real axis. The inversion mapping thus maps the points interior to the unit circle about the origin of the z-plane onto the exterior of the unit circle about the origin of the w-plane, and vice-versa. The two unit circles map onto one another.

Algebraically, we write $w = u + iv, z = x + iy$, when

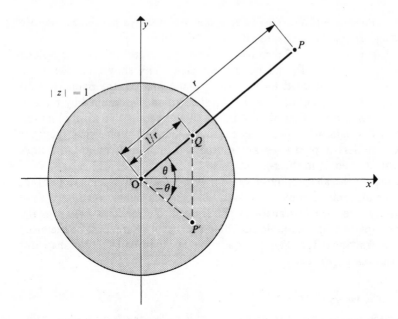

Fig. 10.7 Inversion in unit circle followed by reflection in the x-axis.

$$u = \frac{x}{x^2 + y^2}, \qquad v = \frac{-y}{x^2 + y^2}.$$

To learn how the line $x = \alpha$ in the z-plane maps onto the w-plane we need only set $x = \alpha$ in the expressions for u and v and then eliminate y to obtain the equation

$$u^2 + v^2 - \frac{u}{\alpha} = 0.$$

Similarly, the line $y = \beta$ in the z-plane maps onto the curve in the w-plane defined by the equation

$$u^2 + v^2 + \frac{v}{\beta} = 0.$$

When these equations are rewritten in the form

$$\left(u - \frac{1}{2\alpha}\right)^2 + v^2 = \left(\frac{1}{2\alpha}\right)^2 \qquad \text{and} \qquad u^2 + \left(v + \frac{1}{2\beta}\right)^2 = \left(\frac{1}{2\beta}\right)^2,$$

it is easily seen that the line $x = \alpha$ in the z-plane has for its image in the w-plane a circle of radius $\frac{1}{2}\alpha$ with its centre at $(\frac{1}{2}\alpha, 0)$, whilst the line $y = \beta$ in the z-plane has for its image in the w-plane a circle of radius $\frac{1}{2}\beta$ with its centre at $(0, -\frac{1}{2}\beta)$. We may conclude that lines parallel to the x- and y-axes map onto circles in the w-plane which pass through the origin and have their centres on the u- and v-axes.

Had the general straight line $y = mx + c$ in the z-plane been mapped, then this same form of argument would have shown that any such line not passing through the origin will transform into a circle through the origin in the w-plane. Lines through the origin in the z-plane transform into lines through the origin in the w-plane. The verification of these remarks is left as an exercise for the reader.

10·3 (d) The bilinear transformation

Any mapping of the general form

$$w = \frac{az + b}{cz + d} \tag{10·11}$$

is called a *bilinear* transformation or a *linear fractional* transformation. The general linear transformation and the inversion mapping are special cases of the bilinear transformation. We now show that bilinear transformations are characterized by the property that they map circles and straight lines in the z-plane onto circles and straight lines in the w-plane, though not necessarily in this order.

Let us now write the transformation (10·11) in the form

$$w = \frac{a}{c} - \frac{ad - bc}{c^2[z + (d/c)]}. \tag{10·12}$$

We assume $c \neq 0$ and $ad - bc \neq 0$; this is justified since if $c = 0$ the transformation reduces to the general linear transformation, whereas if $ad - bc = 0$, then w reduces to a constant. So, if we define new variables z_1 and z_2 by

$$z_1 = z + \frac{d}{c}, \qquad z_2 = \frac{1}{z_1}, \tag{10·13}$$

then (10·22) becomes

$$w = \frac{a}{c} - \left(\frac{ad - bc}{c^2}\right) z_2. \tag{10·14}$$

We must now consider the sequential effect of the mappings that transform from the z-plane to the w-plane via the intermediate planes z_1 and z_2. The mapping from the z-plane to the z_1-plane is a pure translation and thus leaves the shape and size of all curves invariant. The mapping from the z_1-plane to the z_2-plane is an inversion and, as we have just seen, maps straight lines not passing through the origin onto circles, and straight lines through the origin onto straight lines. Finally, the mapping from the z_2-plane to the w-plane is a general linear transformation and so comprises a rotation and a translation. Hence, in particular, this final mapping will transform straight lines into straight lines and circles into circles. This justifies our earlier statement that the bilinear transformation maps straight lines and circles into straight lines and circles, though not necessarily in this order.

Example 10·5   Find the image in the w-plane of the circle $|z| = 2$ if

$$w = \frac{z - i}{z + i}.$$

Solution   Setting $w = u + iv$, $z = x + iy$ we find that

$$u = \frac{x^2 + y^2 - 1}{x^2 + y^2 + 2y + 1}, \qquad v = \frac{-2x}{x^2 + y^2 + 2y + 1}.$$

Now the circle $|z| = 2$ has the equation $x^2 + y^2 = 4$, which used in the expressions for u, v gives

$$u = \frac{3}{2y + 5}, \qquad v = \frac{-2x}{2y + 5}.$$

Next, solving these for x and y, we find

$$x = \frac{-3v}{2u}, \qquad y = \frac{1}{2}\left(\frac{3}{u} - 5\right)$$

so that on the required circle $x^2 + y^2 = 4$ this pair of equations is equivalent to

$$3(u^2 + v^2) - 10u + 3 = 0.$$

When this equation is expressed in the form

$$\left(u - \frac{5}{3}\right)^2 + v^2 = \frac{16}{9}$$

it can be recognized as the equation of a circle in the w-plane having a radius of $4/3$ and its centre at the point $(5/3, 0)$.

This conclusion could have been obtained more easily by using the following argument. The equation

$$w = \frac{z - i}{z + i}$$

is equivalent to

$$z = i\left(\frac{1 + w}{1 - w}\right).$$

Hence, as $z\bar{z} = x^2 + y^2$, we have

$$x^2 + y^2 = i(-i)\left(\frac{1 + w}{1 - w}\right)\left(\frac{1 + \bar{w}}{1 - \bar{w}}\right) = \frac{1 + \bar{w} + w + w\bar{w}}{1 - w - \bar{w} + w\bar{w}}.$$

In terms of $w = u + iv$, $\bar{w} = u - iv$ this becomes

$$x^2 + y^2 = \frac{1 + 2u + u^2 + v^2}{1 - 2u + u^2 + v^2}.$$

and, on the circle $x^2 + y^2 = 4$, it reduces our previous result

$$3(u^2 + v^2) - 10u + 3 = 0.$$

PROBLEMS

Section 10·1

10·1 Sketch each of the following curves defined in the complex plane:
  (a) $x = s, y = \sqrt{(1 - s^2)}$ for $-1 \leq s \leq 1$;
  (b) $x = a \sin s, y = b \cos s$ for $0 \leq s \leq 2\pi$ (a, b real);
  (c) $x = \cosh s, y = \sinh s$ for $-\infty < s < \infty$;
  (d) $|z + 2 - i| = 3$;
  (e) $z\bar{z} = 4$.

Sketch the region defined by each of the following sets of inequalities and indicate when the boundary points belong to the region so defined.

10·2 $\text{Im}(z + iz) \geq 0$ and $\text{Re } z \geq 0$.

10·3 $2 < |z| < 3$ with $0 \leq \arg z \leq \frac{1}{2}\pi$.

10·4 $1 \leq |z - 1| \leq 2$ and $1 \leq |z + 1| \leq 2$.

10·5 Sketch the region that lies inside the curve defined by
$$\arg (z + 2) - \arg (z + 3) = \frac{1}{4}\pi$$
and is such that $\text{Im } z \geq \frac{1}{2}$.

Give an alternative representation of this region.

10·6 Draw the curve C defined by
$$\arg (z - i) - \arg (z - 1) = \frac{1}{2}\pi.$$

Section 10·2

10·7 For what values of z are the following complex functions defined:
  (a) $w = z^2 + iz + 1$;   (b) $w = (z - 1)/(z - 2)$;
  (c) $(z + 1)(z - i)(z^2 + 4)$;   (d) $w = \sinh z$.

10·8 If $f(z) = u + iv$, find the expressions for the functions u, v in terms of x, y given that:
  (a) $f(z) = z^2 + z\bar{z} + 1$;   (b) $f(z) = \dfrac{z + i}{z + 2}$;

  (c) $f(z) = \cosh z$;   (d) $f(z) = \cos z$.

10·9 Given the following forms of $f(z)$ deduce their value if $z = 1 + 2i$:
  (a) $f(z) = x^2 + 3xy + iy^2$;

  (b) $f(z) = \dfrac{x^2 + 2iy + 1}{x^2 + y^2}$;

  (c) $f(z) = \sin \dfrac{\pi x}{2} (x^2 - iy^2) + i \cos \dfrac{3\pi y}{2} (x^2 + iy^2)$.

10·10 Give reasons to justify the assertion that

$$f(z) = z \sin (z^2 + 3z + 2) + 1/(z + 2 - i)$$

is continuous everywhere except at $z = -2 + i$.

10·11 Determine which of the following functions $f(z)$ satisfy the Cauchy–Riemann equations:

(a) $f(z) = z^3 - iz^2 + 3$;
(b) $f(z) = \cosh (z + 3i)$;
(c) $f(z) = z \sin z + z\bar{z}$;
(d) $f(z) = (x^3 - 3xy^2) + i(3x^2y - y^3)$;
(e) $f(z) = z(z + \bar{z})/2$;
(f) $f(z) = \sinh 3x \cos y + i \cosh 3x \sin y$.

10·12 Find the points, if any, at which the following functions are not analytic:

(a) $f(z) = 3z + \sinh z$;
(b) $f(z) = z/(z + 2)$;
(c) $f(z) = \cos 1/z$;

(d) $f(z) = \dfrac{\sin z}{z^2 + 1}$.

10·13 Find the values of the constants a and b in order that the functions w should satisfy the Cauchy–Riemann equations:

(a) $w = a \sin x \cosh by + i2 \cos x \sinh y$;
(b) $w = x^3 - axy^2 - x + 1 + i(3x^2 - by^3 - 1)$.

10·14 Verify that the following functions w satisfy the Cauchy–Riemann equations and in each case express the derivative of w as a function of z:

(a) $w = (x^3 - 3xy^2 + y) + i(3x^2y - y^3 - x)$;
(b) $w = (x \sinh x \cos y - y \cosh x \sin y) + i(y \sinh x \cos y$
$$+ x \cosh x \sin y);$$
(c) $w = e^{ax} (\cos ay + i \sin ay)$.

Section 10·3

10·15 Sketch the images in the w-plane of the line $y = 2x - 1$ in the z-plane that result from the mappings:

(a) $w = iz - (2 + i)$;
(b) $w = 2z + 3$;
(c) $w = (1 + i)z + 1$.

10·16 Determine the images in the w-plane of the circle $| z - 1 | = 1$ in the z-plane that result from the mappings:

(a) $w = 3z - i$;
(b) $w = (i - 1)z + 2$.

In each case shade the regions in the w-plane that correspond to the interior of the circle $| z - 1 | = 1$.

10·17 Sketch the region in the w-plane corresponding to the region $x \geq 2, y \leq x$ in the z-plane given that

$$w = (2i - 1)z + (1 + i).$$

10·18 Determine the equation of the line in the w-plane which is the image of the line $x = 1$ in the z-plane under the mapping

$w = z^3$.

10·19 Give an algebraic proof that if $c \neq 0$, then the general straight line $y = mx + c$ in the z-plane is mapped by the transformation $w = 1/z$ onto a circle in the w-plane.

11 Scalars, vectors, and fields

11·1 Curves in space

If the coordinates (x, y, z) of a point P in space are described by

$$x = f(t), \qquad y = g(t), \qquad z = h(t), \tag{11·1}$$

where f, g, h are continuous functions of t, then as t increases so the point P moves in space tracing out some curve. It follows that Eqns (11·1) represent a *parametric* description of a curve Γ in space and, furthermore, that they define a direction along the curve Γ corresponding to the direction in which P moves as t increases. For example, the parametric equations

$$x = 2 \cos 2\pi t, \qquad y = 2 \sin 2\pi t, \qquad z = 2t,$$

for $0 \le t \le 1$ describe one turn of a helix, as may be seen by noticing that the projection of the point P on the (x, y)-plane traces one revolution of the circle $x^2 + y^2 = 4$ as t increases from $t = 0$ to $t = 1$, whilst the z-coordinate of P steadily increases from $z = 0$ to $z = 2$.

If we now denote by $\mathbf{r}$ the position vector OP of a point P on Γ relative to the origin O of our coordinate system, and introduce the triad of orthogonal unit vectors $\mathbf{i}$, $\mathbf{j}$, $\mathbf{k}$ used in Chapter 4, it follows that (Fig. 11·1)

$$\mathbf{r} = f(t)\mathbf{i} + g(t)\mathbf{j} + h(t)\mathbf{k}. \tag{11·2}$$

Expressions of this form are called *vector functions* of one real variable, in which the dependence on the parameter t is often displayed concisely by writing $\mathbf{r} = \mathbf{r}(t)$. The name vector function arises because $\mathbf{r}$ is certainly a vector and, as it depends on the real independent variable t, it must also be a function in the sense that to each t there corresponds a vector $\mathbf{r}(t)$. Knowledge of the vector function $\mathbf{r}(t)$ implies knowledge of the three scalar functions f, g, and h, and conversely.

The geometrical analogy used here to interpret a general vector function $\mathbf{r}(t)$ is particularly valuable in dynamics where the point $P(t)$ with position vector $\mathbf{r}(t)$ usually represents a moving particle, and the curve Γ its trajectory in space. Under these conditions it is frequently most convenient if the parameter t is identified with the time, though in some circumstances identification with the distance s to P measured along Γ from some fixed point on Γ is preferable. Useful though these geometrical and dynamical analogies are, we shall in the main use them only to help further our understanding of general vector functions.

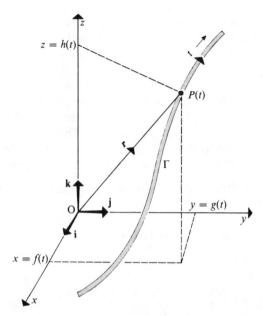

Fig. 11·1 Vector function of one variable interpreted as a curve in space.

The name vector function suggests, correctly, that it is possible to give satisfactory meanings to the terms limit, continuity, and derivative when applied to $\mathbf{r}(t)$. As in the ordinary calculus, the key concept is that of a limit. Intuitively the idea of a limit is clear: when we say $\mathbf{u}(t)$ tends to a limit $\mathbf{v}$ as $t \to t_0$, we mean that when t is close to t_0, the vector function $\mathbf{u}(t)$ is in some sense close to the vector $\mathbf{v}$. In what sense though can the two vectors $\mathbf{u}(t)$ and $\mathbf{v}$ be said to be close to one another? Ultimately, all that is necessary is to interpret this as meaning that $|\mathbf{u}(t) - \mathbf{v}|$ is small.

So, we shall say that $\mathbf{u}(t)$ tends to the limit $\mathbf{v}$ as $t \to t_0$ if, by taking t sufficiently close to t_0, it is possible to make $|\mathbf{u}(t) - \mathbf{v}|$ arbitrarily small. As with our previous notion of continuity we shall then say that $\mathbf{u}(t)$ is continuous at t_0 if $\lim_{t \to t_0} \mathbf{u}(t) = \mathbf{v}$ and, in addition, $\mathbf{u}(t_0) = \mathbf{v}$. We incorporate these ideas into a formal definition as follows:

DEFINITION 11·1 (vector functions—limits and continuity) Let $\mathbf{u}(t) = u_1(t)\mathbf{i} + u_2(t)\mathbf{j} + u_3(t)\mathbf{k}$ and $\mathbf{v} = v_1\mathbf{i} + v_2\mathbf{j} + v_3\mathbf{k}$, then if for any $\varepsilon > 0$ there is some number δ such that

$$|\mathbf{u}(t) - \mathbf{v}| < \varepsilon \qquad \text{when} \qquad |t - t_0| < \delta,$$

we shall say that $\mathbf{u}(t)$ tends to the limit $\mathbf{v}$ as $t \to t_0$, and write

$$\lim_{t \to t_0} \mathbf{u}(t) = \mathbf{v}.$$

If in addition $\mathbf{u}(t_0) = \mathbf{v}$, then $\mathbf{u}(t)$ will be said to be continuous at $t = t_0$. A vector function that is continuous at all points in the interval $a \le t \le b$ will be said to be continuous throughout that interval.

As usual, a vector function that is not continuous at $t = t_0$ will be said to be discontinuous. It is obvious from this definition that $\mathbf{u}(t)$ can only tend to the limit $\mathbf{v}$ as $t \to t_0$ if the limit of each component of $\mathbf{u}(t)$ is equal to the corresponding component of the vector $\mathbf{v}$. Thus the limit of a vector function of one variable is directly related to the limits of the three scalar functions of one variable $u_1(t)$, $u_2(t)$, and $u_3(t)$. This is proved by writing

$$| \mathbf{u}(t) - \mathbf{v} | = [(u_1(t) - v_1)^2 + (u_2(t) - v_2)^2 + (u_3(t) - v_3)^2]^{1/2},$$

showing that $| \mathbf{u}(t) - \mathbf{v} | < \varepsilon$ as $t \to t_0$ is only possible if

$$\lim_{t \to t_0} (u_i(t) - v_i) = 0 \qquad \text{for } i = 1, 2, 3,$$

or

$$\lim_{t \to t_0} u_1(t) = v_1, \qquad \lim_{t \to t_0} u_2(t) = v_2, \qquad \lim_{t \to t_0} u_3(t) = v_3.$$

A systematic application of these arguments enables the following theorem to be proved.

THEOREM 11·1 (continuous vector functions) If the vector functions $\mathbf{u}(t)$, $\mathbf{v}(t)$ are defined and continuous throughout the interval $a \le t \le b$, then the vector functions $\mathbf{u}(t) + \mathbf{v}(t)$, $\mathbf{u}(t) \times \mathbf{v}(t)$, and the scalar function $\mathbf{u}(t) . \mathbf{v}(t)$ are also defined and continuous throughout that same interval.

Example 11·1 At what points are the vector functions $\mathbf{u}(t)$, $\mathbf{v}(t)$ discontinuous if

$$\mathbf{u}(t) = \sin t \mathbf{i} + \sec t \mathbf{j} + \frac{1}{t - 1} \mathbf{k},$$

$$\mathbf{v}(t) = t \mathbf{i} + (1 + t^2)\mathbf{j} + e^t \mathbf{k}.$$

Verify by direct calculation that $\mathbf{u}(t) + \mathbf{v}(t)$, $\mathbf{u}(t) . \mathbf{v}(t)$, and $\mathbf{u}(t) \times \mathbf{v}(t)$ are continuous functions in any interval not containing a point of discontinuity of $\mathbf{u}(t)$ or $\mathbf{v}(t)$.

Solution The $\mathbf{i}$ component of $\mathbf{u}(t)$ is defined and continuous for all t, whereas the $\mathbf{j}$ component is discontinuous for $t = (2n + 1)\frac{1}{2}\pi$ with $n = 0, \pm 1, \pm 2,$ . . . and the $\mathbf{k}$ component is discontinuous for the single value $t = 1$. All three components of $\mathbf{v}(t)$ are continuous for all t. We have by vector addition

$$\mathbf{u}(t) + \mathbf{v}(t) = (t + \sin t)\mathbf{i} + (1 + t^2 + \sec t)\mathbf{j} + \left(e^t + \frac{1}{t - 1}\right)\mathbf{k},$$

showing that the components of $\mathbf{u}(t) + \mathbf{v}(t)$ give rise to the same points of discontinuity as the function $\mathbf{u}(t)$. We may thus conclude that the vector sum is continuous throughout any interval not containing one of these points. For example, $\mathbf{u}(t) + \mathbf{v}(t)$ is continuous in both the open interval $(\tfrac{1}{2}\pi, 3\pi/2)$ and the closed interval $[5, 7]$ but it is discontinuous in $(0, \pi)$.

The scalar product $\mathbf{u}(t) \cdot \mathbf{v}(t)$ is given by

$$\mathbf{u}(t) \cdot \mathbf{v}(t) = t \sin t + (1 + t^2) \sec t + \frac{e^t}{(t - 1)},$$

which is, of course, a scalar. Again we see by inspection that the scalar product is continuous in any interval not containing a point of discontinuity of $\mathbf{u}(t)$.

The vector product $\mathbf{u}(t) \times \mathbf{v}(t)$ is

$$\mathbf{u}(t) \times \mathbf{v}(t) = \begin{vmatrix} \mathbf{i} & \mathbf{j} & \mathbf{k} \\ \sin t & \sec t & 1/(t - 1) \\ t & 1 + t^2 & e^t \end{vmatrix}$$

giving,

$$\mathbf{u}(t) \times \mathbf{v}(t) = \left(e^t \sec t - \frac{1 + t^2}{t - 1} \right) \mathbf{i} + \left(\frac{t}{t - 1} - e^t \sin t \right) \mathbf{j}$$
$$+ [(1 + t^2) \sin t - t \sec t]\mathbf{k}.$$

Here also inspection of the components shows that the vector product is continuous in any interval not containing a point of discontinuity of $\mathbf{u}(t)$.

The following definition (interpreted later) shows that, as might be expected, the idea of a derivative can also be applied to vector functions of one variable.

DEFINITION 11·2 (derivative of vector function)   Let $\mathbf{u}(t)$ be a continuous vector function throughout some interval $a \le t \le b$ at each point of which the limit

$$\lim_{\Delta t \to 0} \frac{\mathbf{u}(t + \Delta t) - \mathbf{u}(t)}{\Delta t}$$

is defined. Then $\mathbf{u}(t)$ is said to be differentiable throughout that interval with the derivative

$$\frac{d\mathbf{u}}{dt} = \lim_{\Delta t \to 0} \frac{\mathbf{u}(t + \Delta t) - \mathbf{u}(t)}{\Delta t}.$$

The geometrical interpretation of the derivative of a vector function of a real variable is apparent in Fig. 11·2. In that figure the curve Γ is described

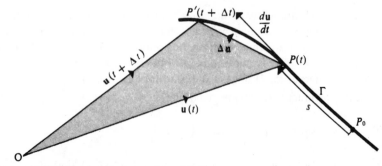

Fig. 11·2 Geometrical interpretation of du/dt.

by a point P(t) with position vector **u**(t) relative to O. The point denoted by P'($t + \Delta t$) is the position assumed by **u** at time $t + \Delta t$, so that $\underline{\text{OP}} = \mathbf{u}(t)$, $\underline{\text{OP}'} = \mathbf{u}(t + \Delta t)$, and $\underline{\text{PP}'} = \Delta\mathbf{u}$ is the increment in **u**(t) consequent upon the increment Δt in t.

It is obvious that as $\Delta t \to 0$, so the vector $\Delta\mathbf{u}$ tends to the line of the tangent to the curve Γ at P(t) with $\Delta\mathbf{u}$ being directed from P to P'. To interpret du/dt in terms of components when $\mathbf{u}(t) = u_1(t)\mathbf{i} + u_2(t)\mathbf{j} + u_3(t)\mathbf{k}$, we need only observe that

$$\frac{d\mathbf{u}}{dt} = \lim_{\Delta t \to 0} \frac{\mathbf{u}(t + \Delta t) - \mathbf{u}(t)}{\Delta t}$$

$$= \lim_{\Delta t \to 0} \left[\frac{u_1(t + \Delta t) - u_1(t)}{\Delta t}\right]\mathbf{i} + \lim_{\Delta t \to 0} \left[\frac{u_2(t + \Delta t) - u_2(t)}{\Delta t}\right]\mathbf{j}$$

$$+ \lim_{\Delta t \to 0} \left[\frac{u_3(t + \Delta t) - u_3(t)}{\Delta t}\right]\mathbf{k},$$

from which it follows that

$$\frac{d\mathbf{u}}{dt} = \frac{du_1}{dt}\mathbf{i} + \frac{du_2}{dt}\mathbf{j} + \frac{du_3}{dt}\mathbf{k}. \tag{11·3}$$

The unit vector **T** that is tangent to Γ at P(t) and points in the direction in which P(t) will move with increasing t is obviously

$$\mathbf{T} = \frac{d\mathbf{u}}{dt} \Big/ \left|\frac{d\mathbf{u}}{dt}\right|. \tag{11·4}$$

If s is the distance to P measured positively in the sense P to P' along Γ from some fixed point on that curve (Fig. 11·2), then we know from our work with differentials that $du_1 = u'_1 dt$, $du_2 = u'_2 dt$, $du_3 = u'_3 dt$. Now as the differentials du_1, du_2, du_3 are mutually orthogonal and represent the increments in the coordinates [$u_1(t)$, $u_2(t)$, $u_3(t)$] of P to an adjacent point distant ds away along Γ with coordinates [$u_1(t + dt)$, $u_2(t + dt)$, $u_3(t + dt)$], we may apply Pythagoras' theorem to obtain

$$(ds)^2 = (u'_1 dt)^2 + (u'_2 dt)^2 + (u'_3 dt)^2,$$

whence

$$\frac{ds}{dt} = \sqrt{\left[\left(\frac{du_1}{dt}\right)^2 + \left(\frac{du_2}{dt}\right)^2 + \left(\frac{du_3}{dt}\right)^2\right]}. \tag{11·5}$$

Comparison of Eqns (11·3) and (11·5) then gives the result

$$\left|\frac{du}{dt}\right| = \frac{ds}{dt}, \tag{11·6}$$

from which we see that if t is regarded as time, then the vector function $\mathbf{v} = du/dt$ is the *velocity vector* of P(t) as it moves with *speed* ds/dt along Γ in the direction of **T**. These results merit recording as a theorem.

THEOREM 11·2   Let $\mathbf{u}(t) = u_1(t)\mathbf{i} + u_2(t)\mathbf{j} + u_3(t)\mathbf{k}$ be a differentiable vector function of the real variable t, then

$$\frac{d\mathbf{u}}{dt} = \frac{du_1}{dt}\mathbf{i} + \frac{du_2}{dt}\mathbf{j} + \frac{du_3}{dt}\mathbf{k}.$$

If Γ denotes the curve traced out by the point P(t) with position vector $\mathbf{u}(t)$ as t increases, and s is the distance to P(t) measured along Γ from some fixed point, then

$$\frac{ds}{dt} = \left|\frac{d\mathbf{u}}{dt}\right|$$

and the unit tangent **T** to the curve Γ at P(t) oriented in the sense of increasing t is

$$\mathbf{T} = \left(\frac{d\mathbf{u}}{dt}\right)\Big/\left|\frac{d\mathbf{u}}{dt}\right|.$$

As a consequence of this theorem we may write

$$\frac{d\mathbf{u}}{dt} = \frac{ds}{dt}\left(\frac{d\mathbf{u}}{dt}\right)\Big/\left|\frac{d\mathbf{u}}{dt}\right| = \frac{ds}{dt}\mathbf{T}, \tag{11·7}$$

which is a result of considerable use in dynamics when t is identified with time.

Higher order derivatives such as $d^2\mathbf{u}/dt^2$ and $d^3\mathbf{u}/dt^3$ may also be defined in the obvious fashion as $d^2\mathbf{u}/dt^2 = (d/dt)(d\mathbf{u}/dt)$, $d^3\mathbf{u}/dt^3 = (d/dt)(d^2\mathbf{u}/dt^2)$ provided only that the components of $\mathbf{u}(t)$ have suitable differentiability properties. Thus, for example, if the second derivatives of the components of $\mathbf{u}(t)$ exist we have

$$\frac{d^2\mathbf{u}}{dt^2} = \frac{d^2u_1}{dt^2}\mathbf{i} + \frac{d^2u_2}{dt^2}\mathbf{j} + \frac{d^2u_3}{dt^2}\mathbf{k}. \tag{11·8}$$

We have seen that if t is identified with time and $\mathbf{u}(t)$ is the position vector of a point P, then $d\mathbf{u}/dt$ is the *velocity vector* of P. It follows from this same argument that $d^2\mathbf{u}/dt^2$ is the *acceleration vector* of P.

Example 11·2   The position vector $\mathbf{r}$ of a particle at time t is given by

$$\mathbf{r} = a \cos \omega t \mathbf{i} + a \sin \omega t \mathbf{j} + \alpha t^2 \mathbf{k},$$

where $\mathbf{i}$, $\mathbf{j}$, $\mathbf{k}$ have their usual meanings and a, ω, and α are constants. Find the acceleration vector at time t, and deduce the times at which it will be perpendicular to the position vector. Hence deduce the unit tangent to the particle trajectory at these times.

Solution   By making the identifications $\mathbf{u} = \mathbf{r}$, $u_1(t) = a \cos \omega t$, $u_2(t) = a \sin \omega t$ and $u_3(t) = \alpha t^2$ and then applying Theorem 11·2 we find that the velocity vector is

$$\frac{d\mathbf{r}}{dt} = -a\omega \sin \omega t \mathbf{i} + a\omega \cos \omega t \mathbf{j} + 2\alpha t \mathbf{k}.$$

A further differentiation yields the required acceleration vector

$$\frac{d^2\mathbf{r}}{dt^2} = -a\omega^2 \cos \omega t \mathbf{i} - a\omega^2 \sin \omega t \mathbf{j} + 2\alpha \mathbf{k}.$$

Expressed vectorially, the condition that $\mathbf{r}$ and $d^2\mathbf{r}/dt^2$ should be perpendicular is simply that $\mathbf{r} \cdot (d^2\mathbf{r}/dt^2) = 0$. Hence to find the time at which this condition is satisfied we must solve the equation

$$(a \cos \omega t \mathbf{i} + a \sin \omega t \mathbf{j} + \alpha t^2 \mathbf{k}) \cdot (-a\omega^2 \cos \omega t \mathbf{i} - a\omega^2 \sin \omega t \mathbf{j} + 2\alpha \mathbf{k}) = 0.$$

Forming the required scalar product gives

$$-a^2\omega^2 \cos^2 \omega t - a^2\omega^2 \sin^2 \omega t + 2\alpha^2 t^2 = 0$$

which immediately simplifies to

$$a^2\omega^2 = 2\alpha^2 t^2,$$

showing that the desired times are

$$t = \pm \frac{a\omega}{\alpha\sqrt{2}}.$$

To deduce the unit tangent $\mathbf{T}$ at these times we use the fact that

$$\mathbf{T} = \left(\frac{d\mathbf{r}}{dt}\right) \Big/ \left|\frac{d\mathbf{r}}{dt}\right|,$$

where here

$$\left|\frac{d\mathbf{r}}{dt}\right| = \sqrt{(a^2\omega^2 + 4\alpha^2 t^2)}.$$

Denoting by $\mathbf{T}_\pm$, the unit tangent to the trajectory at $t = \pm a\omega/\alpha\sqrt{2}$, we find by substitution of these values of t in the above expression that

$$\mathbf{T}_+ = \frac{1}{\sqrt{3}}\left(-\sin\frac{a\omega^2}{\alpha\sqrt{2}}\mathbf{i} + \cos\frac{a\omega^2}{\alpha\sqrt{2}}\mathbf{j} + \sqrt{2}\,\mathbf{k}\right)$$

and

$$\mathbf{T}_- = \frac{1}{\sqrt{3}}\left(\sin\frac{a\omega^2}{\alpha\sqrt{2}}\mathbf{i} + \cos\frac{a\omega^2}{\alpha\sqrt{2}}\mathbf{j} - \sqrt{2}\,\mathbf{k}\right).$$

With the obvious differentiability requirements, if $\mathbf{u}(t)$ and $\mathbf{v}(t)$ are differentiable vector functions with respect to t, then so also are $\mathbf{u} + \mathbf{v}$, $\mathbf{u}.\mathbf{v}$, $\mathbf{u} \times \mathbf{v}$, and $\phi\mathbf{u}$, where $\phi = \phi(t)$ is a scalar function of t. As the following theorem is easily proved by resolution of the vector functions involved into component form, it is stated without proof.

THEOREM 11·3 (differentiation, sums and products of vector functions)   If $\mathbf{u}(t)$ and $\mathbf{v}(t)$ are differentiable vector functions throughout some interval $a \leq t \leq b$ and $\phi(t)$ is a differentiable scalar function throughout that same interval then,

(a) $\dfrac{d}{dt}(\mathbf{u} + \mathbf{v}) = \dfrac{d\mathbf{u}}{dt} + \dfrac{d\mathbf{v}}{dt}$;

(b) $\dfrac{d}{dt}(\phi\mathbf{u}) = \phi\dfrac{d\mathbf{u}}{dt} + \mathbf{u}\dfrac{d\phi}{dt}$;

(c) $\dfrac{d}{dt}(\mathbf{u}.\mathbf{v}) = \mathbf{u}\cdot\dfrac{d\mathbf{v}}{dt} + \dfrac{d\mathbf{u}}{dt}\cdot\mathbf{v}$;

(d) $\dfrac{d}{dt}(\mathbf{u} \times \mathbf{v}) = \mathbf{u} \times \dfrac{d\mathbf{v}}{dt} + \dfrac{d\mathbf{u}}{dt} \times \mathbf{v}$;

and, if $\mathbf{c}$ is a constant vector,

(e) $\dfrac{d}{dt}\mathbf{c} = \mathbf{0}$;

where the order of the vector products on the right-hand side of (d) must be strictly observed.

11·2   Antiderivatives and integrals of vector functions

The notion of an antiderivative, already encountered in Chapter 8, extends naturally to a vector function of a real variable.

DEFINITION 11·3 (antiderivative—vector function) The vector function $F(t)$ of the real variable t will be said to be the *antiderivative* of the vector function $f(t)$ if

$$\frac{d}{dt} F(t) = f(t).$$

Naturally, an antiderivative $F(t)$ is indeterminate so far as an additive arbitrary constant vector C is concerned, because by Theorem 11·3 (e), $dC/dt = 0$. Continuing the convention adopted in Chapter 8, the operation of antidifferentiation with respect to a vector function of the single real variable t will be denoted by $\int$, so that

$$\int f(t)dt = F(t) + C,\tag{11·9}$$

where C is an arbitrary constant vector.

It is obvious that Eqn (11·9), when taken in conjunction with Theorem 11·2, implies the following result.

THEOREM 11·4 (antiderivative of vector function) If

$$\int f(t)dt = F(t) + C,$$

where $f(t) = f_1(t)i + f_2(t)j + f_3(t)k$, $F(t) = F_1(t)i + F_2(t)j + F_3(t)k$ and $C = C_1i + C_2j + C_3k$ is an arbitrary constant vector, then

$$\int f_i(t)dt = F_i(t) + C_i, \qquad i = 1, 2, 3$$

with

$$\frac{dF_i}{dt} = f_i(t).$$

Expressed in words, the antiderivative of $f(t)$ has components equal to the antiderivatives of the components of $f(t)$. As with the scalar case, in many books the entire right-hand side of Eqn (11·9) is loosely referred to as the indefinite integral of the vector function $f(t)$, rather than as here using this term to refer only to its first member.

Example 11·3 Find the antiderivative of $f(t)$ given that

$$f(t) = \cos t i + (1 + t^2)j + e^{-t}k.$$

Solution It follows immediately from Theorem 11·4 that,

$$\int f(t)dt = i \int \cos t \, dt + j \int (1 + t^2)dt + k \int e^{-t} \, dt$$

$$= \sin t i + \left(t + \frac{t^3}{3}\right)j - e^{-t}k + C.$$

The obvious modification to Theorem 11·4 to enable us to work with definite integrals of vector functions of a single real variable comprises the next theorem. Because it is strictly analogous to the scalar case it is offered without proof.

THEOREM 11·5 (definite integral of vector function)   If $\mathbf{F}(t)$ is an anti-derivative of $\mathbf{f}(t)$, then

$$\int_a^b \mathbf{f}(t)\mathrm{d}t = \mathbf{F}(b) - \mathbf{F}(a).$$

Example 11·4   Evaluate the definite integral

$$\int_0^{\frac{1}{4}\pi} (t^2\mathbf{i} + \sec^2 t\mathbf{j} + \mathbf{k})\mathrm{d}t.$$

Solution   From Theorem 11·5 we have the result

$$\int_0^{\frac{1}{4}\pi} (t^2\mathbf{i} + \sec^2 t\mathbf{j} + \mathbf{k})\mathrm{d}t = \left(\frac{t^3}{3}\mathbf{i} + \tan t\mathbf{j} + \mathbf{k}t\right)\Big|_0^{\frac{1}{4}\pi} = \frac{\pi^3}{192}\mathbf{i} + \mathbf{j} + \tfrac{1}{4}\pi\mathbf{k}.$$

A slightly more interesting application of a definite integral is provided by the following example concerning the motion of a particle in space.

Example 11·5   A point moving in space has acceleration

$$\sin 2t\mathbf{i} - \cos 2t\mathbf{k}.$$

Find the equation of its path if it passes through the point with position vector $\mathbf{r}_0 = \mathbf{j} + 2\mathbf{k}$ with velocity $2\mathbf{j}$ at time $t = 0$.

Solution   If $\mathbf{r}$ is the general position vector of the point at time t, then the velocity $\mathbf{v}(t) = \mathrm{d}\mathbf{r}/\mathrm{d}t$ and the acceleration $\mathbf{a}(t) = \mathrm{d}^2\mathbf{r}/\mathrm{d}t^2$. Hence

$$\frac{\mathrm{d}^2\mathbf{r}}{\mathrm{d}t^2} = \sin 2t\mathbf{i} - \cos 2t\mathbf{k},$$

so that integrating the acceleration equation from 0 to t and replacing t in the integrand by the dummy variable τ gives

$$\int_0^t \left(\frac{\mathrm{d}^2\mathbf{r}}{\mathrm{d}\tau^2}\right)\mathrm{d}\tau = \int_0^t (\sin 2\tau\mathbf{i} - \cos 2\tau\mathbf{k})\mathrm{d}\tau.$$

Hence

$$\left(\frac{\mathrm{d}\mathbf{r}}{\mathrm{d}\tau}\right)\Big|_0^t = -\tfrac{1}{2}(\cos 2\tau\mathbf{i} + \sin 2\tau\mathbf{k})\Big|_0^t,$$

and so

$$\mathbf{v}(t) = \mathbf{v}_0 - \tfrac{1}{2}(1 - \cos 2t)\mathbf{i} - \tfrac{1}{2}\sin 2t\mathbf{k}.$$

Now from the initial conditions of the problem $v_0 = 2j$, so that the velocity equation becomes

$$v(t) = \tfrac{1}{2}(1 - \cos 2t)i + 2j - \tfrac{1}{2}\sin 2t k.$$

To find the equation of the path a further integration is required so, setting $v(t) = dr/dt$, integrating the velocity equation from 0 to t gives

$$\int_0^t \left(\frac{dr}{d\tau}\right) d\tau = \int_0^t (\tfrac{1}{2}(1 - \cos 2\tau)i + 2j - \tfrac{1}{2}\sin 2\tau k)d\tau.$$

Hence

$$\left. r(\tau) \right|_0^t = \tfrac{1}{2}(\tau - \tfrac{1}{2}\sin 2\tau)i + 2\tau j + \tfrac{1}{4}\cos 2\tau k \Big|_0^t,$$

and so

$$r(t) = r_0 + \tfrac{1}{2}(t - \tfrac{1}{2}\sin 2t)i + 2tj + \tfrac{1}{4}(\cos 2t - 1)k.$$

Again appealing to the initial conditions of the problem we find that $r_0 = j + 2k$, so that, finally, the particle path must be

$$r(t) = \tfrac{1}{2}(t - \tfrac{1}{2}\sin 2t)i + (1 + 2t)j + \tfrac{1}{4}(7 + \cos 2t)k.$$

11·3 Application to Kinematics

Kinematics, an important branch of mechanics, is essentially concerned with the geometrical aspect of the motion of particles along curves. Of particular

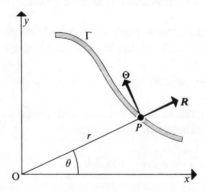

Fig. 11·3 Planar motion of particle in terms of polar coordinates.

importance is that class of motions that occur entirely in one plane, and so are called planar motions. In many of these situations, for example, particle motion in an orbit, the position of a particle is best defined in terms of the polar coordinates (r, θ) in the plane of the motion. Let us then determine expressions for the velocity and acceleration of a particle in terms of polar coordinates.

We first appeal to Fig. 11·3, which represents a particle P moving in the indicated direction along the curve Γ. The unit vectors $\mathbf{R}$, $\mathbf{\Theta}$ are normal to each other and are such that $\mathbf{R}$ is directed from O to P along the radius vector OP, and $\mathbf{\Theta}$ points in the direction of increasing θ. Then clearly $\mathbf{R}$ and $\mathbf{\Theta}$ are vector functions of the single variable θ, with

$$\mathbf{R} = \cos\theta\mathbf{i} + \sin\theta\mathbf{j} \quad \text{and} \quad \mathbf{\Theta} = -\sin\theta\mathbf{i} + \cos\theta\mathbf{j}. \tag{11·10}$$

It follows from these relationships that

$$\frac{d\mathbf{R}}{d\theta} = \mathbf{\Theta} \quad \text{and} \quad \frac{d\mathbf{\Theta}}{d\theta} = -\mathbf{R}. \tag{11·11}$$

In terms of the unit vectors $\mathbf{R}$, $\mathbf{\Theta}$ the point P has the position vector

$$\mathbf{r} = r\mathbf{R}, \tag{11·12}$$

so that the velocity $d\mathbf{r}/dt$ must be

$$\frac{d\mathbf{r}}{dt} = \frac{dr}{dt}\mathbf{R} + r\frac{d\mathbf{R}}{dt}$$

$$= \frac{dr}{dt}\mathbf{R} + r\frac{d\mathbf{R}}{d\theta}\frac{d\theta}{dt},$$

showing that the velocity vector of P is

$$\dot{\mathbf{r}} = \dot{r}\mathbf{R} + r\dot{\theta}\mathbf{\Theta}, \tag{11·13}$$

where differentiation with respect to time has been denoted by a dot.

Here the quantity $\dot{r}$ is called the radial component of velocity and $r\dot{\theta}$ is called the transverse component of velocity. A further differentiation with respect to time yields for the acceleration vector $\ddot{\mathbf{r}} = d^2\mathbf{r}/dt^2$ the expression

$$\ddot{\mathbf{r}} = \ddot{r}\mathbf{R} + \dot{r}\dot{\mathbf{R}} + \dot{r}\dot{\theta}\mathbf{\Theta} + r\ddot{\theta}\mathbf{\Theta} + r\dot{\theta}\dot{\mathbf{\Theta}}$$

or

$$\ddot{\mathbf{r}} = \ddot{r}\mathbf{R} + \dot{r}\dot{\theta}\frac{d\mathbf{R}}{d\theta} + (\dot{r}\dot{\theta} + r\ddot{\theta})\mathbf{\Theta} + r\dot{\theta}^2\frac{d\mathbf{\Theta}}{d\theta}.$$

Hence by Eqn (11·11) this is seen to be equivalent to

$$\ddot{\mathbf{r}} = (\ddot{r} - r\dot{\theta}^2)\mathbf{R} + (2\dot{r}\dot{\theta} + r\ddot{\theta})\mathbf{\Theta}. \tag{11·14}$$

The quantity $\ddot{r} - r\dot{\theta}^2$ is called the radial component of acceleration, and $2\dot{r}\dot{\theta} + r\ddot{\theta}$ is called the transverse component of acceleration.

Example 11·6 A particle is constrained to move with constant speed v along the cardioid $r = a(1 + \cos\theta)$. Prove that

$$v = 2a\dot\theta \cos\left(\frac{\theta}{2}\right),$$

and show that the radial component of the acceleration is constant.

Solution From Eqn (11·13) and the expression $r = a(1 + \cos\theta)$, it follows that the velocity vector **r** is given by

$$\dot{\mathbf{r}} = -a\sin\theta\dot\theta\mathbf{R} + a(1 + \cos\theta)\dot\theta\mathbf{\Theta}.$$

Now as $v^2 = \dot{\mathbf{r}}^2 = \dot{\mathbf{r}} \cdot \dot{\mathbf{r}}$, we have

$$v^2 = a^2\dot\theta^2\sin^2\theta + a^2\dot\theta^2(1 + \cos\theta)^2 = 2a^2\dot\theta^2(1 + \cos\theta).$$

Using the identity $1 + \cos\theta = 2\cos^2(\theta/2)$ in this expression and taking the square root yields the required result

$$v = 2a\dot\theta\cos(\theta/2).$$

To complete the problem we now make appeal to the fact that the radial acceleration component is $\ddot{r} - r\dot\theta^2$, whilst by supposition $v = \text{constant}$. From our previous working we know that

$$v^2 = 2a^2\dot\theta^2(1 + \cos\theta),$$

so that differentiating with respect to t and cancelling $\dot\theta$ gives

$$\ddot\theta = \frac{\dot\theta^2\sin\theta}{2(1 + \cos\theta)}$$

or, as

$$\dot\theta^2 = \frac{v^2}{2a^2(1 + \cos\theta)},$$

$$\ddot\theta = \frac{v^2\sin\theta}{4a^2(1 + \cos\theta)^2}.$$

Hence as $\ddot{r} = -a(\cos\theta\dot\theta^2 + \sin\theta\ddot\theta)$, substituting for r, $\dot\theta^2$, and $\ddot\theta$ in the radial component of acceleration we find, as required, that

$$\ddot{r} - r\dot\theta^2 = \frac{-3v^2}{4a} = \text{constant}.$$

11·4 Fields, gradient, and directional derivative

The scalar function $\phi = \sqrt{(1 - x^2)} + \sqrt{(1 - y^2)} + \sqrt{(1 - z^2)}$ is defined within and on the cube shaped domain $|x| \le 1$, $|y| \le 1$, $|z| \le 1$ and assigns a specific number ϕ to every point within that region. In the language of vector analysis, ϕ is said to define a *scalar field* throughout the cube. In general, any scalar function ϕ of position will define a scalar field within its domain of definition. A typical physical example of a scalar field is provided by the temperature at each point of a body.

Similarly, if **F** is a vector function of position, we say that **F** defines a *vector field* throughout its domain of definition in the sense that it assigns a specific vector to each point. Thus the vector function $\mathbf{F} = \sin x\mathbf{i} + xy\mathbf{j} + ye^z\mathbf{k}$ defines a vector field throughout all space.

As heat flows in the direction of decreasing temperature, it follows that associated with the scalar temperature field within a body there must also be a vector field which assigns to each point a vector describing the direction and maximum rate of flow of heat. Other physical examples of vector fields are provided by the velocity field **v** throughout a fluid, and the magnetic field **H** throughout a region.

To examine more closely the nature of a scalar field, and to see one way in which a special type of vector field arises, we must now define what is called the gradient of a scalar function. This is a vector differentiation operation that associates a vector field with every continuously differentiable scalar function.

DEFINITION 11·4 (gradient of scalar function) If the scalar function $\phi(x, y, z)$ is a continuously differentiable function with respect to the independent variables x, y, and z then the *gradient* of ϕ, written grad ϕ, is defined to be the vector

$$\text{grad } \phi = \frac{\partial \phi}{\partial x}\mathbf{i} + \frac{\partial \phi}{\partial y}\mathbf{j} + \frac{\partial \phi}{\partial z}\mathbf{k}.$$

For the moment let it be understood that $\mathbf{r} = x\mathbf{i} + y\mathbf{j} + z\mathbf{k}$ is a specific point, and consider a displacement from it $d\mathbf{r} = dx\mathbf{i} + dy\mathbf{j} + dz\mathbf{k}$. Then it follows from the definition of grad ϕ that

$$d\mathbf{r} \cdot \text{grad } \phi = \frac{\partial \phi}{\partial x} dx + \frac{\partial \phi}{\partial y} dy + \frac{\partial \phi}{\partial z} dz,$$

in which it is supposed that grad ϕ is evaluated at $\mathbf{r} = x\mathbf{i} + y\mathbf{j} + z\mathbf{k}$. Theorem 5·19 then asserts that the right-hand side of this expression is simply the total differential $d\phi$ of the scalar function ϕ, so that we have the result

$$d\phi = d\mathbf{r} \cdot \text{grad } \phi. \tag{11·15}$$

If we set $ds = |d\mathbf{r}|$, then $d\mathbf{r}/ds$ is the unit vector in the direction of $d\mathbf{r}$. Writing $\mathbf{a} = d\mathbf{r}/ds$, Eqn (11·15) is thus seen to be equivalent to

$$\frac{d\phi}{ds} = \mathbf{a} \cdot \text{grad } \phi. \tag{11·16}$$

Because $\mathbf{a} \cdot \text{grad } \phi$ is the projection of grad ϕ along the unit vector $\mathbf{a}$, expression (11·16) is called the *directional derivative* of ϕ in the direction of $\mathbf{a}$.

In other words, $\mathbf{a} \cdot \text{grad } \phi$ is the rate of change of ϕ with respect to distance measured in the direction of **a**. We have already utilized the notion of a directional derivative in connection with the derivation of the Cauchy–Riemann equations, though at that time neither the term nor vector notation was employed.

As the largest value of the projection **a** . grad ϕ at a point occurs when **a** is taken in the same direction as grad ϕ, it follows that grad ϕ points in the direction in which the maximum change of the directional derivative of ϕ occurs.

In more advanced treatments of the gradient operator it is this last property that is used to define grad ϕ, since it is essentially independent of the coordinate system that is utilized. From this more general point of view our Definition 11·4 then becomes the interpretation of grad ϕ in terms of rectangular Cartesian coordinates.

The vector differential operator ∇, pronounced either 'del' or 'nabla', is defined in terms of rectangular Cartesian coordinates as

$$\nabla \equiv \mathbf{i}\,\frac{\partial}{\partial x} + \mathbf{j}\,\frac{\partial}{\partial y} + \mathbf{k}\,\frac{\partial}{\partial z}. \tag{11·17}$$

As the name implies, ∇ is a vector differential operator, not a vector. It only generates a vector when it acts on a suitably differentiable scalar function. We have the obvious result that

$$\text{grad }\phi \equiv \frac{\partial \phi}{\partial x}\mathbf{i} + \frac{\partial \phi}{\partial y}\mathbf{j} + \frac{\partial \phi}{\partial z}\mathbf{k} \equiv \left(\mathbf{i}\,\frac{\partial}{\partial x} + \mathbf{j}\,\frac{\partial}{\partial y} + \mathbf{k}\,\frac{\partial}{\partial z}\right)\phi \equiv \nabla\phi. \tag{11·18}$$

Example 11·7 Determine grad ϕ if $\phi = z^2 \cos(xy - \tfrac{1}{4}\pi)$, and hence deduce its value at the point $(1, \tfrac{1}{2}\pi, 1)$.

Solution We have

$$\frac{\partial \phi}{\partial x} = -yz^2 \sin(xy - \tfrac{1}{4}\pi), \quad \frac{\partial \phi}{\partial y} = -xz^2 \sin(xy - \tfrac{1}{4}\pi)$$

and

$$\frac{\partial \phi}{\partial z} = 2z \cos(xy - \tfrac{1}{4}\pi).$$

Hence,

$$\text{grad }\phi = \frac{\partial \phi}{\partial x}\mathbf{i} + \frac{\partial \phi}{\partial y}\mathbf{j} + \frac{\partial \phi}{\partial z}\mathbf{k}$$

$$= -yz^2 \sin(xy - \tfrac{1}{4}\pi)\mathbf{i} - xz^2 \sin(xy - \tfrac{1}{4}\pi)\mathbf{j} + 2z \cos(xy - \tfrac{1}{4}\pi)\mathbf{k}.$$

At the point $(1, \tfrac{1}{2}\pi, 1)$ we thus have

$$(\text{grad }\phi)_{(1,\,\frac{1}{2}\pi,\,1)} = \frac{1}{\sqrt{2}}\left(-(\tfrac{1}{2}\pi)\mathbf{i} - \mathbf{j} + 2\mathbf{k}\right).$$

Example 11·8   If $\mathbf{r} = x\mathbf{i} + y\mathbf{j} + z\mathbf{k}$, and $r = |\mathbf{r}|$, deduce the form taken by grad r^n.

Solution   As $r = (x^2 + y^2 + z^2)^{1/2}$, it follows from Eqn (11·18) and the chain rule that

$$\text{grad } r^n = \left(\mathbf{i} \frac{\partial}{\partial x} + \mathbf{j} \frac{\partial}{\partial y} + \mathbf{k} \frac{\partial}{\partial z} \right) r^n$$

$$= \left(\mathbf{i} \frac{\partial r}{\partial x} \cdot \frac{\partial}{\partial r} + \mathbf{j} \frac{\partial r}{\partial y} \cdot \frac{\partial}{\partial r} + \mathbf{k} \frac{\partial r}{\partial z} \cdot \frac{\partial}{\partial r} \right) r^n$$

$$= n r^{n-1} \left(\frac{\partial r}{\partial x} \mathbf{i} + \frac{\partial r}{\partial y} \mathbf{j} + \frac{\partial r}{\partial z} \mathbf{k} \right).$$

However,

$$\frac{\partial r}{\partial x} = \frac{x}{r}, \qquad \frac{\partial r}{\partial y} = \frac{y}{r}, \qquad \frac{\partial r}{\partial z} = \frac{z}{r}$$

and so

$$\text{grad } r^n = n r^{n-2}(x\mathbf{i} + y\mathbf{j} + z\mathbf{k}) = n r^{n-2}\mathbf{r}.$$

The following theorem is an immediate consequence of the definition of the gradient operator and of the operation of partial differentiation.

THEOREM 11·6 (properties of gradient operator)   If ϕ and ψ are two continuously differentiable scalar functions in some domain D, and a, b are scalar constants, then

(a) grad $a = \mathbf{0}$;
(b) grad $(a\phi + b\psi) = a$ grad $\phi + b$ grad ψ;
(c) grad $(\phi \psi) = \phi$ grad $\psi + \psi$ grad ϕ.

The surfaces $\phi(x, y, z) = $ constant associated with a scalar function ϕ are called *level surfaces* of ϕ. If we form the total differential of ϕ at a point on a specific level surface $\phi = $ constant then $\mathrm{d}\phi = 0$ and, as in Eqn (5·23), we obtain the result

$$\frac{\partial \phi}{\partial x} \mathrm{d}x + \frac{\partial \phi}{\partial y} \mathrm{d}y + \frac{\partial \phi}{\partial z} \mathrm{d}z = 0.$$

This is equivalent to

$$\mathbf{dr} \cdot \text{grad } \phi = 0, \tag{11·19}$$

where now $\mathbf{dr}$ is constrained to lie in the level surface.

This vector condition shows that grad ϕ must be normal to $\mathbf{dr}$, and as $\mathbf{dr}$ is constrained to be an arbitrary tangential vector to the level surface at the point in question, it follows that the vector grad ϕ must be normal to the level surface. The unit normal $\mathbf{n}$ to the surface is thus $\mathbf{n} = $ grad $\phi / |$ grad $\phi |$.

Notice that this normal is unique apart from its sign. This simple argument has proved the following general result.

THEOREM 11·7 (normal to level surface)   If ϕ is a continuously differentiable scalar function, the unit normal $\mathbf{n}$ to any point of the level surface $\phi =$ constant is determined by

$$\mathbf{n} = \frac{\text{grad } \phi}{|\text{grad } \phi|}.$$

Example 11·9   If $\phi = x^2 + 3xy^2 + yz^3 - 12$, find the unit normal $\mathbf{n}$ to the level curve $\phi = 3$ at the point $(1, 2, 1)$. Deduce the equation of the tangent plane to the level surface at this point.

Solution   The level surface $\phi = 3$ is defined by the equation $\psi = 0$, where

$$\psi = x^2 + 3xy^2 + yz^3 - 15 = 0.$$

Hence

$$\text{grad } \psi = (2x + 3y^2)\mathbf{i} + (6xy + z^3)\mathbf{j} + 3yz^2\mathbf{k}$$

which, at $(1, 2, 1)$, becomes

$$(\text{grad } \psi)_{(1,2,1)} = 14\mathbf{i} + 13\mathbf{j} + 6\mathbf{k}.$$

As $\psi = 0$ is the desired level surface, it follows from Theorem 11·7 that the unit normal to this surface at the point $(1, 2, 1)$ must be,

$$\mathbf{n} = \frac{14\mathbf{i} + 13\mathbf{j} + 6\mathbf{k}}{\sqrt{[(14)^2 + (13)^2 + (6)^2]}} = \frac{14\mathbf{i} + 13\mathbf{j} + 6\mathbf{k}}{\sqrt{401}}.$$

Now the equation of a plane is $\mathbf{n} \cdot \mathbf{r} = p$, where $\mathbf{r} = x\mathbf{i} + y\mathbf{j} + z\mathbf{k}$ is a general point on the plane, $\mathbf{n}$ is the unit normal to the plane, and p is its perpendicular distance from the origin. The point $\mathbf{r}_0 = \mathbf{i} + 2\mathbf{j} + \mathbf{k}$ is a point on the plane so that $\mathbf{n} \cdot \mathbf{r} = \mathbf{n} \cdot \mathbf{r}_0 \, (=p)$. Hence

$$\left(\frac{14\mathbf{i} + 13\mathbf{j} + 6\mathbf{k}}{\sqrt{401}}\right) \cdot (x\mathbf{i} + y\mathbf{j} + z\mathbf{k}) = \left(\frac{14\mathbf{i} + 13\mathbf{j} + 6\mathbf{k}}{\sqrt{401}}\right) \cdot (\mathbf{i} + 2\mathbf{j} + \mathbf{k}),$$

showing that the required equation is

$$14x + 13y + 6z = 46.$$

PROBLEMS

Section 11·1

11·1 Sketch and give a brief description of the curves described by the following vector functions of a single real variable t:

(a) $\mathbf{r} = a \cos 2\pi t \mathbf{i} + b \sin 2\pi t \mathbf{j} + t \mathbf{k}$;

(b) $\mathbf{r} = a \cos 2\pi t \mathbf{i} + b \sin 2\pi t \mathbf{j} + t^2 \mathbf{k}$;

(c) $\mathbf{r} = t \mathbf{i} + t^2 \mathbf{j} + t^3 \mathbf{k}$.

11·2 Form the vector functions $\mathbf{u}(t) + \mathbf{v}(t)$, $\mathbf{u}(t) \times \mathbf{v}(t)$, and the scalar function $\mathbf{u}(t) \cdot \mathbf{v}(t)$ given that:

$$\mathbf{u}(t) = t^2 \mathbf{i} + \sinh t \mathbf{j} + \left(\frac{1 - t^2}{1 + t^2} \right) \mathbf{k}$$

and

$$\mathbf{v}(t) = 2t \mathbf{i} + \cosh t \mathbf{j} + \sin t \mathbf{k}.$$

11·3 Determine $d\mathbf{u}/dt$ and $d^2\mathbf{u}/dt^2$ for the vectors $\mathbf{u}$ defined in (a) and (c) of Problem 11·1, and find $d\mathbf{u}/dt$ for the vector

$$\mathbf{u} = \frac{\mathbf{r}}{r^2} + (\mathbf{a} \cdot \mathbf{r})\mathbf{b} + \mathbf{a} \times \frac{d^2\mathbf{r}}{dt^2},$$

where $\mathbf{r} = \mathbf{t}(t)$, $r = |\mathbf{r}|$ and $\mathbf{a}$, $\mathbf{b}$ are constant vectors.

11·4 The position vector of a particle at time t is

$$\mathbf{r} = \cos(t - 1)\mathbf{i} + \sinh(t - 1)\mathbf{j} + \alpha t^3 \mathbf{k}.$$

Find the condition imposed on α by requiring that at time $t = 1$ the acceleration vector is normal to the position vector.

11·5 Find the unit tangent $\mathbf{T}$ to the curve

$$\mathbf{r} = t\mathbf{i} + t^2 \mathbf{j} + t^3 \mathbf{k}$$

at the points corresponding to $t = 0$ and $t = 1$.

11·6 Prove results (a) to (c) of Theorem 11·3.

Section 11·2

11·7 Find the antiderivative of the following two functions $\mathbf{f}(t)$:

(a) $\mathbf{f}(t) = \cosh 2t \mathbf{i} + \dfrac{1}{t}\mathbf{j} + t^3 \mathbf{k}$;

(b) $\mathbf{f}(t) = t^2 \sin t \mathbf{i} + e^t \mathbf{j} + \log t \mathbf{k}$.

11·8 Verify the following antiderivatives using Definition 11·3:

(a) $\displaystyle\int \left(\mathbf{r} \cdot \frac{d\mathbf{r}}{dt} \right) dt = \tfrac{1}{2}(\mathbf{r} \cdot \mathbf{r}) + C = \tfrac{1}{2}r^2 + C$;

(b) $\displaystyle\int \left(\frac{d\mathbf{r}}{dt} \cdot \frac{d^2\mathbf{r}}{dt^2} \right) dt = \frac{1}{2} \frac{d\mathbf{r}}{dt} \cdot \frac{d\mathbf{r}}{dt} + C = \frac{1}{2} \left(\frac{d\mathbf{r}}{dt} \right)^2 + C$;

(c) $\displaystyle\int \mathbf{r} \times \frac{d^2\mathbf{r}}{dt^2} = \mathbf{r} \times \frac{d\mathbf{r}}{dt} + \mathbf{C}$,

where C, $\mathbf{C}$ are arbitrary constants.

11·9 The displacement of a particle P is given in terms of the time t by

$$\mathbf{r} = \cos 2t \mathbf{i} + \sin 2t \mathbf{j} + t^2 \mathbf{k}.$$

If v and f are the magnitudes of the velocity and acceleration respectively, show that

$$5v^2 = f^2(1 + t^2).$$

11·10 A point moving in space has acceleration $\cos t\mathbf{i} + \sin t\mathbf{j}$. Find the equation of its path if it passes through the point $(-1, 0, 0)$ with velocity $-\mathbf{j} + \mathbf{k}$ at time $t = 0$.

11·11 Use the result of Problem 11·8 to express $d\mathbf{r}/dt$ in terms of $\mathbf{r}$, given that $\mathbf{r}$ satisfies the vector differential equation

$$\frac{d^2\mathbf{r}}{dt^2} + \Omega^2\mathbf{r} = 0.$$

Section 11·3

11·12 A particle moves in a curve given by

$$r = a(1 - \cos\theta) \quad \text{with} \quad \frac{d\theta}{dt} = 3.$$

Find the components of velocity and acceleration. Show that the velocity is zero when $\theta = 0$. Find the acceleration when $\theta = 0$.

11·13 A particle moves on that portion of the curve $r = ae^\theta \cos\theta$ ($a =$ constant) for which $0 < \theta < \frac{1}{4}\pi$, so that its radial velocity u remains constant. Find its transverse velocity and its radial and transverse components of acceleration as functions of u and θ.

Section 11·4

11·14 Find the gradient of the following functions ϕ:
(a) $\phi = \cosh xyz$;
(b) $\phi = x^2 + y^2 + z^2$;
(c) $\phi = xy \tanh(x - z)$.

11·15 Find the directional derivative of the following functions ϕ in the direction of the vector $(\mathbf{i} + 2\mathbf{j} - 2\mathbf{k})$:
(a) $\phi = 3x^2 + xy^2 + yz$;
(b) $\phi = x^2yz + \cos y$;
(c) $\phi = 1/xyz$.

11·16 If $\mathbf{a}$ is a constant vector and $\mathbf{r} = x\mathbf{i} + y\mathbf{j} + z\mathbf{k}$, $r = |\mathbf{r}|$ prove that
(a) $\operatorname{grad}(\mathbf{a}.\mathbf{r}) = \mathbf{a}$;
(b) $\operatorname{grad} r = \mathbf{r}$;
(c) $\operatorname{grad}\left(\dfrac{1}{r}\right) = -\dfrac{\mathbf{r}}{r^3}$.

11·17 Find the unit normal $\mathbf{n}$ to the surface $x^2 + 2y^2 - z^2 - 8 = 0$ at the point $(1, 2, 1)$. Deduce the equation of the tangent plane to the surface at this point.

11·18 Find the unit normal $\mathbf{n}$ to the surface $x^2 - 4y^2 + 2z^2 = 6$ at the point $(2, 2, 3)$. Deduce the equation of the plane which has $\mathbf{n}$ as its normal and which passes through the origin.

12 Series, Taylor's theorem and its uses

12·1 Series

The term *series* denotes the sum of the members of a sequence of numbers $\{a_n\}$, in which a_n represents the general term. The number of terms added may be finite or infinite, according as the sequence used is finite or infinite in the sense of Chapter 3. The sum to N terms of the infinite sequence $\{a_n\}$ is written

$$a_1 + a_2 + \cdots + a_N = \sum_{n=1}^{N} a_n,$$

and it is called a *finite* series because the number of terms involved in the summation is finite. The so called *infinite* series derived from the infinite sequence $\{a_n\}$ by the addition of all its terms is written

$$a_1 + a_2 + \cdots + a_r + \cdots = \sum_{n=1}^{\infty} a_n.$$

The following are specific examples of numerical series of essentially different types:

(a) $\displaystyle\sum_{n=1}^{N} n^2 = 1^2 + 2^2 + \cdots + N^2,$

in which the general term $a_n = n^2$;

(b) $\displaystyle\sum_{n=0}^{\infty} \frac{1}{n!} = 1 + 1 + \frac{1}{2!} + \frac{1}{3!} + \cdots + \frac{1}{r!} + \cdots,$

in which the general term $a_n = 1/n!$;

(c) $\displaystyle\sum_{n=1}^{\infty} \frac{1}{n} = 1 + \frac{1}{2} + \frac{1}{3} + \cdots + \frac{1}{r} + \cdots,$

in which the general term $a_n = 1/n$;

(d) $\displaystyle\sum_{n=1}^{\infty} \frac{2n^2 + 1}{4n + 2} = \frac{1}{2} + \frac{9}{10} + \frac{19}{14} + \cdots + \frac{2r^2 + 1}{4r + 2} + \cdots,$

in which the general term $a_n = (2n^2 + 1)/(4n + 2)$;

(e) $\displaystyle\sum_{n=1}^{\infty} (-1)^{n+1} = 1 - 1 + 1 - 1 + \cdots + (-1)^{r+1} + \cdots,$

in which the general term $a_n = (-1)^{n+1}$.

Only (a) is a finite series; the remainder are infinite.

There is obviously no difficulty in assigning a sum to a finite series, but how are we to do this in the case of an infinite series? A practical approach would be to attempt to approximate the infinite series by means of a finite series comprising only its first N terms. To justify this it would be necessary to show in some way that the sum of the remainder R_N of the series after N terms tends to zero as N increases and, even better if possible, to obtain an upper bound for R_N. This was, of course, the approach adopted in Chapter 6 when discussing the exponential series which comprises example (b). In the event of an upper bound for R_N being available, this could be used to deduce the number of terms that need be taken in order to determine the sum to within a specified accuracy.

The spirit of this practical approach to the summation of series is exactly what is adopted in a rigorous discussion of series. The first question to be determined is whether or not a given series has a unique sum; the estimation of the remainder term follows afterwards, and usually proves to be more difficult.

To assist us in our formal discussion of series we use the already familiar notion of the *nth partial sum* S_n of the series $\sum_{n=1}^{\infty} a_n$, which is defined to be the finite sum

$$S_n = \sum_{r=1}^{n} a_r = a_1 + a_2 + \cdots + a_n.$$

Then, in terms of S_n, we have the following definition of convergence, which is in complete agreement with the approach we have just outlined.

DEFINITION 12·1 (convergence of series)   The series $\sum_{n=1}^{\infty} a_n$ will be said to be *convergent* to the finite sum S if its nth partial sum S_n is such that

$$\lim_{n \to \infty} S_n = S.$$

If the limit of S_n is not defined, or is infinite, the series will be said to be *divergent*.

The remainder after n terms, R_n, is given by

$$R_n = a_{n+1} + a_{n+2} + \cdots + a_{n+r} + \cdots,$$

so that if $\{S_n\}$ converges to the limit S, then $R_n = S - S_n$ and Definition 12·1 is obviously equivalent to requiring that

$$\lim_{n \to \infty} (S - S_n) = \lim_{n \to \infty} R_n = 0.$$

Example 12·1   Find the nth partial sum of the series

$$1 + \frac{1}{3} + \frac{1}{9} + \frac{1}{27} + \cdots + \frac{1}{3^n} + \cdots,$$

and hence show that it converges to the sum 3/2. Find the remainder after n terms and deduce how many terms need be summed in order to yield a result in which the error does not exceed 0·01.

Solution   This series is a geometric progression with initial term unity and common ratio 1/3. Its sum to n terms, which is the desired nth partial sum S_n, may be determined by a well known formula (see Problem 12·2) which gives

$$S_n = \frac{1 - (1/3)^n}{1 - 1/3} = \frac{3}{2} [1 - (1/3)^n].$$

We have

$$\lim_{n \to \infty} S_n = \lim_{n \to \infty} \frac{3}{2} \left[1 - \left(\frac{1}{3}\right)^n \right] = 3/2,$$

showing that the series is convergent to the sum 3/2.

As S_n is the sum to n terms, the remainder after n terms, R_n, must be given by $R_n = 3/2 - S_n$, and so

$$R_n = \frac{1}{2} \left(\frac{1}{3}\right)^{n-1}.$$

If the remainder must not exceed 0·01, $R_n \leq 0·01$, from which it is easily seen that the number n of terms needed is $n \geq 5$. The determination of R_n was simple in this instance because we were fortunate enough to have available an explicit formula for S_n. In general such a formula is seldom available.

The definition of convergence has immediate consequences as regards the addition and subtraction of series. Suppose Σa_n and Σb_n are convergent series with sums α, β. (It is customary to omit summation limits when they are not important.) Let their respective partial sums be $S_n = a_1 + a_2 + \cdots + a_n$, $S_n' = b_1 + b_2 + \cdots + b_n$ and consider the series $\Sigma(a_n + b_n)$ which has the partial sum $S_n'' = S_n + S_n'$. Then

$$\lim_{n \to \infty} S_n'' = \lim_{n \to \infty} (S_n + S_n')$$

$$= \lim_{n \to \infty} S_n + \lim_{n \to \infty} S_n' = \alpha + \beta,$$

showing that

$$\sum_{n=1}^{\infty} (a_n + b_n) = \alpha + \beta.$$

A corresponding result for the difference of two series may be proved in similar fashion. We have established the following general result.

THEOREM 12·1 (sum and difference of convergent series) If the series $\sum_{n=1}^{\infty} a_n$ and $\sum_{n=1}^{\infty} b_n$ are convergent to the respective sums α and β, then

$$\sum_{n=1}^{\infty} (a_n + b_n) = \alpha + \beta; \qquad \sum_{n=1}^{\infty} (a_n - b_n) = \alpha - \beta.$$

Example 12·2 Suppose that $a_n = (1/2)^n$ and $b_n = (1/3)^n$, so that the series involved are again geometric progressions with $\sum_{n=1}^{\infty} (1/2)^n = 2$ and $\sum_{n=1}^{\infty} (1/3)^n = 3/2$. Then it follows from Theorem 12·1 that

$$\sum_{n=1}^{\infty} [(1/2)^n + (1/3)^n] = 7/2 \qquad \text{and} \qquad \sum_{n=1}^{\infty} [(1/2)^n - (1/3)^n] = 1/2.$$

Let us now derive a number of standard tests by which the convergence or divergence of a series may be established. We begin with a test for divergence.

Suppose first that a series Σa_n with nth partial sum S_n converges to the sum S. Then from our discussion of the convergence of a sequence given in Chapter 3, we know that for any $\varepsilon > 0$ there must exist some integer N such that

$$|S_n - S| < \varepsilon \qquad \text{for} \qquad n > N.$$

This immediately implies the additional result

$$|S_{n+1} - S| < \varepsilon.$$

Hence,

$$\varepsilon + \varepsilon > |S_{n+1} - S| + |S_n - \overset{.}{S}| = |S_{n+1} - S| + |S - S_n|$$
$$\geq |S_{n+1} - S_n|.$$

However, as $S_{n+1} - S_n = a_{n+1}$, we have proved that

$$|a_{n+1}| < 2\varepsilon \qquad \text{for} \qquad n > N.$$

As ε was arbitrary, this shows that for a series to be convergent, it is necessary that

$$\lim_{n \to \infty} |a_n| = 0$$

or, equivalently,

$$\lim_{n \to \infty} a_n = 0.$$

If this is not the case then the series $\sum_{n=1}^{\infty} a_n$ must diverge. This condition thus provides us with a positive test for divergence.

THEOREM 12·2 (a) (test for divergence) The series $\sum_{n=1}^{\infty} a_n$ diverges if

$$\lim_{n \to \infty} a_n \neq 0.$$

This theorem shows, for example, that the series (d) is divergent, because $a_n = (2n^2 + 1)/(4n + 2)$, and hence it increases without bound as n increases.

It is important to take note of the fact that this theorem gives *no* information in the event that $\lim_{n \to \infty} a_n = 0$. Although we have shown that this is a *necessary* condition for convergence, it is not a *sufficient* condition because divergent series exist for which the condition is true.

Theorem 12·2 (a) gives no information about either series (b) or (c) as in each case $\lim_{n \to \infty} a_n = 0$. In fact, by using another argument, we have already proved that the series representation for e in (b) is convergent, whereas we shall prove shortly that the harmonic series (c) is divergent. Series (e) must also be divergent according to our definition, because a_n oscillates finitely between 1 and -1, and also S_n does not tend to any limit.

The terms of series are not always of the same sign, and so it is useful to associate with the series Σa_n the companion series $\Sigma \, | \, a_n \, |$. If this latter series is convergent, then the series Σa_n is said to be *absolutely convergent*. It can happen that although Σa_n is convergent, $\Sigma \, | \, a_n \, |$ is divergent. When this occurs the series Σa_n is said to be *conditionally* convergent. Now when terms of differing signs are involved, the sum of the absolute values of the terms of a series clearly exceeds the sum of the terms of the series, and so it seems reasonable to expect that absolute convergence implies convergence. Let us prove this fact.

THEOREM 12·2 (b) (absolute convergence implies convergence) If the series $\sum_{n=1}^{\infty} | \, a_n \, |$ is convergent, then so also is the series $\sum_{n=1}^{\infty} a_n.$

Proof The proof of this result is simple. Let $S_n = | \, a_1 \, | + | \, a_2 \, | + \cdots + | \, a_n \, |$ and $S_n' = a_1 + a_2 + \cdots + a_n$ be the nth partial sums, respectively, of the series in Theorem 12·2. Then, as $a_r + | \, a_r \, |$ is either zero or $2 \, | \, a_r \, |$, it follows that

$$0 \leq S_n + S_n' \leq 2S_n'.$$

Now by supposition $\lim_{n \to \infty} S_n' = S'$ exists, so that taking limits we arrive at

$$0 \leq \lim_{n \to \infty} (S_n + S_n') \leq 2S'.$$

This implies that the series with nth term $a_n + |a_n|$ must be convergent and hence, using Theorem 12·1, that $\sum_{n=1}^{\infty} a_n$ must be convergent.

Example 12·3 Consider the series

$$\sum_{n=0}^{\infty} \frac{(-1)^n}{n!} = 1 - 1 + \frac{1}{2!} - \frac{1}{3!} + \cdots.$$

As $a_n = (-1)^n/n!$, we have $|a_n| = 1/n!$, which is the general term of the exponential series defining e. Thus Theorem 12·2, and the convergence of the exponential series, together imply the convergence of $\sum_{n=0}^{\infty} (-1)^n/n!$ In fact this is the series representation of $1/e$.

Suppose Σb_n is a convergent series of positive terms, and that Σa_n is a series with the property that if N is some positive integer, then $|a_n| \leq b_n$ for $n > N$. Then clearly the convergence of Σb_n implies the convergence of $\sum |a_n|$ and, by Theorem 12·2, also the convergence of Σa_n. By a similar argument, if for $n > N$, $0 \leq b_n \leq a_n$, and Σb_n is known to be divergent, then clearly Σa_n must also be divergent. We incorporate these results into a useful comparison test.

THEOREM 12·3 (comparison test)

(a) *Convergence test* Let Σb_n be a convergent series of positive terms, and let Σa_n be a series with the property that there exists a positive integer N such that

$$|a_n| \leq b_n \quad \text{for} \quad n > N.$$

Then Σa_n is an absolutely convergent series.

(b) *Divergence test* Let Σb_n be a divergent series of positive terms, and let Σa_n be a series of positive terms with the property that there exists a positive integer N such that

$$0 \leq b_n \leq a_n \quad \text{for} \quad n > N.$$

Then Σa_n is a divergent series.

Example 12·4

(a) Consider the series $\sum_{n=1}^{\infty} [2 + (-1)^n]/2^n$. We have

$$a_n = \frac{2 + (-1)^n}{2^n} \leq \frac{3}{2^n},$$

and as $\sum_{n=1}^{\infty} 3/2^n = 3 \sum_{n=1}^{\infty} 1/2^n = 9/2$, the conditions of Theorem 12·3 (a) are satisfied if we set $b_n = 3/2^n$. It thus follows that the series Σa_n is convergent.

(b) Consider the series $\sum_{n=1}^{\infty} (n + 1)/n^2$. Here we have

$$a_n = \frac{n + 1}{n^2} = \frac{1}{n}\left(\frac{n + 1}{n}\right) > \frac{1}{n},$$

and as the harmonic series $\Sigma 1/n$ is divergent, the conditions of Theorem 12·3 (b) are satisfied when we set $b_n = 1/n$. Hence Σa_n is divergent.

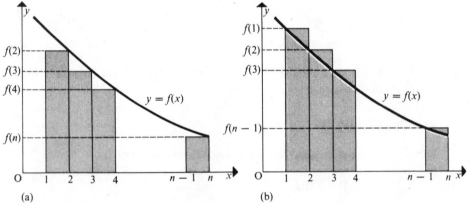

(a)   (b)

Fig. 12·1   Comparison between series and integral.

A powerful test for the convergence or divergence of a series Σa_n of positive terms follows by a comparison of the shaded rectangles in Fig. 12.1.

Let $f(x)$ be a non-increasing function defined for $1 \le x < \infty$ which decreases to zero as x tends to infinity, and let $f(n) = a_n$, where n is an integer. Then we have the obvious inequality

$$\sum_{r=2}^{n} f(r) \le \int_{1}^{n} f(x)\,dx \le \sum_{r=1}^{n} f(r)$$

or, equivalently,

$$\sum_{r=2}^{n} a_r \le \int_{1}^{n} f(x)\,dx \le \sum_{r=1}^{n} a_r.$$

As the right-hand side of this inequality only exceeds the left-hand side by the single term a_1, it must follow that in the limit, the infinite series Σa_r and the integral

$$\lim_{n \to \infty} \int_{1}^{n} f(x)\,dx$$

converge or diverge together. This conclusion may be incorporated into a test as follows.

THEOREM 12·4 (integral test)   Let $f(x)$ be a positive non-increasing function defined on $1 \leq x < \infty$ with $\lim_{x \to \infty} f(x) = 0$. Then, if $a_n = f(n)$, the series $\sum_{n=1}^{\infty} a_n$ converges or diverges according as

$$\lim_{n \to \infty} \int_1^n f(x)\, dx$$

is finite or infinite.

Corollary 12·4 (R_N deduced from integral test).   Let $f(x)$ be a positive non-increasing function defined on $1 \leq x < \infty$ with $\lim_{n \to \infty} f(x) = 0$, and let $\sum_{n=1}^{\infty} a_n$ be convergent, where $a_n = f(n)$. Then the remainder R_N after N terms satisfies the inequality

$$R_N \leq \int_N^{\infty} f(x)\, dx.$$

Proof   The result follows at once from the obvious inequality

$$\sum_{r=N+1}^{N'} a_r \leq \int_N^{N'} f(x)\, dx \leq \sum_{r=N}^{N'} a_r$$

by taking the limit as $N' \to \infty$. This is possible because, by hypothesis, Σa_n is convergent so that the improper integral involved exists.

Example 12·5

(a) Consider the series $\sum_{n=1}^{\infty} 1/n^k$, where $k > 0$. Then the function $f(x) = 1/x^k$ satisfies the conditions of Theorem 12·4. Hence this series converges or diverges according as

$$\lim_{n \to \infty} \int_1^n \frac{dx}{x^k}$$

is finite or infinite. If $k \neq 1$ we have

$$\lim_{n \to \infty} \int_1^n \frac{dx}{x^k} = \left(\frac{1}{1-k} \right) \lim_{n \to \infty} \left[\frac{1}{n^{k-1}} - 1 \right].$$

Hence for $0 < k < 1$ this limit is infinite, showing that the series is divergent for k in this range, whereas for $k > 1$ this limit has the finite value $1/(k-1)$, showing that the series is convergent for $k > 1$. Applying Corollary 12·4 shows that when $k > 1$, the remainder R_N after N terms must satisfy the inequality

$$R_N \leq N^{(1-k)}/(k - 1).$$

When $k = 1$ we obtain the harmonic series, which must be treated separately. As it follows that

$$\lim_{n \to \infty} \int_1^n \frac{dx}{x} = \lim_{n \to \infty} \log n \to \infty,$$

we have proved that the harmonic series is divergent.

(b) Consider the series $\sum_{n=1}^{\infty} n/(1 + n^2)$. Here we set $f(x) = x/(1 + x^2)$, so we must examine

$$L = \lim_{n \to \infty} \int_1^n \frac{x\,dx}{1 + x^2}.$$

Setting $x^2 = u$ we find

$$L = \lim_{n \to \infty} \tfrac{1}{2}[\log(1 + x^2) - \log 2] \to \infty.$$

Hence the series is divergent.

Another useful test known as the ratio test for convergence may be derived from Theorem 12·3, essentially using a geometric progression for purposes of comparison. The idea involved in this test is that a series is tested against itself, and that its convergence or divergence is then deduced from the rate at which successive terms decrease or increase.

Suppose that Σa_n is a series for which the ratio a_{n+1}/a_n is always defined and that $\lim_{n \to \infty} | a_{n+1}/a_n | = L$, where $L < 1$. Let r be some fixed number such that $L < r < 1$. Then the existence of the limit L implies that there exists an integer N such that

$$| a_{n+1} | < r | a_n | \qquad \text{for} \qquad n > . N.$$

Hence it follows that

$$| a_{N+2} | < r | a_{N+1} |, \quad | a_{N+3} | < r | a_{N+2} | < r^2 | a_{N+1} |, \ldots,$$

and in general

$$| a_{N+m+1} | < r^m | a_{N+1} |.$$

Thus if R_N is the remainder after N terms we have

$$R_N = \sum_{n=N+1}^{\infty} a_n \leq \sum_{n=N+1}^{\infty} | a_n | < | a_{N+1} | (1 + r + r^2 + \cdots). \qquad (*)$$

The expression in brackets is a convergent geometric progression because, by hypothesis, $r < 1$. As the remainder term R_N is finite, and is less than the sum of the absolute values of the terms comprising the tail of the series, it is easily seen that the series Σa_n must be absolutely convergent. If $L > 1$ the terms grow in size, and the series Σa_n is divergent. Nothing may be deduced if $L = 1$ for then the series may either be convergent or divergent as illustrated by Example 12·5 (a). In that case $a_{n+1}/a_n = n^k/(n + 1)^k$, giving $\lim |a_{n+1}/a_n| = 1$; and the series was seen to be divergent for $0 < k \leq 1$ and convergent for $k > 1$.

Expressed formally, as follows, these results are called the *ratio test*.

THEOREM 12·5 (ratio test)   If the series $\sum\limits_{n=1}^{\infty} a_n$ is such that $a_n \neq 0$ and

$$\lim_{n \to \infty} \left| \frac{a_{n+1}}{a_n} \right| = L,$$

then

(a) the series Σa_n converges absolutely if $L < 1$,
(b) the series Σa_n diverges if $L > 1$,
(c) the test fails if $L = 1$.

Example 12·6

(a) Consider the series

$$\sum_{n=1}^{\infty} \frac{(-1)^n n!}{n^n}.$$

Then $a_n \neq 0$ and

$$\frac{a_{n+1}}{a_n} = (-1)^{2n+1} \frac{(n + 1)! n^n}{(n + 1)^{n+1} n!} = (-1)^{2n+1} \left(1 + \frac{1}{n} \right)^{-n}.$$

Hence

$$\lim_{n \to \infty} \left| \frac{a_{n+1}}{a_n} \right| = \lim_{n \to \infty} 1 \bigg/ \left(1 + \frac{1}{n} \right)^n = 1/e,$$

where the final result follows by virtue of the work of Section 3·3. As $e > 1$, the ratio test proves the absolute convergence of this series.

(b) Consider the series $\sum\limits_{n=1}^{\infty} 1/n!$. Here $a_n = 1/n! \neq 0$ and

$$\frac{a_{n+1}}{a_n} = \frac{n!}{(n + 1)!} = \frac{1}{n + 1} = \left| \frac{a_{n+1}}{a_n} \right|.$$

Hence

$$\lim_{n\to\infty}\left|\frac{a_{n+1}}{a_n}\right| = \lim_{n\to\infty}\frac{1}{n+1} = 0,$$

and as $0 < 1$ the ratio test proves the series to be convergent.

(c) Consider the series $\sum_{n=1}^{\infty} 3^n/n$.

Then $a_n \neq 0$ and

$$\frac{a_{n+1}}{a_n} = \left(\frac{3^{n+1}}{n+1}\right)\left(\frac{n}{3^n}\right) = 3\left(\frac{n}{n+1}\right) = \left|\frac{a_{n+1}}{a_n}\right|.$$

Now

$$\lim_{n\to\infty}\left|\frac{a_{n+1}}{a_n}\right| = \lim_{n\to\infty}\frac{3n}{n+1} = 3,$$

and as $3 > 1$ the ratio test proves the series to be divergent.

(d) Consider the series $\sum_{n=1}^{\infty} 1/(2n+1)^2$.

Then $a_n \neq 0$ and

$$\frac{a_{n+1}}{a_n} = \left(\frac{2n+1}{2n+3}\right)^2 = \left|\frac{a_{n+1}}{a_n}\right|.$$

Now

$$\lim_{n\to\infty}\left|\frac{a_{n+1}}{a_n}\right| = \lim_{n\to\infty}\left(\frac{2n+1}{2n+3}\right)^2 = 1,$$

so that the ratio test fails in this case. In fact the series is convergent, as may readily be proved by use either of the comparison test, with $b_n = 1/n^2$, or the integral test.

As the remainder term R_N used in the proof of the ratio test may be either positive or negative, the estimate (*) is equivalent to

$$|R_N| \leq |a_{N+1}|(1 + r + r^2 + \cdots)$$

or, summing the geometric progression, to

$$|R_N| \leq \frac{|a_{N+1}|}{1-r}.$$

This simple result provides an estimate of the error if the summation is terminated after N terms and comprises our next result.

Corollary 12·5 (R_N deduced from ratio test) Let the series $\sum_{n=1}^{\infty} a_n$ be convergent, and let the ratio test be applicable with

$$\lim_{n \to \infty} \left| \frac{a_{n+1}}{a_n} \right| = L.$$

Then, if r is a number such that $L < r < 1$, the remainder R_N after N terms is such that

$$| R_N | \leq \frac{| a_{N+1} |}{1 - r}.$$

Let us use Example 12·6 (a) to illustrate this and to compute $| R_5 |$. We have $L = 1/e$ and, as $e = 2·7182 . . .$, we could take $r = 0·5$. Then $1/(1 - r) = 2$, whence

$$| R_N | \leq \frac{2N!}{N^N}.$$

Hence $| R_5 | \leq 48/625$.

For our final result we prove that all series in which the signs of terms alternate, whilst the absolute values of successive terms decrease monotonically to zero, are convergent. Such series are called *alternating* series and are of the general form

$$\sum_{n=1}^{\infty} (-1)^{n+1} a_n = a_1 - a_2 + a_3 - a_4 + \cdots,$$

where $a_n > 0$ for all n.

To prove our assertion of convergence we assume $a_1 > a_2 > a_3 > \cdots$, and $\lim a_n = 0$ and first consider the partial sum S_{2r} corresponding to an even number of terms $2r$. We write S_{2r} in the form

$$S_{2r} = (a_1 - a_2) + (a_3 - a_4) + \cdots + (a_{2r-1} - a_{2r}).$$

Then, because $a_1 > a_2 > a_3 > \cdots$, it follows that $S_{2r} > 0$. By a slight rearrangement of the brackets we also have

$$S_{2r} = a_1 - (a_2 - a_3) - (a_4 - a_5) - \cdots - (a_{2r-2} - a_{2r-1}) - a_{2r},$$

showing that as all the brackets and quantities are positive, $S_{2r} < a_1$. Hence, as S_{2r} is a bounded monotonic decreasing sequence, we know from Chapter 3 that it must tend to a limit S, where

$$0 < S < a_1.$$

Next consider the partial sum S_{2r+1} corresponding to an odd number of terms $2r + 1$. We may write $S_{2r+1} = S_{2r} + a_{2r+1}$. Then, taking the limit of S_{2r+1} we have

$$\lim_{r \to \infty} S_{2r+1} = \lim_{r \to \infty} S_{2r} + \lim_{r \to \infty} a_{2r+1} = S,$$

because by supposition $\lim a_{2r+1} = 0$. Thus both the partial sums S_{2r} and the partial sums S_{2r+1} tend to the same limit S. Hence we have proved that for n both even and odd

$$\lim_{n \to \infty} S_n = S,$$

thereby showing that the series converges.

THEOREM 12·6 (alternating series test)   The series $\sum_{n=1}^{\infty}(-1)^{n+1}a_n$ converges if $a_n > 0$ and $a_{n+1} \leq a_n$ for all n and, in addition,

$$\lim_{n \to \infty} a_n = 0.$$

Example 12·7

(a) Consider the alternating series

$$\sum_{n=1}^{\infty} \frac{(-1)^n}{2^n} = 1 - \frac{1}{2} + \frac{1}{2^2} - \frac{1}{2^3} + \cdots,$$

in which the absolute value of the general term $a_n = \frac{1}{2}^n$. Then, as it is true that $a_{n+1} < a_n$ and $\lim a_n = 0$, the test shows that the series is convergent.

(b) Consider the alternating series

$$\sum_{n=1}^{\infty}(-1)^{n+1}\,\sqrt[2n-1]{2} = \sqrt{2} - \sqrt[3]{2} + \sqrt[4]{2} - \sqrt[5]{2} + \cdots,$$

in which the absolute value of the general term $a_n = \sqrt[2n-1]{2}$. Now it is true that $a_{n+1} < a_n$, but $\lim a_n = 1$, so that the last condition of the theorem is violated rendering it inapplicable. Theorem 12·2 shows the series to be divergent.

The form of argument that was used to show $0 < S_{2r} < a_1$ also shows that

$$0 < \sum_{r=2m+1}^{\infty}(-1)^{n+1}a_r < a_{2m+1}$$

and, by a slight modification, that

$$-a_{2m} < \sum_{r=2m}^{\infty}(-1)^{n+1}a_r < 0.$$

As $R_{2m} = \sum_{r=2m+1}^{\infty}(-1)^r a_r$ is the remainder after an even number $2m$ of terms, and $R_{2m-1} = \sum_{r=2m}^{\infty}(-1)^r a_r$ is the remainder after an odd number $2m - 1$ of terms, it follows that if N is either even or odd, then

$$0 < |R_N| < a_{N+1}.$$

Expressed in words this asserts that when an alternating series is terminated after the Nth term, the absolute value of the error involved is less than the magnitude a_{N+1} of the next term.

Corollary 12·6 (R_N for alternating series) If the alternating series $\sum_{n=1}^{\infty}(-1)^{n+1}a_n$ converges, and R_N is the remainder after N terms, then

$$0 < |R_N| < a_{N+1}.$$

Using the convergent alternating series in Example 12·7 (a) for purposes of illustration we see that $a_n = 1/2^n$, and so the remainder R_N must be such that

$$0 < |R_N| < 1/2^{N+1}.$$

For example, termination of the summation of this series after five terms would result in an error whose absolute magnitude is less than $1/64$.

A calculation involving the summation of a finite number of terms is often facilitated by grouping and interchanging their order. Although these operations are legitimate when the number of terms involved is finite, we must question their validity when dealing with an infinite number of terms. Later we shall show that the grouping of terms is permissible for any convergent series, but that rearrangement of terms is only permissible in a series when it is absolutely convergent, for only then does this operation leave the sum unaltered.

An example will help here to indicate the dangers of manipulating a series without first questioning the legitimacy of the operations to be performed upon it. Consider the alternating series

$$1 - \tfrac{1}{2} + \tfrac{1}{3} - \tfrac{1}{4} + \tfrac{1}{5} - \tfrac{1}{6} + \cdots,$$

which is seen to be convergent by virtue of our last theorem, and denote its sum by S. Then we have

$$S = 1 - \tfrac{1}{2} + \tfrac{1}{3} - \tfrac{1}{4} + \tfrac{1}{5} - \tfrac{1}{6} + \tfrac{1}{7} - \tfrac{1}{8} + \tfrac{1}{9} - \tfrac{1}{10} + \tfrac{1}{11} - \tfrac{1}{12} + \cdots$$

or, on rearranging the terms,

$$S = 1 - \tfrac{1}{2} - \tfrac{1}{4} + \tfrac{1}{3} - \tfrac{1}{6} - \tfrac{1}{8} + \tfrac{1}{5} - \tfrac{1}{10} - \tfrac{1}{12} + \cdots$$
$$= (1 - \tfrac{1}{2}) - \tfrac{1}{4} + (\tfrac{1}{3} - \tfrac{1}{6}) - \tfrac{1}{8} + (\tfrac{1}{5} - \tfrac{1}{10}) - \tfrac{1}{12} + \cdots$$
$$= \tfrac{1}{2}(1 - \tfrac{1}{2} + \tfrac{1}{3} - \tfrac{1}{4} + \tfrac{1}{5} - \tfrac{1}{6} + \cdots)$$
$$= \tfrac{1}{2}S.$$

This can only be true if $S = 0$, but clearly this is impossible because Corollary 12·6 above shows that the error in the summation after only one term is less than $\tfrac{1}{2}$ and therefore S is certainly positive with $\tfrac{1}{2} < S < 1$.

What has gone wrong. The answer is that in a sense we are 'robbing Peter to pay Paul'. This occurs because both the series $\Sigma 1/(2n + 1)$ and the series

$\Sigma 1/2n$ from which are derived the positive and negative terms in our series are divergent, and we have so rearranged the terms that they are weighted in favour of the negative ones. Other rearrangements could in fact be made to yield any sum that was desired. In other words, we are working with a series that is only conditionally convergent, and not absolutely convergent. It would seem from this that perhaps if a series Σa_n is absolutely convergent, then its terms should be capable of rearrangement and grouping without altering the sum. Let us prove the truth of this conjecture, but first we prove the simpler result that the grouping or bracketing of the terms of a convergent series leaves its sum unaltered.

Suppose that Σa_n is a convergent series with sum S. Take as representative of the possible groupings of its terms the series derived from Σa_n by the insertion of parentheses (brackets) as indicated below:

$$(a_1 + a_2) + (a_3 + a_4 + a_5) + a_6 + (a_7 + a_8) + \cdots .$$

Now denote the bracketed terms by b_1, b_2, . . ., where $b_1 = a_1 + a_2$, $b_2 = a_3 + a_4 + a_5$, . . ., so that we have associated a new series Σb_n with the original series Σa_n. If the nth partial sums of Σa_n and Σb_n are S_n and S'_n, respectively, then the partial sums S'_1, S'_2, S'_3, S'_4, . . . of Σb_n are, in reality, the partial sums S_2, S_5, S_6, S_8, . . . of Σa_n. As Σa_n is convergent to S by hypothesis, any subsequence of its partial sums $\{S_n\}$ must also converge to S. In particular this applies to the sequence S_2, S_5, S_6, S_8, . . ., derived by the inclusion of parentheses. Hence Σb_n is also convergent to the sum S, which proves our result.

We now examine the effect of rearranging the terms of a series. Let Σa_n be absolutely convergent so that $\Sigma \mid a_n \mid$ must be convergent, and let Σb_n be a rearrangement of Σa_n. Then, as the terms of $\Sigma \mid b_n \mid$ are in one-to-one correspondence with those of $\Sigma \mid a_n \mid$, it is clear that $\Sigma \mid b_n \mid = \Sigma \mid a_n \mid$, from which we deduce that Σb_n is also absolutely convergent.

Next we must show that Σa_n and Σb_n have the same sum. If S_n is the nth partial sum of Σa_n which has the sum S, then by taking n sufficiently large we may make $\mid S_n - S \mid$ as small as we wish; say less than an arbitrarily small positive number ε. Now let S'_m be the mth partial sum of Σb_n. Then, as S_n contains the first n terms of Σa_n, with their suffixes in sequential order, by taking m large enough we can obviously make S'_m contain all the terms of S_n together with $m - n$ additional terms a_p, a_q, . . ., a_r, where $n < p < q < \cdots < r$.

Hence we may write

$$S'_m = S_n + a_p + a_q + \cdots + a_r,$$

whence

$$S'_m - S = S_n - S + a_p + a_q + \cdots + a_r.$$

Taking absolute values gives

$$| S'_m - S | \leq | S_n - S | + | a_p | + | a_q | + \cdots + | a_r |.$$

Now, n was chosen such that $| S_n - S | < \varepsilon$, so that

$$| S'_m - S | \leq \varepsilon + | a_p | + | a_q | + \cdots + | a_r |.$$

However, the remaining terms on the right-hand side of this inequality all occur after a_n in the series Σa_n, and as $| S_n - S | < \varepsilon$, it must follow that their total contribution cannot exceed ε, and thus

$$| S'_m - S | < 2\varepsilon.$$

This shows that the mth partial sum of Σb_n converges to the sum S, so that rearrangement of the terms of an absolutely convergent series is permissible and does not affect its sum.

THEOREM 12·7 (grouping and rearrangement of series) If the series $\sum\limits_{n=1}^{\infty} a_n$ is convergent, then parentheses may be inserted into the series without affecting its sum. If, in addition, the series $\sum\limits_{n=1}^{\infty} a_n$ is absolutely convergent, then its terms may be rearranged without altering its sum.

Example 12·8

Consider the series

$$\sum_{m=1}^{\infty} \frac{1}{m(m + 1)},$$

which is easily seen to be absolutely convergent by use of the comparison test with $b_m = 1/m^2$. As absolute convergence obviously implies convergence, the first part of Theorem 12·7 asserts that we may group terms by inserting parentheses as we wish. So, using the identity

$$\frac{1}{m(m + 1)} = \frac{1}{m} - \frac{1}{m + 1},$$

we find for the nth partial sum S_n the expression

$$S_n = \sum_{m=1}^{n} \left(\frac{1}{m} - \frac{1}{m + 1} \right).$$

Now successive terms in this summation cancel, or *telescope* as the process is sometimes called, leaving only the first and the last. This is best seen by writing out the expression for S_n in full as follows:

$$S_n = \left(\frac{1}{1} - \frac{1}{2} \right) + \left(\frac{1}{2} - \frac{1}{3} \right) + \cdots + \left(\frac{1}{n - 1} - \frac{1}{n} \right) + \left(\frac{1}{n} - \frac{1}{n + 1} \right)$$

$$= 1 - \frac{1}{n + 1}.$$

Hence, if the sum of the series is S, we have

$$S = \lim_{n \to \infty} S_n = \lim_{n \to \infty} \left[1 - \frac{1}{n+1} \right] = 1.$$

12·2   Power series

Up to now we have been concerned entirely with series that did not contain the variable x. A more general type of series called a *power series* in $(x - x_0)$ has the general form

$$\sum_{n=0}^{\infty} a_n(x - x_0)^n = a_0 + a_1(x - x_0) + a_2(x - x_0)^2 + \cdots, \qquad (12·1)$$

in which the coefficients $a_0, a_1, \ldots, a_n, \ldots$ are constants. When x is assigned some fixed value ξ, say, the power series Eqn (12·1) reduces to an ordinary series of the kind discussed in the previous section, and so may be tested for convergence by any appropriate test mentioned there.

For simplicity we now apply the *ratio test* to series Eqn (12·1), allowing x to remain a free variable, in order to try to deduce the interval for x in which the series is absolutely convergent. If $\alpha_n(x)$ is the absolute value of the ratio of the $(n + 1)$th term to the nth term as a function of x, we have

$$\alpha_n(x) = \left| \frac{a_{n+1}(x - x_0)^{n+1}}{a_n(x - x_0)^n} \right| = \left| \frac{a_{n+1}}{a_n} \right| |x - x_0|.$$

Now for any specific value of x, the ratio test asserts that the series will be convergent if $\lim_{n \to \infty} \alpha_n(x) < 1$, whence we must require

$$\lim_{n \to \infty} \left| \frac{a_{n+1}}{a_n} \right| |x - x_0| < 1.$$

Thus the largest value r, say, of $|x - x_0|$ for which this is true is given by

$$r = \lim_{n \to \infty} \left| \frac{a_n}{a_{n+1}} \right|, \qquad (12·2)$$

provided that this limit exists.

The inequality

$$|x - x_0| < r \qquad (12·3)$$

thus defines the x-interval $(x_0 - r, x_0 + r)$ within which the power series Eqn (12·1) is *absolutely convergent*. For x outside this interval the ratio test shows that the power series must be *divergent*. (See Fig. 12·2.) The interval itself is called the *interval of convergence* of the power series, and the number r is called the *radius of convergence* of the power series. The interval of convergence has been deliberately displayed in the form of an open interval because the ratio test can offer no information about the behaviour of the series at the end points. In fact the power series may either be convergent or divergent at these points.

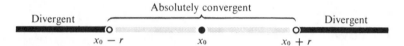

Absolutely convergent

Divergent   Divergent

$x_0 - r$ x_0 $x_0 + r$

Fig. 12·2 Interval of convergence.

DEFINITION 12·2 (radius of convergence of power series) The radius of convergence r of the power series $\sum_{n=0}^{\infty} a_n(x - x_0)^n$ is defined as:

$$r = \lim_{n \to \infty} \left| \frac{a_n}{a_{n+1}} \right|,$$

provided that this limit exists.

Example 12·9

(a) Let us show that the series for the exponential function is absolutely convergent for all real x. We have

$$e^x = 1 + x + \frac{x^2}{2!} + \frac{x^3}{3!} + \cdots + \frac{x^n}{n!} + \cdots,$$

in which the general term $a_n = 1/n!$.

Now

$$\left| \frac{a_n}{a_{n+1}} \right| = \left| \frac{(n + 1)!}{n!} \right| = (n + 1),$$

so that

$$r = \lim_{n \to \infty} (n + 1) \to \infty.$$

We have thus proved that the power series for e^x is absolutely convergent for all real x. This was an example of a power series with an infinite radius of convergence.

(b) Consider the series

$$x - \frac{x^2}{2} + \frac{x^3}{3} - \frac{x^4}{4} + \cdots,$$

which reduces to the illustrative example following Corollary 12·6, when $x = 1$. We shall see later that this is the power series expansion of $\log(1 + x)$. Then, again applying limit (12·2), we have $a_n = (-1)^{n+1}/n$, and so

$$\left| \frac{a_n}{a_{n+1}} \right| = \left(\frac{n+1}{n} \right).$$

Thus we have

$$r = \lim_{n \to \infty} \left(\frac{n+1}{n} \right) = 1.$$

Hence, the series is absolutely convergent for $|x| < 1$. As we already know the series is convergent for $x = 1$, and divergent for $x = -1$ for then it becomes the harmonic series with the signs of all terms reversed, we have proved that the power series for $\log(1 + x)$ is absolutely convergent for $-1 < x \le 1$. This was an example of a power series with radius of convergence unity.

As a power series is yet another example of the representation of a function of the variable x, it is reasonable to enquire how we may differentiate and integrate functions that are so defined. For simplicity we will take $x_0 = 0$, and work with the power series about the origin

$$f(x) = \sum_{n=0}^{\infty} a_n x^n. \tag{12·4}$$

This is no restriction because Eqn (12·1) can be brought into this form by shifting the origin by means of the change of variable $t = x - x_0$. We will assume that Eqn (12·4) has a radius of convergence $r > 0$.

Intuition suggests that the derivative of $f(x)$ could be obtained by differentiating the right-hand side of Eqn (12·4) term by term and, similarly, that $\int_0^x f(t)\mathrm{d}t$ could be obtained by term by term integration. However, extreme caution must be exercised in such matters for we have already seen that what is legitimate for the sum of a finite number of terms is not necessarily legitimate for an infinite series. Furthermore, we are now dealing with an infinite series of functions, and not just an ordinary series. In fact, although we shall not prove it, termwise differentiation of a power series is always permissible when x lies within the interval of convergence $-r < x < r$ of Eqn (12·4). As differentiability implies continuity we have, as an incidental result, that a power series is continuous within its interval of convergence.

The termwise integrability of power series is easier to prove. Denote by $H(x)$ the series

$$H(x) = \sum_{n=0}^{\infty} \frac{a_n}{n+1} x^{n+1}, \tag{12·5}$$

which is obtained by termwise integration of Eqn (12·4). That is

$$H(x) = \int_0^x f(t) \, \mathrm{d}t.$$

Now the ratio of the nth to the $(n + 1)$th coefficients of Eqn (12·5) is $(n + 1)a_{n-1}/na_n$, whence

$$\lim_{n \to \infty} \left| \frac{(n + 1)}{n} \cdot \frac{a_{n-1}}{a_n} \right| = \lim_{n \to \infty} \left(\frac{n + 1}{n} \right) \lim_{n \to \infty} \left| \frac{a_{n-1}}{a_n} \right| = r.$$

This shows that the power series Eqn (12·5) also has radius of convergence r. We have just remarked that a power series is differentiable for x within its interval of convergence, so that $H'(x) = f(x)$ for $-r < x < r$. Thus by the fundamental theorem of calculus

$$\int_0^x f(t)\mathrm{d}t = H(x) - H(0) = H(x),$$

which was to be proved. Let us collect together these results into the form of a theorem.

THEOREM 12·8 (differentiation and integration of power series) Let the function $f(x)$ be defined by the power series

$$f(x) = \sum_{n=0}^{\infty} a_n x^n,$$

with radius of convergence $r > 0$. Then, within the common interval of convergence $-r < x < r$,

(a) $f(x)$ is a continuous function;

(b) $f'(x) = \sum_{n=1}^{\infty} n a_n x^{n-1}$;

(c) $\int_0^x f(t)\mathrm{d}t = \sum_{n=0}^{\infty} \frac{a_n}{n + 1} x^{n+1}$.

Example 12·10 Find the radius and interval of convergence of

$$f(x) = \sum_{n=1}^{\infty} \frac{x^n}{n(n + 1)}.$$

Deduce $f'(x)$ and find its interval of convergence.

Solution   The nth coefficient a_n of the power series for $f(x)$ is $a_n = 1/n(n + 1)$, and so the radius of convergence r is given by

$$r = \lim_{n \to \infty} \left| \frac{a_n}{a_{n+1}} \right| = \lim_{n \to \infty} \left| \frac{n + 2}{n} \right| = 1.$$

To specify the complete interval of convergence it remains to examine the behaviour of the power series at the end points of the interval $-1 < x < 1$.

The series may be seen to be convergent at $x = 1$ by using the comparison test with $b_n = 1/n^2$. When $x = -1$ the series becomes an alternating series and is seen to be convergent by Theorem 12·6. Thus the complete interval of convergence for $f(x)$ is $-1 \leq x \leq 1$.

Under the conditions of Theorem 12·8 (b) we may differentiate the power series for $f(x)$ term by term within $-1 < x < 1$, so that

$$f'(x) = \sum_{n=1}^{\infty} \frac{x^{n-1}}{n+1}.$$

To specify the complete interval of convergence for this new series which, by Theorem 12·8 (b), is certainly convergent in $-1 < x < 1$, we must again examine the end points of the interval $-1 < x < 1$. The series for $f'(x)$ becomes an alternating series when $x = -1$, and is convergent by Theorem 12·6. At $x = 1$ it becomes the harmonic series, and so is divergent. The complete interval of convergence for $f'(x)$ is thus $-1 \leq x < 1$. The effect of termwise differentiation has been to produce divergence of the differentiated series at the right-hand end point of an interval of convergence at which $f(x)$ is convergent.

Example 12·11 Find the power series representation of arctan x by considering the integral

$$\arctan x = \int_0^x \frac{dt}{1 + t^2}.$$

Deduce a series expansion for $\frac{1}{4}\pi$.

Solution An application of the Binomial Theorem to the function $(1 + a)^{-1}$ gives the result

$$\frac{1}{1 + a} = 1 - a + a^2 - a^3 + a^4 - \cdots,$$

for $-1 < a < 1$. Setting $a = t^2$ we arrive at the power series representation of $(1 + t^2)^{-1}$,

$$\frac{1}{1 + t^2} = 1 - t^2 + t^4 - t^6 + t^8 - \cdots. \tag{A}$$

The conditions of Theorem 12·8 (c) apply, and we may integrate this power series term by term to obtain

$$\arctan x = \int_0^x \frac{dt}{1 + t^2} = \int_0^x (1 - t^2 + t^4 - t^6 + t^8 - \cdots) dt$$

or,

$$\arctan x = x - \frac{x^3}{3} + \frac{x^5}{5} - \frac{x^7}{7} + \cdots. \tag{B}$$

This is the desired power series for arctan x and by the conditions of Theorem 12·8 (b) it is certainly convergent within the interval $-1 < x < 1$, which is the interval of convergence of the original power series Eqn (A).

At each of the end points $x = \pm 1$ of this interval, the power series Eqn (B) becomes an alternating series which is seen to be convergent by Theorem 12·6. Hence the interval of convergence of the integrated series Eqn (B) is $-1 \leq x \leq 1$. Using the fact that arctan $1 = \frac{1}{4}\pi$, we find

$$\tfrac{1}{4}\pi = 1 - \tfrac{1}{3} + \tfrac{1}{5} - \tfrac{1}{7} + \cdots$$

12·3   Taylor's theorem

So far we have discussed the convergence properties of a function $f(x)$ which is defined by a given power series. Let us now reverse this idea and enquire how, when given a specific function $f(x)$, its power series representation may be obtained. Otherwise expressed, we are asking how the coefficients a_n in the power series

$$f(x) = \sum_{n=0}^{\infty} a_n x^n \tag{12·6}$$

may be determined when $f(x)$ is some given function.

First, by setting $x = 0$, we discover that $f(0) = a_0$. Then, on the assumption that the power series Eqn (12·6) has a radius of convergence $r > 0$, differentiate it term by term to obtain

$$f'(x) = \sum_{n=1}^{\infty} n a_n x^{n-1}, \tag{12·7}$$

for $-r < x < r$.

Again setting $x = 0$ shows that $f'(0) = a_1$. Differentiating Eqn (12·7) again with respects to x yields

$$f''(x) = \sum_{n=2}^{\infty} n(n-1) a_n x^{n-2}, \tag{12·8}$$

from which we conclude $f''(0) = 2!a_2$.

Proceeding systematically in this manner gives the general result

$$f^{(m)}(x) = \sum_{n=m}^{\infty} m(m-1)\ldots(m-n+1) a_m x^{n-m}, \tag{12·9}$$

so that $f^{(n)}(0) = n!a_n$.

Thus the coefficients in power series Eqn (12·6) are determined by the formula

$$a_n = \frac{f^{(n)}(0)}{n!} \tag{12·10}$$

for $n \geq 1$ and $a_0 = f(0)$.

Substituting these coefficients into Eqn (12·6) we finally arrive at the power series

$$f(x) = f(0) + xf'(0) + \frac{x^2}{2!} f''(0) + \cdots + \frac{x^n}{n!} f^{(n)}(0) + \cdots. \qquad (12\cdot11)$$

The expression on the right-hand side of this equation is known as the *Maclaurin series* for $f(x)$, and it presupposes that $f(x)$ is differentiable an infinite number of times. To justify the use of the equality sign in Eqn (12·11) it is, of course, necessary to test the series for convergence to verify that its radius of convergence $r > 0$, and to show that $|f(x) - S_n(x)| \to 0$ as $n \to \infty$, where $S_n(x)$ is the sum of the first n terms of the Maclaurin series. We shall return to this matter later.

To transform Eqn (12·11) into a power series in $(x - x_0)$ we set $x = x_0 + h$ and let $f(x_0 + h) = \phi(h)$. Then $\phi'(h) = f'(x_0 + h)$, $\phi''(h) = f''(x_0 + h)$, . . ., $\phi^{(n)}(h) = f^{(n)}(x_0 + h)$, ... It thus follows that $\phi^{(n)}(0) = f^{(n)}(x_0)$ for $n \geq 1$ and $\phi(0) = f(x_0)$. The Maclaurin series for $\phi(h)$ is

$$\phi(h) = \phi(0) + h\phi'(0) + \frac{h^2}{2!} \phi''(0) + \cdots + \frac{h^n}{n!} \phi^{(n)}(0) + \cdots,$$

or, reverting to the function f,

$$f(x) = f(x_0) + (x - x_0)f'(x_0) + \frac{(x - x_0)^2}{2!} f''(x_0) + \cdots$$

$$+ \frac{(x - x_0)^n}{n!} f^{(n)}(x_0) + \cdots. \qquad (12\cdot12)$$

Expressed in this form the expression on the right-hand side is called the *Taylor series* for $f(x)$ about the point $x = x_0$.

Example 12·12 Find the Maclaurin series for $\log(1 + x)$ and $\log(1 - x)$. Deduce the expansion for $\log[(1 + x)/(1 - x)]$.

Solution Setting $f(x) = \log(1 + x)$ we find

$$f'(x) = \frac{1}{1 + x}, f''(x) = \frac{-1}{(1 + x)^2}, \cdots, f^{(n)}(x) = \frac{(-1)^{n-1}(n - 1)!}{(1 + x)^n}, \cdots,$$

and so

$$f^{(n)}(0) = (-1)^{n-1}(n - 1)!$$

for $n \geq 1$ and $f(0) = 0$. Combining this expression for $f^{(n)}(0)$ with Eqn (12·11) gives for the Maclaurin series for $\log(1 + x)$,

$$\log(1 + x) = x - \frac{x^2}{2} + \frac{x^3}{3} - \frac{x^4}{4} + \cdots.$$

This has already been examined for convergence in Example 12·9 (b) and found to be absolutely convergent in the interval $-1 < x \leq 1$.

In the case of the function $\log (1 - x)$ the same argument shows that

$$f^{(n)}(0) = -(n - 1)!$$

for $n \geq 1$ and $f(0) = 0$, so that the Maclaurin series for $\log (1 - x)$ has the form

$$\log (1 - x) = -x - \frac{x^2}{2} - \frac{x^3}{3} - \frac{x^4}{4} - \cdots .$$

This can readily be seen to have $-1 \leq x < 1$ for its interval of convergence. Using the fact that $\log \{(1 + x)/(1 - x)\} = \log (1 + x) - \log (1 - x)$ gives the desired result

$$\log \left(\frac{1 + x}{1 - x}\right) = 2\left\{x + \frac{x^3}{3} + \frac{x^5}{5} + \frac{x^7}{7} + \cdots\right\},$$

for $-1 < x < 1$.

Strictly speaking, we are not yet entitled to use the equality sign between the function and its Maclaurin series, as we have not yet established the convergence of the nth partial sum of the series to the function it represents.

To make further progress it now becomes necessary for us to settle the question of when a Maclaurin or Taylor series is really equal to the function with which it is associated. Let the function $f(x)$ be infinitely differentiable and have the Taylor series representation Eqn (12·12), and let $P_{n-1}(x)$ be the sum of the first n terms of the series terminating at the power $(x - x_0)^{n-1}$, so that

$$P_{n-1}(x) = f(x_0) + (x - x_0)f(x_0) + \frac{(x - x_0)^2}{2!} f''(x_0) + \cdots +$$

$$\frac{(x - x_0)^{n-1}}{(n - 1)!} f^{(n-1)}(x_0).$$

Then a necessary and sufficient condition that the Taylor series should converge to $f(x)$ is obviously that

$$\lim_{n \to \infty} | f(x) - P_{n-1}(x) | = 0.$$

This suggests that to establish convergence we must examine the behaviour of the remainder of the series after n terms. To achieve this we now prove *Taylor's theorem*, one form of which is stated below.

THEOREM 12·9 (Taylor's theorem with a remainder) Let $f(x)$ be a function which is differentiable n times in the interval $a \leq x \leq b$. Then there exists a number ξ, strictly between a and b, such that

$$f(b) = f(a) + (b - a)f'(a) + \frac{(b - a)^2}{2!}f''(a) + \cdots$$

$$+ \frac{(b - a)^{n-1}}{(n - 1)!}f^{(n-1)}(a) + \frac{(b - a)^n}{n!}f^{(n)}(\xi).$$

Proof The proof of Taylor's theorem we now offer will be based on Rolle's theorem. Let k be defined such that

$$f(b) = f(a) + (b - a)f'(a) + \cdots + \frac{(b - a)^{n-1}}{(n - 1)!}f^{(n-1)}(a) + \frac{(b - a)^n}{n!}k,$$

and define the function $F(x)$ by the expression

$$F(x) = f(b) - f(x) - (b - x)f'(x) - \cdots - \frac{(b - x)^{n-1}}{(n - 1!)}f^{(n-1)}(x)$$

$$- \frac{(b - x)^n}{n!}k.$$

Then $F(b) = F(a) = 0$, and a simple calculation shows that

$$F'(x) = \frac{(b - x)^{n-1}}{(n - 1)!}\{f^{(n)}(x) - k\}.$$

Since, by hypothesis, $f^{(n-1)}(x)$ is differentiable in $a \le x \le b$, the function $F(x)$ satisfies the conditions of Rolle's theorem, which asserts that there must be a number ξ, strictly between a and b, for which $F'(\xi) = 0$. As $a < \xi < b$, the factor $(b - \xi)^{n-1} \ne 0$, so that we must have $k = f^{(n)}(\xi)$. This completes the proof of Taylor's theorem.

If we identify b with x and a with x_0, Taylor's theorem with a remainder takes the form

$$f(x) = f(x_0) + (x - x_0)f'(x_0) + \cdots + \frac{(x - x_0)^{n-1}}{(n - 1)!}f^{(n-1)}(x_0)$$

$$+ \frac{(x - x_0)^n}{n!}f^{(n)}(\xi), \qquad (12\cdot13)$$

where $x_0 < \xi < x$. For obvious reasons the last term of this expression is called the remainder term and is usually denoted by $R_n(x)$. The form stated here in which

$$R_n(x) = \frac{(x - x_0)^n}{n!}f^{(n)}(\xi), \qquad (12\cdot14)$$

with $x_0 < \xi < x$ is known as the *Lagrange* form of the remainder term.

When $x_0 = 0$ Eqn (12·13) reduces to *Maclaurin's theorem with a Lagrange remainder,*

$$f(x) = f(0) + xf'(0) + \frac{x^2}{2!}f''(0) + \cdots + \frac{x^{n-1}}{(n-1)!} f^{(n-1)}(0)$$

$$+ \frac{x^n}{n!}f^{(n)}(\xi), \quad (12\cdot15)$$

where $0 < \xi < x$.

Example 12·13 Find the Lagrange remainders $R_n(x)$ after n terms in the Maclaurin series expansions of e^x, $\sin x$, and $\cos x$. By showing that in each case $R_n(x) \to 0$ as $n \to \infty$, prove that these functions are equal to their Maclaurin series expansions.

Solution If $f(x) = e^x$, it is easily shown that Eqn (12·15) takes the form

$$e^x = 1 + x + \frac{x^2}{2!} + \frac{x^3}{3!} + \cdots + \frac{x^{n-1}}{(n-1)!} + R_n(x),$$

where $R_n(x) = (x^n/n!)e^\xi$, and $0 < \xi < x$. Now $e^\xi < e^{|x|}$, and in connection with Eqn (6·15) we proved that

$$\frac{x^n}{n!} < \frac{x^{R-1}}{(R-1)!} (\tfrac{1}{2})^{n-R+1},$$

where R is an integer greater than $2x$. Hence for any fixed x, $e^{|x|}$ is a finite constant and $x^n/n! \to 0$ as $n \to \infty$. It follows from this that $R_n(x) \to 0$ as $n \to \infty$. This provides an alternative verification of the results of Section 6·1.

If $f(x) = \sin x$, then the Maclaurin series with a Lagrange remainder Eqn (12·20) becomes

$$\sin x = x - \frac{x^3}{3!} + \frac{x^5}{5!} - \cdots + \frac{x^n}{n!} \sin \left(\xi + \frac{n\pi}{2}\right),$$

where $0 < \xi < x$. The Lagrange remainder Eqn (12·14) is the last term

$$R_n(x) = \frac{x^n}{n!} \sin \left(\xi + \frac{n\pi}{2}\right).$$

Since $|\sin [\xi + (n\pi/2)]| \leq 1$ we must have

$$|R_n(x)| \leq \left|\frac{x^n}{n!}\right|,$$

showing that $R_n(x) \to 0$ as $n \to \infty$. This establishes the convergence of $\sin x$ to its Maclaurin series, and the argument for the cosine function is exactly similar.

Example 12·14 Establish that $\log(1 + x)$ converges to its Maclaurin series in the interval $-1 < x \leq 1$.

Solution   The Maclaurin series with a remainder is (see Example 12·12)

$$\log (1 + x) = x - \frac{x^2}{2} + \frac{x^3}{3} - \frac{x^4}{4} + \cdots \frac{(-1)^{n-2}x^{n-1}}{n - 1} + R_n(x),$$

where the Lagrange remainder is

$$R_n(x) = \frac{(-1)^{n-1}x^n}{n(1 + \xi)^n},$$

with $\xi < x$. For the interval $0 \le x \le 1$, we must have $0 < \xi < 1$ so that $1 + \xi > 1$, and hence $(1 + \xi)^n > 1$. Thus $| R_n(x) | < x^n/n < 1/n \to 0$ as $n \to \infty$, thereby proving convergence of the Maclaurin series to $\log (1 + x)$ for $0 \le x \le 1$.

We must proceed differently to prove convergence for the interval $-1 < x < 0$. Set $y = -x$ and consider the interval $0 < y < 1$, in which we may write

$$\log (1 + x) = \log (1 - y) = - \int_0^y \frac{dt}{1 - t}.$$

Using the identity

$$\frac{1}{1 - t} = 1 + t + t^2 + \cdots + t^{n-1} + \frac{(-t)^n}{1 - t},$$

we have, after integration,

$$\log (1 - y) = -y - \frac{y^2}{2} - \frac{y^3}{3} - \cdots - \frac{y^n}{n} + \int_0^y \frac{(-t)^n \, dt}{1 - t}.$$

Thus our remainder term is now expressed in the form of the integral

$$R_n(y) = (-1)^n \int_0^y \frac{t^n}{1 - t} \, dt.$$

Now,

$$| R_n(y) | = \int_0^y \frac{t^n}{1 - t} \, dt < \left(\frac{1}{1 - y} \right) \int_0^y t^n \, dt = \frac{y^{n+1}}{(1 - y)(n + 1)}$$

$$< \frac{1}{(1 - y)(n + 1)},$$

so that $| R_n(y) | \to 0$ as $n \to \infty$. This establishes convergence in the interval $-1 < x < 0$. Taken together with the first result we have succeeded in showing that the Maclaurin series of $\log (1 + x)$ converges to the function itself in the interval $-1 < x \le 1$. This provides the justification for our final result in Example 12·12.

When performing numerical calculations with Taylor series, the remainder term provides information on the number of terms that must be retained in order to attain any specified accuracy. Suppose, for example, we wished to calculate $\sin 31°$ correct to five decimal places by means of Eqn (12·13). Then first we would need to set $f(x) = \sin x$ to obtain

$$\sin x = \sin x_0 + (x - x_0) \cos x_0 - \frac{(x - x_0)^2}{2!} \sin x_0 + \cdots$$

$$+ \frac{(x - x_0)^{n-1}}{(n - 1)!} \sin \left(x_0 + \frac{n\pi}{2} \right) + R_n(x),$$

where the remainder

$$R_n(x) = \frac{(x - x_0)^n}{n!} \sin \left(\xi + \frac{n\pi}{2} \right),$$

with $x_0 < \xi < x$.

As the arguments of trigonometric functions must be specified in radian measure it is necessary to set x equal to the radian equivalent of $31°$ and then to choose a convenient value for x_0. We have $31°$ is equivalent to $\pi/6 + \pi/180$ radians, so that a convenient value for x_0 would be $x_0 = \pi/6$. This is, of course, the radian equivalent of $30°$. The remainder term $R_n(x)$ now becomes

$$R_n(x) = \left(\frac{\pi}{180} \right)^n \frac{1}{n!} \sin \left(\xi + \frac{n\pi}{2} \right),$$

whence

$$| R_n(x) | \leq \left(\frac{\pi}{180} \right)^n \cdot \frac{1}{n!}.$$

For our desired accuracy we must have $| R_n(x) | < 5 \times 10^{-6}$. Hence n must be such that

$$\left(\frac{\pi}{180} \right)^n \cdot \frac{1}{n!} < 5 \times 10^{-6}.$$

A short calculation soon shows this condition is satisfied for $n \geq 3$, so that the expansion need only contain powers as far as $(x - x_0)^2$.

The polynomial

$$P_{n-1}(x) = f(x_0) + (x - x_0)f'(x_0) + \cdots + \frac{(x - x_0)^{n-1}}{(n - 1)!} f^{(n-1)}(x_0)$$

$$(12·16)$$

associated with Taylor's theorem as expressed in Eqn (12·13) is called a *Taylor polynomial* of degree $(n - 1)$ about the point $x = x_0$. It is obviously an *approximating* polynomial for the function $f(x)$ in the sense that $| f(x) - P_{n-1}(x) | \to 0$ as $n \to \infty$ for all x within the interval of convergence. Hence

$P_{n-1}(x)$ is strictly analogous to the nth partial sum used in the previous section. By way of example, the Taylor polynomial $P_3(x)$ for the exponential function e^x about the point $x = 0$ is

$$P_3(x) = 1 + x + \frac{x^2}{2!} + \frac{x^3}{3!},$$

whilst its general Taylor polynomial $P_n(x)$ about the point $x = 0$ is

$$P_n(x) = 1 + x + \frac{x^2}{2!} + \cdots + \frac{x^n}{n!}.$$

In many books Theorem 12·9 is called the *generalized mean value theorem*, since when $n = 1$ it reduces to the already familiar mean value theorem derived in Chapter 5 (Theorem 5·12). Let us now derive the analogue of Taylor's theorem with a remainder for a function of two variables.

Suppose that $f(x, y)$ has continuous partial derivatives up to those of nth order, and consider the function

$$F(t) = f(a + ht, b + kt), \tag{12·17}$$

in which a, b, h, and k are constants. Then $F(t) = f(x, y)$, where $x = a + ht$, $y = b + kt$, and in the neighbourhood of (a, b) we have

$$\frac{dF}{dt} = \frac{df}{dt} = \frac{\partial f}{\partial x}\frac{dx}{dt} + \frac{\partial f}{\partial y}\frac{dy}{dt}$$

$$= h\frac{\partial f}{\partial x} + k\frac{\partial f}{\partial y}.$$

Write this result in the form

$$\frac{df}{dt} = \left(h\frac{\partial}{\partial x} + k\frac{\partial}{\partial y} \right) f,$$

where the expression in parentheses is a *partial differential operator* with respect to x and y and is *not* a function. It only generates a function when it acts on a suitably differentiable function f. In consequence, differentiating r times, we have

$$\left(\frac{d}{dt} \right)^r f = \left(h\frac{\partial}{\partial x} + k\frac{\partial}{\partial y} \right)^r f \quad \text{for } r = 1, 2, \ldots, \tag{12·18}$$

with the understanding that:

$$\left(h\frac{\partial}{\partial x} + k\frac{\partial}{\partial y} \right) f = h\frac{\partial f}{\partial x} + k\frac{\partial f}{\partial y},$$

$$\left(h\frac{\partial}{\partial x} + k\frac{\partial}{\partial y} \right)^2 f = h^2\frac{\partial^2 f}{\partial x^2} + 2hk\frac{\partial^2 f}{\partial x\partial y} + k^2\frac{\partial^2 f}{\partial y^2},$$

$$\left(h\frac{\partial}{\partial x} + k\frac{\partial}{\partial y}\right)^3 f = h^3\frac{\partial^3 f}{\partial x^3} + 3h^2 k\frac{\partial^3 f}{\partial x^2 \partial y} + 3hk^2\frac{\partial^3 f}{\partial x \partial y^2} + k^3\frac{\partial^3 f}{\partial y^3}.$$

Now $F(0) = f(a, b)$, $F(1) = f(a + h, b + k)$, and $F(t)$ is differentiable n times for $0 \le t \le 1$. Consequently, by applying Theorem 12·9 to the function $F(t)$ we obtain

$$F(1) = F(0) + F'(0) + \frac{1}{2!}F''(0) + \cdots + \frac{1}{(n-1)!}F^{(n-1)}(0)$$

$$+ \frac{1}{n!}F^{(n)}(\xi), \qquad (12\cdot19)$$

where $0 < \xi < 1$.

However, we also have

$$F^{(r)}(0) = \left(h\frac{\partial}{\partial x} + k\frac{\partial}{\partial y}\right)^r f\bigg|_{\substack{x=a \\ y=b}}$$

and

$$F^{(n)}(\xi) = \left(h\frac{\partial}{\partial x} + k\frac{\partial}{\partial y}\right)^n f\bigg|_{\substack{x=a+\xi h \\ y=b+\xi k}}, \qquad (12\cdot20)$$

whence by substitution of Eqn (12·20) into Eqn (12·19) we obtain:

$$f(a + h, b + k) = f(a, b) + hf_x(a, b) + kf_y(a, b)$$

$$+ \frac{1}{2!}\left(h\frac{\partial}{\partial x} + k\frac{\partial}{\partial y}\right)^2 f\bigg|_{\substack{x=a \\ y=b}} + \cdots + \frac{1}{(n-1)!}\left(h\frac{\partial}{\partial x} + k\frac{\partial}{\partial y}\right)^{n-1} f\bigg|_{\substack{x=a \\ y=b}}$$

$$+ \frac{1}{n!}\left(h\frac{\partial}{\partial x} + k\frac{\partial}{\partial y}\right)^n f\bigg|_{\substack{x=a+\xi h \\ y=b+\xi k}}, \qquad (12\cdot21)$$

where $0 < \xi < 1$.

This result is Taylor's theorem for a function $f(x, y)$ of two variables and it is terminated with a Lagrange remainder term involving nth partial derivatives. The result is also often known as the generalized mean value theorem for a function of two variables. In particular, by taking $n = 1$ we obtain the result

$$f(a + h, b + k) = f(a, b) + hf_x(a + \xi h, b + \xi k) + kf_y(a + \xi h, b + \xi k), \qquad (12\cdot22)$$

where $0 < \xi < 1$. This is the two variable analogue of Theorem 5·12 to which it obviously reduces when $f = f(x)$, for then $f_y \equiv 0$. Result Eqn (12·21) is of such importance that it merits stating in the form of a theorem.

THEOREM 12·10 (generalized mean value theorem in two variables) Let $f(x, y)$ have continuous partial derivatives up to those of order n in some neighbourhood of the point (a, b). Then if (x, y) is any point within this neighbourhood,

$$f(x, y) = f(a, b) + \left((x - a) \frac{\partial}{\partial x} + (y - b) \frac{\partial}{\partial y} \right) f \bigg|_{(a,b)} + \frac{1}{2!} \left((x - a) \frac{\partial}{\partial x} \right.$$

$$\left. + (y - b) \frac{\partial}{\partial y} \right)^2 f \bigg|_{(a,b)} + \cdots$$

$$+ \frac{1}{(n - 1)!} \left((x - a) \frac{\partial}{\partial x} + (y - b) \frac{\partial}{\partial y} \right)^{n-1} f \bigg|_{(a,b)} + R_n(x, y),$$

where the Lagrange remainder

$$R_n(x, y) = \frac{1}{n!} \left((x - a) \frac{\partial}{\partial x} + (y - b) \frac{\partial}{\partial y} \right)^n f \bigg|_{(\eta,\zeta)},$$

in which $\eta = a + \xi(x - a)$,　$\zeta = b + \xi(y - b)$,　and $0 < \xi < 1$.

Example 12·15 Use the generalized mean value theorem in two variables to expand the function

$$f(x, y) = e^{x+2xy}$$

about the point $(0, 0)$. Terminate the expansion with the Lagrange remainder term $R_3(x, y)$ and display its form.

Solution　As the expansion is required about the point $(0, 0)$ we must set $a = 0$, $b = 0$ in Theorem 12·10 and take $n = 3$. Routine calculation shows that:

$$f(0, 0) = 1, \quad f_x(0, 0) = 1, \quad f_y(0, 0) = 0, \quad f_{xx}(0, 0) = 1, \quad f_{xy}(0, 0) = 2,$$
$$f_{yy}(0, 0) = 0,$$

whilst

$$f_{xxx}(x, y) = (1 + 2y)^3 e^{x+2xy},$$
$$f_{xxy}(x, y) = 2(1 + 2y)[2 + x(1 + 2y)] e^{x+2xy},$$
$$f_{yyx}(x, y) = 4x[2 + x(1 + 2y)] e^{x+2xy},$$
$$f_{yyy}(x, y) = 8x^3 e^{x+2xy}.$$

From Theorem 12·10 we find

$$e^{x+2xy} = 1 + x + \tfrac{1}{2}x^2 + 2xy + R_3(x, y),$$

where

$$R_3(x, y) = \frac{1}{3!} (x^3 f_{xxx}(x, y) + 3x^2 y f_{xxy}(x, y) + 3xy^2 f_{yyx}(x, y)$$
$$+ y^3 f_{yyy}(x, y))_{(\eta, \zeta)}$$

with $\eta = \xi x$, $\zeta = \xi y$, and $0 < \xi < 1$.

12·4 Application of Taylor's theorem

The applications of Taylor's theorem with a remainder are so numerous that we can do no more here than describe some of the most common. It is hoped that these illustrations will indicate the power of this theorem and the fact that its use is not confined exclusively to the estimation of errors in the series expansion of functions.

12·4 (a) Indeterminate forms

The form of L'Hospital's rule given in Theorem 5·14 is capable of immediate extension as follows.

THEOREM 12·11 (extended L'Hospital's rule) Let $f(x)$ and $g(x)$ be n times differentiable functions which are such that $f(a) = g(a) = 0$ and $f^{(r)}(a) = g^{(r)}(a) = 0$ for $r = 1, 2, \ldots, n - 1$, but $\lim_{x \to a} f^{(n)}(x)$ and $\lim_{x \to a} g^{(n)}(x)$ are not both zero.

Then

$$\lim_{x \to a} \frac{f(x)}{g(x)} = \frac{\lim_{x \to a} f^{(n)}(x)}{\lim_{x \to a} g^{(n)}(x)}.$$

Proof Using Taylor's theorem with a remainder to expand numerator and denominator separately gives

$$\frac{f(a + h)}{g(a + h)} = \frac{f(a) + hf'(a) + \cdots + \dfrac{h^n}{n!} f^{(n)}(\xi_1)}{g(a) + hg'(a) + \cdots + \dfrac{h^n}{n!} g^{(n)}(\xi_2)} = \frac{f^{(n)}(\xi_1)}{g^{(n)}(\xi_2)},$$

where $a < \xi_1 < a + h$, $a < \xi_2 < a + h$. If now $h \to 0$, then $\xi_1, \xi_2 \to a$ and we obtain the result of the theorem

$$\lim_{h \to 0} \frac{f(a + h)}{g(a + h)} = \frac{\lim_{x \to a} f^{(n)}(x)}{\lim_{x \to a} g^{(n)}(x)}.$$

Example 12·16 Find the value of the expression

$$\lim_{x \to 0} \frac{x \sin x}{(a^x - 1)(b^x - 1)}.$$

Solution This is an indeterminate form. Setting $f(x) = x \sin x$, $g(x) = (a^x - 1)(b^x - 1)$, we first compute $f'(x)$ and $g'(x)$. We find $f'(x) = \sin x + x \cos x$ and $g'(x) = a^x \log a(b^x - 1) + b^x \log b(a^x - 1)$, and clearly $\lim_{x \to 0} f'(x) = \lim_{x \to 0} g'(x) = 0$. The earlier form of L'Hospital's rule thus fails, and we must make appeal to Theorem 12·11 and compute $f''(x)$ and $g''(x)$. We find $f''(x) = 2 \cos x - x \sin x$ and $g''(x) = 2a^x b^x \log a \log b + a^x(\log a)^2(b^x - 1) + b^x(\log b)^2(a^x - 1)$, from which we see that $\lim_{x \to 0} f''(x) = 2$, $\lim_{x \to 0} g''(x) = 2 \log a \log b$. By the conditions of Theorem 12·11 we have

$$\lim_{x \to 0} \frac{x \sin x}{(a^x - 1)(b^x - 1)} = \frac{1}{\log a \log b}.$$

12·4 (b) Local behaviour of functions of one variable

In Chapter 5 we repeatedly turned to the problem of the local behaviour of a function of one variable in order to identify local maxima, local minima, and points of inflection. Here again Taylor's theorem with a remainder helps to identify such points when not only the first derivative, but also successive higher order derivatives vanish at a point.

Suppose that $f(x)$ is n times differentiable near $x = a$ and that $f^{(1)}(a) = f^{(2)}(a) = \cdots = f^{(n-1)}(a) = 0$, but that $f^{(n)}(a) \neq 0$. Then by Taylor's theorem

$$f(a + h) = f(a) + hf^{(1)}(a) + \cdots + \frac{h^{n-1}}{(n-1)!}f^{(n-1)}(a) + \frac{h^n}{n!}f^{(n)}(\xi),$$

where $a < \xi < a + h$, but because of the vanishing of the first $(n - 1)$ derivatives at $x = a$ this simplifies to

$$f(a + h) - f(a) = \frac{h^n}{n!}f^{(n)}(\xi).$$

The behaviour of the left-hand side of this expression was used in Chapter 5 to identify the nature of the extrema involved so that we see its sign is now determined solely by the sign of $h^n f^{(n)}(\xi)$ or, for suitably small h, by the sign of $h^n f^{(n)}(a)$. It is left to the reader to verify that the following theorem is an immediate consequence of this simple result when taken in conjunction with Definition 5·4.

THEOREM 12·12 (identification of local extrema—one independent variable) A necessary and sufficient condition that a suitably differentiable function $f(x)$ have a local $\begin{Bmatrix} \text{maximum} \\ \text{minimum} \end{Bmatrix}$ at $x = a$ is that the first derivative $f^{(n)}(x)$ with a non-zero value at $x = a$ shall be of even order and $\begin{Bmatrix} f^{(n)}(a) < 0 \\ f^{(n)}(a) > 0 \end{Bmatrix}$. If the first derivative other than $f^{(1)}(a)$ with a non-zero value at $x = a$ is of odd order, then $f(x)$ has a point of inflection with an associated zero gradient at $x = a$.

12·4 (c)   Newton's method

Newton's method is a simple and powerful method for the accurate determination of the roots of an equation $f(x) = 0$, and is based on Taylor's theorem with the Lagrange remainder $R_2(x)$.

Suppose x_0 is an approximate root of $f(x) = 0$ and h is such that $x = x_0 + h$ is an exact root. Then by Taylor's theorem

$$f(x_0 + h) = f(x_0) + hf'(x_0) + \frac{h^2}{2} f''(\xi),$$

where $x_0 < \xi < x_0 + h$.

As, by supposition, $f(x_0 + h) = 0$ we find

$$0 = f(x_0) + hf'(x_0) + \frac{h^2}{2} f''(\xi).$$

Now ξ is not known, but on the assumption that h is small we may define a first approximation h_1 to h by neglecting the third term and writing

$$h_1 = -\frac{f(x_0)}{f'(x_0)}.$$

The next approximation to the root itself must be $x_1 = x_0 + h_1$, whence by the same argument, the approximation h_2 to the correction needed to make x_1 an exact root is

$$h_2 = -\frac{f(x_0 + h_1)}{f'(x_0 + h_1)}.$$

Proceeding in this manner we find that the nth approximation x_n to the exact root of $f(x) = 0$ is, in terms of the $(n - 1)$th approximation x_{n-1},

$$x_n = x_{n-1} - \frac{f(x_{n-1})}{f'(x_{n-1})}.$$

The successive calculation of improved approximations in this manner is called *iteration*, and x_n itself is called the nth *iterate*.

If the sequence $\{x_n\}$ tends to a limit x^*, it follows that this limit must be the desired root, for then the numerator of the correction term vanishes. The choice of an approximate root x_0 with which to start the process may be made in any convenient manner. The most usual method is to seek to show that the root lies between two fairly close values $x = a$, $x = b$ and then to take for x_0 any value that is intermediate between them. The numbers a, b are usually found by direct calculation, which is used to prove that $f(a)$ and $f(b)$ are of opposite sign, so that by the intermediate value theorem a zero of $y = f(x)$ must occur in the interval $a < x < b$.

The reasons for both the success and failure of Newton's method are best appreciated in geometrical terms. The calculation of x_n from x_{n-1} amounts to tracing back the tangent to the curve $y = f(x)$ at x_{n-1} until it intersects the x-axis at the point x_n. If x_n lies between x_{n-1} and x^* for all n then the process converges; otherwise it diverges. Fig. 12·3 (a) illustrates a convergent iteration and Fig. 12·3 (b) a divergent one.

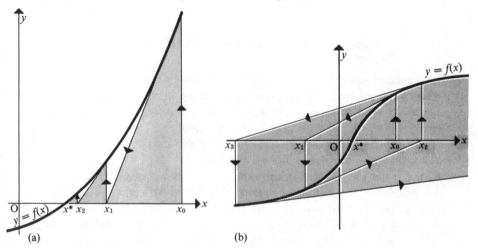

(a)   (b)

Fig. 12·3   (a) Convergent Newton iteration process; (b) divergent Newton iteration process.

Example 12·17 Locate the real root of the cubic

$$x^3 + x^2 + 2x + 1 = 0.$$

Use the result to find the remaining roots.

Solution Setting $f(x) = x^3 + x^2 + 2x + 1$ we see that $f(0) = 1 > 0$ and $f(-1) = -1 < 0$, so that by the intermediate value theorem a root of the equation $f(x) = 0$ must lie in the interval $-1 < x < 0$. Take $x_0 = -0·5$, since this lies within the desired interval.

Now $f'(x) = 3x^2 + 2x + 2$ so that Newton's method requires us to employ the relation

$$x_n = x_{n-1} - \frac{x_{n-1}^3 + x_{n-1}^2 + 2x_{n-1} + 1}{3x_{n-1}^2 + 2x_{n-1} + 2},$$

starting with $x_0 = -0·5$.

A straightforward calculation shows that to four decimal places $x_1 = -0.5714$, $x_2 = -0.5698$, and $x_3 = -0.5698$. The iteration process has thus converged to within the required accuracy in only three iterations. The real root is $x^* = -0.5698$, and the remaining two roots can now be found by dividing $f(x) = 0$ by the factor $(x + 0.5698)$ and then solving the remaining quadratic in the usual manner. If this is done, long division gives

$$\frac{x^3 + x^2 + 2x + 1}{x + 0.5698} = x^2 + 0.4302x + 1.7549,$$

from which we find the other two roots are

$$x = -0.2151 + i\,1.3071 \qquad \text{and} \qquad x = -0.2151 - i\,1.3071.$$

12·5 Applications of the generalized mean value theorem

The applications of the extension of Taylor's theorem to functions of two or more variables are perhaps even more extensive than those of Taylor's theorem itself. This section illustrates a few of the simplest and most used, connected mainly with functions of two variables. The final application, connected with the least squares fitting of a polynomial, is the only one concerning functions of more than two variables.

12·5 (a)   Stationary points of functions of two variables

Consider the function $z = f(x, y)$ of the two real independent variables x, y which is defined in some region D of the (x, y)-plane bounded by the curve γ. The notion of its graph is already familiar to us and it comprises a surface S with points $(x, y, f(x, y))$, the projection of the boundary Γ of which onto the (x, y)-plane is the curve γ. A typical situation is shown in Fig. 12·4 (a, b) where the point P is obviously a maximum and the point Q is obviously a minimum.

Intuitively, and by analogy with the single variable case, it would seem that all that is necessary to locate extrema such as P, Q is to find those points (x_0, y_0) at which $f_x(x_0, y_0) = f_y(x_0, y_0) = 0$. This is, in effect, saying that the tangent plane at either a maximum or a minimum must be parallel to the (x, y)-plane. Unfortunately, this is not a sufficiently stringent condition, for the point R in Fig. 12·5 is neither a maximum, nor a minimum, yet the tangent plane at that point is certainly parallel to the (x, y)-plane. Because of the shape of the surface it is called a *saddle point*. It is characterized by the fact that if the surface is sectioned through R by different planes parallel to the z-axis, then for some the curve of section has a minimum at R and for others a maximum.

Each of these points P, Q, R is called a *stationary* point of the function $z = f(x, y)$ because f_x and f_y vanish at these points.

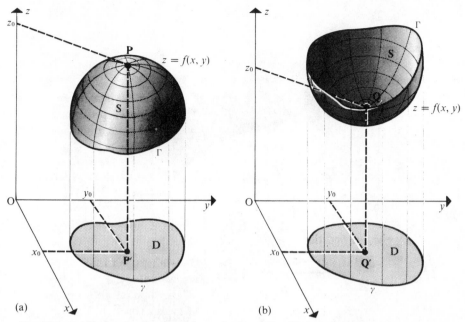

Fig. 12·4 (a) Surface having maximum at P; (b) surface having minimum at Q.

DEFINITION 12·3 (stationary points of $f(x, y)$) Let $f(x, y)$ be a differentiable function in some region of the (x, y)-plane. Then any point (x_0, y_0) in D for which $f_x(x_0, y_0) = 0$ and $f_y(x_0, y_0) = 0$ is called a stationary point of the function $f(x, y)$ in D.

If for all (x, y) near (x_0, y_0) it is true that $f(x, y) < f(x_0, y_0)$, then $f(x, y)$ will be said to have a *local maximum* at (x_0, y_0). If for all (x, y) near to (x_0, y_0) it is true that $f(x, y) > f(x_0, y_0)$, then $f(x, y)$ will be said to have a *local minimum* at (x_0, y_0). In the event that $f(x, y)$ assumes values both greater and less than $f(x_0, y_0)$ for (x, y) near to a stationary point (x_0, y_0), then $f(x, y)$ will be said to have a *saddle point* at (x_0, y_0).

We now use the generalized mean value theorem to prove the following result.

THEOREM 12·13 (identification of extrema of $f(x, y)$) Let $f(x, y)$ be a function with continuous first and second order partial derivatives. Then a sufficient condition that (x_0, y_0) is a local $\begin{Bmatrix} \text{maximum} \\ \text{minimum} \end{Bmatrix}$ for $f(x, y)$ is that:

(a) $f_x(x_0, y_0) = f_y(x_0, y_0) = 0$;

(b) $f_{xx}(x_0, y_0)f_{yy}(x_0, y_0) > f_{xy}{}^2(x_0, y_0)$;

(c) $\begin{cases} f_{xx}(x_0, y_0) < 0 \\ f_{xx}(x_0, y_0) > 0. \end{cases}$

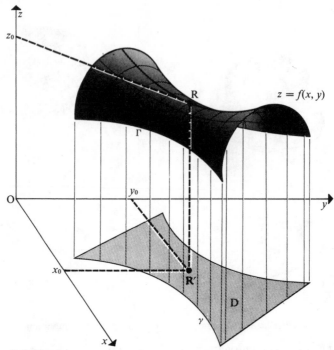

Fig. 12·5 Saddle point.

A sufficient condition that $f(x, y)$ should have a saddle point at (x_0, y_0) is that in addition to condition (a) above being satisfied, it is also true that:

(d) $f_{xx}(x_0, y_0)f_{yy}(x_0, y_0) < f_{xy}^2(x_0, y_0)$.

Proof Note first that (b) implies either that $f_{xx}(x_0, y_0) > 0$ and $f_{yy}(x_0, y_0) > 0$ or that $f_{xx}(x_0, y_0) < 0$ and $f_{yy}(x_0, y_0) < 0$. Consider the case $f_{xx}(x_0, y_0) > 0$. Then by the generalized mean value theorem with $n = 2$,

$$f(x_0 + h, y_0 + k) - f(x_0, y_0) = \tfrac{1}{2}[h^2 f_{xx}(\eta, \zeta) + 2hk f_{xy}(\eta, \zeta) \\ + k^2 f_{yy}(\eta, \zeta)],$$

where $\eta = x_0 + \xi h$, $\zeta = y_0 + \xi k$ with $0 < \xi < 1$. Now as f_{xx}, f_{xy}, and f_{yy} are assumed continuous, it follows from (b) that for sufficiently small h and k, $f_{xx}(\xi, \eta)f_{yy}(\xi, \eta) - f_{xy}^2(\xi, \eta) > 0$. Thus we have

$$f(x_0 + h, y_0 + k) - f(x_0, y_0) = \tfrac{1}{2}(Ah^2 + 2Bhk + Ck^2),$$

where $A = f_{xx}(\eta, \zeta)$, $B = f_{xy}(\eta, \zeta)$, $C = f_{yy}(\eta, \zeta)$. Completing the square on the right-hand side of this equation allows us to write it as

$$f(x_0 + h, y_0 + k) - f(x_0, y_0) = \tfrac{1}{2}A\left[\left(h + \frac{B}{A}k\right)^2 + \left(\frac{AC - B^2}{A^2}\right)k^2\right].$$

Clearly, $(h + (B/A)k)^2 \geq 0$ and $[(AC - B^2)/A^2]k^2 > 0$ if $k > 0$ since, by hypothesis, $AC - B^2 > 0$. In the event $k = 0$, then $Ah^2 + 2Bhk + Ck^2 = Ah^2 > 0$ provided $h \neq 0$.

Thus, if not both h and $k = 0$, since we are assuming $A > 0$ we have shown that

$$f(x_0 + h, y_0 + k) - f(x_0, y_0) > 0$$

for small h, k or, equivalently,

$$f(x, y) > f(x_0, y_0),$$

for all (x, y) near (x_0, y_0). This is the condition that $f(x, y)$ should have a local minimum at (x_0, y_0).

The verification of the condition for a local maximum at (x_0, y_0) follows from the above argument by setting $g(x, y) = -f(x, y)$ and then supposing that $f_{xx}(x_0, y_0) < 0$. This establishes that $g(x, y)$ has a local minimum at (x_0, y_0) so that $f(x, y)$ must have a local maximum at that point.

The verification of the condition for a saddle point follows directly from consideration of the result

$$f(x_0 + h, y_0 + k) - f(x_0, y_0) = \tfrac{1}{2}A \left[\left(h + \frac{B}{A} k \right)^2 + \left(\frac{AC - B^2}{A^2} \right) k^2 \right]$$

which was derived above. For now, by hypothesis, $AC - B^2 < 0$, so that the terms within the large bracket are of opposite signs. This implies that $f(x_0 + h, y_0 + k) - f(x_0, y_0)$ can be made either positive or negative near (x_0, y_0) by a suitable choice of h, k. This is the condition for a saddle point and completes the proof of the theorem.

Example 12·18 Find the stationary points of the function

$$f(x, y) = 2x^3 - 9x^2y + 12xy^2 - 60y$$

and identify their nature.

Solution We have,

$$f_x = 6x^2 - 18xy + 12y^2 \qquad \text{and} \qquad f_y = -9x^2 + 24xy - 60.$$

The conditions $f_x = f_y = 0$ are equivalent to

$$(f_x = 0) \qquad (x - y)(x - 2y) = 0$$

and

$$(f_y = 0) \qquad 3x^2 - 8xy + 20 = 0.$$

From the first condition we may either have $x = y$ or $x = 2y$. Substituting $x = y$ in the second condition gives rise to the equation $y^2 = 4$, so that the stationary points corresponding to $x = y$ are $(2, 2)$ and $(-2, -2)$. Substituting $x = 2y$ in the second condition gives rise to the condition $y^2 = 5$, so that the stationary points corresponding to $x = 2y$ are $(2\sqrt{5}, \sqrt{5})$ and $(-2\sqrt{5}, -\sqrt{5})$.

There are thus four stationary points associated with the function in question and we must apply the tests given in Theorem 12·13 to identify their nature. We have

$$f_{xx} = 12x - 18y, \quad f_{xy} = -18x + 24y, \quad f_{yy} = 24x,$$

and it is easily verified that $f_{xx}f_{yy} - f_{xy}^2 < 0$ at both of the points $(2, 2)$ and $(-2, -2)$, showing that they must be saddle points. A similar calculation shows that $f_{xx}f_{yy} - f_{xy}^2 > 0$ at each of the other stationary points, though $f_{xx} > 0$ at $(2\sqrt{5}, \sqrt{5})$, showing that it must be a minimum, whereas $f_{xx} < 0$ at $(-2\sqrt{5}, -\sqrt{5})$, showing that it must be a maximum.

12·6   Fourier series on $[-\pi, \pi]$

The fundamental idea underlying Fourier series is that all functions $f(x)$ of practical importance which are defined on the interval $-\pi \leq x \leq \pi$ can be expressed in terms of a convergent *trigonometric series* of the form

$$f(x) = \frac{a_0}{2} + \sum_{n=1}^{\infty} (a_n \cos nx + b_n \sin nx), \tag{12·23}$$

in which the constant coefficients a_n, b_n are related to $f(x)$ in a special way. The apparent restriction of $f(x)$ to the interval $[-\pi, \pi]$ is unimportant, since an elementary change of variable will always reduce an arbitrary interval $[a, b]$ to $[-\pi, \pi]$.

Notice here that because of the periodicity properties of the sine and cosine functions, the right-hand side of (12·23) must of necessity be periodic with period 2π. This implies that the best we can expect of such a representation is that, at each point of the interval $[-\pi, \pi]$, the trigonometric series has for its sum the function $f(x)$. Naturally, although the trigonometric series will assign functional values to $f(x)$ for *all* real x, it does not follow that these need agree with the actual functional values of $f(x)$ outside the *fundamental interval* $[-\pi, \pi]$. In fact, the series will provide a *periodic extension* of the functional behaviour of $f(x)$ over the fundamental interval $[-\pi, \pi]$ to every interval of the form $[(2r - 1)\pi, (2r + 1)\pi]$, in the sense that $f(x) = f(x + 2r\pi)$ for $r = 0, \pm 1, \pm 2, \ldots$.

To deduce the relationship between $f(x)$ and the coefficients a_n, b_n let us first reinterpret results (8·28) to (8·30) of Chapter 8 in terms of definite integrals taken over the interval $[-\pi, \pi]$. We find at once that for any integers $m, n = 0, 1, \ldots$,

$$\int_{-\pi}^{\pi} \sin mx \cos nx \, dx = 0 \text{ for all } m, n, \qquad (12\cdot24)$$

$$\int_{-\pi}^{\pi} \sin mx \sin nx \, dx = \begin{cases} 0 \text{ for } m \neq n \\ \pi \text{ for } m = n \neq 0, \end{cases} \qquad (12\cdot25)$$

$$\int_{-\pi}^{\pi} \cos mx \cos nx \, dx = \begin{cases} 0 \text{ for } m \neq n \\ \pi \text{ for } m = n \neq 0 \\ 2\pi \text{ for } m = n = 0. \end{cases} \qquad (12\cdot26)$$

In mathematical terms, the facts expressed by Eqn (12·24) and by the first results of Eqns (12·25) and (12·26) are described by saying that the functions belonging to the system

$$1, \cos x, \sin x, \cos 2x, \sin 2x, \ldots, \cos nx, \sin nx, \ldots, \qquad (12\cdot27)$$

are *orthogonal* over the interval $[-\pi, \pi]$. In words, these equations say that the product of any two different functions of this sequence when integrated over the interval $[-\pi, \pi]$ will yield zero.

The significance os the orthogonality property of system (12·27) is seen when Eqn (12·23) is multiplier by $\cos mx$ and the result is then integrated over the interval $[-\pi, \pi]$. We find that

$$\int_{-\pi}^{\pi} f(x) \cos mx \, dx = \frac{a_0}{2} \int_{-\pi}^{\pi} \cos mx \, dx + \sum_{n=1}^{\infty} \left(a_n \int_{-\pi}^{\pi} \cos mx \cos nx \, dx \right.$$

$$\left. + b_n \int_{-\pi}^{\pi} \cos mx \sin nx \, dx \right),$$

which on account of the above results immediately reduces to

$$\int_{-\pi}^{\pi} f(x) \cos mx \, dx = \pi a_m \text{ for } m = 0, 1, \ldots. \qquad (12\cdot28)$$

Had Eqn (12·23) been multiplied by $\sin mx$ and the result been integrated over the interval $[-\pi, \pi]$, an exactly similar argument would have yielded the result

$$\int_{-\pi}^{\pi} f(x) \sin mx \, dx = \pi b_m \text{ for } m = 1, 2, \ldots. \qquad (12\cdot29)$$

Thus we have found that for $f(x)$ to have the trigonometric series representation (12·23), we must define the constant coefficients a_n, b_n by the relationships

$$a_n = \frac{1}{\pi} \int_{-\pi}^{\pi} f(x) \cos nx \, dx \qquad (12\cdot30)$$

for $n = 0, 1, \ldots$, and

$$b_n = \frac{1}{\pi} \int_{-\pi}^{\pi} f(x) \sin nx \, dx \tag{12·31}$$

for $n = 1, 2, \ldots$.

The coefficients a_n, b_n so defined are called the *Fourier coefficients* of $f(x)$, and the corresponding right-hand side of (12·23) is then called the *Fourier series* of $f(x)$.

In principle, at least, Fourier series would appear to offer the possibility of representation of discontinuous as well as continuous functions, because whereas for a Taylor series expansion a function needs to be differentiable, for a Fourier series expansion it would appear that it only needs to be integrable. This assertion follows because in Eqn (7·17) we have already seen that the integration of piecewise continuous functions presents no difficulty, and so the Fourier coefficients a_n, b_n can even be computed when $f(x)$ is piecewise continuous. Naturally, we must examine the functional value which a Fourier series attributes to a point of discontinuity of the function which it represents, since at such points it is reasonable to expect the behaviour of the series to differ from that of the function itself.

Another important feature of a Fourier series is that it offers a method of synthesis of a function in terms of simple harmonic components having periodicities which are sub-multiples of 2π. This is particularly valuable when an oscillatory problem is being studied since, in effect, it describes the function involved in terms of the simple harmonic oscillatory modes which occur naturally in the problem.

Example 12·19 Determine the Fourier series expansion of the function

$$f(x) = \pi^2 - x^2 \text{ for } -\pi \le x \le \pi.$$

Solution As $f(x) = \pi^2 - x^2$, we see from Eqn (12·30) that the Fourier coefficients a_n are determined by the integral

$$a_n = \frac{1}{\pi} \int_{-\pi}^{\pi} (\pi^2 - x^2) \cos nx \, dx,$$

where $n = 0, 1, \ldots$. When $n = 0$ this yields

$$a_0 = \frac{1}{\pi} \int_{-\pi}^{\pi} (\pi^2 - x^2) \, dx = \frac{4}{3} \pi^2.$$

For the case $n \ne 0$ we have

$$a_n = \pi \int_{-\pi}^{\pi} \cos nx \, dx - \frac{1}{\pi} \int_{-\pi}^{\pi} x^2 \cos nx \, dx,$$

$$= \pi \left(\frac{1}{n}\sin nx\right)\bigg|_{-\pi}^{\pi} - \frac{1}{\pi}\left(\frac{2x\cos nx}{n^2} + \frac{(n^2x^2 - 2)\sin nx}{n^3}\right)\bigg|_{-\pi}^{\pi}$$

$$= (-1)^{n+1}\frac{4}{n^2}.$$

To determine the Fourier coefficients b_n we must use Eqn (12·31) which shows that

$$b_n = \frac{1}{\pi}\int_{-\pi}^{\pi}(\pi^2 - x^2)\sin nx\,dx,$$

where $n = 1, 2, \ldots$. Instead of evaluating this integral directly, let us divide the interval of integration and rewrite the result as the sum of two integrals. First we write

$$b_n = \frac{1}{\pi}\int_{-\pi}^{0}(\pi^2 - x^2)\sin nx\,dx + \frac{1}{\pi}\int_{0}^{\pi}(\pi^2 - x^2)\sin nx\,dx,$$

and then, setting $x = -z$ in the first integral, this becomes

$$b_n = -\frac{1}{\pi}\int_{\pi}^{0}(\pi^2 - z^2)\sin(-nz)\,dz + \frac{1}{\pi}\int_{0}^{\pi}(\pi^2 - x^2)\sin nx\,dx.$$

However, $\sin(-nz) = -\sin nz$, and the minus sign in front of the first integral may be utilized to reverse the order of the limits of integration, so that finally we arrive at

$$b_n = -\frac{1}{\pi}\int_{0}^{\pi}(\pi^2 - z^2)\sin nz\,dx + \frac{1}{\pi}\int_{0}^{\pi}(\pi^2 - x^2)\sin nx\,dx,$$

for $n = 1, 2, \ldots$. As the variable in a definite integral is only a dummy variable, we may replace z by x in the first integral to deduce that

$$b_n = 0 \text{ for all } n.$$

This result could, of course, have been obtained by direct evaluation of the definite integral for b_n using integration by parts, though the argument would have been more tedious.

Inserting the Fourier coefficients a_n, b_n into Eqn (12·23) then gives the Fourier series of $f(x) = \pi^2 - x^2$ for $-\pi \le x \le \pi$. The result obtained is

$$f(x) = \frac{2}{3}\pi^2 + 4\left(\cos x - \frac{1}{2^2}\cos 2x + \frac{1}{3^2}\cos 3x - \ldots\right.$$

$$\left. + \frac{(-1)^{n+1}}{n^2}\cos nx + \ldots\right).$$

The relationship between the Fourier series representation of $f(x) = \pi^2 - x^2$ in the fundamental interval $[-\pi, \pi]$, the periodic extension it assigns to $f(x)$ outside the fundamental interval, and the actual behaviour of $f(x)$ both inside and outside the fundamental interval are illustrated in Fig. 12·6 (a). The curve denotes both the functional behaviour of $f(x)$ and that of its Fourier series in the fundamental interval, the dotted curve denotes the periodic extension of $f(x)$ and the chain-dotted curve denotes the actual behaviour of $f(x)$ outside the fundamental interval. Setting $x = 0$ and $x = \pm \pi$ in this result gives us the two interesting series

$$\frac{\pi^2}{12} = 1 - \frac{1}{2^2} + \frac{1}{3^2} - \frac{1}{4^2} + \cdots,$$

and

$$\frac{\pi^2}{6} = 1 + \frac{1}{2^2} + \frac{1}{3^2} + \frac{1}{4^2} + \cdots .$$

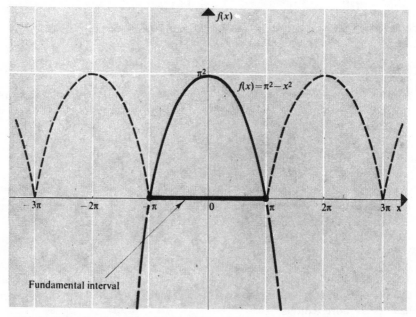

Fig. 12·6 (a) Fourier series representation of $f(x) = \pi^2 - x^2$.

Let us define the mth partial sum $S_m(x)$ of this Fourier series to be

$$S_m(x) = \tfrac{2}{3}\pi^2 + 4 \sum_{n=1}^{m-1} \frac{(-1)^{n-1}}{n^2} \cos nx.$$

Then, when working numerically with the Fourier series, the function will need to be approximated by a partial sum. The behaviour of the second, third, and fourth partial sums

$$S_2(x) = \tfrac{2}{3}\pi^2 + 4\cos x,$$

$$S_3(x) = \tfrac{2}{3}\pi^2 + 4\cos x - \cos 2x,$$

$$S_4(x) = \tfrac{2}{3}\pi^2 + 4\cos x - \cos 2x + \tfrac{4}{9}\cos 3x,$$

is shown in Fig. 12·6 (b).

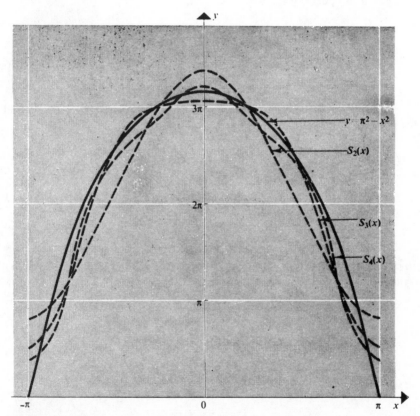

Fig. 12·6 (b) Approximation of $\pi^2 - x^2$ by partial sums.

Example 12·20 Determine the Fourier series expansion of the function

$$f(x) = \begin{cases} a \text{ for } -\pi \leq x < 0 \\ b \text{ for } 0 \leq x \leq \pi. \end{cases}$$

Solution Proceeding as before and using the notation of Eqn (7·17),

$$a_n = \frac{1}{\pi}\int_{-\pi}^{0-} a\cos nx \, \mathrm{d}x + \frac{1}{\pi}\int_{0+}^{\pi} b\cos nx \, \mathrm{d}x$$

for $n = 0, 1, 2, \ldots$, and

$$b_n = \frac{1}{\pi} \int_{-\pi}^{0-} a \sin nx \, \mathrm{d}x + \frac{1}{\pi} \int_{0+}^{\pi} b \sin nx \, \mathrm{d}x$$

for $n = 1, 2, \ldots$.

A simple calculation shows that

$$a_0 = a + b, \; a_n = 0 \text{ for } n = 1, 2, \ldots,$$

and

$$b_{2n} = 0, \; b_{2n+1} = \frac{2(b - a)}{(2n + 1)\pi} \text{ for } n = 1, 2, \ldots.$$

Substitution of these Fourier coefficients into Eqn (12·23) then shows that the Fourier series of $f(x)$ in $[-\pi, \pi]$ is

$$f(x) = \left(\frac{a + b}{2}\right) + \frac{2}{\pi}(b - a)(\sin x + \tfrac{1}{3} \sin 3x + \tfrac{1}{5} \sin 5x + \ldots$$

$$\frac{1}{(2n + 1)} \sin(2n + 1)x + \ldots).$$

This Fourier series certainly converges to the function $f(x)$ when $x = \pm\frac{1}{2}\pi$, as can be seen by assigning these values to x and employing the final result of Example 12·11 to sum the series. Observe though that in this case

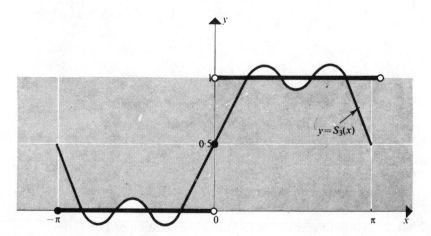

Fig. 12·7 Approximation of $f(x) = 0$ for $-\pi \leq x < 0$ and $f(x) = 1$ for $0 \leq x \leq \pi$ by the partial sum $S_9(x)$.

the Fourier series assumes the value $(a + b)/2$ for $x = 0$ and $x = \pm\pi$ which is not in agreement with the actual functional values at those points. It is, in fact, the *average* of the functional values to the immediate left and right of the discontinuity at $x = 0$. This result is not coincidental, and it can be shown to be true of all Fourier series at jump discontinuities. The approximation of $f(x)$ by the third partial sum

$$S_3(x) = \left(\frac{a + b}{2}\right) + \frac{2}{\pi}(b - a)(\sin x + \tfrac{1}{3}\sin 3x)$$

in the case $a = 0$, $b = 1$ is shown in Fig. 12·7 in which a circle denotes an end point not included and a dot an end point that is included.

12·7 Different forms of Fourier series

A number of special forms of Fourier series occur depending on whether or not the fundamental interval is $[-\pi, \pi]$, or the function $f(x)$ is even or odd. These are of sufficient importance to merit recording them formally, though we leave their proof as a simple exercise for the reader.

(a) Change of interval

THEOREM 12·14 (*Change of origin of fundamental interval*)   If $f(x)$ is defined arbitrarily in the fundamental interval $[-\pi, \pi]$ and by periodic extension outside it then, for any α, the Fourier coefficients a_n, b_n of $f(x)$ are given by

$$a_n = \frac{1}{\pi}\int_{\alpha - \pi}^{\alpha + \pi} f(x)\cos nx\, dx \text{ for } n = 0, 1, \ldots,$$

$$b_n = \frac{1}{\pi}\int_{\alpha - \pi}^{\alpha + \pi} f(x)\sin nx\, dx \text{ for } n = 1, 2, \ldots..$$

The Fourier series of $f(x)$ then has the form

$$f(x) = \frac{a_0}{2} + \sum_{n=1}^{\infty}(a_n\cos nx + b_n\sin nx) \text{ for } \alpha - \pi \leq x \leq \alpha + \pi.$$

It often happens that the fundamental interval to be used is $[-L, L]$ instead of $[-\pi, \pi]$. When this occurs, the simple variable change $t = \pi x/L$ maps the interval $-L \leq x \leq L$ onto the interval $-\pi \leq t \leq \pi$ for which we already have a Fourier expansion theorem.

THEOREM 12·15 (*Change of interval length*)   The Fourier series of an integrable function $f(x)$ defined on the interval $[-L, L]$ is the series

$$f(x) = \frac{a_0}{2} + \sum_{n=1}^{\infty} \left(a_n \cos \frac{n\pi x}{L} + b_n \sin \frac{n\pi x}{L} \right)$$

with

$$a_n = \frac{1}{L} \int_{-L}^{L} f(x) \cos \frac{n\pi x}{L} \, dx \text{ for } n = 0, 1, \ldots,$$

and

$$b_n = \frac{1}{L} \int_{-L}^{L} f(x) \sin \frac{n\pi x}{L} \, dx \text{ for } n = 1, 2, \ldots .$$

Example 12·21 Deduce the Fourier series of the function

$$f(x) = x^3 \text{ for } -1 \le x \le 1.$$

Solution In this case the fundamental interval is $[-1, 1]$ so that setting $L = 1$ in Theorem 12·15 gives for the Fourier coefficients of $f(x) = x^3$ the values

$$a_n = \int_{-1}^{1} x^3 \cos n\pi x \, dx \quad \text{and} \quad b_n = \int_{-1}^{1} x^3 \sin n\pi x \, dx.$$

Routine integration then shows $a_n = 0$ for $n = 0, 1, \ldots,$ and

$$b_n = (-1)^n \frac{2(6 - n^2\pi^2)}{n^3\pi^3} \text{ for } n = 1, 2, \ldots .$$

Hence the Fourier series of $f(x) = x^3$ in the fundamental interval $-1 \le x \le 1$ is

$$f(x) = \frac{2}{\pi^3} \sum_{n=1}^{\infty} (-1)^n \frac{(6 - n^2\pi^2)}{n^3} \sin n\pi x.$$

Since the periodic extension of $f(x)$ is discontinuous at $x = \pm 1$, the Fourier series at these points will converge to the value $\frac{1}{2}[f(1-) + f(1+)]$. In this case this value is zero, since $f(1-) = 1$ and $f(1+) = -1$.

(b) Fourier sine and cosine series

THEOREM 12·16 (*Fourier sine and cosine series*) If $f(x)$ is an arbitrary function defined and integrable on $[0, \pi]$, then it may either be expanded as a *Fourier cosine series*

$$f(x) = \frac{a_0}{2} + \sum_{n=1}^{\infty} a_n \cos nx \text{ for } 0 \le x \le \pi,$$

in which

$$a_n = \frac{2}{\pi} \int_0^\pi f(x) \cos nx \, dx \text{ for } n = 0, 1, \ldots,$$

or as a *Fourier sine series*

$$f(x) = \sum_{n=1}^\infty b_n \sin nx \text{ for } 0 \leq x \leq \pi,$$

in which

$$b_n = \frac{2}{\pi} \int_0^\pi f(x) \sin nx \, dx \text{ for } n = 1, 2, \ldots ..$$

Example 12·22   Deduce the Fourier cosine and sine series of the function $f(x) = x$ in $[0, \pi]$.

Solution   From the first part of Theorem 12·16 we have

$$a_n = \frac{2}{\pi} \int_0^\pi \cos nx \, dx \text{ for } n = 0, 1, \ldots ..$$

A simple integration then shows

$$a_0 = \pi, \quad a_{2n-1} = \frac{-4}{\pi(2n-1)^2}, \quad a_{2n} = 0.$$

The Fourier cosine series of $f(x) = x$ thus has the form

$$f(x) = \frac{\pi}{2} - \frac{4}{\pi} \sum_{n=1}^\infty \frac{\cos(2n-1)x}{(2n-1)^2} \text{ for } 0 \leq x \leq \pi.$$

From the second part of Theorem 12·16 we find

$$b_n = \frac{2}{\pi} \int_0^\pi x \sin nx \, dx \text{ for } n = 1, 2, \ldots ..$$

Evaluating this integral gives

$$b_n = (-1)^{n+1} \frac{2}{n} \text{ for } n = 1, 2, \ldots,$$

so that the Fourier sine series of $f(x) = x$ is seen to have the form

$$f(x) = 2 \sum_{n=1}^\infty (-1)^{n+1} \frac{\sin nx}{x}.$$

From the practical point of view the cosine series is preferable to the sine series in this instance because it converges more rapidly.

PROBLEMS

Section 12·1

12·1 Write down the general term in each of the following series:

(a) $\dfrac{3}{4} + \dfrac{5}{4^2} + \dfrac{7}{4^3} + \cdots$;

(b) $\dfrac{2}{5} + \dfrac{4}{8} + \dfrac{6}{11} + \dfrac{8}{14} + \cdots$;

(c) $1 + \dfrac{1\cdot3}{1\cdot4} + \dfrac{1\cdot3\cdot5}{1\cdot4\cdot7} + \dfrac{1\cdot3\cdot5\cdot7}{1\cdot4\cdot7\cdot10} + \cdots$

12·2 The series $a + ar + ar^2 + \cdots + ar^n + \cdots$ is called either the *geometric progression* or the *geometric series* with initial term a and common ratio r. Denote by S_n the sum of its first n terms so that

$$S_n = \sum_{m=0}^{n-1} ar^m.$$

By considering the difference $S_n - rS_n$ prove that

$$S_n = a\left(\frac{1-r^n}{1-r}\right).$$

If $r < 1$ deduce that

$$\sum_{m=0}^{\infty} ar^m = \frac{a}{1-r}.$$

What is the remainder R_n of the series after n terms.

12·3 Sum the following infinite series and find their remainders after n terms:

(a) $2 + 1 + \dfrac{2}{4} + \dfrac{1}{5} + \dfrac{2}{4^2} + \dfrac{1}{5^2} + \dfrac{2}{4^3} + \dfrac{1}{5^3} + \cdots$;

(b) $2 - 1 + \dfrac{2}{4} - \dfrac{1}{5} + \dfrac{2}{4^2} - \dfrac{1}{5^2} + \dfrac{2}{4^3} - \dfrac{1}{5^3} + \cdots$.

12·4 State which of the following series is divergent by Theorem 12·2:

(a) $\left(1 - \dfrac{1}{2}\right) + \left(1 + \dfrac{1}{2^2}\right) + \left(1 - \dfrac{1}{2^3}\right) + \cdots + \left(1 + \dfrac{(-1)^n}{2^n}\right) + \cdots$;

(b) $\dfrac{1}{3} + \dfrac{4}{9} + \dfrac{9}{19} + \cdots + \dfrac{n^2}{2n^2 + 1} + \cdots$;

(c) $\dfrac{1}{1} + \dfrac{2^2}{2^2} + \dfrac{3^2}{6^2} + \cdots + \dfrac{n^2}{(n!)^2} + \cdots$;

12·5 Prove the divergence of the harmonic series by obtaining simple underestimates for the sums of each of the indicated groupings of its terms and showing that they themselves form a series which is obviously divergent.

$$1 + \tfrac{1}{2} + \underbrace{\tfrac{1}{3} + \tfrac{1}{4}}_{\text{2 terms}} + \underbrace{\tfrac{1}{5} + \tfrac{1}{6} + \tfrac{1}{7} + \tfrac{1}{8}}_{\text{4 terms}} + \underbrace{\tfrac{1}{9} + \tfrac{1}{10} + \cdots + \tfrac{1}{16}}_{\text{8 terms}} + \cdots.$$

12·6 Use the comparison test to classify the following series as convergent or divergent:

(a) $1 + (\tfrac{2}{3}) + (\tfrac{2}{3})^4 + (\tfrac{2}{3})^9 + \cdots + (\tfrac{2}{3})^{n^2} + \cdots$;

(b) $\tfrac{3}{4} + \tfrac{4}{7} + \tfrac{5}{12} + \cdots + \left(\dfrac{n+2}{n^2+3}\right) + \cdots$;

(c) $1 + \dfrac{1}{3^2} + \dfrac{1}{5^3} + \dfrac{1}{7^4} + \cdots + \dfrac{1}{(2n-1)^n} + \cdots$

12·7 Use the integral test to determine the convergence or divergence of the following series:

(a) $\dfrac{1}{2 \log 2} + \dfrac{1}{3 \log 3} + \dfrac{1}{4 \log 4} + \cdots + \dfrac{1}{n \log n} + \cdots$;

(b) $\dfrac{1}{2 \log^2 2} + \dfrac{1}{3 \log^2 3} + \dfrac{1}{4 \log^2 4} + \cdots + \dfrac{1}{n \log^2 n} + \cdots$.

Where appropriate, estimate the remainder after six terms.

12·8 Classify the following convergent series as conditionally convergent or absolutely convergent:

(a) $1 - \dfrac{1}{\sqrt{2}} + \dfrac{1}{\sqrt{3}} - \cdots + \dfrac{(-1)^{n+1}}{\sqrt{n}} + \cdots$;

(b) $\dfrac{-3}{5} + \left(\dfrac{5}{9}\right)^2 - \left(\dfrac{7}{13}\right)^3 + \cdots + (-1)^n \left(\dfrac{2n+1}{4n+1}\right)^n + \cdots$;

(c) $1 - \dfrac{1}{3^2} + \dfrac{1}{5^2} - \dfrac{1}{7^2} + \cdots + (-1)^{n+1} \dfrac{1}{(2n-1)^2} + \cdots$.

12·9 Test the following series for convergence by the ratio test:

(a) $\dfrac{2}{1} + \dfrac{2 \cdot 5}{1 \cdot 5} + \dfrac{2 \cdot 5 \cdot 8}{1 \cdot 5 \cdot 9} + \cdots + \dfrac{2 \cdot 5 \cdot 8 \ldots (3n-1)}{1 \cdot 5 \cdot 9 \ldots (4n-3)} + \cdots$;

(b) $\dfrac{1}{\sqrt{3}} + \dfrac{3}{3} + \dfrac{5}{(\sqrt{3})^3} + \cdots + \dfrac{(2n-1)}{(\sqrt{3})^n} + \cdots$;

(c) $\dfrac{2}{5} + \dfrac{5}{14} + \dfrac{10}{29} + \cdots + \dfrac{n^2+1}{3n^2+2} + \cdots$;

(d) $1 + \dfrac{2^{30}}{2!} + \dfrac{3^{30}}{3!} + \cdots + \dfrac{n^{30}}{n!} + \cdots$;

(e) $\dfrac{1}{10} + \dfrac{2!}{10^2} + \dfrac{3!}{10^3} + \cdots + \dfrac{n!}{10^n} + \cdots$.

Where appropriate, estimate the remainder after four terms.

12·10 Test the following alternating series for convergence:

(a) $\dfrac{1}{2^2} - \dfrac{1}{4^2} + \dfrac{1}{6^2} - \dfrac{1}{8^2} + \cdots + (-1)^{n+1}\dfrac{1}{(2n)^2} + \cdots$;

(b) $\dfrac{1}{\sqrt{5}} - \dfrac{1}{3\sqrt{5}} + \dfrac{1}{4\sqrt{5}} - \cdots + (-1)^{n+1}\dfrac{1}{n+1\sqrt{5}} + \cdots$;

(c) $\dfrac{1}{3} - \dfrac{1}{2}\left(\dfrac{1}{3}\right)^2 + \dfrac{1}{3}\left(\dfrac{1}{3}\right)^3 - \cdots + (-1)^{n+1}\dfrac{1}{n}\left(\dfrac{1}{3}\right)^n + \cdots$.

Where appropriate, estimate the remainder after ten terms.

Section 12·2

12·11 Find the interval of convergence of each of the following power series:

(a) $x + \dfrac{2!}{2^2}x^2 + \dfrac{3!}{3^3}x^3 + \cdots + \dfrac{n!}{n^n}x^n + \cdots$;

(b) $x + \dfrac{(2!)^2}{4!}x^2 + \dfrac{(3!)^2}{6!}x^3 + \cdots + \dfrac{(n!)^2}{(2n)!}x^n + \cdots$;

(c) $1 + 5x + \dfrac{5^2x^2}{2!} + \dfrac{5^3x^3}{3!} + \cdots + \dfrac{5^nx^n}{n!} + \cdots$;

(d) $1 + \dfrac{x}{1} + \dfrac{x^2}{2^2} + \dfrac{x^3}{3^3} + \cdots + \dfrac{x^n}{n^n} + \cdots$;

(e) $\dfrac{x+7}{9} + \dfrac{(x+7)^3}{2 \cdot 9^2} + \dfrac{(x+7)^5}{3 \cdot 9^3} + \cdots + \dfrac{(x+7)^{2n+1}}{n \cdot 9^n} + \cdots$.

12·12 Find the radius of convergence of each of the following power series and verify that the differentiated series has the same radius of convergence:

(a) $\displaystyle\sum_{n=1}^{\infty} \dfrac{(x-3)^n}{(2n+1)2^n}$;

(b) $\displaystyle\sum_{n=0}^{\infty} \dfrac{(-1)^n(x-2)^n}{(2n+3)\sqrt{(n+1)}}$;

(c) $\displaystyle\sum_{n=1}^{\infty} \dfrac{(x+9)^n}{(n+1)^2}$.

12·13 Find the power series representation of arcsin x by considering the integral

$$\arcsin x = \int_0^x \frac{dx}{\sqrt{(1-x^2)}},$$

and find its radius of convergence. Does the power series converge at the end points of its interval of convergence.

12·14 By using the power series representation for log $(1-x)$, find a power series representation for the definite integral

$$\int_0^x \frac{\log (1 - x)}{x} \, dx.$$

For what values of x is the resulting power series convergent.

Section 12·3

12·15 Derive the Maclaurin series expansion of each of the following functions together with its interval of convergence.

(a) $x e^{-2x}$.

(b) $\cosh (x^2/2)$.

(c) $(1 + e^x)^3$;

(d) $\dfrac{1}{1 + x - 2x^2}$. [Hint: Use partial fractions.]

(e) $\log [x + \sqrt{(1 + x^2)}]$.

12·16 Taking $n = 1$, use Taylor's theorem with a remainder to determine whether the following functions increase or decrease with x for $x > 0$:

(a) $x - \tanh x$; (b) $\arctan x - x$; (c) $\log (1 + x) - x$.

12·17 Write down the Lagrange remainder $R_3(x)$ for each of the following functions:

(a) $f(a + x) = \sinh (a + x)$;

(b) $f(a + x) = \sin (a + x)$.

12·18 Write down the Taylor polynomial $P_4(x)$ for each of the following functions:

(a) $\log (1 + 2x)$;

(b) $\cos (\alpha + x)$;

(c) $1/\sqrt{(4 - x^2)}$;

(d) $a^x (a > 0)$.

12·19 Estimate the error if e^x is represented by its Taylor polynomial

$$P_5(x) = 1 + x + \frac{x^2}{2!} + \frac{x^3}{3!} + \frac{x^4}{4!}$$

in the interval $0 \leq x \leq \frac{1}{2}$.

12·20 Expand $f(x, y) = x^2 + 3xy^2 + y$ about the point $(1, 1)$.

12·21 Write down the first three terms of the Taylor series expansion of $f(x, y) = e^x \sin y$ about the origin.

12·22 Write down the first three terms of the Taylor series expansion of $f(x, y) = e^{x+y}$ about the point $(1, -1)$.

Section 12·4

12·23 Evaluate the limit

$$\lim_{x \to 2} \frac{\sin^2 \pi x}{2 e^{x/2} - x e}.$$

12·24 Given that $f(x, y) = e^{xy} - 1 - y(e^x - 1)$, and that b is neither equal to 1 nor 0, evaluate

$$\lim_{x \to 0} \frac{f(x, a)}{f(x, b)}.$$

12·25 Evaluate the limit

$$\lim_{x \to 0} \frac{(4^x - 2^x) - x(\log 4 - \log 2)}{x^2}.$$

12·26 Evaluate the limit

$$\lim_{x \to 0} \left(\frac{\cot x - 1/x}{\coth x - 1/x} \right).$$

12·27 Use Newton's method to calculate $\sqrt{21}$ accurately to four decimal places by seeking the zero of the function $f(x) = 21 - x^2$. Start your iteration with $x_0 = 4$.

12·28 Locate the pair of integers between which lies the one real root of the equation

$$x^3 - x - 1 = 0.$$

Determine the value of this root to four decimal places.

12·29 Find the positive root of $\sin x - \frac{1}{2}x = 0$ to three decimal places.

Section 12·5

12·30 Locate and identify the nature of the stationary points of the function

$$f(x, y) = x^3 + 3xy^2 - 15x - 12y.$$

Find the functional value of any maxima or minima.

12·31 Locate and identify the stationary points of the function

$$f(x, y) = 2x^4 - 3x^2y^2 + y^4 + 8x^2 + 3y^2.$$

12·32 Locate and identify the stationary point of the function

$$f(x, y) = x^3y^2(6 - x - y)$$

which lies in the first quadrant $x > 0$, $y > 0$.

12·33 Locate and identify the stationary points of the function

$$f(x, y) = x^3 + y^3 - 3xy.$$

12·34 By considering the proof of Theorem 12·13 show that the conditions stated in (c) are equivalent to, and may be replaced by,

(c) $\begin{cases} f_{yy}(x_0, y_0) < 0 \\ f_{yy}(x_0, y_0) > 0. \end{cases}$

Section 12·6

Find the Fourier series of each of the following functions.

12·35 $f(x) = x$ for $-\pi \le x \le \pi$.

12·36 $f(x) = \begin{cases} 0 \text{ for } -\pi \leq x < 0 \\ -x \text{ for } 0 \leqslant x \leqslant \pi. \end{cases}$

12·37 $f(x) = |\sin x|$ for $-\pi \leq x \leq \pi.$

12·38 $f(x) = \sin^2 x$ for $-\pi \leq x \leq \pi.$

12·39 $f(x) = \begin{cases} 0 \text{ for } -\pi \leq x < 0 \\ 1 \text{ for } 0 \leq x \leq \frac{1}{2}\pi \\ 0 \text{ for } \frac{1}{2}\pi < x \leq \pi. \end{cases}$

Section 12·7

12·40 Find the Fourier series of the function $f(x) = x$ for $0 \leq x \leq 2\pi.$

12·41 Find the Fourier series of the function
$$f(x) = \begin{cases} x^2 \text{ for } -\frac{1}{2}\pi \leq x \leq \pi \\ (2\pi - x)^2 \text{ for } \pi \leq x \leq \frac{3}{2}\pi \end{cases}$$

12·42 Find the Fourier series of the function $f(x) = x$ for $-1 \leq x \leq 1.$

12·43 Find the Fourier series of the function
$f(x) = |x|$ for $-3 \leq x \leq 3.$

12·44 Find the Fourier series of the function
$f(x) = \sin x$ for $0 \leq x \leq \pi.$

12·45 Find the Fourier sine series for the function
$f(x) = \cos x$ in $0 \leq x \leq \pi.$

12·46 Find the Fourier cosine series for the function
$f(x) = e^z$ for $0 \leq x \leq \pi.$

13 Differential equations and geometry

13·1 Introductory ideas

Special examples of differential equations have already been encountered; for example, those that gave rise to the exponential function and to the sine and cosine functions. It is now appropriate to make a systematic study of certain differential equations that are both useful and of frequent occurrence. We shall begin by examining a number of simple examples to illustrate the basic ideas.

Any equation involving one or more derivatives of a differentiable function of a single independent variable is called an *ordinary* differential equation. The following related equations taken from elementary dynamics are familiar examples.

$$\frac{d^2x}{dt^2} = g \qquad \text{(acceleration equation)}$$

$$\frac{dx}{dt} = u + gt \qquad \text{(velocity equation)}.$$

They describe, respectively, the acceleration and velocity of a particle falling freely under the action of gravity. Here g is the acceleration due to gravity, x is the distance of the particle from a fixed origin in its line of motion at time t, and u is the initial velocity of the particle. In these simple equations the dependent variable is represented by the displacement x and the independent variable by the time t. The integration of these equations is elementary and already familiar to the reader, who will also recognize that the velocity equation is in fact the integral of the acceleration equation with the arbitrary constant of integration set equal to the initial velocity u, since the velocity equation must describe the velocity at the start of the motion (when $t = 0$).

The first step in a systematic study of useful ordinary differential equations, aimed at producing general methods of solution wherever possible, is a straightforward classification of the equations. This we achieve by associating two numbers with each equation which we shall refer to as its *order* and its *degree*. We define the order of an ordinary differential equation to be the order of the highest derivative appearing in the equation, and the degree to be the exponent to which this highest derivative is raised when fractions and radicals involving y or its derivatives have been removed from the equation. Clearly, the notion of degree is only applicable when a differential equation has a simple algebraic structure allowing such a classification to be made. Thus both the simple dynamical equations just described are of degree 1, but the acceleration equation is of second order whereas the velocity equation is of first order. These are, in fact, examples of a specially important class of equations known as *linear differential equations*.

All linear differential equations are characterized by the fact that the dependent variable and its derivatives only occur with degree 1, whilst the coefficients multiplying them are either constants or functions of the independent variable. Thus of the following three second order differential equations, only the first two are linear, since the last involves the non-linear product $y(dy/dx)$. In general, differential equations that are not linear are termed non-linear.

$$\frac{d^2y}{dx^2} + 3\frac{dy}{dx} + 2y = 0,$$

$$x^2\frac{d^2y}{dx^2} + x\frac{dy}{dx} + (x^2 - n^2)y = 0,$$

$$\frac{d^2y}{dx^2} + y\frac{dy}{dx} - y = 0.$$

The classification of a more complicated differential equation is illustrated by the following example, involving both fractions and radicals, in which k is a constant:

$$\frac{(y'')^{3/2}}{y + (y'')^2} = k.$$

Clearing the fractions and radicals gives rise to the ordinary differential equation

$$k^2y''^4 - y''^3 + 2k^2yy''^2 + k^2y^2 = 0,$$

showing that the order is 2 and the degree is 4.

If y', y'', . . ., $y^{(n)}$ respectively denote successive derivatives, up to order n, of a differentiable function $y(x)$ with independent variable x, then a general nth order ordinary differential equation has the form

$$F(x, y, y', \ldots, y^{(n)}) = 0, \tag{13·1}$$

where F is an arbitrary function of the variables involved.

DEFINITION 13·1 (solution of differential equation) A solution of the ordinary differential Eqn (13·1) is a function $y = \phi(x)$ that is differentiable a suitable number of times in some interval I containing the independent variable x, and which has the property that

$$F(x, \phi(x), \phi'(x), \ldots, \phi^{(n)}(x)) = 0 \tag{13·2}$$

for all x belonging to I.

Notice that it is important to define the interval I since the differential equation does not necessarily describe the solution for unrestricted values of the argument x.

Thus a solution of the velocity equation just used as an example would be a differentiable function $x = \phi(t)$ defined for some interval I of time t with the property that

$$\phi'(t) - gt - u = 0, \tag{13·3}$$

for all t in the interval I. In this case I would be of finite size since the particle could not fall for an unlimited time without being arrested by contact with the ground, after which the ordinary differential equation giving rise to solution Eqn (13·3) would no longer be valid.

The prefix ordinary is used to describe differential equations involving only one dependent and one independent variable, in contrast with *partial* differential equations, which involve partial derivatives, and so have at least two independent variables and may also contain more than one dependent variable. Normally, when the type of differential equation being discussed is clear from the context, the adjectives 'ordinary' and 'partial' are omitted.

It is possible to develop the theory of differential equations in considerable generality, but our approach, as mentioned before, will be to examine a number of useful special forms of equation. We shall, however, first examine a few of the ways in which important forms of ordinary differential equation may arise.

13·2   Possible physical origin of some equations

At this stage it will be useful to illustrate some typical forms of differential equation, showing their manner of derivation from physical situations. We shall consider a number of essentially different physical problems and in each case take the discussion as far as the derivation of the governing differential equation.

Example 13·1   Experiment has shown that certain objects falling freely in air from a great height experience an air resistance that is proportional to the square of the velocity of the body. Let us determine the differential equation that describes this motion, and for convenience take our origin for the time t at the start of the motion. We shall assume that the body has a constant mass m and that at time t the velocity of fall is v, so that the air resistance at time t becomes λv^2 units of force, where λ is a constant of proportionality.

Now by definition, the acceleration a is the rate of change of velocity, so that $a = \mathrm{d}v/\mathrm{d}t$ and, since the body has constant mass m, it immediately follows from Newton's second law that the force accelerating the body is

$m(dv/dt)$. To obtain the equation of motion this force must now be equated to the other forces acting vertically downwards which are, taking account of the sign, the weight mg and the resistance $-\lambda v^2$. The equation of motion is thus

$$m\frac{dv}{dt} = mg - \lambda v^2$$

or, dividing throughout by the constant m,

$$\frac{dv}{dt} = g - \frac{\lambda}{m}v^2,$$

which is a special case of a differential equation in which the variables are *separable*. A general differential equation of this form involving the independent variable x and the dependent variable y can be written in either of the two general forms

$$\frac{dy}{dx} = M(x) . N(y) \tag{13·4}$$

or

$$P(x)Q(y)dx + R(x)S(y)dy = 0. \tag{13·5}$$

Example 13·2   In many simple chemical reactions the conversion of a raw material to the desired product proceeds under constant conditions of temperature and pressure at a rate directly proportional to the mass of raw material remaining at any time. If the initial mass of the raw material is Q, and the mass of the product chemical at time t is q, then the unconverted mass remaining at time t is $Q - q$. Then, if $-k(k > 0)$ denotes the proportionality factor governing the rate of the reaction, the reaction conversion rate $d(Q - q)/dt$ must be equal to $-k$ times the unconverted mass $Q - q$. The desired reaction rate equation thus has the form

$$\frac{d}{dt}(Q - q) = -k(Q - q),$$

where the minus sign has been introduced into the definition of k to allow for the fact that $Q - q$ decreases as t increases.

Example 13·3   A simple closed electrical circuit contains an inductance L and a resistance R in series, and a current i is caused to flow by the application of a voltage $V_0 \sin \omega t$ across two terminals located between the resistance and inductance. The equation governing this current i may be obtained by a simple application of Kirchhoff's second law, which tells us that the algebraic sum of the drops in potential around the circuit must be zero. Thus, since the driving potential is $V_0 \sin \omega t$ and the changes in potential across the inductance and resistance are in the opposite sense to i and so are, respectively, $-L(\mathrm{d}i/\mathrm{d}t)$ and $-Ri$, it follows that

$$V_0 \sin \omega t - L\frac{\mathrm{d}i}{\mathrm{d}t} - Ri = 0$$

or,

$$\frac{\mathrm{d}i}{\mathrm{d}t} + \frac{R}{L}i = \frac{V_0}{L}\sin \omega t.$$

The final equations of Examples 13·2 and 13·3 are both specially simple cases of linear first order differential equations. If the dependent variable is denoted by y and the independent variable by x, then all linear first order differential equations have the general form

$$\frac{\mathrm{d}y}{\mathrm{d}x} + P(x)y = Q(x). \tag{13·6}$$

Example 13·4 Mechanical vibrations occur frequently in physics and engineering and they are usually controlled by the introduction of some suitable dissipative force. A typical situation might involve a mass m on which acts a restoring force proportional to the displacement x of the mass from an equilibrium position, and a resistance to motion that is proportional to the velocity of the mass. Such a system, which to a first approximation could represent a vehicle suspension involving a spring and damper, is often tested by subjecting it to a periodic external force $F \cos \omega t$ in order to simulate varying road conditions. In this situation the displacement x would represent the movement of the centre of gravity of the vehicle about an equilibrium position as a result of passage of the vehicle along a road with a sinusoid profile. If the resisting force F_d has a proportionality constant k, and the restoring force F_r has a proportionality constant λ, then $F_d = k(\mathrm{d}x/\mathrm{d}t)$ and $F_r = \lambda x$. Applying Newton's second law, as in Example 13·1, and equating forces acting on the system we obtain the equation of motion

$$m\frac{\mathrm{d}^2x}{\mathrm{d}t^2} = F\cos \omega t - k\frac{\mathrm{d}x}{\mathrm{d}t} - \lambda x$$

or,

$$\frac{\mathrm{d}^2x}{\mathrm{d}t^2} + \frac{k}{m}\frac{\mathrm{d}x}{\mathrm{d}t} + \frac{\lambda}{m}x = \frac{F}{m}\cos \omega t.$$

This is a particular case of a linear constant coefficient second order differential equation, all of which have the general form

$$\frac{\mathrm{d}^2y}{\mathrm{d}x^2} + a\frac{\mathrm{d}y}{\mathrm{d}x} + by = f(x), \tag{13·7}$$

where x is the independent variable, y the dependent variable, and a and b are constants. Equations (13·6) and (13·7) are said to be *inhomogeneous* when they contain a term involving only the independent variable; otherwise they are said to be *homogeneous*. The differential equation of Example 13·2 is thus homogeneous of order 1 with dependent variable $(Q - q)$, whilst that of Example 13·4 is inhomogeneous of order 2; both are linear and involve constant coefficients. If in Example 13·2 the temperature of the reaction were allowed to vary with time, then in general the velocity constant k of the reaction would become a function of the time t and the equation would assume the homogeneous form of Eqn (13·6) with a variable coefficient.

The special importance of the types of differential equation singled out here lies in their frequent occurrence throughout the physical sciences. We shall later proceed with a systematic study of solution methods for these standard forms, together with other common cases of interest.

13·3   Arbitrary constants and initial conditions

If we consider the simple differential equation

$$\frac{\mathrm{d}^2 x}{\mathrm{d}t^2} = g, \tag{13·8}$$

then a single integration with respect to time gives $\mathrm{d}x/\mathrm{d}t = gt$ as a possible first integral. This is certainly a solution of Eqn (13·8) in the sense defined in Eqn (13·2), but it is not the most general solution since

$$\frac{\mathrm{d}x}{\mathrm{d}t} = c_1 + gt, \tag{13·9}$$

where c_1 is an arbitrary constant, is also a solution. This specific example illustrates the general result that in order to obtain the most complete form of solution, each integral involved in the solution of a differential equation must be interpreted as an antiderivative or, more loosely, as an indefinite integral. When maximum generality is sought the result is termed the *general* or *complete* solution of the differential equation. It is, therefore, important that when obtaining the general solution of a differential equation, an arbitrary constant should be introduced immediately after each integration. Thus the general solution of Eqn (13·8), which is obtained after two integrations, is

$$x = c_2 + c_1 t + \tfrac{1}{2} g t^2, \tag{13·10}$$

where c_2 is another arbitrary constant.

These arbitrary constants may be given definite values, and a *particular* solution obtained, if the solution is required to satisfy a set of conditions, at some starting time $t = t_0$, equal in number to the order of the differential equation. If, for example, Eqn (13·8) describes the acceleration of a body falling under the influence of gravity, and air resistance may be neglected, then Eqn (13·10) is the general solution of the problem of the position of the body at time t. In the event that the body started to fall with an initial velocity u at time $t = 0$, it follows from Eqn (13·9) that $c_1 = u$. Similarly, if the body was at position $x = x_0$ at time $t = 0$, it follows from Eqn (13·10) that $c_2 = x_0$, and so the particular solution corresponding to the initial conditions $x = x_0$, $dx/dt = u$ at $t = 0$ is

$$x = x_0 + ut + \tfrac{1}{2}gt^2. \tag{13·11}$$

General starting conditions of this type are known as initial conditions by analogy with time dependent problems such as this in which the solution evolves away from some known initial state. On occasion it is convenient to write initial conditions in an abbreviated form which we illustrate by repeating the initial conditions that gave rise to solution Eqn (13·11):

$$x\Big|_{t=0} = x_0, \qquad \frac{dx}{dt}\Big|_{t=0} = u.$$

Something more of the role of arbitrary constants may be appreciated if they are eliminated by differentiation from a general expression describing a family of curves in order that the differential equation describing the family may be obtained. Suppose, for example, that a general two-parameter family of curves is defined by the expression

$$y(x) = A \cosh 2x + B \sinh 2x,$$

where A and B are arbitrary constants (which we now regard as parameters). Then differentiation shows that

$$y' = 2(A \sinh 2x + B \cosh 2x) \quad \text{and} \quad y'' = 4(A \cosh 2x + B \sinh 2x),$$

from which it follows that elimination of A and B gives the differential equation

$$y'' = 4y.$$

This is the differential equation that has the two-parameter family of curves $y(x)$ as its general solution.

We should now see whether, having found a particular solution of a differential equation with given initial conditions, this is indeed the only possible solution. This is called the *uniqueness* problem for the solution and is obviously important in physical applications. To answer uniqueness questions for general classes of differential equations is difficult, but in the case of the dynamical problem just discussed a simple argument will suffice.

Let $v = dx/dt$ denote the velocity of the body so that Eqn (13·8) takes the form $dv/dt = g$. Now suppose that some other function w is also a solution of Eqn (13·8), satisfying the same initial conditions, so that $dw/dt = g$ and $v = w = u$ at $t = 0$. Then, setting $V = v - w$, it is easily established by subtraction of the two linear differential equations that the differential equation satisfied by the difference between the two postulated solutions is $dV/dt = 0$, thereby showing that $V = $ constant. However, as the initial conditions require that $V = 0$, it follows at once that $w \equiv v$. The velocity is thus uniquely determined by the differential equation and the initial condition. To complete the proof that the position is also uniquely determined it is only necessary to apply the foregoing argument to the velocity equation $dx/dt = u + gt$ obtained by direct integration of $dv/dt = g$. This matter of uniqueness will be taken up again in the next section in connection with some useful geometrical ideas.

It is not always necessary, or indeed possible, to prescribe only initial data for a differential equation, as we now illustrate by reformulating the previous example.

We have seen that the velocity v of the body is uniquely determined by Eqn (13·9) once it has been specified at some given instant of time. Similarly, when the velocity is known, the position is uniquely determined by Eqn (13·10) once it has been specified at some given instant of time. Velocity and position were specified at the same instant of time in the initial value problem just discussed; we now illustrate an alternative problem that could equally well have been considered.

As the solution (13·10) implies the result (13·9), it would be quite permissible to determine a particular solution by requiring the body to be at the positions x_0 and x_1 at the respective times t_0 and t_1. These conditions would enable the determination of the arbitrary constants c_1 and c_2 and would, of course, completely determine the velocity. Conditions such as these that are imposed on the solution of a differential equation at two different values of the independent variable are called *two-point boundary* conditions. This name is derived from the fact that in many important applications the conditions to be imposed are prescribed at two physical boundaries associated with the problem.

In the simple initial value problem discussed here the question of the existence of a solution was never in doubt since we were able to find the general solution by direct integration. This is not usually the case; with more complicated differential equations the first question to be asked is 'does a solution exist' and, if so, 'is it unique'.

To illustrate this let us again consider Eqn (13·8), but this time with different two-point boundary conditions. At first sight it might appear reasonable to specify the velocity rather than the position at two different times, but a moment's reflection shows that this is not possible. This arises because Eqn (13·9) determines the velocity, and unless the two pre-assigned velocities were in agreement with this equation there could obviously be no solution satisfying such two-point boundary conditions. Furthermore, even if they were in agreement with Eqn (13·9), only the arbitrary constant c_1 would be so determined, leaving an infinity of solutions Eqn (13·10) of the original differential equation corresponding to arbitrary values of c_2.

Having made this point we shall not pursue it further in this first course on differential equations.

13·4 Properties of solutions—isoclines

Before discussing methods of solution of differential equations the general notion of a solution already outlined must be more fully discussed. Then, after examining the differentiability of solutions in this context, we shall, as a preliminary to developing special methods of solution, illustrate how the properties of particular differential equations may be explored by the application of some useful geometrical ideas.

The concept of a solution given in Definition 13·1 implies that for a differential equation of order n, the solution must be differentiable at least n times within the interval I of its definition. This is easily seen by writing the equation of order n displayed implicitly in Eqn (13·1) in the explicit form

$$y^{(n)} = f(x, y, y', \ldots, y^{(n-1)}), \tag{13·12}$$

and then using the fact that it is necessary for y to be differentiable $n - 1$ times for the arguments $y', y'', \ldots, y^{(n-1)}$ of the function f to exist.

To illustrate these ideas we shall consider the solution of the differential equation

$$\frac{d^2y}{dx^2} + y = 0$$

within the interval I determined by $-1 \leq x \leq 1$. It is easily verified by differentiation that

$$y_1(x) = \cos x + \sin x \quad \text{and} \quad y_2(x) = \cos x + 2 \sin x$$

are both solutions of the differential equation over the interval I. However, the function $y(x)$ defined on I by

$$y(x) = \begin{cases} \cos x + \sin x & (-1 \leq x \leq 0) \\ \cos x + 2 \sin x & (0 < x \leq 1) \end{cases}$$

is not a solution of the differential equation, because although $y(x)$ is continuous over I, dy/dx has a discontinuity at $x = 0$ and, since the equation is of second order, we must require of its solution that at least y and y' be continuous over I. This example introduces into the context of differential equations the idea of left- and right-handed derivatives already encountered in connection with continuity and differentiability. Here $y(x)$ has the property that $y'(0-) = 1$ and $y'(0+) = 2$.

A solution of a differential equation must also be finite in its interval of definition and sometimes it is necessary to restrict the interval to achieve this. The simple differential equation

$$\frac{dy}{dx} = y^2 + 1$$

clearly has as its solution the function $y = \tan x$ whenever $x \neq (2n + 1)\frac{1}{2}\pi$. Thus $y = \tan x$ is a solution in any *open* interval contained in $I_n = [(2n - 1)\frac{1}{2}\pi \leq x \leq (2n + 1)\frac{1}{2}\pi]$. In particular, a solution exists in the interval $(-\frac{1}{2}\pi, \frac{1}{2}\pi)$ but not in the interval $(-\frac{1}{2}\pi, \frac{1}{2}\pi]$. In examples of this type, the behaviour of the solution in the neighbourhood of the end points of the intervals of definition means that the solutions in adjacent intervals cannot be connected across the points of discontinuity.

These ideas are greatly clarified if the notion of an *isocline* is introduced, which we now do in relation to the first order differential equation $y' = f(x, y)$. From the ideas already formulated about the nature of a solution on a specified interval I, it is clear that when initial conditions $x = x_0$, $y = y_0$ are specified, the differential equation then assigns to the derivative of the solution at $x = x_0$ the value $(y')_0 = f(x_0, y_0)$. Alternatively, representing the solution by a curve in the (x, y)-plane, we may use geometrical terminology and speak instead of the initial point (x_0, y_0), and of the derivative of the solution $(y')_0 = f(x_0, y_0)$ at that point. From the geometrical interpretation of a derivative this implies that the tangent to the solution curve through the point (x_0, y_0) is inclined to the x-axis at an angle $\theta = \arctan f(x_0, y_0)$. For any particular solution y_p through an initial point (x_0, y_0), the derivative $y_p'(x)$ will, in general, have different values associated with different values of the independent variable x, and so far we have no way of determining how y_p varies with x. However, if instead of a particular solution the general properties of all solutions of the differential equation $y' = f(x, y)$ are to be considered, then the equation may be regarded as assigning to each point of the (x, y)-plane a specific value of y', and hence also a specific angle of inclination θ for the tangent to the solution curve through that point. If the derivative y' is set equal to a constant value K, then the equation $K = f(x, y)$ determines those curves in the (x, y)-plane along which the tangents to the solutions all have a constant angle of inclination $\theta_0 = \arctan K$ to the x-axis. Because of the equal angle of inclination of the tangents to the solution curves that pass through points on these lines, the lines themselves are termed isoclines. Different values of K determine different isoclines, and it is quite possible that no isoclines exist for some values of K.

It is, of course, important to recognize that an isocline is only a curve characterizing a special property of all the solutions of the differential equation and that it is not itself a curve representing a solution.

In principle, the simple idea of an isocline provides an approximate procedure for the determination of a numerical solution of a general first order

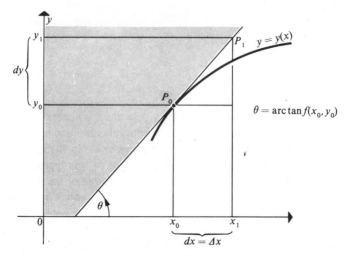

Fig. 13·1 Graphical solution of $y' = f(x, y)$, $y(x_0) = y_0$.

differential equation. If we know that the solution passes through the initial point (x_0, y_0), and it is required to determine the solution as far as $x = d$, then we start by sub-dividing the interval $[x_0, d]$ into n sub-intervals. Letting $x_l = x_0 + l\Delta x$, where $l = 1, 2, \ldots, n$ and $\Delta x = (d - x_0)/n$, we determine y_1, the approximate solution at $x = x_1$, by drawing through the point $P_0(x_0, y_0)$ a line with gradient $f(x_0, y_0)$, and setting y_1 equal to the value of y at point P_1 where the tangent line intersects the ordinate through $x = x_1$ (Fig. 13·1). This process is repeated until x becomes equal to $x_n = d$, when the numbers $y_1, y_2, \ldots, y_n$ represent approximations to the true solution y at the points $x_1, x_2, \ldots, x_n$. Naturally, the accuracy of these numbers depends, in part, on the number n of sub-divisions that has been chosen.

The polygonal line obtained by joining adjacent points (x_i, y_i) by straight line elements is called the *Cauchy polygon* approximation to the solution, after A. L. Cauchy who was the first to establish its convergence to the true solution as $\Delta x \to 0$.

This process may be mechanized by the use of differentials together with a suitable notation. To see this let us introduce the differentials dx and dy_i through the equation

$$dy_i = f(x_i, y_i)dx,$$

with $dx = \Delta x$, so that at (x_i, y_i) the ratio dy_i/dx of the finite quantities dx and dy_i is equal to the value of $y'(x_i)$. This leads to the general result

$$y_{i+1} = y_i + dy_i$$

or, equivalently,

$$y_{i+1} = y_i + f(x_i, y_i)dx.$$

As a particular case of this we have

$$y_1 = y_0 + f(x_0, y_0)dx,$$

which is just the first step of our earlier argument alternatively expressed. Other arguments show that the magnitude of the error arising at each step is of the order of $(dx)^2$.

The following example applies this result to the numerical integration of a simple differential equation over five equal steps $\Delta x = 0·1$ though, if desired, the interval Δx may be varied from step to step.

Example 13·5 Let us find the approximate value of y at $x = 0·5$, given that $y' = xy$ and $y(0) = 1$.

Solution If we take five equal sub-intervals, so that $\Delta x = 0·1$, the results of the calculations based on the relation $dy_i = 0·1 \, x_i y_i$ will be as follows:

i	x_i	y_i	dy_i	$e^{\frac{1}{2}x^2}$ ← exact sol⁻
0	0	1	0	1
1	0·1	1	0·01	1·0050
2	0·2	1·01	0·0202	1·0202
3	0·3	1·0302	0·0309	1·0460
4	0·4	1·0611	0·0424	1·0833
5	0·5	1·1035		1·1331

(handwritten annotations:)
$\rightarrow$ e.g. $x_3 = 3 \times 0·1 = 0·3$,

$y_3 = y_2 + dy_2$

$= 1·01 + 0·0202 = 1·0302$

$dy_3 = x_3 y_3 \Delta x = ·3 \times 1·0302 \times 0·1$
$= 0·030906$

A comparison of the third column with the final column, which tabulates the exact solution $y = e^{\frac{1}{2}x^2}$, demonstrates the relatively poor accuracy obtainable by this simple approach, known as *Euler's method* for the numerical integration of a differential equation. The approximate value $y(0·5) = 1·1035$ obtained by Euler's method is seen to be already 2·6 per cent low, and attempts to determine y for values of $x > 0·5$ would result in a very rapid growth of error. The Cauchy polygon is compared with the exact solution in Fig. 13·2. Later we shall show how a simple modification to this method will produce a considerable improvement.

Returning to the subject of isoclines we shall now utilize several examples to illustrate some typical situations. As a solution curve arises as an integral of the original differential equation, it is customary to refer to the solution curves as *integral curves*.

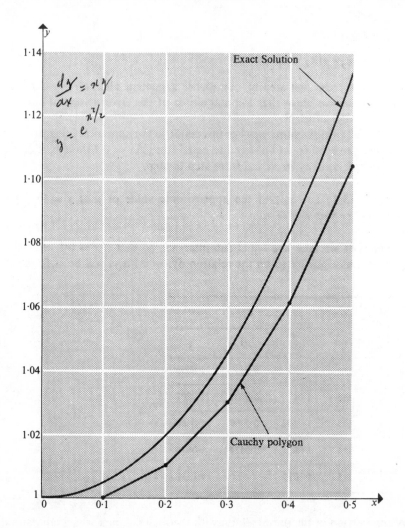

Fig. 13·2 Comparison of exact solution and Cauchy polygon.

Example 13·6 Consider the simple differential equation $y' = x + 1$, which is easily seen to have the general solution $y = \frac{1}{2}x^2 + x + C$. Setting $y' = K$ then shows that the isoclines of this differential equation are the lines $x = K - 1$. Representative isoclines are illustrated in Fig. 13·3 as the full vertical lines. Short inclined lines have been added to these isoclines to indicate the direction of the tangents to the integral curves that intersect the isoclines; their angles of inclination have the magnitude arctan K. Three integral curves, represented by curved full lines, have been drawn to show the relationship between isoclines, the tangents or gradients associated with

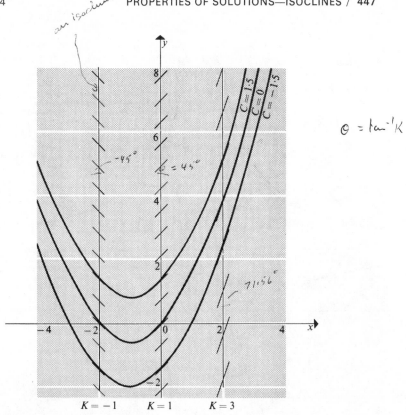

an isocline

$\theta = \tan^{-1}K$

$C = 1.5$
$C = 0$
$C = -1.5$

$-45°$

$= 45°$

$71.56°$

$K = -1$ $K = 1$ $K = 3$

Fig. 13·3 Isoclines, direction field, and integral curves.

isoclines, and the integral curves themselves. The pattern of these tangents associated with the isoclines shows the direction taken by integral curves and is accordingly termed the *direction field* associated with the integral curves.

Figure 13·3 also serves to illustrate the geometrical analogue of Euler's method; namely, to use a map of the isoclines, each marked with their associated tangents indicating the direction field of the integral curves, in order to trace a solution that starts from a given point and always intersects each isocline at an angle equal to the gradient associated with it.

It is easily seen that with the simple equation $y' = x + 1$ there are no points in the finite (x, y)-plane at which the gradient is either infinite or ambiguous. The next two examples show more complicated situations involving characteristic behaviour of direction fields and integral curves at special points.

Example 13·7 In the case of the differential equation $y' = (1 - y)/(1 + x)$, the general solution determining the integral curves is $y = 1 + C/(1 + x)$. As always, the isoclines are determined by setting $y' = K$ in the differential

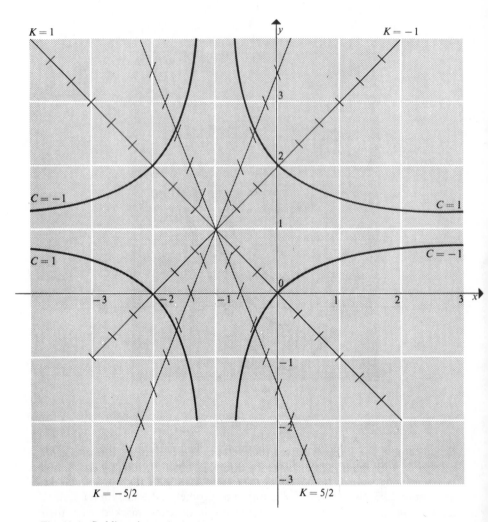

Fig. 13·4 Saddle point at $(-1, 1)$.

equation, thereby giving rise to the equation $1 - y = K(1 + x)$. This is simply a family of straight lines through the point P in the (x, y)-plane with the coordinates $(-1, 1)$. Integral curves for $C = \pm 1$ are shown in Fig. 13·4 together with representative isoclines. It is easily seen from the differential equation that the line $y = 1$ is both a degenerate integral curve and an isocline corresponding to $K = 0$. As all the isoclines pass through the point P it is obviously a special point in the direction field. We shall call such a point P at which the derivative y' is indeterminate a *singular* point of the direction field of the differential equation in question. The hyperbola-like pattern of the integral curves in the vicinity of P is characteristic of a certain important form of behaviour, and any family of integral curves having this property are said to have a *saddle point* at P.

Example 13·8   A direction field of a different kind is provided by the differential equation $y' = 2y/x$, which has the lines $y = \frac{1}{2}Kx$ as its isoclines and the curves $y = Cx^2$ as its integral curves. Their inter-relationship is illustrated· in Fig. 13·5, which also shows quite clearly that the singular point at the origin is of an essentially different kind to that of the previous example. Again the isoclines all pass through this point but, whereas in Example 13·7 there was only one degenerate integral curve through the point P, in the present case every integral curve passes through the singular point. The parabola-like behaviour of the integral curves in the vicinity of the origin is characteristic of a different form of singularity, and integral curves with this general property are said to have a *node* at the common point.

The last two examples also serve to illustrate that initial conditions to differential equations may not always be prescribed arbitrarily without reference to the equation in question, since there may either be no solution or an infinity of solutions satisfying a differential equation and arbitrarily prescribed initial conditions. For example, no integral curve passes through the point $(-1, 2)$ in Fig. 13·4, whereas every integral curve passes through the point $(0, 0)$ in Fig. 13·5. Since, in the first case, solutions have infinities along the line $x = -1$, and in the second case, the direction field is indeterminate at $(0, 0)$, this suggests that for a unique solution to exist the isoclines must be well behaved and free both from points at which infinite gradients occur and points of intersection giving rise to indeterminacies of gradient in the direction field.

To make these ideas a little more precise let us use the following simple argument to suggest the form of a general existence theorem for solutions of the general first order differential equation

$$y' = f(x, y), \tag{13·13}$$

in some small interval $[a, b]$ containing the point $x = x_0$ at which we require $y = y_0$.

Setting $K = f(x_0, y_0)$ and assuming K to be finite, the corresponding isocline is then defined by the implicit functional relationship $K = f(x, y)$ or, alternatively, by $F(x, y) = 0$, where $F(x, y) = f(x, y) - K$.

By our earlier work on implicit functions we know that a unique relationship $y = \phi(x)$ defining the isocline may be obtained in the neighbourhood of some point (x_0, y_0), provided the partial derivatives F_x and F_y are continuous in the neighbourhood of (x_0, y_0) and $F_y(x_0, y_0) \neq 0$. However, since K is constant for the particular isocline in question, $F_x \equiv f_x$ and $F_y \equiv f_y$, and so we may conclude that the continuity of f_x and f_y in the neighbourhood

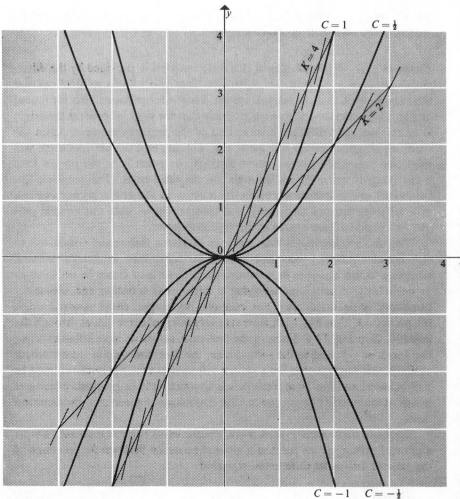

Fig. 13·5 Node at origin.

of (x_0, y_0), together with the condition $f_y(x_0, y_0) \neq 0$, will ensure that locally there is a unique isocline with the associated gradient K. Consequently, there is no singularity of the direction field near (x_0, y_0), and so an argument such as the Euler method will yield a solution in the neighbourhood of (x_0, y_0). In reality the simple argument used here has resulted in conditions to be applied to the function $f(x, y)$ that will certainly ensure the existence of a unique solution, so they are *sufficient* conditions; nevertheless, we shall show that they are too restrictive, and so are not all *necessary* conditions.

That the conditions are sufficient, but not necessary, is easily demonstrated by appealing to Example 13·6, in which $f(x, y) = x + 1$. We already know that the general solution is $y = \frac{1}{2}x^2 + x + C$ and so always exists, but although $f_x = 1$ and $f_y = 0$ are both continuous functions, the result $f_y = 0$ violates the supplementary condition that $f_y(x_0, y_0) \neq 0$. Thus this condition is clearly not a necessary one.

More subtle methods of analysis give rise to the following less restrictive theorem which, although satisfactory for most practical purposes, is still only a statement of sufficient conditions.

THEOREM 13·1   If the functions $f(x, y)$ and $f_y(x, y)$ are continuous in a rectangle R of the (x, y)-plane containing the point (x_0, y_0) then, for some sufficiently small positive number h, there exists a unique solution $y = y(x)$ of the differential equation

$$y' = f(x, y)$$

that is defined on the interval $x_0 - h \leq x \leq x_0 + h$ and is such that $y(x_0) = y_0$.

In effect this theorem asserts that when the stated conditions are satisfied, a unique integral curve passes through each point of the rectangle R. We shall not pursue these arguments further, but they are obviously of importance when used in connection with discussions involving differential equations of unfamiliar type to determine whether solutions, once obtained, are unique.

An application of the conditions of the theorem to the three previous examples shows that the first satisfies them everywhere in the finite plane, the second has infinities in f and f_y along $x = -1$, and the third has infinities in f and f_y along $x = 0$. Consequently Example 13·6 has a unique integral curve through every point of the finite plane, whereas in Examples 13·7 and 13·8 the respective lines $x = -1$ and $x = 0$ must be omitted from the (x, y)-plane; a unique integral curve then passes through all the remaining points of the finite plane.

Example 13·9   The use of isoclines in the determination of properties of solutions of differential equations can often be supplemented by other useful information obtainable directly from the equation. We illustrate this by considering the differential equation

$$y' = y + \tfrac{1}{2}x + e^{-x},$$

which is seen to have isoclines determined by the equation

$$y = K - \tfrac{1}{2}x - e^{-x}.$$

Having constructed a set of isoclines together with the associated direction field of tangents, we notice first that the *extrema* of the integral curves will occur along the isocline $y = -\tfrac{1}{2}x - e^{-x}$ corresponding to $y' = K = 0$. This isocline, together with several others, is shown in Fig. 13·6 (a), in which

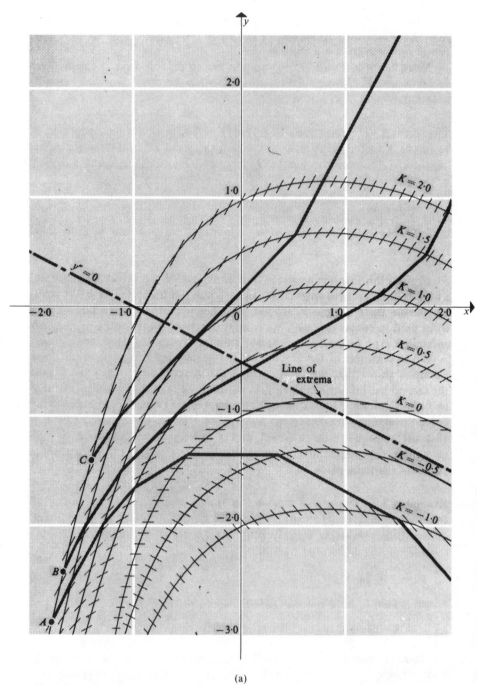

(a)

Fig. 13·6 Integral curves and curves characterizing extrema of solutions: (a) approximate integral curves; (b) exact integral curves.

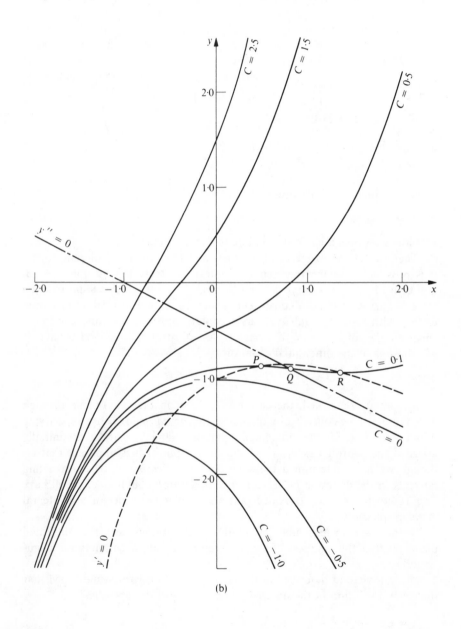

(b)

short inclined lines have again been used to indicate the direction field associated with the isoclines.

Additional information may be obtained by seeking the location of *points of inflection* of the integral curves which, when they occur, must coincide with the vanishing of y''. This information may be obtained directly from the differential equation itself if we first differentiate it with respect to x to obtain

$$y'' = y' + \tfrac{1}{2} - e^{-x},$$

and then substitute for y' to obtain

$$y'' = y + \tfrac{1}{2}(1 + x).$$

Hence the points of inflection will lie along the line

$$y = -\tfrac{1}{2}(1 + x),$$

which is shown as a chain-dotted line in Fig. 13·6 (a).

Then, using the property of isoclines and the associated direction field, it is possible to sketch representative integral curves. Taking points A, B, and C in Fig. 13·6 (a) as typical points in the (x, y)-plane, three approximate integral curves have been constructed using the graphical method discussed earlier. Although these integral curves contain substantial errors, due to the small number of isoclines, they nevertheless illustrate the general behaviour of solutions of the differential equation

$$y' = y + \tfrac{1}{2}x + e^{-x}.$$

As already remarked, the choice of a point in the (x, y)-plane through which to begin the construction of an integral curve is equivalent to specifying initial conditions for the differential equation. Namely, x and y are initially assigned the coordinates x_0, y_0 of the chosen point. It is apparent that although we have determined the solution for *increasing* x, by constructing tangents in the direction of *decreasing* x, a solution could equally well have been found for $x < x_0$, provided that no singular point lies on the integral curve in question.

In this case no ambiguity or infinity of derivatives occurs in the finite plane, so that the solution of this differential equation contains no singularities.

Using a method described in the next chapter it is easily established that the general solution of the differential equation just discussed is

$$y = Ce^x - \tfrac{1}{2}e^{-x} - \tfrac{1}{2}(1 + x),$$

and representative curves are shown in Fig. 13·6 (b) corresponding to the indicated values of C. These curves illustrate, as do those of Fig. 13·6 (a), that the nature of the extrema differ from curve to curve. Thus the lower three integral curves in Fig. 13·6 (b) possess absolute maxima but no points of inflection, whereas the upper three integral curves possess points of inflection but neither maxima nor minima. However, the integral curve corresponding to $C = 0·1$ possesses a local maximum at P, a point of inflection at Q, and a local minimum at R. The line of points of inflection is again shown as a chain-dotted line, whilst the line of extrema (the isocline for which $K = 0$) is shown as a dotted line.

13·5 Modified Euler method

The Euler method for the numerical solution of a first order differential equation provides a means of determining the solution of an initial value problem but, as we have already seen in an example and several problems, the accuracy is poor. We now show that attention to the geometrical implications of the method can greatly improve its accuracy. In Fig. 13·1 the gradient appropriate to point P_0 was used to determine the change dy_0 in the functional value over the entire interval $dx = \Delta x$. This is obviously only a first approximation to the true situation, and a better approximation to the increment in y consequent upon a step Δx would be provided by using the average of the gradients at P_0 and P_1 in place of $f(x_0, y_0)$ in the Euler method. This simple refinement applied to the previous argument is known as the *modified Euler method* in which the error at each step is of the order of $(dx)^3$.

The proposed modification is shown diagrammatically in Fig. 13·7, in which the full straight lines passing through points P_0 and P_1 have respective gradients $m_0 = f(x_0, y_0)$ and $m_1 = f(x_0 + dx, y_0 + dy_0)$. Then, if the dotted line through P_0 has gradient $m_0' = \frac{1}{2}(m_0 + m_1)$, the improved approximation dy_0' to the increment in y is simply $dy_0' = m_0' dx$. In terms of the angles θ, θ_0, and θ_1 defined in the figure, $\tan \theta = \frac{1}{2}\{\tan \theta_0 + \tan \theta_1\}$.

The improved accuracy is best illustrated by repeating the numerical Example 13·5 to determine the value of y at $x = 0·5$, given that $y' = xy$ and $y(0) = 1$. To simplify the headings on the tabulation we set $m_i = f(x_i, y_i)$ and $m_{i+1} = f(x_i + dx, y_i + dy_i)$ and, as before, use increments $\Delta x = 0·1$.

i	x_i	y_i	m_i	dy_i	m_{i+1}	$m_i' = \frac{1}{2}(m_i + m_{i+1})$	dy_i'	$e^{\frac{1}{2}x^2}$
0	0·0	1·0	0·0	0·0	0·1	0·05	0·005	1·0
1	0·1	1·0050	0·1005	0·0101	0·2030	0·1517	0·0152	1·0050
2	0·2	1·0202	0·2040	0·0204	0·3122	0·2581	0·0258	1·0202
3	0·3	1·0460	0·3138	0·0314	0·4310	0·3724	0·0372	1·0460
4	0·4	1·0832	0·4333	0·0433	0·5633	0·4983	0·0498	1·0833
5	0·5	1·1330						1·1331

The approximate value $y(0·5) = 1·1330$ shown in the third column is now only 0·0001 low, demonstrating the superiority of the modified Euler method over its predecessor.

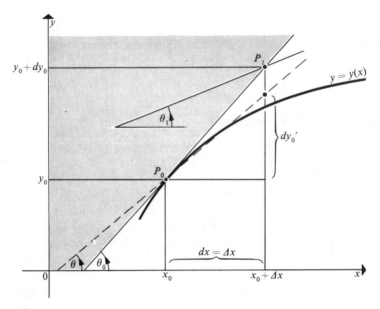

Fig. 13·7 Euler's modified method.

PROBLEMS

Section 13·1

13·1 Determine the order and degree of each of the following equations:

(a) $x^2 y''' + y'^2 + y = 0$;

(b) $y'^2 + 2xy = 0$;

(c) $\dfrac{(2y'' + x)^2}{(xy + 1)} = 3$;

(d) $\left(\dfrac{y'' + xy}{y'' + 3}\right)^{3/2} = (2y'' + xy' + y)$;

(e) $y''' + 2y''^2 + 6xy = e^x$.

Section 13·2

13·2 Determine the differential equation of the curve that has the property that the length of the interval of the x-axis contained between the intercepts of the tangent and ordinate to a general point on the curve has a constant value k.

13·3 Obtain the differential equation governing the motion of a particle of mass m that is projected vertically upwards in a medium in which the resistance is λ times the square of the particle velocity v.

13·4 Derive the differential equation which describes the rate of cooling of a body at a temperature T on the assumption that the rate of cooling is k times the excess of the body temperature above the ambient temperature T_0 of the surrounding air. This is known as *Newton's Law of Cooling* and it is a good approximation for small temperature differences.

Section 13·3

13·5 Eliminate the arbitrary constants in the following expressions to determine the differential equations for which they are the general solutions:

(a) $\frac{1}{2}x^2 + y^2 = C^2$;

(b) $y = Cx + C^3$;

(c) $x^3 = C(x + y)^2$;

(d) $\log\left(\dfrac{1 + x}{y}\right) = Cy$;

(e) $y = Ae^x + Be^{2x}$;

(f) $y = (C + Dx)e^{2x}$.

13·6 Determine whether the following expressions satisfy the associated differential equations for all real x:

(a) $y = x^2 - 2x$; $\quad xy'' + y = x^2$;

(b) $y = \dfrac{1}{x}$; $\quad y' = y^2 + \dfrac{2}{x^2}$;

(c) $y = \sin 3x + \cos 3x$; $\quad y'' + 9y = 0$;

(d) $y = e^{-x}(A \cos 2x + B \sin 2x)$; $\quad y'' + 2y' + 5y = 0$;

(e) $y = 2x(e^x + C)$; $\quad xy' - y = x^2e^x$;

(f) $y = A \cos x + B \sin x - \frac{1}{2}x \cos x$; $\quad y'' + y = \sin x$.

Section 13·4

13·7 Determine whether the following differential equations have the associated functions as their solution over the stated intervals:

(a) $y' = x$, $y(x) = \begin{cases} \frac{1}{2}x^2 + 1, & x \le 1 \\ \frac{1}{2}x^2 + 2, & 1 < x \end{cases}$ $(-1 \le x \le 1)$;

(b) $y'' - 9y = 0$, $y(x) = A \cosh 3x + B \sinh 3x$ $(-\infty < x < \infty)$;

(c) $y' = 4$, $y(x) = \begin{cases} 4x + 2, & x < 0 \\ 4x - 2, & x > 0 \end{cases}$ $(-2 \le x \le -\frac{1}{2})$;

(d) $y' + y^2 = 0$, $y(x) = \dfrac{1}{1 + x}$, with $y(-1) = 1$ $(-2 \le x \le 0)$;

(e) $x^4y'' + y = 0$, $y(x) = x \sin \dfrac{1}{x}$, with $y(0) = 1$ $(0 \le x \le 1)$;

(f) $y' - 3x^2y = 0$, $y(x) = e^{x^3}$ $(-\infty < x \le 10)$.

13·8 Taking intervals $\Delta x = 0{\cdot}2$, use Euler's method to determine $y(1)$, given that $y' + y = 0$ and $y(0) = 1$. Compare your results with the exact solution $y = e^{-x}$. Construct the Cauchy polygon.

13·9 Taking intervals $\Delta x = 0{\cdot}1$, use Euler's method to determine $y(1)$, given that $dy/dx = (x^2 + y)/x$ and $y(0{\cdot}5) = 0{\cdot}5$. Compare your results with the exact solution $y = \frac{1}{2}x + x^2$. Construct the Cauchy polygon.

13·10 Taking intervals $\Delta x = 0{\cdot}2$, use Euler's method to determine $y(1)$, given that $y' = y + e^{-x}$ and $y(0) = 0$. Compare your results with the exact solution $y = \sinh x$. Construct the Cauchy polygon.

13·11 Repeat Problem 13·10, taking $\Delta x = 0.1$ and determine the improvement in accuracy.

13·12 Draw the isoclines and sketch the direction fields for the differential equations in Problems 13·8 to 13·10.

13·13 Use the isoclines and direction field for the differential equation $dy/dx = (x^2 + y)/x$ to deduce the behaviour of the integral curves close to the origin. What form of singularity occurs at the origin?

13·14 Using isoclines and the associated direction field, determine the approximate value of $y(1)$, given that $y' = y + e^{-x}$ and $y(0) = -1$. Compare your result with the exact solution $y = -\cosh x$.

Section 13·5

13·15 Repeat Problem 13·8, using the modified Euler method and compare with the previous numerical solution.

13·16 Repeat Problem 13·9, using the modified Euler method and compare with the previous numerical solution.

14 First order differential equations

14·1 Equations with separable variables

The class of differential equations in which the variables are separable was identified in Section 13·2, where it was remarked that either of the two forms

$$\frac{dy}{dx} = M(x)N(y) \tag{14·1}$$

or

$$P(x)Q(y)dx + R(x)S(y)dy = 0 \tag{14·2}$$

may arise. Here, Eqn (14·2) must, of course, be interpreted in the sense of differentials already defined elsewhere.

As the name implies, such equations are solved by rewriting so that functions of the variable x, together with its differential dx, and functions of the variable y, together with its differential dy, occur on opposite sides of the equation. The general solution may then be obtained by direct integration and the introduction of an arbitrary constant.

Written in symbolic form the solutions of Eqns (14·1) and (14·2) may be expressed as

$$\int \frac{dy}{N(y)} = \int M(x)\,dx + C, \tag{14·3}$$

provided $N(y)$ is non-vanishing, and

$$\int \frac{S(y)}{Q(y)}\,dy = -\int \frac{P(x)}{R(x)}\,dx + C, \tag{14·4}$$

provided $Q(y)$ and $R(x)$ are non-vanishing where, of course, C is an arbitrary constant.

Example 14·1 Suppose, as in Example 13·7, that

$$\frac{dy}{dx} = \frac{1-y}{1+x}.$$

Solution Divide the equation by $(1 - y)$ and multiply by the differential dx to obtain

$$\frac{1}{1 - y} \left(\frac{dy}{dx}\right) dx = \frac{dx}{1 + x}.$$

From our definition of a differential we recognize that $(dy/dx)\,dx = dy$, so that in differential form the equation becomes

$$\frac{dy}{1 - y} = \frac{dx}{1 + x}.$$

Henceforth, to shorten discussions, we shall proceed directly from differential equations to results of this form, omitting the formal introduction of the differentials. That is, when convenient, we shall regard dy/dx either as the single entity denoting a derivative or as the ratio of the two differentials dy and dx.

On integrating this last result we obtain

$$\int \frac{dy}{1 - y} + \log C = \int \frac{dx}{1 + x}$$

or

$$-\log |\,1 - y\,| + \log C = \log |\,1 + x\,|,$$

where, for convenience, we write the arbitrary constant in the form $\log C$. The general solution is thus $|(1 - y)(1 + x)| = C$ where, by virtue of the form of this expression, C is, of course, an essentially positive constant. Alternatively, the modulus signs may be removed and the general solution written as

$$y = 1 \pm \frac{C}{1 + x}.$$

In arriving at this solution we divided by the factor $(1 - y)$, which it was assumed was non-zero. To complete the solution we must now recognize that if this factor vanishes then the method of solution just outlined will fail. Clearly, when this happens we must also enquire whether the vanishing of the factor itself will give rise to a solution. Now the factor $(1 - y)$ vanishes when $y = 1$, and it is simple to substitute $y = 1$ into the original differential equation to verify that it is in fact a degenerate solution.

Example 14·2   Next let us solve the following equation, already expressed in differential form:

$$x \cos y \, dx - e^{-x} \sec y \, dy = 0.$$

Solution   This is of the form shown in Eqn (14·2) and, as implied by Eqn (14·4), after division by $e^{-x} \cos y$ the solution may be written

$$\int \sec^2 y \, dy = \int x e^x \, dx + C.$$

Performing the indicated integrations then gives

$$\tan y = e^x(x - 1) + C \qquad \text{or} \qquad y = \arctan[e^x(x - 1) + C].$$

Again we must check to see if the vanishing of the divisor gives rise to another solution. Here the divisor was $e^{-x} \cos y$, which only has zeros when $y = (2n + 1)\frac{1}{2}\pi$ with n an integer. Substitution of these values of y into the original differential equation shows that they are not solutions.

14·2 Homogeneous equations

A function $F(x, y)$ is said to be *homogeneous of degree n* in the algebraic sense if $F(kx, ky) = k^n F(x, y)$ for every real number k. Thus the function $F(x, y) = ax^2 + bxy + cy^2$ is homogeneous of degree 2, since $F(kx, ky) = k^2 F(x, y)$. If a change of variable $y = sx$ is made then any homogeneous function $F(x, y)$ of degree n may be written in the form $F(x, y) = F(x \times 1, x \times s) = x^n F(1, s)$, since here x is now cast in the role of k. This variable change has resulted in $F(x, y)$ being expressed as the product of x^n, which is only a function of x, and $F(1, s)$, which is only a function of s.

This facilitates the integration of a differential equation of the form

$$M(x, y)dx + N(x, y)dy = 0, \tag{14·5}$$

in which the functions $M(x, y)$ and $N(x, y)$ are both homogeneous functions of the same degree. Such equations are called *homogeneous equations*. 'Homogeneous' here refers to the algebraic property shared by M and N and is not to be confused with the same term introduced in Section 13·2.

The change of variable implies that $dy = s\,dx + x\,ds$, and if M and N are of degree n, then $M(x, y) = x^n M(1, s)$ and $N(x, y) = x^n N(1, s)$.

If, for simplification, we write $P(s) = M(1, s)$ and $Q(s) = N(1, s)$, Eqn (14·5) becomes

$$x^n P(s)dx + x^n Q(s)(s\,dx + x\,ds) = 0 \tag{14·6}$$

or, on cancelling x^n and rearranging,

$$[P(s) + s\,Q(s)]dx + xQ(s)ds = 0.$$

This equation is of the form variables separable and has the general solution

$$\int \frac{Q(s)ds}{P(s) + sQ(s)} = \log\left|\frac{C}{x}\right|, \tag{14·7}$$

where C is an arbitrary constant.

The final solution in terms of x and y may be obtained by using the relation $s = y/x$. The vanishing of the divisor x^n also implies a possible solution $x = 0$ which must be tested for validity against the original differential equation.

Example 14·3 Let us integrate the following homogeneous differential equation both directly and by an application of Eqn (14·7),

$$(y - x)dx - x\,dy = 0.$$

Solution

Direct method: The functions multiplying dx and dy are both of degree 1, so the equation is homogeneous. Setting $y = sx$ produces the equation

$$x(s - 1)dx - x(s\,dx + x\,ds) = 0.$$

Cancelling the factor x and rearranging we see that

$$\int ds = -\int \frac{dx}{x} + C,$$

whence

$$s = C - \log |x| \qquad \text{or} \qquad y = x(C - \log |x|).$$

The original equation shows that the vanishing of the cancelled factor x gives rise to an additional solution $x = 0$.

Solution by Equation (14·7):   Here $M(x, y) = y - x$ and $N(x, y) = -x$ so that $M(1, s) = s - 1$ and $N(1, s) = -1$. Hence $P(s) = s - 1$ and $Q(s) = -1$. Substituting in Eqn (14·7) gives

$$\int ds = \log \left| \frac{C}{x} \right| \qquad \text{or} \qquad s = \log \left| \frac{C}{x} \right|.$$

Again using $s = y/x$ we find that

$$y = x(\log |C| - \log |x|),$$

which agrees with the previous answer apart from the form of the arbitrary constant.

14·3   Exact equations

DEFINITION 14·1 (exact differential equation)   A differential equation of the form

$$M(x, y)dx + N(x, y)dy = 0 \tag{14·8}$$

with the property that $M(x, y)$ and $N(x, y)$ are related to a differentiable function $F(x, y)$ by the equations

$$M(x, y) = \frac{\partial F}{\partial x}, \qquad N(x, y) = \frac{\partial F}{\partial y}, \tag{14·9}$$

is said to be *exact*.

There are various ways of solving this simple but important type of differential equation. Here we choose to display the relationship of its solution to the familiar ideas of a total derivative and to the parametric representation of a variable y which is a function of x. Since the original differential equation (14·8) implies a functional relationship between x and y, we may suppose that the parametric representation $x = x(t)$, $y = y(t)$ describes the solution, where t is some parameter. Then the differentials dx, dy, and dt are related to the derivatives dx/dt and dy/dt by the expressions $dx = (dx/dt)dt$, $dy = (dy/dt)dt$. Using these results in Eqn (14.8), substituting for $M(x, y)$ and $N(x, y)$ from Eqn (14·9), and cancelling the differential dt, we find that

$$\frac{\partial F}{\partial x}\frac{dx}{dt} + \frac{\partial F}{\partial y}\frac{dy}{dt} = 0. \tag{14·10}$$

However, since the total derivative of $F[x(t), y(t)]$ with respect to t is

$$\frac{dF}{dt} = \frac{\partial F}{\partial x}\frac{dx}{dt} + \frac{\partial F}{\partial y}\frac{dy}{dt}, \tag{14·11}$$

Eqns (14·10) and (14·11) together imply that $dF/dt = 0$ or, equivalently, that $F(x, y) = $ constant, where now the parameter t may be disregarded. Thus we have established that the general solution of Eqns (14·8), (14·9) is

$$F(x, y) = C. \tag{14·12}$$

Now we already know that if $F(x, y)$ is a differentiable function with continuous first and second order derivatives then there must everywhere be equality between the mixed derivatives $\partial^2 F/\partial x \partial y$ and $\partial^2 F/\partial y \partial x$. Applying this argument to Eqns (14·9) shows that the condition for a differential equation of type (14·8) to be exact must be that everywhere $\partial M/\partial y = \partial N/\partial x$.

We have thus proved the following theorem, by means of which we may test a differential equation to discover if it is exact.

THEOREM 14·1 If the functions $M(x, y)$ and $N(x, y)$ and their derivatives $\partial M/\partial y$ and $\partial N/\partial x$ are continuous and, furthermore, $\partial M/\partial y = \partial N/\partial x$ everywhere, then the differential equation

$$M(x, y)dx + N(x, y)dy = 0$$

is exact.

Equations (14·9) also provide the means by which the function $F(x, y)$ may be obtained, since integrating the first equation with respect to x gives

$$F(x, y) = \int M(x, y)dx + A(y), \tag{14·13}$$

where $A(y)$ is some function of y and acts as though it were a constant as regards partial differentiation with respect to x. Thus $A(y)$ takes the place of the constant of integration that would occur were M to be only a function of x. The determination of $A(y)$ then follows immediately by differentiating Eqn (14·13) partially with respect to y, and using the second of Eqns (14·9) to determine dA/dy, from which $A(y)$ follows by integration in the form

$$A(y) = \int \left\{ N(x, y) - \frac{\partial}{\partial y} \left[\int M(x, y)dx \right] \right\} dy + C'. \tag{14·14}$$

Clearly, C' can only be an ordinary constant of integration and not a function of x since, by its manner of construction, the integrand of Eqn (14·14) is only a function of y so that to introduce a further function of x in place of C' would be inconsistent with the equation. This arbitrary constant C' can be combined with the constant C appearing in Eqn (14·12) so that there is in fact only one constant of integration appearing in the general solution, as would be expected with a first order equation.

Example 14·4 Let us solve the differential equation

$$[2x + 3 \cos y]dx + [2y - 3x \sin y]dy = 0.$$

The equation is exact since, as $M(x, y) = 2x + 3 \cos y$ and $N(x, y) = 2y - 3x \sin y$, it follows that $\partial M/\partial y = \partial N/\partial x = -3 \sin y$, satisfying the conditions of Theorem 14·1. Now as $\partial F/\partial x = 2x + 3 \cos y$,

$$F(x, y) = \int (2x + 3 \cos y)dx + A(y)$$

or

$$F(x, y) = x^2 + 3x \cos y + A(y).$$

Differentiating partially with respect to y we find that

$$\frac{\partial F}{\partial y} = -3x \sin y + \frac{dA}{dy}$$

which, since $\partial F/\partial y = N(x, y)$, is equivalent to $dA/dy = 2y$ and hence $A(y) = y^2 + C'$. The function $F(x, y)$ is thus

$$F(x, y) = x^2 + 3x \cos y + y^2 + C'$$

and so the general solution is

$$x^2 + 3x \cos y + y^2 = C.$$

If a particular solution is required then the value of the constant C must be determined by requiring the integral curve to pass through a specified point. For example, suppose that our equation is required to have the initial value $x = 0$, $y = \frac{1}{2}\pi$, the clearly $C = \frac{1}{4}\pi^2$ and the integral curve representing the solution is

$$x^2 + 3x \cos y + y^2 = \frac{1}{4}\pi^2.$$

Although not all differential equations of the form (14·8) are exact, it sometimes happens that they can be made exact by multiplying by some simple function $\mu(x, y)$, called an *integrating factor*, which is not identically zero. So, if $\mu(x, y)$ is an integrating factor for Eqn (14·8), then

$$\mu(x, y) M(x, y)dx + \mu(x, y) N(x, y)dy = 0 \qquad (14·15)$$

is exact, and by virtue of Theorem 14·1 the integrating factor must satisfy the equation

$$\frac{\partial}{\partial y} (\mu M) = \frac{\partial}{\partial x} (\mu N). \qquad (14·16)$$

It is possible to establish that an integrating factor always exists, provided only that M and N are differentiable functions, but unfortunately there is no general method by which it may be found.

Two special forms of differential equation in which an integrating factor may always be found have already been encountered in Eqns (14·2) and (14·5). The integrating factor for Eqn (14·2), in which the variables are separable, is $\mu(x, y) = 1/[R(x)Q(y)]$ and leads to the result (14·4). Similarly, the integrating factor for Eqn (14·5), with homogeneous coefficients, was $\mu(x, s) = 1/[P(s) + sQ(s)]$, where $s = y/x$ and gave rise to the solution shown in Eqn (14·7).

When the integrating factor is believed to be of a simple form in which only certain constants need to be determined, direct substitution in Eqn (14·16) would confirm this, and also indicate conditions by which these constants may be determined. The arguments used in this trial and error method will now be applied to a search for an integrating factor having the form $\mu(x, y) = x^m y^n$ in the following simple example.

Example 14·5 Solve the differential equation

$$(2xy + y^2)dx + (2x^2 + 3xy)dy = 0.$$

Solution First we notice that $\partial M/\partial y \neq \partial N/\partial x$, so the equation is not exact and an integrating factor μ is required. As M and N are simple algebraic functions we shall try an expression of the form $\mu(x, y) = x^m y^n$, in which the constants m and n must be determined so that

$$x^m y^n(2xy + y^2)dx + x^m y^n(2x^2 + 3xy)dy = 0$$

is exact. By condition (14·16) this implies that if $\mu(x, y) = x^m y^n$ is in actual fact an integrating factor, then m and n must be chosen so that

$$\frac{\partial}{\partial y} [x^m y^n(2xy + y^2)] = \frac{\partial}{\partial x} [x^m y^n(2x^2 + 3xy)].$$

This condition gives rise to the equation

$$nx^m y^{n-1}(2xy + y^2) + x^m y^n(2x + 2y) = mx^{m-1}y^n(2x^2 + 3xy)$$

$$+ x^m y^n(4x + 3y),$$

from which we must determine m and n if the chosen form of integrating factor is correct. Since this expression must be an identity, we now equate coefficients of terms of equal degrees in x and y and, if possible, select m and n such that all conditions are satisfied. In this case only two conditions arise:

(a) terms involving $x^m y^{n+1}$:

$$n + 2 = 3m + 3;$$

(b) terms involving $x^{m+1} y^n$:

$$2n + 2 = 2m + 4.$$

These conditions are satisfied if $m = 0$, $n = 1$, so an integrating factor of the type assumed does exist, and in this case $\mu = y$. The exact differential equation is thus

$$(2xy^2 + y^3)dx + (2x^2y + 3xy^2)dy = 0,$$

which is easily seen to have the general solution

$$x^2 y^2 + xy^3 = C.$$

When values of m and n cannot be found that will produce an identity from condition (14·16), then the integrating factor is not of the form $\mu(x, y) = x^m y^n$.

14·4 The linear equation of first order

The general linear equation of first order already encountered in Eqn (13·6) has the form

$$\frac{dy}{dx} + P(x)y = Q(x). \tag{14·17}$$

To solve it let us seek an integrating factor μ that will make $\mu(dy/dx) + \mu Py$ a derivative; namely the derivative $(d/dx)(\mu y)$. Since the equation is linear, μ must be independent of y, so we need only consider μ to be of the form $\mu = \mu(x)$. Thus the integrating factor μ is required to be a solution of the equation $(d/dx)(\mu y) = \mu(dy/dx) + \mu Py$. Expanding the left-hand side and simplifying then leads to the simple differential equation $y[(d\mu/dx) - P\mu] = 0$. As the solution y of Eqn (14·17) is not identically zero, it follows that μ must be the solution of

$$\frac{d\mu}{dx} = P(x)\mu. \tag{14·18}$$

The variables x and μ are separable, giving

$$\frac{d\mu}{\mu} = P(x)dx,$$

showing that $\log | \mu | = \int P(x)\mathrm{d}x + C'$, where C' is an arbitrary constant. Taking exponentials we find that the most general integrating factor is $\mu = e^{C'} \cdot e^{\int P(x)\mathrm{d}x}$.

However, as the arbitrary factor $e^{C'}$ is always non-zero, and so may be cancelled when this expression is used as a multiplier in Eqn (14·17), we may always take as the integrating factor the expression

$$\mu = e^{\int P(x)\mathrm{d}x}. \tag{14·19}$$

Multiplying Eqn (14·17) by μ and using its properties then gives

$$\frac{\mathrm{d}}{\mathrm{d}x} \left(y e^{\int P(x)\mathrm{d}x} \right) = Q(x) \, e^{\int P(x)\mathrm{d}x}.$$

After a final integration and simplification we obtain the general solution

$$y = e^{-\int P(x)\mathrm{d}x} \{ C + \int Q(x) \, e^{\int P(x)\mathrm{d}x} \, \mathrm{d}x \}, \tag{14·20}$$

where C is the arbitrary constant of the final integration and must be retained.

Although this general solution is useful, in applications it is usually better to use the fact that expression (14·19) is an integrating factor and to proceed directly from that point without recourse to Eqn (14·20).

An important general point illustrated by Eqn (14·20) which, as we shall see later, characterizes all linear equations is that the general solution comprises the sum of two parts. The first part, $Ce^{-\int P(x)\mathrm{d}x}$, is the solution of the homogeneous equation (corresponding to $Q(x) \equiv 0$) and the second part $e^{-\int P(x)\mathrm{d}x} \int Q(x) \, e^{\int P(x)\mathrm{d}x} \, \mathrm{d}x$ is particular to the form of the inhomogeneous term $Q(x)$. This second part is called the *particular integral*, whilst the first part is the *complementary function*. Notice that the two parts of the solution are additive and that the arbitrary constant is associated with the complementary function. These observations characterize linear differential equations of any order and will be encountered again in the next chapter.

Example 14·6 Solve

$$\frac{\mathrm{d}y}{\mathrm{d}x} + ky = a \sin mx$$

subject to the initial condition $y = 1$ when $x = 0$.

Solution In this case $P(x) = k$, so that the integrating factor $\mu = e^{kx}$. Hence

$$\frac{\mathrm{d}}{\mathrm{d}x} \left(y e^{kx} \right) = a e^{kx} \sin mx,$$

giving rise to

$$ye^{kx} = a \int e^{kx} \sin mx \, dx + C$$

or,

$$y = e^{-kx} \left(C + a \int e^{kx} \sin mx \, dx \right).$$

Performing the indicated integration gives the general solution

$$y = Ce^{-kx} + \frac{a}{k^2 + m^2} (k \sin mx - m \cos mx),$$

the first term being the complementary function and the second term the particular integral.

To determine the constant C we now utilize the initial conditions $y(0) = 1$ by writing

$$1 = C - \frac{am}{k^2 + m^2}.$$

The particular solution is thus

$$y = \left(\frac{k^2 + m^2 + am}{k^2 + m^2} \right) e^{-kx} + \frac{a}{k^2 + m^2} (k \sin mx - m \cos mx).$$

If we make the identifications $y \equiv i$, $x \equiv t$, $k \equiv R/L$, $a \equiv V_0/L$, and $m \equiv \omega$, we discover that we have just obtained the solution of Example 13·3 (concerning the electric circuit), subject to the initial condition that a unit current flows when the circuit is closed.

Example 14·7   Find the general solution of the linear equation

$$\frac{dy}{dx} + xy = x.$$

Solution   The integrating factor $\mu = e^{\int x \, dx}$ so that $\mu = e^{x^2/2}$. Hence

$$\frac{d}{dx}(ye^{x^2/2}) = xe^{x^2/2}$$

or

$$ye^{x^2/2} = C + \int xe^{x^2/2} \, dx.$$

Since the indefinite integral on the right-hand side is $e^{x^2/2}$, it follows that

$$y = Ce^{-x^2/2} + 1.$$

The particular integral in this case is simply the constant unity.

PROBLEMS

Section 14·1

14·1 Find the general solution of the following problems by separating the variables:

(a) $\tan x \sin^2 y \, dx + \cos^2 x \cot y \, dy = 0$;

(b) $xy' - y = y^3$;

(c) $xyy' = 1 - x^2$;

(d) $y - xy' = a(1 + x^2y')$;

(e) $3e^x \tan y \, dx + (1 - e^x) \sec^2 y \, dy = 0$;

(f) $y' \tan x = y$.

Section 14·2

14·2 Find the solutions of the following homogeneous equations:

(a) $(x - y)y \, dx - x^2 \, dy = 0$ if initially $x = 1$, $y = 2$;

(b) $(x^2 + y^2)dx - 2xy \, dy = 0$;

(c) $(x^2 - 3y^2)dx + 2xy \, dy = 0$;

(d) $\left(y + x \sinh^2 \dfrac{y}{x} \right) dx - x \, dy = 0$.

Section 14·3

14·3 Show that the stated functions μ are integrating factors for their associated differential equations and find the general solutions:

(a) $\left(\dfrac{2x + y + 3}{x} \right) dx + \left(\dfrac{x + 2y + 3}{y} \right) dy = 0 \quad (\mu = xy)$;

(b) $y \, dx + x \, dy + \left(\dfrac{2x}{1 + \cos xy} \right) dx = 0 \quad (\mu = 1 + \cos xy)$;

(c) $(ye^{-y/2} + e^{y/2})dx + 2x \cosh \tfrac{1}{2}y \, dy \quad (\mu = e^{y/2})$.

14·4 Determine if $\mu = x^m y^n$ is an integrating factor for the following differential equations and, if so, deduce appropriate values for m and n. Use your results to determine the solution in each case:

(a) $(9y^2 + 4xy^2 + 3y)dx + (6xy + 2x^2y + x)dy = 0$;

(b) $y(2x + \cosh x)dx + 2(x^2 + \sinh x)dy = 0$;

(c) $(x + y)dx + x(x + 3y)dy = 0$;

(d) $2e^y dx + \left(xe^y + \dfrac{2y}{x} \right) dy = 0$ if initially $x = 1$, $y = 0$;

(e) $y^2 \, dx + (x + \cos y)dy = 0$;

(f) $y(2 + 3xy^2)dx + 2x(1 + 2xy^2)dy = 0$.

Section 14·4

14·5 Find the general solutions of the following differential equations, and when initial conditions are given determine the particular solution appropriate to them:

(a) $\dfrac{dy}{dx} - \dfrac{y}{x} = 2x$;

(b) $\dfrac{dy}{dx} + \dfrac{2y}{x} = x^3$ $(y = 0,\ x = 1)$;

(c) $\dfrac{dy}{dx} - \dfrac{2}{x+1} y = 3(x+1)^3$;

(d) $xy' + y - e^x = 0$ $(y = 1,\ x = 2)$;

(e) $y' - y \tan x = \sec x$ $(y = 1,\ x = 0)$;

(f) $y\,dx - [y^2 + (y+1)x]dy = 0$;

(g) $y' + xy = xy^2$;

(h) $y' + \dfrac{y}{x} = xy^2 \sin x$.

14·6 Show that the solution of the homogeneous form of the differential equation

$$\frac{dy}{dx} + P(x)y = Q(x)$$

is

$$y = Ae^{\int P(x)dx}.$$

If A is now regarded as a function of x show, by substituting back into the inhomogeneous equation, that $A(x)$ is the solution of the equation

$$\frac{dA}{dx} = Q(x)e^{\int P(x)dx},$$

in which the variables are separable. Integrate this result and find the general solution of the original differential equation.

14·7 By applying the previous method to the differential equation

$$\frac{dy}{dx} + \frac{y}{x} = e^x,$$

show that $A(x) = C + e^x(x - 1)$ and hence find the solution that satisfies the initial condition $y = 1$ when $x = 1$.

15 Higher order differential equations

15·1 Linear equations with constant coefficients—homogeneous case

Thus far our encounter with differential equations of order greater than unity has been essentially confined to the second order equation which served to introduce the sine and cosine functions in Chapter 6. We now take up the solution of higher order differential equations in more general terms and, amongst other matters, extend systematically the notion of a complementary function and a particular integral first introduced in connection with a linear first order differential equation in Chapter 14.

Differential equations of the form

$$\frac{d^n y}{dx^n} + a_1 \frac{d^{n-1} y}{dx^{n-1}} + \cdots + a_n y = f(x), \tag{15·1}$$

in which $a_1, a_2, \ldots, a_n$ are constants (usually real), are called *linear constant coefficient* equations. We begin by studying the homogeneous form of the equation (that is, $f(x) = 0$) which in this context is usually called the *reduced* equation:

$$\frac{d^n y}{dx^n} + a_1 \frac{d^{n-1} y}{dx^{n-1}} + \cdots + a_n y = 0. \tag{15·2}$$

Equation (15·2) may be solved by using the fact that $y = Ce^{\lambda x}$ is obviously a solution for arbitrary constant C, provided λ satisfies the equation that results when this expression is substituted into Eqn (15·2) giving

$$(\lambda^n + a_1\lambda^{n-1} + \cdots + a_n)e^{\lambda x} = 0. \tag{15·3}$$

The substitution $y = Ce^{\lambda x}$ has thus associated a characteristic polynomial in λ,

$$P(\lambda) \equiv \lambda^n + a_1\lambda^{n-1} + \cdots + a_n, \tag{15·4}$$

with the differential Eqn (15·1) and, as $e^{\lambda x} \neq 0$, it follows that the permissible values of λ in the solutions $y = Ce^{\lambda x}$ must be the roots of $P(\lambda) = 0$. The values of λ are thus determined by solving the characteristic equation

$$P(\lambda) \equiv \lambda^n + a_1\lambda^{n-1} + \cdots + a_n = 0. \tag{15·5}$$

This equation is also known as the *indicial* or *auxiliary* equation.

Since $P(\lambda)$ is a polynomial of degree n, it follows that $P(\lambda) = 0$ has precisely n roots $\lambda = \lambda_i$ and, furthermore, we know that when the coefficients $a_1, a_2, \ldots, a_n$ are real; these roots must therefore either be real or must occur in complex conjugate pairs. Each expression $y_i = C_i e^{\lambda_i x}$, $i = 1, 2, \ldots, n$ is a solution of Eqn (15·2) for arbitrary C_i, and direct substitution into Eqn (15·2) verifies that

$$y = C_1 e^{\lambda_1 x} + C_2 e^{\lambda_2 x} + \cdots + C_n e^{\lambda_n x} \tag{15·6}$$

is also a solution. The additive property of solutions of higher order linear differential equations expressed by this result is often referred to as the *linear superposition* of solutions.

This solution of the reduced Eqn (15·2) is called the *complementary function* and if the λ_i are distinct, the n individual solutions y_i are linearly independent. We shall now prove this assertion of linear independence indirectly, by assuming the result to be invalid and producing a contradiction, thereby showing that the assumption of linear dependence is incorrect. This contradiction proves the result.

Consider the case in which the roots λ_i are all distinct, but assume that the y_i are linearly dependent. Then there must exist some constants C_1, $C_2, \ldots, C_n$, not all zero, such that

$$C_1 e^{\lambda_1 x} + C_2 e^{\lambda_2 x} + \cdots + C_n e^{\lambda_n x} = 0. \tag{15·7}$$

Successive differentiation of expression Eqn (15·7) shows that also

$$\left. \begin{array}{l} C_1 \lambda_1 e^{\lambda_1 x} + C_2 \lambda_2 e^{\lambda_2 x} + \cdots + C_n \lambda_n e^{\lambda_n x} = 0, \\ \cdot \quad \cdot \quad \cdot \quad \cdot \quad \cdot \quad \cdot \quad \cdot \quad \cdot \quad \cdot \quad \cdot \\ C_1 \lambda_1^{n-1} e^{\lambda_1 x} + C_2 \lambda_2^{n-1} e^{\lambda_2 x} + \cdots + C_n \lambda_n^{n-1} e^{\lambda_n x} = 0. \end{array} \right\} \tag{15·8}$$

Now if Eqns (15·7) and (15·8) are to be true for a non-trivial set of constants C_i, the determinant $|W|$ of the coefficients of the C_i must vanish for all x, thereby giving rise to the condition

$$|W| \equiv \begin{vmatrix} e^{\lambda_1 x} & e^{\lambda_2 x} & \cdots & e^{\lambda_n x} \\ \lambda_1 e^{\lambda_1 x} & \lambda_2 e^{\lambda_2 x} & \cdots & \lambda_n e^{\lambda_n x} \\ \cdot & \cdot & \cdot & \cdot \\ \lambda_1^{n-1} e^{\lambda_1 x} & \lambda_2^{n-1} e^{\lambda_2 x} & \cdots & \lambda_n^{n-1} e^{\lambda_n x} \end{vmatrix} = 0. \tag{15·9}$$

The determinant $|W|$ formed in this manner is known as the *Wronskian* of the n solutions y_i, and plays an important role in more general studies of differential equations. In this case $|W|$ has a simple form and, as the common exponential factor in each column is non-zero, it follows that these may be removed as factors of $|W|$, showing that condition Eqn (15·9) implies the vanishing of an alternant determinant $|A|$ (see Problem 9·20), where

$$|A| \equiv \begin{vmatrix} 1 & 1 & \cdots & 1 \\ \lambda_1 & \lambda_2 & \cdots & \lambda_n \\ \cdot & \cdot & \cdot & \cdot \\ \lambda_1^{n-1} & \lambda_2^{n-1} & \cdots & \lambda_n^{n-1} \end{vmatrix} = 0. \tag{15·10}$$

Now we know from the theory of determinants that the value of $|A|$ is simply the product of all possible factors of the form $(\lambda_i - \lambda_j)$ with a suitable sign appended, so that if the roots λ_i are all distinct $|A|$ cannot vanish. This contradiction thus proves that Eqn (15·7) can be true only if all the C_i are zero, and hence the n solutions y_i must be linearly independent. We have thus proved:

THEOREM 15·1 (linear independence of solutions—single roots)   A differential equation

$$y^{(n)} + a_1 y^{(n-1)} + \cdots + a_n y = 0$$

which has n distinct roots λ_i of its characteristic equation $P(\lambda) = 0$ has n linearly independent solutions $y_i = C_i e^{\lambda_i x}$. Its general solution, the complementary function, is of the form

$$y = C_1 e^{\lambda_1 x} + C_2 e^{\lambda_2 x} + \cdots + C_n e^{\lambda_n x}.$$

Example 15·1   Suppose that $y'' + 3y' + 2y = 0$, then $P(\lambda) \equiv \lambda^2 + 3\lambda + 2$ and the roots of $P(\lambda) = 0$ are $\lambda = -1$, $\lambda = -2$. The linearly independent solutions are $y_1 = C_1 e^{-x}$ and $y_2 = C_2 e^{-2x}$, and the general solution or complementary function is $y = C_1 e^{-x} + C_2 e^{-2x}$. A simple calculation shows that the Wronskian $|W| = -e^{-3x}$.

When r of the roots of the characteristic polynomial $P(\lambda)$ coincide and equal λ^*, say, then $\lambda = \lambda^*$ is said to be a root of *multiplicity* r. The form of the general solution Eqn (15·6) is then inapplicable because r of its terms are linearly dependent. In this situation an additional $(r - 1)$ linearly independent solutions need to be determined to complete the general solution.

THEOREM 15·2 (linear independence of solutions—repeated roots)   When $\lambda = \lambda_1$ is a root of multiplicity r of the characteristic equation $P(\lambda) = 0$ belonging to

$$y^{(n)} + a_1 y^{(n-1)} + \cdots + a_n y = 0,$$

then $e^{\lambda_1 x}$, $xe^{\lambda_1 x}$, $x^2 e^{\lambda_1 x}$, . . ., $x^{r-1} e^{\lambda_1 x}$ are linearly independent solutions of the differential equation corresponding to the r-fold root λ_1.

Proof   Because the stated form of the linearly independent solutions may be established more easily by a different technique, which we shall discuss later, we only prove the result for a root of multiplicity 2 (a double root). The assertion of linear independence, however, will be proved for any value of r.

When $\lambda = \lambda_1$ is a root of multiplicity 2 the characteristic polynomial $P(\lambda)$ may be written

$$P(\lambda) \equiv \lambda^n + a_1\lambda^{n-1} + \cdots + a_{n-1}\lambda + a_n = (\lambda - \lambda_1)^2 Q(\lambda), \qquad (15\cdot11)$$

where $Q(\lambda)$ is a polynomial of degree $(n - 2)$. Clearly, by definition, $P(\lambda_1) = 0$, and by differentiation of $P(\lambda) = 0$ with respect to λ it also follows that

$$n\lambda_1^{n-1} + (n - 1)a_1\lambda_1^{n-2} + \cdots + a_{n-1} = 0, \qquad (15\cdot12)$$

which is a result that will be needed shortly.

If now we write $y = xe^{\lambda_1 x}$, then $y^{(r)} = r\lambda_1^{r-1}e^{\lambda_1 x} + \lambda_1^r xe^{\lambda_1 x}$ and, as $P(\lambda_1) = 0$, substitution into $y^{(n)} + a_1 y^{(n-1)} + \cdots + a_n y$ gives $(n\lambda_1^{n-1} + (n - 1)a_1\lambda_1^{n-2} + \cdots + a_{n-1})e^{\lambda_1 x}$. This vanishes by virtue of Eqn (15·12), so we have shown that $y = xe^{\lambda_1 x}$ is actually a solution of differential Eqn (15·2) when $\lambda = \lambda_1$ is a double root. As $y = e^{\lambda_1 x}$ is obviously also a solution, we have established that when $\lambda = \lambda_1$ is a double root, the general solution must take the form

$$y = C_1 e^{\lambda_1 x} + C_2 xe^{\lambda_1 x} + C_3 e^{\lambda_3 x} + \cdots + C_n e^{\lambda_n x}, \qquad (15\cdot13)$$

where the remaining $(n - 2)$ roots $\lambda_3, \lambda_4, \ldots, \lambda_n$ are assumed to be distinct.

Whether or not the λ_i are multiple roots, n initial conditions must be specified in order to construct a particular solution from the appropriate form of the general solution. Used in conjunction with the general solution they enable n simultaneous equations to be formed for the determination of the n arbitrary constants $C_1, C_2, \ldots, C_n$. The usual initial conditions for an nth order differential equation are the specification of the values of y, $y^{(1)}$, $y^{(2)}, \ldots, y^{(n-1)}$ at some initial point $x = x_0$.

Now in connection with the Wronskian, we have already seen that a set of exponential functions is linearly independent provided the exponents are distinct. Hence, to show that the functions in Eqn (15·13) are linearly independent, it will be sufficient to show that $e^{\lambda_1 x}$ and $xe^{\lambda_1 x}$ are linearly independent. This result is self evident, because removal of the common factor $e^{\lambda_1 x}$ leaves the functions 1 and x, which are obviously linearly independent.

For completeness, and for application to roots of multiplicity greater than 2, we prove the more general result that the functions $1, x, x^2, \ldots, x^m$ are linearly independent.

Assuming first that this is not true and that these functions are linearly dependent, it follows that there must exist a non-trivial set of constants $C_0, C_1, \ldots, C_m$ such that, for all x,

$$C_0 + C_1 x + C_2 x^2 + \cdots + C_m x^m = 0.$$

However, we know that as this is an algebraic equation of degree m, there can at most be only m distinct values of x for which it can be true. The expression cannot thus be true for all x, and so the assumption of linear dependence is false. This establishes the result.

Example 15·2 Suppose $y''' + 4y'' + 5y' + 2y = 0$, then $P(\lambda) \equiv \lambda^3 + 4\lambda^2 + 5\lambda + 2 = (\lambda + 1)^2(\lambda + 2)$ and the roots of $P(\lambda) = 0$ are $\lambda = -1$, $\lambda = -1$, and $\lambda = -2$. The root $\lambda = -1$ has multiplicity 2 and so is a double root. The linearly independent solutions corresponding to $\lambda = -1$ are $y_1 = C_1 e^{-x}$ and $y_2 = C_2 x e^{-x}$, the remaining linearly independent solution being $y_3 = C_3 e^{-2x}$. The general solution is $y = C_1 e^{-x} + C_2 x e^{-x} + C_3 e^{-2x}$. To determine the particular solution appropriate to, say, the initial conditions $y = 1$, $y' = 0$, $y'' = 1$ when $x = 0$ we proceed as follows. From the general solution we find by differentiation that $y' = -C_1 e^{-x} + C_2(1 - x)e^{-x} - 2C_3 e^{-2x}$ and $y'' = C_1 e^{-x} - C_2(2 - x)e^{-x} + 4C_3 e^{-2x}$. Substituting the initial conditions gives rise to the three simultaneous equations $1 = C_1 + C_3$, $0 = -C_1 + C_2 - 2C_3$, and $1 = C_1 - 2C_2 + 4C_3$, which have as their solution $C_1 = -1$, $C_2 = 3$, $C_3 = 2$. Hence the required particular solution is $y = (3x - 1)e^{-x} + 2e^{-2x}$.

Example 15·3 Suppose $y^{(5)} + 3y^{(4)} - y^{(3)} - 7y^{(1)} + 4y = 0$, then $P(\lambda) \equiv \lambda^5 + 3\lambda^4 - \lambda^3 - 7\lambda + 4 = (\lambda - 1)^2(\lambda + 2)^2(\lambda + 1)$ and the roots of $P(\lambda) = 0$ are $\lambda = 1$, $\lambda = 1$, $\lambda = -2$, $\lambda = -2$, and $\lambda = -1$. The roots $\lambda = 1$ and $\lambda = -2$ are double roots and the root $\lambda = -1$ is a single root. The general solution is $y = C_1 e^x + C_2 x e^x + C_3 e^{-2x} + C_4 x e^{-2x} + C_5 e^{-x}$.

Example 15·4 Suppose $y^{(5)} - 3y^{(4)} + 3y^{(3)} - y^{(2)} = 0$, then $P(\lambda) \equiv \lambda^5 - 3\lambda^4 + 3\lambda^3 - \lambda^2 = \lambda^2(\lambda - 1)^3$, and the roots of $P(\lambda) = 0$ are $\lambda = 0$, $\lambda = 0$, $\lambda = 1$, $\lambda = 1$, and $\lambda = 1$. The root $\lambda = 0$ is a double root and the root $\lambda = 1$ is a triple root. The general solution is $y = C_1 + C_2 x + C_3 e^x + C_4 x e^x + C_5 x^2 e^x$. Here the terms C_1 and $C_2 x$ are the linearly independent solutions corresponding to the double root $\lambda = 0$.

Finally we must give consideration to the situation in which the roots λ_i occur in complex conjugate pairs. Suppose that $\lambda_s = \mu + i\nu$ and its complex conjugate $\bar{\lambda}_s = \mu - i\nu$ are roots of the characteristic polynomial $P(\lambda) = 0$ of Eqn (15·2). Then, by analogy with the case of real roots, $y_s = \exp[(\mu + i\nu)x]$ and $y_s^* = \exp[(\mu - i\nu)x]$ must be solutions of Eqn (15·2). Linear combinations of y_s and y_s^* will also be solutions of Eqn (15·2) and so, in particular, $u = \frac{1}{2}(y_s + y_s^*)$ and $v = (1/2i)(y_s - y_s^*)$ will be solutions. A simple calculation then shows that $u = e^{\mu x} \cos \nu x$ and $v = e^{\mu x} \sin \nu x$. Hence the combination of terms corresponding to the complex root λ_s and its complex conjugate $\bar{\lambda}_s$ in the general solution gives rise to the solution

$$e^{\mu x}(A \cos \nu x + B \sin \nu x),$$

where A and B are arbitrary real constants. We have established the following general result:

THEOREM 15·3 (form of solution—complex roots)　When $\lambda = \mu + i\nu$ and its complex conjugate $\bar{\lambda}$ are single roots of the characteristic equation $P(\lambda) = 0$ of the differential equation

$$y^{(n)} + a_1 y^{(n-1)} + \cdots + a_n y = 0,$$

and the remaining roots $\lambda_3, \lambda_4, \ldots, \lambda_n$ are real and distinct, the general solution has the form

$$y = e^{\mu x}(C_1 \cos \nu x + C_2 \sin \nu x) + C_3 e^{\lambda_3 x} + C_4 e^{\lambda_4 x} + \cdots + C_n e^{\lambda_n x}.$$

The extension of Theorem 15·3 when other complex conjugate pairs of roots occur in the characteristic polynomial, or when there are multiple real roots, is obvious and immediate. When a complex root has multiplicity greater than 1, the results of Theorem 15·2 may be incorporated into Theorem 15·3 to modify the constants C_1 and C_2. Thus, for example, if complex root λ_s had multiplicity 2, the general solution stated in Theorem 15·3 would take the form

$$y = e^{\mu x}\{(C_1 + C_2 x) \cos \nu x + (C_3 + C_4 x) \sin \nu x\} + C_5 e^{\lambda_5 x}$$
$$+ C_6 e^{\lambda_6 x} + \cdots + C_n e^{\lambda_n x}.$$

Example 15·5　The differential equation $y'' + 4y' + 13y = 0$ has the characteristic polynomial $P(\lambda) \equiv \lambda^2 + 4\lambda + 13 = (\lambda + 2 + 3i)(\lambda + 2 - 3i)$ and the roots of $P(\lambda) = 0$ are $\lambda = -2 - 3i$ and $\lambda = -2 + 3i$. The general solution is $y = e^{-2x}(C_1 \cos 3x + C_2 \sin 3x)$.

Example 15·6　The differential equation $y^{(5)} + 3y^{(4)} + 10y^{(3)} + 6y^{(2)} + 5y^{(1)} - 25y = 0$ has the characteristic polynomial $P(\lambda) \equiv \lambda^5 + 3\lambda^4 + 10\lambda^3 + 6\lambda^2 + 5\lambda - 25 = (\lambda - 1)(\lambda + 1 + 2i)^2(\lambda + 1 - 2i)^2$. The complex roots $\lambda = -1 - 2i$ and $\lambda = -1 + 2i$ of $P(\lambda) = 0$ are double roots, and the single root $\lambda = 1$ is the only real root. The general solution is $y = e^{-x}[(C_1 + C_2 x) \cos 2x + (C_3 + C_4 x) \sin 2x] + C_5 e^x$.

15·2　Linear equations with constant coefficients— inhomogeneous case

We now examine methods of solution of the inhomogeneous differential Eqn (15·1). Our approach will be to progress from a semi-intuitive method known as the method of undetermined coefficients through a rather more systematic treatment using the operator D, which will be introduced later, and thence to the method of variation of parameters. To complete the chapter, a brief introduction is given to the solution of linear differential equations by means of the Laplace transform.

It is an easily verified fact that $y = C_1 \cos x + C_2 \sin x + \frac{1}{2}e^{-x}$ is a solution of the inhomogeneous equation $y'' + y = e^{-x}$. The first two terms of this solution obviously comprise the complementary function of the reduced equation $y'' + y = 0$, whilst the last term is a function which, when substituted into the differential equation, gives rise to the inhomogeneous term. There thus appear to be two distinct parts to this solution, the first being the general solution to the reduced equation and the second, which is additive, being a solution particular to the form of the inhomogeneous term. We now prove a theorem that establishes that this is in fact the pattern of solution that applies to all inhomogeneous linear equations. The sum of the two parts is termed the *general solution* or the *complete primitive* of the inhomogeneous equation.

To simplify manipulation it will be convenient to introduce a concise notation for the left-hand side of differential Eqn (15·1) and we achieve this by defining $L[y] \equiv y^{(n)} + a_1 y^{(n-1)} + \cdots + a_n y$. In terms of this notation, in which $a_1, a_2, \ldots, a_n$ are understood to be constants, we now state:

THEOREM 15·4 (form of general solution of linear inhomogeneous equations) The general solution of the inhomogeneous equation $L[y] = f(x)$ is of the form $y(x) = y_c(x) + y_p(x)$, where $y_c(x)$ is the general solution or complementary function of the reduced equation $L[y] = 0$, and $y_p(x)$ is a particular solution of $L[y] = f(x)$.

Proof The proof is straightforward. Firstly, $y(x) = y_c(x) + y_p(x)$ does satisfy the equation since $L[y_c(x)] = 0$ and $L[y_p(x)] = f(x)$, and as $(d^r/dx^r)(y_c + y_p) = (d^r y_c/dx^r) + (d^r y_p/dx^r)$ for $r = 1, 2, \ldots, n$, it follows that

$$L(y) = L[y_c(x) + y_p(x)] = L[y_c(x)] + L[y_p(x)]$$
$$= 0 + f(x) = f(x).$$

As $y_c(x)$ contains n arbitrary constants we choose to write it in the form $y_c(x; C_1, C_2, \ldots, C_n)$ to make this explicit. Then, clearly, adding two complementary functions with differing constants C_i and C_i' gives

$$y_c(x; C_1, C_2, \ldots, C_n) + y_c(x; C_1', C_2', \ldots, C_n')$$
$$= y_c(x; C_1 + C_1', C_2 + C_2', \ldots, C_n + C_n'),$$

which is simply the same form of complementary function but with modified constants. Suppose next that $y_{1p}(x)$ and $y_{2p}(x)$ are two particular solutions of $L[y] = f(x)$. Then $L[y_{1p}(x) - y_{2p}(x)] = L[y_{1p}(x)] - L[y_{2p}(x)] = f(x) - f(x) = 0$ and so $y_{1p}(x) - y_{2p}(x)$ is a solution of $L[y] = 0$. Hence $y_{1p}(x) - y_{2p}(x) = y_c(x; C_1^0, C_2^0, \ldots, C_n^0)$, which is again the same form of complementary function but with some other set of constants.

Now

$$y_c(x; C_1, C_2, \ldots, C_n) + y_{1p}(x) = y_c(x; C_1, C_2, \ldots, C_n) + y_{1p}(x)$$
$$- y_{2p}(x) + y_{2p}(x)$$
$$= y_c(x; C_1, C_2, \ldots, C_n)$$
$$+ y_c(x; C_1{}^0, C_2{}^0, \ldots, C_n{}^0)$$
$$+ y_{2p}(x)$$
$$= y_c(x; C_1 + C_1{}^0, C_2 + C_2{}^0, \ldots,$$
$$C_n + C_n{}^0) + y_{2p}(x),$$

and so we have shown that any two particular solutions give rise to the same form of general solution. The arbitrary constants in this general solution must be determined by applying the initial conditions to the complete primitive $y = y_c(x) + y_p(x)$.

We have already seen the simplest example of this theorem in connection with Eqn (14·20), which clearly displays the two parts of the general solution of an inhomogeneous linear equation of first order. As in that section, we shall call the particular solution $y_p(x)$ of the inhomogeneous equation a *particular integral* of the differential equation.

15·2 (a)   The method of undetermined coefficients

The determination of simple particular integrals by means of the *method of undertermined coefficients* is best illustrated by example. In essence, the method is based on the fact that simple forms of inhomogeneous term $f(x)$ in Eqn (15·1) can only arise as the result of differentiation of obvious functions in which the values of certain constants are the only things that need determination. A solution is achieved by substitution of a trial function into the inhomogeneous equation and subsequent comparison of coefficients of corresponding terms.

Case (a): $f(x)$ *a polynomial in* x.   Suppose $y'' + y = x^2$, then by inspection we see that the particular integral y_p can only be a polynomial in x and, furthermore, that it cannot be of degree higher than 2 since the equation contains an undifferentiated term y. Let us set $y_p = ax^2 + bx + c$, then $y_p'' = 2a$ and substitution into the original equation gives $2a + ax^2 + bx + c = x^2$. Equating the coefficients of corresponding powers of x shows that $a = 1$, $b = 0$, $2a + c = 0$, so that $c = -2$, and the required particular integral must be $y_p = x^2 - 2$. The general solution, or complete primitive, is $y = C_1 \cos x + C_2 \sin x + x^2 - 2$. To determine the solution appropriate to the initial conditions $y = -2$, $y' = 0$, say, at $x = \frac{1}{2}\pi$ we notice that the condition on y gives $-2 = C_1 \cos \frac{1}{2}\pi + C_2 \sin \frac{1}{2}\pi + \frac{1}{4}\pi^2 - 2$, or $C_2 = -\frac{1}{4}\pi^2$, whilst the condition on y' gives $0 = -C_1 \sin \frac{1}{2}\pi + C_2 \cos \frac{1}{2}\pi + \pi$, or

$C_1 = \pi$. Hence the solution appropriate to these initial conditions is $y = \pi(\cos x - \frac{1}{4}\pi \sin x) + x^2 - 2$.

The method extends directly to linear constant coefficient equations of any order and to polynomials $f(x)$ of any degree. For example, using the same argument to determine the particular integral of $y'' + 3y' + y = 1 + x^3$ we would try a particular integral of the form $y_p = ax^3 + bx^2 + cx + d$. Substitution in the equation and comparison of the coefficients of corresponding powers of x would then show that $a = 1$, $9a + b = 0$, $6a + 6b + c = 0$, and $2b + 3c + d = 1$. These equations have the solution $a = 1$, $b = -9$, $c = 48$, and $d = -125$. The particular integral must be $y_p = x^3 - 9x^2 + 48x - 125$.

Case (b): $f(x)$ an exponential function.   Suppose now that $y'' + 3y' + 2y = 3e^{2x}$. As e^{2x} does not appear in the complementary function $y_c = C_1 e^{-x} + C_2 e^{-2x}$ (in which case it would be a solution of the homogeneous equation), it can only arise in the inhomogeneous term as a result of differentiation of a function of the form $y_p = ke^{2x}$. Substituting this in the equation and cancelling the common factor e^{2x} shows that $4k + 6k + 2k = 3$ or $k = \frac{1}{4}$. The required particular integral must thus be $y_p = \frac{1}{4}e^{2x}$. A sum of exponentials occurring in the inhomogeneous term would be treated analogously, the constant multiplier of each being determined separately by the above method.

A complication arises if an exponential in the inhomogeneous term also occurs in the complementary function, as is the case with $y'' + y' - 2y = 2e^x$, which has the complementary function $y_c = C_1 e^x + C_2 e^{-2x}$. Attempting to find a solution by substituting $y_p = ke^x$ would fail here since e^x is a solution of $y'' + y' - 2y = 0$. A moment's reflection and consideration along lines similar to those concerning Theorem 15·2 shows that in this case we must try $y_p = kxe^x$. Then $y_p' = k(1 + x)e^x$ and $y_p'' = k(2 + x)e^x$, so that substitution into the differential equation and cancellation of the common factor e^x gives the condition $3k = 2$. In this case the required particular integral is $y_p = \frac{2}{3}xe^x$. By an obvious extension, if an exponential term $e^{\lambda x}$ appears in the inhomogeneous term $f(x)$, and also occurs in the complementary function as a result of a root of the characteristic equation which has multiplicity r, then the particular integral will be of the form $y_p = kx^r e^{\lambda x}$. Suppose, for example, that $y'' - 2y' + y = 2e^x$, then the complementary function $y_c = (C_1 + C_2 x)e^x$ arises as a result of the double root $\lambda = 1$ of the characteristic equation $P(\lambda) \equiv \lambda^2 - 2\lambda + 1 = 0$. Hence we must seek a particular integral of the form $y_p = kx^2 e^x$. Substituting this in the equation and cancelling the common factor e^x shows that $k = 1$; hence the particular integral is $y_p = x^2 e^x$. The general solution is $y = (C_1 + C_2 x)e^x + x^2 e^x$.

Case (c): $f(x)$ a trigonometric function.   The same method may be applied to an inhomogeneous trigonometric term involving $\sin mx$ or $\cos mx$. In this case, provided neither function occurs solely as a term in the complementary function, a particular integral of the form $y_p = a \cos mx + b \sin mx$ must be sought. Let us consider $y'' - 2y' + 2y = \sin x$, which has the complementary function $y_c = e^x(C_1 \cos x + C_2 \sin x)$. Substituting $y_p = a \cos x + b \sin x$ gives the result $-a \cos x - b \sin x + 2a \sin x - 2b \cos x + 2a \cos x + 2b \sin x = \sin x$. For this to be an identity we must equate the coefficients of $\sin x$ and $\cos x$ on each side of the equation. For the term $\sin x$ this gives rise to the equation $2a + b = 1$, and for the term $\cos x$, the equation $a - 2b = 0$. Hence $a = 2/5$, $b = 1/5$, so that $y_p = \frac{1}{5}(2 \cos x + \sin x)$. The general solution is of course $y = e^x(C_1 \cos x + C_2 \sin x) + \frac{1}{5}(2 \cos x + \sin x)$.

If, however, the trigonometric function in the inhomogeneous term occurs as part of the complementary function then considerations similar to those in the latter part of Case (b) will apply. We must then try to find a particular integral of the form $y_p = x^r(a \cos mx + b \sin mx)$, where r is the multiplicity of the root of the characteristic polynomial giving rise to the term $\cos mx$ or $\sin mx$ in the complementary function.

Suppose, for example, that $y'' + y = \cos x$, then the characteristic polynomial is $P(\lambda) \equiv \lambda^2 + 1 = (\lambda - i)(\lambda + i)$, showing that terms in the complementary function $y_c = C_1 \cos x + C_2 \sin x$ arise from roots of $P(\lambda) = 0$ having multiplicity 1. We must thus attempt a particular solution of the form $y_p = ax \cos x + bx \sin x$. Substitution into the equation gives $-2a \sin x + 2b \cos x - ax \cos x - bx \sin x + ax \cos x + bx \sin x = \cos x$. Equating coefficients of $\cos x$ and $\sin x$ as before shows that $a = 0$, $b = \frac{1}{2}$, and hence $y_p = \frac{1}{2}x \sin x$.

Case (d): $f(x)$ a product of exponential and trigonometric functions. The previous methods extend to allow the determination of a particular integral when the inhomogeneous term is of the form $e^{kx} \cos mx$ or $e^{kx} \sin mx$. In this case a solution of the form $y_p = e^{kx}(a \cos mx + b \sin mx)$ must be sought. In the case of the equation $y'' - 3y' + 2y = e^{-x} \sin x$, as the complementary function $y_c = C_1 e^x + C_2 e^{2x}$ does not contain the inhomogeneous term, we seek a solution of the form $y_p = e^{-x}(a \cos x + b \sin x)$. Substituting and proceeding as before it is easily established that $a = b = 1/10$, from which we deduce that $y_p = \frac{1}{10}e^{-x}(\cos x + \sin x)$.

15·2 (b) The operator D

Let us now introduce a new method of solution of constant coefficient equations, making use of the *differentiation operator D*. By the operator D we are to understand the operation of differentiation already used in Leibnitz's theorem, so that $D \equiv d/dx$, $D^2 \equiv d^2/dx^2$ and, in general, $D^n \equiv d^n/dx^n$, where for the moment n is a positive integer.

DEFINITION 15·1 (polynomial operator) We define the polynomial operator

$$P(D) \equiv D^n + a_1 D^{n-1} + \cdots + a_n \qquad (15\cdot14)$$

by

$$P(D)f(x) \equiv (D^n + a_1 D^{n-1} + \cdots + a_n)f(x) = f^{(n)}(x)$$
$$+ a_1 f^{(n-1)}(x) + \cdots + a_n f(x),$$

where $f(x)$ is any suitably differentiable function.

The number a_n in the operator $P(D)$ is, of course, to be understood to mean a_n times the *identity differentiation operator*, which we may write as D^0, with the understanding that $D^0 y \equiv y$. It is conventional, as in Eqn (15·14), to omit this identity differentiation operator since we shall see later that no confusion arises if we write a_n in place of $a_n D^0$.

DEFINITION 15·2 (Sum and product of operators)   Given two polynomial operators $P(D)$ and $Q(D)$, we define the operators $P(D) + Q(D)$ and $P(D)Q(D)$ by

$$[P(D) + Q(D)]f(x) = P(D)f(x) + Q(D)f(x)$$

and

$$[P(D)\, Q(D)]f(x) = P(D)[Q(D)f(x)],$$

where $f(x)$ is any suitably differentiable function.

We shall say two operators $P(D)$ and $Q(D)$ are equal, and will write $P(D) \equiv Q(D)$, if for all suitable $f(x)$, $P(D)f(x) \equiv Q(D)f(x)$. It is important to recognize that $P(D)$ and $Q(D)$ so defined are operators and not functions, in the sense that they only give rise to a function when they operate on some suitably differentiable function.

Example 15·7   In operator D form the differential equation $y'' - 3y' + 2y = xe^{-x}$ may be written $(D^2 - 3D + 2)y = xe^{-x}$, where here $P(D) \equiv D^2 - 3D + 2$. By our first definition we could insert parentheses and equally well write $P(D) \equiv D^2 - (3D - 2) \equiv (D^2 - 3D) + 2$.

It is an immediate consequence of this notation that if a is a constant, and f and g are suitably differentiable functions, then

$$D(aD^r)f \equiv aD^{r+1}f \qquad (15\cdot15)$$

and

$$P(D)[f + g] \equiv P(D)f + P(D)g. \qquad (15\cdot16)$$

The first result follows because $D[(aD^r)f] = \dfrac{\mathrm{d}}{\mathrm{d}x}\left[a\dfrac{\mathrm{d}^r f}{\mathrm{d}x^r}\right] = a\dfrac{\mathrm{d}^{r+1}f}{\mathrm{d}x^{r+1}}$ $= aD^{r+1}f.$

Here we have used the fact that a constant multiplier of a function commutes with the operation of differentiation so that $(d/dx)(a\phi(x)) = a(d\phi/dx)$. The second result follows because if $P(D) = \Sigma a_r D^{n-r}$, then

$$P(D)(f + g) = \Sigma a_r(f + g)^{(n-r)}(x) = \Sigma a_r f^{(n-r)}(x) + \Sigma a_r g^{(n-r)}(x)$$
$$= P(D)f + P(D)g.$$

Other important results that may be established in similar fashion are:

$$P(D) + Q(D) = Q(D) + P(D) \quad \text{(Commutative law)}; \tag{15·17}$$

$$[P(D) Q(D)] R(D) = P(D) [Q(D) R(D)] \quad \text{(Associative law)}; \tag{15·18}$$

$$P(D)[Q(D) + R(D)] = P(D)Q(D) + P(D) R(D) \quad \text{(Distributive law)}. \tag{15·19}$$

A particular case of the last result is

$$(D - \lambda)Q(D) = DQ(D) - \lambda Q(D). \tag{15·20}$$

THEOREM 15·5 (the factorization theorem)　Suppose that the polynomial $\lambda^n + a_1\lambda^{n-1} + \cdots + a_n$ has factors $\lambda - \lambda_1$, $\lambda - \lambda_2$, . . ., $\lambda - \lambda_n$. Then $D^n + a_1 D^{n-1} + \cdots + a_n = (D - \lambda_1)(D - \lambda_2) . . . (D - \lambda_n)$.

Proof　The proof of this result is by induction. Suppose that $\lambda^n + a_1\lambda^{n-1} + \cdots + a_n = (\lambda - \lambda_1)(\lambda^{n-1} + b_1\lambda^{n-2} + \cdots + b_{n-1})$. Then $a_1 = b_1 - \lambda_1$, $a_2 = b_2 - b_1\lambda_1$, . . ., $a_{n-1} = b_{n-1} - b_{n-2}\lambda_1$, $a_n = -b_{n-1}\lambda_1$ and $\lambda^{n-1} + b_1\lambda^{n-2} + \cdots + b_{n-1} = (\lambda - \lambda_2)(\lambda - \lambda_3) . . . (\lambda - \lambda_n)$.

Assume the result to be true for polynomial operators of degree $n - 1$. Then

$$D^{n-1} + b_1 D^{n-2} + \cdots + b_{n-1} = (D - \lambda_2)(D - \lambda_3) . . . (D - \lambda_n).$$

Now $(D - \lambda_1)(D^{n-1} + b_1 D^{n-2} + \cdots + b_{n-1})$
$$= D(D^{n-1} + b_1 D^{n-2} + \cdots + b_{n-1}) - \lambda_1(D^{n-1} + b_1 D^{n-2}$$
$$+ \cdots + b_{n-1})$$
$$= D^n + (b_1 - \lambda_1)D^{n-1} + (b_2 - b_1\lambda_1)D^{n-2} + \cdots$$
$$+ (b_{n-1} - b_{n-2}\lambda_1)D - b_{n-1}\lambda_1$$
$$= D^n + a_1 D^{n-1} + a_2 D^{n-2} + \cdots + a_n.$$

But $(D - \lambda_1)(D^{n-1} + b_1 D^{n-2} + \cdots + b_n) = (D - \lambda_1)(D - \lambda_2) . . . (D - \lambda_n)$, and therefore
$$D^n + a_1 D^{n-1} + \cdots + a_{n-1}D + a_n = (D - \lambda_1)(D - \lambda_2) . . . (D - \lambda_n).$$
This proves the hereditary property, and the result is clearly true for $n = 1$, so by induction it is generally true.

Since the choice of the factor $\lambda - \lambda_1$ with which the above proof was started was arbitrary, we have:

Corollary 15·1　If $p_1, p_2, . . ., p_n$ is a permutation of the numbers $1, 2, . . ., n$, then

$$(D - \lambda_1)(D - \lambda_2) \ldots (D - \lambda_n) = (D - \lambda_{p_1})(D - \lambda_{p_2}) \ldots (D - \lambda_{p_n}).$$

It follows that

$$P(D)Q(D) \equiv Q(D)P(D). \tag{15·21}$$

Example 15·8   In D operator notation the equation $y'' + 5y' + 6y = \cos x$ becomes $(D^2 + 5D + 6)y = \cos x$. Because of the factorization theorem and its corollary, it follows that the quadratic polynomial operator $(D^2 + 5D + 6) \equiv (D + 2)(D + 3) \equiv (D + 3)(D + 2)$.

Let us now briefly recapitulate our discussion of the solution of the reduced equation, but this time using the operator D.

Example 15·9 (distinct roots)   Consider the general second order equation which in factorized form may be written

$$(D - \lambda_1)(D - \lambda_2)y = 0,$$

where $\lambda_1 \neq \lambda_2$. Now set $(D - \lambda_2)y = u$ so that the equation becomes $(D - \lambda_1)u = 0$, or,

$$\frac{du}{dx} - \lambda_1 u = 0.$$

This has the solution $u = C_1 e^{\lambda_1 x}$ so that now we must solve $(D - \lambda_2)y = C_1 e^{\lambda_1 x}$, which is simply the familiar first order linear differential equation

$$\frac{dy}{dx} - \lambda_2 y = C_1 e^{\lambda_1 x}$$

with integrating factor $\mu = e^{-\lambda_2 x}$. Hence

$$\frac{d}{dx}(y e^{-\lambda_2 x}) = C_1 e^{(\lambda_1 - \lambda_2)x},$$

so that the general solution is

$$y_c = \left(\frac{C_1}{\lambda_1 - \lambda_2}\right) e^{\lambda_1 x} + C_2 e^{\lambda_2 x},$$

where C_2 is another arbitrary constant of integration.

The extension to a polynomial operator of degree n is immediate provided the roots are distinct. As the constants are arbitrary the divisor $(\lambda_1 - \lambda_2)$ may be omitted from the first coefficient of the general solution (that is, introduce a new constant $C_1' = C_1/(\lambda_1 - \lambda_2)$).

Example 15·10 (repeated roots)   This time let us consider a third order equation but assume that two of the roots are equal, so that in factorized form it may be written

$$(D - \lambda_1)^2(D - \lambda_2)y = 0.$$

Changing the order of the factors and setting $(D - \lambda_1)^2 y = u$, the equation becomes $(D - \lambda_2)u = 0$, with the solution $u = C_1 e^{\lambda_2 x}$. Hence we must now solve $(D - \lambda_1)^2 y = C_1 e^{\lambda_2 x}$. So, writing $v = (D - \lambda_1)y$, this equation simplifies to $(D - \lambda_1)v = C_1 e^{\lambda_2 x}$. The integrating factor is $\mu = e^{-\lambda_1 x}$, and an application of the argument of the previous example with re-definition of constants where necessary brings us to the solution $v = C_1' e^{\lambda_2 x} + C_2 e^{\lambda_1 x}$. Finally, we must solve $(D - \lambda_1)y = C_1' e^{\lambda_2 x} + C_2 e^{\lambda_1 x}$. This also has an integrating factor $\mu = e^{-\lambda_1 x}$, so that

$$\frac{d}{dx}(y e^{-\lambda_1 x}) = C_1' e^{(\lambda_2 - \lambda_1)x} + C_2,$$

and hence

$$y_c = C_1'' e^{\lambda_2 x} + (C_2 x + C_3)e^{\lambda_1 x}.$$

As before, constant divisors of the form $\lambda_2 - \lambda_1$ have been omitted and the arbitrary constant re-defined. This method has thus automatically generated the two linearly independent solutions $e^{\lambda_1 x}$ and $x e^{\lambda_1 x}$ corresponding to the two repeated factors $(D - \lambda_1)$. An application of the method to a factor with multiplicity r generates the linearly independent terms discussed in Theorem 15·2.

Example 15·11 (complex conjugate roots) If a polynomial $P(\lambda)$ with real coefficients is such that $P(\lambda) = 0$ has a complex root $\lambda = \mu + i\nu$, then we know it must also have a root $\lambda = \mu - i\nu$. Consequently, as in our previous study, we know that the corresponding term in the particular integral must be

$$e^{\mu x}(C_1 \cos \nu x + C_2 \sin \nu x).$$

Also, if the roots have multiplicity m, then the corresponding term must be

$$e^{\mu x}[P_{m-1}(x) \cos \nu x + Q_{m-1}(x) \sin \nu x],$$

where $P_{m-1}(x)$ and $Q_{m-1}(x)$ are polynomials in x of degree $m - 1$ having arbitrary coefficients. These terms must be added to the other terms that arise from the real roots of $P(\lambda) = 0$ to obtain the complementary function y_c.

Consider the equation

$$(D^5 - 5D^4 + 12D^3 - 16D^2 + 12D - 4)y = 0.$$

In factorized form this becomes

$$(D - 1)(D - 1 - i)^2(D - 1 + i)^2 y = 0,$$

showing that the real factor $D - 1$ has multiplicity 1 and the complex conjugate factors in which $\mu = 1$, $\nu = 1$ have multiplicity 2. The complementary function y_c is thus

$$y_c = e^x[C_1 + (C_2 + C_3 x) \cos x + (C_4 + C_5 x) \sin x].$$

Although these applications of the operator D are of interest, they offer no real advantage over the first method we discussed for the solution of homogeneous equations. In both cases the same process of factorization is required, after which the complementary function may be written down by inspection. The real advantage of the operator method is in the determination of particular integrals as we now show.

Writing an inhomogeneous equation in the compact form

$$P(D)y = f(x), \tag{15·22}$$

we are tempted to regard this as an algebraic equation and to write

$$y_p = \frac{1}{P(D)} f(x) \tag{15·23}$$

for the particular integral.

Since $P(D)$ is an operator, the expression $1/P(D)$, if it can be given a meaning, must also be an operator such that when applied to $f(x)$ it generates the particular integral. Indeed, substitution into Eqn (15·22) shows that an appropriate name for $1/P(D)$ would be the *inverse* operator. This is so because when $1/P(D)$ is applied to $P(D)$ it obviously generates the identity operator. An inverse operator of this kind can in fact be satisfactorily defined if we approach the problem with care. For example, in the very simple equation $Dy = x$, we know that y must be some function which when differentiated will yield x. It is obvious that in this case $y = \frac{1}{2}x^2 + C$, so that when we write, symbolically, $y = (1/D)x$ we must interpret $1/D$ as implying the determination of an antiderivative, that is to say, as the ordinary operation of integration. Similarly, $1/D^2$ represents the operation of integration twice repeated. It is often convenient to use the properties of indices to write the n-fold repeated operation of integration $1/D^n$ in the form D^{-n}.

Care must always be taken to indicate on what function the inverse operator is to act. An expression of the form fg/D is ambiguous, since it could mean either $D^{-1}(fg)$ or $f(D^{-1}g)$, which are two different functions.

In Theorem 15·4 we saw that particular integrals differ only by terms belonging to the complementary function, so that integration constants may be omitted when using the operator D for the determination of particular integrals. Retaining these constants of integration will generate both the complementary function and the particular integral.

The fundamental equation that gives meaning to the operator $(D - \lambda)$ is the linear first order inhomogeneous equation

$$\frac{dy}{dx} - \lambda y = f(x),$$

which in operator notation takes the form

$$(D - \lambda)y = f(x). \tag{15.24}$$

We already know from Eqn (14.20) that the particular integral is

$$y_p(x) = e^{\lambda x} \int f(x)e^{-\lambda x}\, dx, \tag{15.25}$$

so that it follows from Eqns (15.24) and (15.25) that

$$y_p(x) = \frac{1}{D - \lambda}f(x) = e^{\lambda x} \int f(x)e^{-\lambda x}\, dx. \tag{15.26}$$

The expression on the right-hand side thus gives meaning to the inverse operator $(D - \lambda)^{-1}$ acting on the function $f(x)$ and will be taken as our fundamental definition.

DEFINITION 15.3 (inverse operator) We define the effect of the inverse operator $(D - \lambda)^{-1}$ acting on a function $f(x)$ by the expression

$$(D - \lambda)^{-1}f(x) = e^{\lambda x} \int f(x)e^{-\lambda x}\, dx.$$

Example 15.12 We shall determine the particular integral of

$$(D - 1)(D - 2)y = e^x,$$

which will necessitate two applications of the inverse operator just defined. First, using the inverse operator $(D - 1)^{-1}$, and identifying λ with 1 and $f(x)$ with e^x in Eqn (15.26), we have

$$(D - 2)y = \left(\frac{1}{D - 1}\right)e^x$$

$$= e^x \int e^x . e^{-x}\, dx = xe^x.$$

Then, using the inverse operator $(D - 2)^{-1}$, identical reasoning gives

$$y_p = \left(\frac{1}{D - 2}\right)xe^x$$

$$= e^{2x} \int xe^x . e^{-2x}\, dx = -(x + 1)e^x.$$

Hence the desired particular integral is $y_p = -(x + 1)e^x$. Notice that the fact that the inhomogeneous term e^x also occurs in the complementary function $y_c = C_1e^x + C_2e^{2x}$ has been automatically accounted for by this method, and so no special case need now be distinguished in this respect.

Example 15.13 The application of the operator defined in Eqn (15.26) to complex factors is equally straightforward. Thus if

$$(D^2 - 2D + 2)y = e^x,$$

then after factorization we have

$$(D - 1 - i)(D - 1 + i)y = e^x.$$

Proceeding as before then gives

$$(D - 1 + i)y = \left(\frac{1}{D - 1 - i}\right) e^x$$

$$= e^{(1+i)x} \int e^x \cdot e^{-(1+i)x} \, dx = ie^x.$$

Hence,

$$y_p = \left(\frac{1}{D - 1 + i}\right) ie^x$$

$$= e^{(1-i)x} \int ie^x \cdot e^{-(1-i)x} \, dx = e^x,$$

showing that the required particular integral is $y_p = e^x$.

SOME SIMPLE RULES   In special cases the inverse operator just defined simplifies to give some easy rules for the determination of y_p. Suppose first that $P(D)y = ke^{\alpha x}$, where $P(D) \equiv D^n + a_1 D^{n-1} + \cdots + a_n$, and where α is not a root of the characteristic polynomial $P(\lambda) \equiv \lambda^n + a_1 \lambda^{n-1} + \cdots + a_n = 0$. By direct differentiation it is easily shown that

$$(D^n + a_1 D^{n-1} + \cdots + a_n)e^{\alpha x} = e^{\alpha x}(\alpha^n + a_1 \alpha^{n-1} + \cdots + a_n),$$

which is a result that may be expressed by the relation

$$P(D)e^{\alpha x} = e^{\alpha x}P(\alpha), \tag{15·27}$$

where $P(\alpha)$ is just a number obtained from $P(D)$ by formally replacing D by α. As α is not a root of $P(\lambda) = 0$ we have $P(\alpha) \neq 0$, and since $[P(D)]^{-1}P(D)e^{\alpha x} \equiv e^{\alpha x}$, comparison of this identity with Eqn (15·27) gives:

Rule 1

If $P(\alpha) \neq 0$, then

$$\frac{1}{P(D)} e^{\alpha x} = \frac{e^{\alpha x}}{P(\alpha)}.$$

Applying Rule 1 to Example 15·13, in which $P(D) \equiv D^2 - 2D + 2$, shows that we are required to solve

$$y_p = \frac{1}{D^2 - 2D + 2} e^x.$$

Here $\alpha = 1$, so that $P(\alpha) = 1 - 2 + 2 = 1$, whence $y_p = e^x$. The rule is inapplicable in Example 15·12 because in that case $P(\alpha) = 0$, and so operator Eqn (15·26) must be used instead.

Another simple case occurs on account of the obvious results $D^{2n} \cos mx = (-1)^n m^{2n} \cos mx$ and $D^{2n} \sin mx = (-1)^n m^{2n} \sin mx$. If the operator only contains *even* powers of D, which we denote by writing $P(D^2)$, then, arguing as in Rule 1, it is easy to establish:

Rule 2

If $P(D)$ only contains even powers of D and so may be written in the form $P(D^2)$ then, providing $P(-m^2) \neq 0$,

$$\frac{1}{P(D^2)} \sin mx = \frac{\sin mx}{P(-m^2)} \quad \text{and} \quad \frac{1}{P(D^2)} \cos mx = \frac{\cos mx}{P(-m^2)}.$$

If, for example, $(D^4 - 3D^2 + 2)y = \cos 2x$, then since $m = 2$ and $P(D^2) \equiv D^4 - 3D^2 + 2$, it follows that $P(-2^2) = (-4)^2 - 3(-4) + 2 = 30$. Hence by Rule 2 we find that $y_p = (1/30) \cos 2x$.

Another rule may be deduced by applying formula in Eqn (15·26) to the function x^s and then comparing the result with the effect of the operator

$$\frac{1}{D - \lambda} = \frac{-1}{\lambda - D} = -\frac{1}{\lambda}\left(1 + \frac{D}{\lambda} + \frac{D^2}{\lambda^2} + \cdots\right) \tag{15·28}$$

applied to x^s. This operator, which arises when $(\lambda - D)^{-1}$ is formally expanded by the Binomial Theorem, is seen to give the same result when applied to x^s as an application of Eqn (15·26) and, because $D^{s+1}x^s \equiv 0$, the expansion may be terminated after the term D^s. Applying this operator to a polynomial of degree m establishes:

Rule 3

$$\frac{1}{D - \lambda}(b_0 + b_1 x + \cdots + b_m x^m) = -\frac{1}{\lambda}\left(1 + \frac{D}{\lambda} + \frac{D^2}{\lambda^2} + \cdots + \frac{D^m}{\lambda^m}\right)$$
$$(b_0 + b_1 x + \cdots + b_m x^m).$$

To illustrate this, let us suppose that $(D - 4)y = 1 + x^2$, then $y_p = (D - 4)^{-1}(1 + x^2)$. By Rule 3 we see that we need to set $n = 2$ and $\lambda = 4$, so that $y_p = (D - 4)^{-1}(1 + x^2) = -(1/4)[1 + (D/4) + (D^2/4^2)](1 + x^2)$. Performing the indicated differentiations we find that $y_p = -(9 + 4x + 8x^2)/32$.

The particular integral corresponding to the more general expression

$$P(D)y = b_0 + b_1 x + \cdots + b_m x^m$$

may be deduced by factorizing the polynomial operator $P(D)$ and making repeated use of Rule 3.

Sometimes it is useful to reformulate Rule 3 so that, if desired, it may be applied directly to operators $P(D)$ without resolving them into simple first order factors of the form $(D - \lambda)$. This is certainly necessary if $P(D)$ has any quadratic factors corresponding to pairs of complex conjugate roots of the characteristic polynomial. The desired modification of Rule 3 is easily achieved by using a repetition of the previous arguments to arrive at:

Modified Rule 3

If $P(D) \equiv D^n + a_1 D^{n-1} + \cdots + a_{n-1}D + a_n$, and $P(D)$ is expressed in the form

$$P(D) \equiv a_n[1 - Q(D)],$$

so that

$$Q(D) \equiv -\left(\frac{1}{a_n} D^n + \frac{a_1}{a_n} D^{n-1} + \cdots + \frac{a_{n-1}}{a_n} D\right),$$

then

$$\frac{1}{P(D)} (b_0 + b_1 x + \cdots + b_m x^m) = \frac{1}{a_n} (1 + Q(D) + Q^2(D) + \cdots$$
$$+ Q^m(D))(b_0 + b_1 x + \cdots + b_m x^m).$$

By way of example suppose $(D^2 - 3D + 2)y = x^2 + 1$. Then $P(D) \equiv D^2 - 3D + 2$, and in the above notation $a_1 = -3$, $a_2 = 2$, showing that $Q(D) \equiv -\frac{1}{2}D^2 + \frac{3}{2}D$. Since the polynomial is of degree 2 we have $m = 2$ and the Modified Rule 3 then gives

$$y_p = \left(\frac{1}{D^2 - 3D + 2}\right)(x^2 + 1) = \frac{1}{2}[1 + (-\frac{1}{2}D^2 + \frac{3}{2}D)$$
$$+ (-\frac{1}{2}D^2 + \frac{3}{2}D)^2](x^2 + 1).$$

Performing the indicated differentiations we finally arrive at the result

$$y_p = \frac{1}{2}x^2 + \frac{3}{2}x + \frac{9}{4}.$$

The final rule concerns the operator inverse to $P(D)[e^{\mu x}u(x)]$, where $u(x)$ is any suitably differentiable function. By direct differentiation we have $D^r(e^{\mu x}u) = D^{r-1}D(e^{\mu x}u) = D^{r-1}(\mu e^{\mu x}u + e^{\mu x}Du) = D^{r-1}e^{\mu x}(\mu + D)u$. Similarly, $D^{r-1}e^{\mu x}(\mu + D)u = D^{r-2}D[e^{\mu x}(\mu + D)u] = D^{r-2}e^{\mu x}(\mu + D)^2u = \ldots = e^{\mu x}(\mu + D)^r u$. Hence, applying the argument to a polynomial $P(D)$ gives

$$P(D)[e^{\mu x}u(x)] = e^{\mu x}[P(D + \mu)u(x)]. \tag{15·29}$$

To derive the inverse operator rule we now set $P(D + \mu)u(x) = v(x)$, which may be any differentiable function, since $u(x)$ is otherwise unspecified, so that $u(x) = P^{-1}(D + \mu)v(x)$. Then, using these results in Eqn (15·29) gives:

Rule 4

If $v(x)$ is a suitably differentiable function, then

$$\frac{1}{P(D)}(e^{\mu x}v) = e^{\mu x}\frac{1}{P(D + \mu)}v.$$

We illustrate this rule by considering the determination of the particular integral of the equation $(D^2 + 5D + 6)y = xe^{-x}$. Here $P(D) \equiv D^2 + 5D + 6$ and $\mu = -1$, so that $P(D + \mu) \equiv (D - 1)^2 + 5(D - 1) + 6 \equiv (D^2 + 3D + 2)$. Then by Rule 4, $y_p = e^{-x}(D^2 + 3D + 2)^{-1}x$. Factorizing this result and using Eqn (15·28) then gives

$$y_p = e^{-x}\left[\frac{1}{(1 + D)(2 + D)}\right]x$$

$$= e^{-x}\left[(1)(1 - D + D^2 - \cdots)\left(\frac{1}{2}\right)\left[\left(1 - \frac{D}{2} + \frac{D^2}{4} - \cdots\right)\right]x.$$

Expanding the square bracket as far as terms involving D, because the operator is only acting on x, we find that

$$y_p = \tfrac{1}{2}e^{-x}\left(1 - \frac{3D}{2}\right)x = \tfrac{1}{2}xe^{-x} - \tfrac{3}{4}e^{-x}.$$

15·3 Simultaneous linear differential equations

Many important physical processes are described by systems of simultaneous linear constant coefficient differential equations. Typical examples are the equations describing interacting control systems, interacting electric circuits, and reversible chemical reactions. The following two first order equations involving the dependent variables x and y together with the independent variable t provide a simple illustration of this type of problem and its solution.

Let us determine x and y as functions of t if initially at $t = 0$ we have that $x = -\tfrac{1}{4}, y = 1$ and, subsequently,

$$\dot{x} = x + y + t$$
$$\dot{y} = 3x - y,$$

where the dot denotes differentiation with respect to t.

From the first equation we see that $y = \dot{x} - x - t$, so that eliminating y from the second equation then gives $\dot{y} = 4x + t - \dot{x}$. If now we differentiate the first equation with respect to t we obtain $\ddot{x} = \dot{x} + \dot{y} + 1$, which can be combined with the previous result to give $\ddot{x} - 4x = t + 1$. This is a second order inhomogeneous equation involving only the dependent variable x.

Using any of the methods previously described it is easily established that its general solution is

$$x = A \cosh 2t + B \sinh 2t - \tfrac{1}{4}(1 + t).$$

To determine y we now substitute this expression for x in the first of the simultaneous equations to obtain

$$y = A(2 \sinh 2t - \cosh 2t) + B(2 \cosh 2t - \sinh 2t) - \tfrac{3}{4}t.$$

Inserting the initial conditions for x and y into these expressions shows that $A = 0$, $B = \tfrac{1}{2}$ and so the required particular solution is

$$x = \tfrac{1}{2} \sinh 2t - \tfrac{1}{4}(1 + t),$$
$$y = \tfrac{1}{2}(2 \cosh 2t - \sinh 2t) - \tfrac{3}{4}t.$$

Notice that had we inserted the general solution for x in the second of the simultaneous equations we should have obtained a first order inhomogeneous equation for y which would have apparently necessitated the introduction of a third arbitrary constant into the solution. This anomaly is easily resolved by recalling that the value of y so determined, when taken with x, must be compatible with the first equation. Hence the apparently additional arbitrary constant required by this approach is in fact dependent on A and B. This complication is completely avoided by the method adopted here.

15·4   Runge–Kutta method

The series methods for the solution of differential equations described in the previous section provide analytical techniques for the determination of the numerical behaviour of a particular solution. If the rate of convergence of the series is poor, or if a non-linearity in the equation precludes the use of such methods, some other approach must be used to obtain an accurate numerical solution.

The very useful and flexible numerical method that we now describe was first introduced by C. Runge at the turn of the century and subsequently modified and improved by W. Kutta. It is essentially a generalization of Simpson's rule and it can be shown that the error involved when integrating a step of length Δx is of the order of $(\Delta x)^5$. The method is simple to use and allows adjustment of the length of the integration step from point to point without modification of the method.

We suppose that x and y assume the values x_n, y_n after the nth integration step in the numerical integration of

$$\frac{\mathrm{d}y}{\mathrm{d}x} = f(x, y). \tag{15·30}$$

Then the value y_{n+1} of the dependent variable y that is to be associated with argument $x_{n+1} = x_n + \Delta x$ is computed as follows.

Use an integration step of length Δx and let

$$k_1 = f(x_n, y_n) . \Delta x$$

$$k_2 = f(x_n + \tfrac{1}{2}\Delta x, y_n + \tfrac{1}{2}k_1) . \Delta x$$

$$k_3 = f(x_n + \tfrac{1}{2}\Delta x, y_n + \tfrac{1}{2}k_2) . \Delta x \qquad (15\cdot31)$$

$$k_4 = f(x_n + \Delta x, y_n + k_3) . \Delta x$$

$$\Delta y = \tfrac{1}{6}(k_1 + 2k_2 + 2k_3 + k_4),$$

then the value y_{n+1} of y corresponding to $x = x_n + \Delta x$ is determined by

$$y_{n+1} = y_n + \Delta y. \qquad (15\cdot32)$$

Example 15·14 Let us again determine the value $y(0\cdot5)$ given that $y' = xy$, with $y(0) = 1$ and $\Delta x = 0\cdot1$. In this simple example, already used to illustrate Euler's method and its modification, we have $f(x, y) = xy$. As we must anticipate an error of the order of $(0\cdot1)^5$ we shall work to five decimal places so that we may compare our solution with the exact result $y = e^{\frac{1}{2}x^2}$.

n	x_n	y_n	$f(x_n, y_n)$	k_1	k_2	k_3	k_4	y_{n+1}	$e^{\frac{1}{2}x^2}$
0	0·0	1·0	0·0	0·0	0·0050	0·00501	0·01005	1·00501	1·0
1	0·1	1·00501	0·10050	0·01005	0·01515	0·01519	0·02040	1·02020	1·00501
2	0·2	1·02020	0·20404	0·02040	0·02576	0·02583	0·03138	1·04603	1·02020
3	0·3	1·04603	0·31381	0·03138	0·03716	0·03726	0·04333	1·08329	1·04603
4	0·4	1·08329	0·43332	0·04332	0·04972	0·04987	0·05666	1·13315	1·08329
5	0·5	1·13315							1·13315

Comparison of the results of column three with the analytical solution $y = e^{\frac{1}{2}x^2}$ show that it is in fact accurate to five decimal places, so that in this case our rough error estimate was too severe.

The superiority of the Runge–Kutta method over the Euler and modified Euler methods is clearly demonstrated if the Runge–Kutta solution is compared with the previous solutions. This improvement is uniformly true and not just in this instance, since it may be shown that the errors involved in the Euler and modified Euler methods are, respectively, of the order $(\Delta x)^2$ and $(\Delta x)^3$. No discussion will be offered here of the more subtle finite difference methods that may be used to provide integration formulae having extremely high accuracy.

The Runge–Kutta method readily extends to allow the numerical solution of simultaneous and higher order equations. Suppose the equations involved are

$$\frac{dy}{dx} = f(x, y, z)$$

$$\frac{dz}{dx} = g(x, y, z)$$

(15·33)

subject to the initial conditions $y = y_0$, $z = z_0$ at $x = x_0$.

Then, at the nth step of integration, setting

$$k_1 = f(x_n, y_n, z_n) . \Delta x$$

$$k_2 = f(x_n + \tfrac{1}{2}\Delta x, y_n + \tfrac{1}{2}k_1, z_n + \tfrac{1}{2}K_1) . \Delta x$$

$$k_3 = f(x_n + \tfrac{1}{2}\Delta x, y_n + \tfrac{1}{2}k_2, z_n + \tfrac{1}{2}K_2) . \Delta x$$

$$k_4 = f(x_n + \Delta x, y_n + k_3, z_n + K_3) . \Delta x,$$

(15·34)

and

$$K_1 = g(x_n, y_n, z_n) . \Delta x$$

$$K_2 = g(x_n + \tfrac{1}{2}\Delta x, y_n + \tfrac{1}{2}k_1, z_n + \tfrac{1}{2}K_1) . \Delta x$$

$$K_3 = g(x_n + \tfrac{1}{2}\Delta x, y_n + \tfrac{1}{2}k_2, z_n + \tfrac{1}{2}K_2) . \Delta x$$

$$K_4 = g(x_n + \Delta x, y_n + k_3, z_n + K_3) . \Delta x,$$

(15·35)

we use the following formulae to compute Δy and Δz:

$$\Delta y = \tfrac{1}{6}(k_1 + 2k_2 + 2k_3 + k_4) \quad \text{and} \quad \Delta z = \tfrac{1}{6}(K_1 + 2K_2 + 2K_3 + K_4).$$

(15·36)

The values of y and z at the $(n + 1)$th step of integration are $y_{n+1} = y_n + \Delta y$, $z_{n+1} = z_n + \Delta z$.

These results may also be used to integrate a second order equation by introducing the first derivative as a new dependent variable. Suppose $y'' - 2y' + 2y = 0$ with $y(0) = y'(0) = 1$. Then setting $y' = z$, the second order equation is seen to be equivalent to the two first order simultaneous equations $y' = z$ and $z' = 2(z - y)$, with $y(0) = 1$ and $z(0) = 1$. Applying formulae Eqns (15·34) to (15·36) with $\Delta x = 0·2$, $f \equiv z$, and $g \equiv 2(y - z)$ in order to determine $y(0·2)$ we find

$k_1 = 0·2,$	$K_1 = 0$
$k_2 = 0·2,$	$K_2 = -0·04$
$k_3 = 0·196,$	$K_3 = -0·048$
$k_4 = 0·1904,$	$K_4 = -0·0976,$

so that $\Delta y = 0 \cdot 19706$ and $\Delta z = \Delta(y') = -0 \cdot 04560$. Hence $y(0 \cdot 2) = 1 \cdot 19706$ and $y'(0 \cdot 2) = 0 \cdot 95440$, which are in complete agreement with the analytical solution $y = e^x \cos x$.

15·5 Oscillatory solutions

Although we have already discussed the general method of solution of second order differential equations with constant coefficients, their importance in practical applications merits special mention when the inhomogeneous term is periodic. The second order differential equation

$$a \frac{d^2 x}{dt^2} + b \frac{dx}{dt} + cx = f(t) \tag{15·37}$$

characterizes many important physical situations. For example, when a represents a mass, b a damping force proportional to velocity, and c a restoring force, Eqn (15·37) could represent a mechanical vibration damper. Alternatively, if a represents an inductance, d a resistance, and c a capacitance, Eqn (15·37) would describe an R–L–C circuit and, indeed, many other situations are characterized by this simple equation.

By analogy with a mechanical system in which $f(t)$ represents the input driving the system, the inhomogeneous term is sometimes called the *forcing function*. It is the inhomogeneous term that gives rise to the particular integral, and we again remark that part of the general solution is attributable solely to the function $f(t)$.

We shall confine attention to the following particular form of Eqn (15·37):

$$y'' + 2\zeta y' + \Omega^2 y = a \sin \omega t, \tag{15·38}$$

where a is called the *amplitude* of the forcing function $\sin \omega t$, which has frequency ω rad/s and period $2\pi/\omega$. The number ζ is usually called the *damping* of the system described by Eqn (15·38), and Ω is then called the natural frequency of the system. Several cases must be distinguished and first we assume that $\zeta \neq 0$ with $\zeta^2 < \Omega^2$. Then, setting $\omega_0^2 = \Omega^2 - \zeta^2$, the roots of the characteristic equation $P(\lambda) \equiv \lambda^2 + 2\zeta\lambda + \Omega^2 = 0$ become $\lambda = -\zeta \pm i\omega_0$. In terms of this new notation, the complementary function y_c can be written

$$y_c = A e^{-\zeta t} \sin (\omega_0 t + \varepsilon), \tag{15·39}$$

where A and ε are arbitrary constants. The particular integral y_p can be expressed in the form

$$y_p = P \sin \omega t + Q \cos \omega t, \tag{15·40}$$

where

$$P = \frac{a(\Omega^2 - \omega^2)}{(\Omega^2 - \omega^2)^2 + 4\zeta^2\omega^2} \quad \text{and} \quad Q = \frac{-2\zeta a\omega}{(\Omega^2 - \omega^2)^2 + 4\zeta^2\omega^2}.$$

In this context ε is usually called the *phase angle* of the solution.

Simple manipulation shows that y_p may be expressed in the alternative form

$$y_p = \frac{a}{[(\Omega^2 - \omega^2)^2 + 4\zeta^2\omega^2]^{1/2}} \sin(\omega t + \delta), \tag{15·41}$$

where $\delta = \arctan(Q/P)$. The complete solution can then be written

$$y = Ae^{-\zeta t}\sin(\omega_0 t + \varepsilon) + \frac{a}{[(\Omega^2 - \omega^2)^2 + 4\zeta^2\omega^2]^{1/2}} \sin(\omega t + \delta). \tag{15·42}$$

If $\zeta < 0$, then as time increases the influence of the complementary function on the complete solution will diminish. In these circumstances, after a suitable lapse of time, only the particular integral will remain and will describe what is often called the *steady state behaviour*. This is to be interpreted in the sense that the complementary function, which essentially describes how the solution started, has ceased to influence the solution. It is for this reason that the complementary function is often said to describe the *transient behaviour* of the solution.

If we agree to call a solution stable when it is bounded in magnitude for all time, it can be seen from the form of y_p in Eqn (15·41) and our discussion of y_c, that the solution Eqn (15·42) is stable provided $\zeta > 0$.

Examining the steady state solution Eqn (15·41) for a stable equation, we notice that the sine function has an amplitude $A(\omega)$ which is frequency dependent:

$$A(\omega) = \frac{a}{[(\Omega^2 - \omega^2)^2 + 4\zeta^2\omega^2]^{1/2}}. \tag{15·43}$$

It is readily established that the denominator of $A(\omega)$ has a minimum when $\omega = \omega_c$, where $\omega_c^2 = \Omega^2 - 2\zeta^2$. Hence the maximum amplitude $A_{\max}$ attained by the steady state solution must occur when $\omega = \omega_c$, and it has the value:

$$A_{\max} = \frac{a}{2\zeta\Omega\left(1 - \dfrac{\zeta^2}{\Omega^2}\right)^{1/2}}. \tag{15·44}$$

The frequency ω_c at which $A_{\max}$ occurs is called the *resonant* frequency and it can be seen that when there is zero damping ($\zeta = 0$), the original Eqn (15·38) describes simple harmonic motion for which $A_{\max} \to \infty$ as $\omega \to \Omega$, which is then the *natural* frequency of the system. That is to say Ω is the frequency of oscillations when the forcing function is removed.

If $0 < \zeta < \Omega$ the complementary function is oscillatory, and physical systems having a damping ζ in this range are said to be *normally* damped.

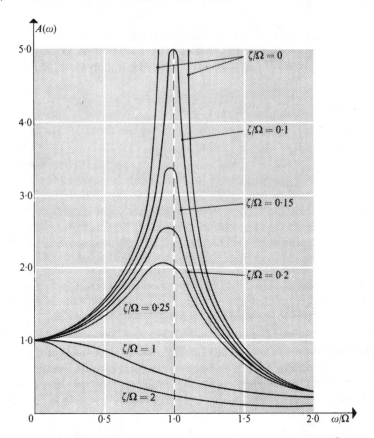

Fig. 15·1 Amplification factor $A(\omega)$ as function of normalized damping ζ/Ω and nondimensional frequency ω/Ω.

If, however, $\zeta > \Omega$ the complementary function or transient solution becomes

$$y_c = C_1 e^{k_1 t} + C_2 e^{k_2 t},$$

where $k_1 = -\zeta + (\zeta^2 - \Omega^2)^{1/2}$ and $k_2 = -\zeta - (\zeta^2 - \Omega^2)^{1/2}$, and is no longer oscillatory. The associated physical system is then said to be *over-damped*.

A critical case occurs when $\zeta = \Omega$, for which the complementary function becomes

$$y_c = (C_1 + C_2 t)e^{-\Omega t}. \tag{15·45}$$

In these circumstances the associated physical system is said to be *critically damped*.

The amplitude $A(\omega)$ is essentially an amplification factor for the forcing function input $a \sin \cdot \omega t$ and it is convenient to summarize the results of this section by constructing a graph of $A(\omega)$ versus ω for different values of the damping ζ.

This is illustrated in Fig. 15·1 for a representative range of values of ζ. The reason for the infinite amplification factor at $\omega = \omega_c$ in the case of zero damping may be readily appreciated by solving the equation

$$y'' + \Omega^2 y = a \sin \Omega t.$$

The complete solution here is

$$y = A \sin (\Omega t + \varepsilon) - \frac{a}{2\Omega} t \cos \Omega t, \tag{15·46}$$

and although the complementary function is finitely bounded for all time, the particular integral is not. A differential equation of this form could, for example, describe the motion of a simple pendulum excited by a periodic disturbance at exactly its natural frequency. The disturbing force would always be in phase with the motion and so would continually reinforce it, thereby causing the amplitude to increase without bound.

PROBLEMS

Section 15·1

15·1 Find the characteristic polynomials and complementary functions of the following differential equations and, where initial conditions are given, find the appropriate particular solution:
  (a) $y'' + 5y' - 14y = 0$;
  (b) $y'' - y = 0$; $y = 1$, $y' = 0$ at $x = 0$;
  (c) $y'' + 4y' + 3y = 0$;
  (d) $y''' + 5y'' + 2y' - 8y = 0$;
  (e) $y''' + 7y'' + 12y' = 0$; $y = 0$, $y' = 9$, $y'' = -39$ at $x = 0$.

15·2 By using the definition of linear dependence state whether the following sets of functions are linearly dependent:
  (a) x^2, x^4, x^6;
  (b) $\cos x$, $-3 \cos x$, $9 \cos x$;
  (c) $\cosh 2x$, $\sinh 2x$, 1;
  (d) $\cosh^2 3x$, $\sinh^2 3x$, 2;
  (e) $x + 1$, $x + 2$, $x + 3$;
  (f) $x + 1$, $x + 2$.

15·3 Obtain the general solution of $y''' - 6y'' + 11y' - 6y = 0$, and by finding the Wronskian of its three constituent functions prove that they are linearly independent.

15·4 By forming the general solution and eliminating the arbitrary constants by differentiation, determine the differential equations that have the following sets of functions as linearly independent solutions:
  (a) e^{2x}, e^{-3x}; (b) e^x, xe^x, x^2e^x; (c) 1, x, e^{2x}.

15·5 Find the general solutions of the following differential equations:
  (a) $y''' - y'' + y' - y = 0$;
  (b) $y'' + y' + y = 0$;
  (c) $y''' - 3ay'' + 3a^2y' - a^3y = 0$;

(d) $y^{iv} + 2y'' + y = 0$;
(e) $y^{iv} + 2y'' + 9y = 0$.

Section 15·2

15·6 Determine the general solutions of the following differential equations using both the method of undetermined coefficients and the operator D method with Rules 1 to 4:

(a) $y'' + 2y' - 3y = x^2 + x + 1$;
(b) $y''' - 3y'' + 3y' - y = 6$;
(c) $y'' + 2y' + y = e^{3x}$;
(d) $y'' + 4y' + 5y = 6e^x(2 \cos 2x + \sin x)$;
(e) $y'' - y = 2e^x$;
(f) $y'' + 4y = \cos 2x$;
(g) $y'' + 9y = \sinh x$;
(h) $y'' + 2y' + 5y = e^x(1 + 2e^x)$;
(i) $y^{iv} + 3y'' + 2y = \cos 3x$;
(j) $y'' - y' - 6y = e^x + \sin x$;
(k) $y'' + 4y' = x + e^x$.

Section 15·3

15·7 Obtain the general solutions of the following simultaneous differential equations:

(a) $\dfrac{dx}{dt} = y, \quad \dfrac{dy}{dt} = x$;

(b) $\dfrac{dx}{dt} = x + y, \quad \dfrac{dy}{dt} = 2x - y$;

(c) $\dfrac{dx}{dt} = x + y + t, \quad \dfrac{dy}{dt} = 2t - 4x - 3y$.

15·8 If $dx/dt = y + z$, $dy/dt = x + z$, $dz/dt = x + y$, obtain the general solution and hence find the particular solution satisfying the initial conditions $x = y = 1$, $z = 0$ at $t = 0$.

Section 15·4

15·9 Use the Runge–Kutta method with $\Delta x = 0·1$ and, working to four decimal places, determine $y(1)$ given that $y' = (x^2 + y)/x$ with $y(0·5) = 0·5$. Compare your results with the exact solution $y = \frac{1}{2}x + x^2$.

15·10 Use the Runge–Kutta method with $\Delta x = 0·2$ and, working to four decimal places, determine $y(1)$ given that $y' = y + e^{-x}$ and $y(0) = 0$. Compare your results with the exact solution $y = \sinh x$.

15·11 Use the Runge–Kutta method with $\Delta x = 0·1$ and, working to four decimal places, determine $y(0·3)$ given that $y'' - 3y' + 2y = 0$ with $y(1) = 0$ and $y'(1) = 0$. Compare your results with the exact solution $y = 2e^x - e^{2x}$.

Section 15·5

15·12 The equation of motion of a forced oscillation is

$$\ddot{y} + 2\dot{y} + 5y = 10 \sin \omega t.$$

Find the complete solution, indicating the difference between the transient and steady state terms. Find also the maximum value of the amplitude of the steady state oscillation that may be obtained by varying ω.

15·13 Sketch the variation of the phase angle δ of the particular integral occurring in Eqn (15·41) as a function of the normalized excitation frequency ω/Ω, for the cases $\zeta = \frac{1}{2}$, $\zeta = 1$, and $\zeta = 2$.

15·14 Derive an expression for x in the case of a critically damped oscillator for which

$$\ddot{x} + 2n\dot{x} + n^2 x = 0,$$

where $\dot{x} = u$ and $x = s$ at time $t = 0$. Show that if this equation describes the motion of a particle, then it will come to rest when $x = u/ne$ if $s = 0$.

15·15 When $\Omega^2 = \zeta^2$, the general solution of the damped harmonic motion described by

$$\ddot{x} + 2\zeta\dot{x} + \Omega^2 x = 0$$

is

$$x = e^{-\zeta t}(A \cos \omega_0 t + B \sin \omega_0 t), \tag{A}$$

where $\omega_0^2 = \Omega^2 - \zeta^2$. Deduce that the extrema of x occur when

$$\tan \omega_0 t = (B\omega_0 - \zeta A)/(A\omega_0 + \zeta B). \tag{B}$$

Denote the positive solutions of this equation by

$$\omega_0 t = \delta_0 + r\pi,$$

where $r = 0, 1, 2, \ldots$, and δ_0 is the smallest positive angle satisfying (B). Thus, defining the sequence of times $\{t_r\}$ by

$$t_r = (\delta_0 + r\pi)/\omega_0, \quad r = 0, 1, 2, \ldots,$$

and the corresponding sequence of displacements $\{x_r\}$ by setting $t = t_r$ in (A), prove that

$$x_{r+1}/x_r = \exp(-\zeta\pi/\omega_0).$$

This establishes that the ratio of the amplitude of successive oscillations decreases by the constant factor $\exp(-\zeta\pi/\omega_0)$. The constant $\zeta\pi/\omega_0$ is called the *logarithmic decrement* of the oscillations.

Probability and statistics

16·1 Probability, discrete distributions and moments

One of the most direct applications of the elements of set theory is to be found in a formal introduction to probability theory. Because the notion of a probability is fundamental to many branches of engineering and science we choose to introduce some basic ideas and definitions now, making full use of the notions of set theory.

In some situations the outcome of an experiment is not determinate, so one of several possible events may occur. Following statistical practice we shall refer to an individual event of this kind as the result or outcome of a *trial*, whereas an agreed number of trials, say N, will be said to constitute an *experiment*. If an experiment comprises throwing a die N times, then a trial would involve throwing it once and the outcome of a trial would be the score that was recorded as a result of the throw. The experiment would involve recording the outcome of each of the N trials.

In general, if a trial has m outcomes we shall denote them by $E_1, E_2, \ldots, E_m$ and refer to each as a *simple event*. Hence a trial involving tossing a coin would have only two simple events as outcomes: namely 'heads', which could be labelled E_1, and 'tails', which would then be labelled E_2. In this instance an experiment would be a record of the outcomes from a given number of such trials. A typical record of an experiment involving tossing a coin eight times would be $E_1, E_2, E_1, E_1, E_1, E_2, E_2, E_1$. With such a simple experiment the E_1, E_2 notation has no apparent advantage over writing H in place of E_1 and T in place of E_2 to obtain the equivalent record H, T, H, H, H, T, T, H. The advantage of the E_i notation accrues from the fact that the subscript attached to the E may be ordered numerically, thereby enabling easier manipulation of the outcomes during analysis.

Events such as the result of tossing a coin or throwing a die are called chance or *random* events, since they are indeterminate and are supposedly the consequence of unbiased chance effects. Experience suggests that the relative frequency of occurrence of each such event averaged over a series of similar experiments tends to a definite value as the number of experiments increases.

The *relative frequency of occurrence* of the simple event E_i in a series of N trials is thus given by the expression

$$\frac{\text{Number of occurrences of event } E_i}{N}.$$

By virtue of its definition, this ratio must either be positive and less than unity, or be zero. For any given N, this ratio provides an estimate of the theoretical ratio that would have been obtained were N to have been made arbitrarily large. This theoretical ratio will be called the *probability* of occurrence of event E_i and will be written $P(E_i)$. In many simple situations its value may be arrived at by making reasonable postulates concerning the mechanisms involved in a trial. Thus when fairly tossing an unbiased coin it would be reasonable to suppose that over a large number of trials the number of 'heads' would closely approximate the number of 'tails' so that $P(H) = P(T) = \frac{1}{2}$. Here, of course, $P(H)$ signifies the probability of occurrence of a 'head' and $P(T)$ signifies the probability of occurrence of a 'tail'.

If there are m outcomes $E_1, E_2, \ldots, E_m$ of a trial, and they occur with the respective frequencies $n_1, n_2, \ldots, n_m$ in a series of N trials, then we have the obvious identity

$$\frac{n_1 + n_2 + \cdots + n_m}{N} = 1.$$

When N becomes arbitrarily large we may interpret each of the relative frequency ratios n_i/N $(i = 1, 2, \ldots, m)$ occurring on the left-hand side as the probability of occurrence $P(E_i)$ of event E_i, thereby giving rise to the general result

$$P(E_1) + P(E_2) + \cdots + P(E_m) = 1. \tag{16·1}$$

By this time a careful reader will have noticed that the definition of probability adopted here has a logical difficulty associated with it, namely, the question whether a relative frequency ratio such as n_i/N can be said to approach a definite number as N becomes arbitrarily large. We shall not attempt to discuss this philosophical point more fully, but rather be content that our simple definition in terms of the relative frequency ratio is in accord with everyday experience.

An examination of Eqn (16·1) and its associated relative frequency ratios is instructive. It shows the obvious results that:

(a) if event E_i never occurs, then $n_i = 0$ and $P(E_i) = 0$;
(b) if event E_i is certain to occur, then $n_i = N$ and $P(E_i) = 1$;
(c) if event E_i occurs less frequently than event E_j, then $n_i < n_j$ and $P(E_i) < P(E_j)$;
(d) if the m possible events $E_1, E_2, \ldots, E_m$ occur with equal frequency, then $n_1 = n_2 = \cdots = n_m = N/m$ and $P(E_1) = P(E_2) = \cdots = P(E_m) = 1/m$.

The relationship between sets and probability begins to emerge once it is appreciated that a trial having m different outcomes is simply a rule by which an event may be classified unambiguously as belonging to one of m different sets. Often a geometrical analogy may be used to advantage when representing the different outcomes of a particular trial.

A convenient example is provided by the simple experiment which involves throwing two dice and recording their individual scores. There will be in all 36 possible outcomes which may be recorded as the ordered number pairs (1, 1), (1, 2), (1, 3), . . ., (2, 1), (2, 2), . . ., (6, 5), (6, 6). Here the first integer in the ordered number pair represents the score on die 1 and the second the score on die 2. These may be plotted as 36 points with integer coordinates as shown in Fig. 16·1 (a).

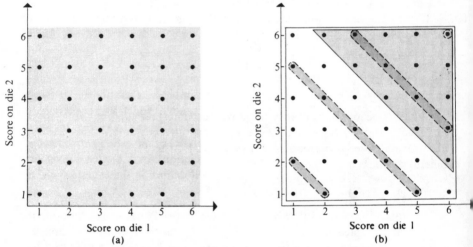

Fig. 16·1 Sample space for two dice: (a) complete sample space; (b) sample space for specific outcome.

Because each of the indicated points in Fig. 16·1 (a) lies in a two-dimensional geometrical space (that is, they are specific points in a plane), and in their totality they describe all possible outcomes, the representation is usually called the *sample space* of events. The probability of occurrence of an event characterized by a point in the sample space is, of course, the probability of occurrence of the simple event it represents.

As a sample space will require a 'dimension' for each of its variables it is immediately apparent that only in simple cases can it be represented graphically.

The points in the sample space may be regarded as defining points in a set D so that specific requirements as to the outcome of a trial will define a subset A of D, at each point of which the required event will occur. Typical of this situation would be the case in which a simple event is the throw of two dice, and the requirement defining the subset is that the combined score after throwing the two dice equals or exceeds 8. Here the set D would be the 36 points within the square in Fig. 16·1 (b) and the set A the 15 points within the triangle. Using set notation we may write $A \subset D$.

The sample space representation becomes particularly valuable when trials are considered whose outcome depends on the combination of events belonging to two different subsets A and B of the sample space. Thus, again using our previous example and taking for A the points within the triangle in Fig. 16·1 (b), the points in B might be determined by the requirement that the combined score be divisible by the integer 3. The set of points B is then those contained within the dotted curves of Fig. 16·1 (b).

A new set C may be derived from two sets A and B in two essentially different ways according as:

(a) C contains points in A or B or both;
(b) C contains points in A and B.

If desired, these statements about sets may be rewritten as statements about events. This is so because there is an unambiguous relationship between an event and the set of points S in the sample space at which that event occurs. Thus, for example, we may paraphrase the first statement by saying, *the event corresponding to points in C denotes the occurrence of the events corresponding to points in A or B, or both.* Because of this relationship it is often convenient to regard an event and the subset of points it defines in the sample space as being synonymous.

The statements provide yet another connection with set theory, since in (a) we may obviously write $C = A \cup B$, whereas in (b) we must write $C = A \cap B$. In terms of the sets A and B defined in connection with Fig. 16·1 (b), the set $C = A \cup B$ contains the points in the triangle together with those within the two dotted curves exterior to the triangle. The set $C = A \cap B$ contains only the five points within the two dotted curves lying inside the triangle.

Here it should be remarked that the statistician usually avoids the set theory symbols $\cup$ and $\cap$, preferring instead to denote the union of A and B by $A + B$ and their intersection by AB. This largely arises because of the duality we have already mentioned that exists between an event and the set of points it defines; the statistician naturally preferring to think in terms of events rather than sets. However, to emphasize the connection with set theory we shall preserve the set theory notation.

Using this duality we now denote by $P(A)$ the probability that an event corresponding to a point in the sample space lies within subset A, and define its value to be as follows:

DEFINITION 16·1

$P(A)$ is the sum of the probabilities associated with every point belonging to the subset A.

In Fig. 16·1 (b) the set A contains the 15 points within the triangle and, since for unbiased dice each point in the sample space is equally probable, it follows at once that the probability 1/36 is to be associated with each of these points. Hence from our definition we see that in this case, $P(A)$ = 15 × (1/36) = 5/12. Similarly, for the set B comprising the 12 points contained within the dotted curves we have $P(B) = 12 × (1/36) = 1/3$.

We can now introduce the idea of a conditional probability through the following definition.

DEFINITION 16·2

$P(A|B)$ is the *conditional probability* that an event known to be associated with set B is also associated with set A.

Clearly we are only interested in the relationship that exists between A and B, with B now playing the part of a sample space. Because in Definition 16·2 B plays the part of a sample space, but is itself only a subset of the complete sample space, it is sometimes given the name of the *reduced sample space*.

In terms of set theory Definition 16·2 is easily seen to be equivalent to

$$P(A|B) = \frac{P(A \cap B)}{P(B)}, \tag{16·2}$$

which immediately shows us how $P(A|B)$ may be computed. Namely, $P(A|B)$ is obtained by dividing the sum of the probabilities at points belonging to the intersection $A \cap B$ of sets A and B by the sum of the probabilities at points belonging to B. This ensures that $P(B|B) = 1$ as would be expected.

We can illustrate this by again appealing to the sets A and B defined in connection with Fig. 16·1 (b). It has already been established that $P(B) = 1/3$, and since there are only five points in $A \cap B$, each with a probability 1/36, it follows that $P(A \cap B) = 5/36$. Hence $P(A|B) = (5/36)/(1/3) = 5/12$. This result expressed in words states that when two dice are thrown and their score is divisible by the integer 3, then the probability that it also equals or exceeds 8 is 5/12.

A direct consequence of Eqn (16·2) is the so-called *probability multiplication rule*:

THEOREM 16·1   If two events define subsets A and B of a sample space, then

$P(A \cap B) = P(B)P(A|B)$.

Sometimes, when it is given that the event corresponding to points in subset B occurs, it is also true that $P(A|B)$ depends only on A, so that $P(A|B) = P(A)$. The events giving rise to subsets A and B will then be said to be independent. The probability multiplication rule then simplifies in an obvious manner which we express as follows:

Corollary 16·1   If the events giving rise to subsets A and B of a sample space are independent, then

$P(A \cap B) = P(A)P(B)$.

Consideration of the interpretation of $P(A \cup B)$ leads to another important result known as the *probability addition rule*:

THEOREM 16·2   If two events define subsets A and B of a sample space, then

$P(A \cup B) = P(A) + P(B) - P(A \cap B)$.

The proof of this theorem is self-evident once it is remarked that when computing $P(A)$ and $P(B)$ from subsets A and B and then forming the expression $P(A) + P(B)$, the sum of probabilities at points in the intersection $A \cap B$ is counted twice. Hence $P(A) + P(B)$ exceeds $P(A \cup B)$ by an amount $P(A \cap B)$.

The probability addition rule also has an important special case when sets A and B are disjoint so that $A \cap B = \phi$. When this occurs the events corresponding to sets A and B are said to be *mutually exclusive* and we express the result as follows:

Corollary 16·2   If the events giving rise to subsets A and B of a sample space are mutually exclusive, then

$P(A \cup B) = P(A) + P(B)$.

As a simple illustration of Theorem 16·2 we again use the sets A and B defined in connection with Fig. 16·1 (b) to compute $P(A \cup B)$. The result is immediate for we have already obtained the results $P(A) = 5/12, P(B) = 1/3$, and $P(A \cap B) = 5/36$, so from Theorem 16·2 follows the result

$P(A \cup B) = 5/12 + 1/3 - 5/36 = 11/18$.

The applications of these theorems and their corollaries are well illustrated by the following simple examples.

Example 16·1   A bag contains a very large number of red and black balls in the ratio 1 red ball to 4 black. If 2 balls are drawn successively from the bag at random, what is the probability of selecting

(a) 2 red balls,
(b) 2 black balls,
(c) 1 red and 1 black ball?

Let A_1 denote the selection of a red ball first (and either colour second), and A_2 the selection of a red ball second (and either colour first). Then $A_1 \cap A_2$ is the selection of 2 red balls and, similarly, $B_1 \cap B_2$ is the selection of 2 black balls. As the balls occur in the ratio 1 red : 4 black it follows that their relative frequency ratios are 1/5 for a red ball and 4/5 for a black ball, so $P(A_1) = 1/5$ and $P(B_1) = 4/5$.

The fact that the bag contains a *large* number of balls implies that the drawing of one or more balls does not materially alter the relative frequency ratio that existed at the start, so $P(A_i) = 1/5$ and $P(B_i) = 4/5$. This, together with the fact that the balls are drawn at random, implies that the drawing of each ball is an *independent* event. The independence of events A and B then allows the use of Corollary 16·1 to determine the required solutions to (a) and (b). We find that

(a) $P(A_1 \cap A_2) = (1/5) \cdot (1/5) = 1/25,$

(b) $P(B_1 \cap B_2) = (4/5) \cdot (4/5) = 16/25.$

Now to answer (c) we notice that there are two mutually exclusive orders in which a red and a black ball may be selected. Namely as the event $C \cup D$ where $C = A_1 \cap B_2$ (red then black) and $D = B_1 \cap A_2$ (black then red). From Corollary 16·2 we then have that $P(C \cup D) = P(C) + P(D)$, where $P(C)$ and $P(D)$ are determined by Corollary 16·1. This shows that $P(C) = P(A_1)P(B_2)$ and $P(D) = P(B_1)P(A_2)$, so that $P(C) = P(D) = (1/5) \cdot (4/5) = 4/25$. The solution to (c) becomes

$$P(C \cup D) = 4/25 + 4/25 = 8/25.$$

The three forms of selection (a), (b), and (c) are themselves mutually exclusive and it must follow that $P(A_1 \cap A_2) + P(B_1 \cap B_2) + P(C \cup D) = 1$, as is readily checked. Indeed, this result could have been used directly to calculate $P(C \cup D)$ from $P(A_1 \cap A_2)$ and $P(B_1 \cap B_2)$ in place of the above argument using Corollary 16·2.

The previous situation becomes slightly more complicated if only a limited number of balls are contained in the bag.

Example 16·2 A bag contains 50 balls of which 10 are red and the remainder black. If 2 balls are drawn successively from the bag at random, what is the probability of selecting

(a) 2 red balls,

(b) 2 black balls,

(c) 1 red and 1 black ball?

This time the approach must be slightly different because, unlike Example 16·1, the removal of a ball from the bag now materially alters the probabilities involved when the next ball is drawn. In fact this is a problem involving conditional probabilities.

Here we shall define A to be the event that the first ball selected is red, and B to be the event that the second ball selected is red. The probability we must now evaluate is the probability of occurrence of event B given that event A has occurred. Expressed in set notation we have to find $P(A \cap B)$, the probability of occurrence of the event associated with $A \cap B$. This is a conditional probability with the set associated with event A playing the role of the reduced sample space. Utilizing this observation we now make use of Theorem 16·2 to write

$$P(A \cap B) = P(A)P(B|A).$$

Now the relative frequency of occurrence of a red ball at the first draw is $10/(10 + 40) = 1/5$, so that $P(A) = 1/5$. (Not till later will we use the fact that the relative frequency of occurrence of a black ball is $40/(10 + 40) = 4/5$.)

Given that a red ball has been drawn, 9 red balls and 40 black balls remain in the bag. If the next ball to be drawn is red then its probability of occurrence is the conditional probability $P(B|A) = 9/(9 + 40) = 9/49$. Hence it follows that the solution to (a) is

$$P(A \cap B) = (1/5) . (9/49) = 9/245.$$

It is interesting to compare this with the value $1/25$ that was obtained in Example 16·1 on the assumption that there was virtually an infinite number of balls in the bag.

If C is defined to be the event that the first ball drawn is black and D the event that the second ball drawn is black, then to answer (b) we must compute $P(C \cap D)$. Obviously, $P(C) = 4/5$, and by using an argument analogous to that above it follows that $P(D|C) = 39/(10 + 39) = 39/49$. Hence the solution to question (b) is

$$P(C \cap D) = (4/5) . (39/49) = 156/245.$$

Again this should be compared with the value $16/25$ obtained in Example 16·1.

The simplest way to answer (c) is to use the fact that events (a), (b), and (c) describe the only possibilities and so are mutually exclusive. Hence the sum of the three probabilities must equal unity. Denoting the probability of event (c) by P we have

$$P = 1 - P(A \cap B) - P(C \cap D),$$

showing that $P = 1 - 9/245 - 156/245 = 16/49$.

To close this section with a brief examination of repeated trials, the ideas of a permutation and a combination must be utilized. The student will already be familiar with these concepts from elementary combinatorial algebra and so we shall only record two definitions.

DEFINITION 16·3   A *permutation* of a set of n mutually distinguishable objects r at a time is an arrangement, or an enumeration of the objects, in which their order of appearance counts.

Thus of the five letters a, b, c, d, e the arrangements a, b, c and a, c, b represent two different permutations of three of the five letters. These are described as permutations of five letters taken three at a time. Other permutations of this kind may be obtained by further re-arrangement of the letters a, b, c and by the replacement of any of them by either or both of the remaining two letters d and e.

The total number of different permutations of n objects r at a time will be denoted by ^nP_r and it is left to the reader to prove as an exercise that

$$^nP_r = \frac{n!}{(n-r)!}, \tag{16.3}$$

where $n!$ (factorial n) $= n(n-1)(n-2)\ldots 3.2.1$, and we adopt the convention that $0! = 1$.

DEFINITION 16·4 A *combination* of a set of n mutually distinguishable objects r at a time is a selection of r objects from the n without regard to their order of arrangement.

It follows from the definition of a permutation that a set of r objects may be arranged in $r!$ different ways so that the number of different combinations of n objects r at a time must be $n!/r!(n-r)!$ which is simply the binomial coefficient $\binom{n}{r}$. We have thus found the result that the number of different combinations of n objects r at a time is

$$\binom{n}{r} = \frac{n!}{r!(n-r)!}. \tag{16.4}$$

In many books it will often be found that the expression ^nC_r is written in place of $\binom{n}{r}$. It will be recalled that the numbers $\binom{n}{r}$ are called *binomial coefficients* because of their occurrence in the binomial expansion

$$(p+q)^n = \sum_{r=0}^{n} \binom{n}{r} p^r q^{n-r}, \text{ with } n \text{ a positive integer.} \tag{16.5}$$

Now consider an experiment involving a series of independent trials in each of which only one of two events A or B may occur. Then if the probabilities of occurrence of events A and B are p and q, respectively, we must obviously have $p + q = 1$. If n such trials constitutes an experiment, we might wish to know with what probability the experiment may be expected to yield r events of type A. The statistician will call such a situation *repeated independent trials*.

An experiment will be deemed to be successful if r events of type A and $n - r$ events of type B occur, irrespective of their order of occurrence. Clearly this can happen in $\binom{n}{r}$ different ways and by Corollary 16·2, since the trials are independent, the probability of occurrence of any one of these events will be $p^r(1 - p)^{n-r}$. Hence, as the results of trials are also mutually exclusive, it follows from Corollary 16·3 that the required probability $P(r)$ of occurrence of r events of A each with probability of occurrence p in n independent trials is

$$P(r) = \binom{n}{r} p^r(1 - p)^{n-r}. \tag{16·6}$$

Identifying the p and q of Eqn (16·5) with the probabilities of occurrence of the events A and B just discussed, we see that $q = 1 - p$, so that Eqn (16·5) takes the form

$$1 = \sum_{r=0}^{n} \binom{n}{r} p^r(1 - p)^{n-r}. \tag{16·7}$$

Each term on the right-hand side of Eqn (16·7) then represents the probability of occurrence of an event of the form just discussed. For example, the first term

$$P(0) = \binom{n}{0}(1 - p)^n$$

is the probability that event A will never occur in a series of n independent trials, whilst the third term

$$P(2) = \binom{n}{2} p^2(1 - p)^{n-2}$$

is the probability that event A will occur exactly twice in a series of n independent trials.

The $n + 1$ numbers $P(r), r = 0, 1, \ldots, n$ have, by definition, the property that

$$P(0) + P(1) + \cdots + P(n) = 1, \tag{16·8}$$

and they are said to define a *discrete probability distribution*. It is conventional to plot them in histogram fashion when they illustrate the probabilities to be associated with the $n + 1$ possible outcomes of an experiment involving n trials. Fig. 16·2 (a) illustrates the case in which $n = 4$ and $p = \frac{1}{4}$ so that

$$P(0) = \binom{4}{0}\left(\frac{1}{4}\right)^0\left(\frac{3}{4}\right)^4 = \frac{81}{256}, \quad P(1) = \binom{4}{1}\left(\frac{1}{4}\right)^1\left(\frac{3}{4}\right)^3 = \frac{27}{64}, \quad \text{and, similarly,}$$

$P(2) = 54/256$, $P(3) = 3/64$, and $P(4) = 1/256$. Because of the origin of this distribution, Eqn (16·6) is said to define the *binomial distribution*. This distribution is historically associated with Jacob Bernoulli (1654–1705) and experiments of the type just examined are sometimes referred to as Bernoullian trials. When the cumulative total

$$U(r) = \sum_{t=0}^{r} P(t), \tag{16·9}$$

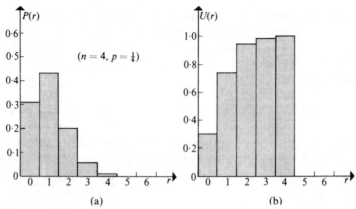

Fig. 16.2 Binomial distribution: (a) binomial probability density function; (b) binomial cumulative distribution function.

is plotted in histogram fashion against r the result is called the *cumulative distribution function*. The cumulative distribution function corresponding to Fig. 16·2 (a) is shown in Fig. 16·2 (b). It is conventional to refer to the $P(r)$ as the probability density function or the frequency function since it describes the proportion of observations appropriate to the value of r.

Example 16·3 If an unbiased coin is tossed six times, what is the probability that only two 'heads' will occur in the sequence of results?

As the coin is unbiased $p = q = \frac{1}{2}$ and so

$$P(2) = \binom{6}{2}\left(\frac{1}{2}\right)^2\left(\frac{1}{2}\right)^4 = \frac{15}{64}.$$

It is an immediate consequence of Eqn (16·6) that

(a) if A occurs with probability p in independent trials then the probability that it will occur *at least* r times in n trials is

$$\sum_{s=r}^{n} \binom{n}{s} p^s(1-p)^{n-s};$$

(b) if A occurs with probability p in independent trials then the probability that it will occur *at most* r times in n trials is

$$\sum_{s=0}^{r} \binom{n}{s} p^s(1-p)^{n-s};$$

and to this we may add Eqn (16·6) in this form:

(c) if A occurs with probability p in independent trials then the probability that it will occur *exactly* r times in n trials is

$$\binom{n}{r} p^r(1-p)^{n-r}.$$

Example 16·4 What is the probability of hitting a target when three shells are fired, assuming each to have a probability $\frac{1}{2}$ of making a hit?

Obviously here $p = \frac{1}{2}$ and we will have satisfied the conditions of the question if *at least* one shell finds the target. Accordingly, using (a) above, the result is

$$\sum_{s=1}^{3} \binom{3}{s}\left(\frac{1}{2}\right)^{s}(1 - \tfrac{1}{2})^{3-s}.$$

Hence the required probability is $\frac{3}{8} + \frac{3}{8} + \frac{1}{8} = \frac{7}{8}$.

The binomial distribution (16·6) is characterized by the two parameters n, p and its form varies considerably with the values they assume. One important case occurs when p is very small and n is large, in such a way that the product np is not negligible. The binomial distribution can then be approximated by a new and simpler distribution characterized by the single parameter $\mu = np$. To see this observe first that Eqn (16·6) can be written in the form

$$P(r) = Q(r, n)\frac{\mu^{r}}{r!}\left(1 - \frac{\mu}{n}\right)^{n}, \tag{16·10}$$

where

$$Q(r, n) = \left(1 - \frac{1}{n}\right)\left(1 - \frac{2}{n}\right) \cdots \left(1 - \frac{r-1}{n}\right)\left(1 - \frac{\mu}{n}\right)^{-r}.$$

However, for any fixed r we have

$$\lim_{n\to\infty} Q(r, n) = 1, \quad \text{while from Section 3·3} \lim_{n\to\infty}\left(1 - \frac{\mu}{n}\right)^{n} = e^{-\mu},$$

so that in the limit as $n \to \infty$,

$$P(r) = \frac{\mu^{r}e^{-\mu}}{r!}. \tag{16·11}$$

This is a probability density function in its own right. We are able to assert this because $P(r) \geq 0$ for all $r \geq 0$ and the probability that events of all frequencies occur, namely

$$\sum_{r=0}^{\infty} P(r) = \sum_{r=0}^{\infty}\frac{\mu^{r}e^{-\mu}}{r!} = e^{-\mu}\sum_{r=0}^{\infty}\frac{\mu^{r}}{r!} = e^{-\mu}e^{\mu} = 1,$$

as is required of a probability density function.

The probability density function in (16·11) is called the *Poisson distribution* and differs from the binomial distribution, for which we have seen it is a limiting approximation, in that it only involves the single parameter μ. It finds useful application in a variety of seemingly different circumstances. Thus, for example, it can be used to describe the particle count rate in a radioactive process, the rate at which telephone calls arrive at an exchange or the rate of arrivals of people in a queue.

The Poisson probability distribution is illustrated in Fig. 16·3 (a) for the choice of parameter $\mu = 3$. This could, for instance, be taken to be an approximation to a binomial distribution in which $n = 150$ and $p = 0·02$.

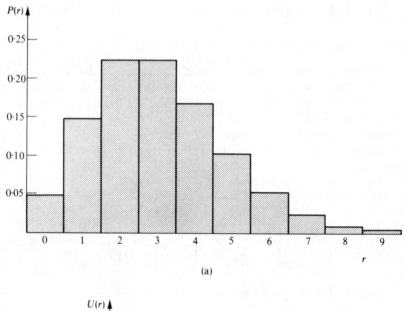

(a)

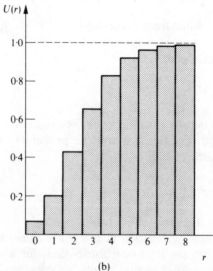

(b)

Fig. 16·3 (a) Poisson distribution with $\mu = 3$ and standard deviation $\sigma = \sqrt{3}$. (b) Corresponding Poisson cumulative distribution.

When the equivalent histogram for a binomial distribution with these values of n and p is shown for comparison on the same diagram, it is virtually indistinguishable from the Poisson distribution. The Poisson distribution in Fig. 16·3 (a) would not be a very good approximation for a binomial distribution with $n = 30$, $p = 0·1$. As a general rule, the approximation of the binomial distribution by the Poisson distribution is satisfactory when $n \geq 100$ and $p < 0·05$. The Poisson cumulative distribution function corresponding to Fig. 16·3 (a) is shown in Fig. 16·3 (b).

Whereas $P(r)$ is the probability that exactly r events will occur, the cumulative probability $U(r)$ is the probability that up to and including r events will occur. Thus $U(3)$ is the probability that 0, 1, 2, or 3 events will occur.

Example 16·5   An ammunition manufacturer knows from experience that 5 out of every 1000 shells manufactured can be expected to be defective. What is the probability that out of a batch of 600, (a) exactly 4 will be defective, (b) no more than 4 will be defective.

Solution   Here $p = 5/1000 = 1/200$ and $n = 600$ so that $\mu = np = 3$. The Poisson distribution is thus an appropriate approximation to the binomial distribution which really describes this situation, and it has the probability density function

$$P(r) = 3^r e^{-3}/r!.$$

Now as $P(r)$ is the probability that exactly r shells are defective, the answer to (a) is $P(4) = 3^4 e^{-3}/4! = 0·1680$. The answer to (b) is just $U(4)$, where

$$U(4) = P(0) + P(1) + P(2) + P(3) + P(4) = 0·8152.$$

By the same form of argument, had we needed to know the probability that at least m but no more than s shells were defective in a batch, then this would have been given by the sum $P(m) + P(m + 1) + P(m + 2) + \ldots + P(s)$.

When a discrete probability density function is known it is often useful to give a simple description of the distribution in terms of two numbers known as the first and second *moments* of the distribution. If the density function is $P(r)$, then when the sums are convergent it is conventional to denote the *first moment* or the *mean* by μ, and the *second moment* or the *variance* by σ^2, where

$$\mu = \sum_{r=0}^{\infty} r P(r), \tag{16·12}$$

and

$$\sigma^2 = \sum_{r=0}^{\infty} (r - \mu)^2 P(r). \tag{16·13}$$

Expanding the expression under the summation sign in (16·13), using (16·12) and the fact that $\sum_{r=0}^{\infty} P(r) = 1$ it is a simple matter to show that an alternative, and often more convenient, expression for the variance σ^2 is the following

$$\sigma^2 = \sum_{r=0}^{\infty} r^2 P(r) - \mu^2. \tag{16·14}$$

Using a mechanical analogy, the mean μ can be thought of as the perpendicular distance of the centre of gravity of the area under the distribution from the $P(r)$ axis. That is, the turning effect about the $P(r)$ axis of a unit mass at a perpendicular distance μ from that axis will be the same as the sum of the individual turning effects of masses $P(r)$ at perpendicular distances r from the axis. It is, indeed, this analogy that gives rise to the term *moment of a distribution* already mentioned. The square root σ of the variance is called the *standard deviation* and provides a measure of the spread of the distribution about the mean.

· We leave to the reader as exercises in manipulation the task of proving the following results:

Binomial distribution

probability density $P(r) = \binom{n}{r} p^r (1 - p)^{n-r}$, $r = 0, 1, \ldots, n$

mean $\mu = np$, variance $\sigma^2 = np(1 - p)$.

Poisson distribution

probability density $P(r) = \dfrac{\mu^r e^{-\mu}}{r!}$, $r = 0, 1, 2, \ldots$

mean $\mu = np$, variance $\sigma^2 = \mu$.

16·2 Continuous distributions—normal distribution

So far the sample spaces we have used have involved discrete points, and it is for this reason that the term discrete has been used in conjunction with the definition of the binomial and Poisson distributions. In other words, in discrete distributions, no meaning is to be attributed to points that are intermediate between the discrete sample space points.

When different situations involving probability are examined it is often necessary to consider sample spaces in which all points are allowed. This happens, for instance, when considering the actual measured lengths of manufactured items having a specific design length, or when considering the actual measured capacity of capacitors, all having the same nominal capacity.

The probability density for situations like these can be thought of as being arrived at as the limit of an arbitrarily large number of discrete observations in the following sense. Suppose first that all the measured quantities x lie within the interval $a \leq x \leq b$. Next, consider the division of $[a, b]$ into n equal intervals of length $\Delta = (b - a)/n$ and number them sequentially from $i = 1$ to n starting from the left-hand side of the interval $[a, b]$. Thus the ith interval corresponds to the inequality $a + (i - 1)\Delta < x < a + i\Delta$. Now in the set or *sample* of N measurements taken from the totality of all possible measurements, hereafter called the *population* from which the measurements are drawn, it will be observed that f_i of these lie within the

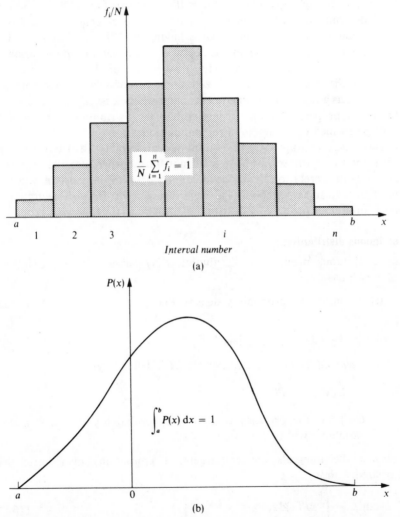

Fig. 16·4 (a) Relative frequency polygon or histogram. (b) Continuous probability distribution obtained as the limit of the frequency polygon as n, $N \to \infty$.

ith interval. This number f_i is the observed *frequency of occurrence* of a measurement in the ith interval and it is obvious that $\sum\limits_{i=1}^{n} f_i = N$. If a *relative frequency polygon* or *histogram* were to be constructed at this stage by plotting the numbers f_i/N against i then it would appear as in Fig. 16·4 (a)

As the number of observations N becomes arbitrarily large, so as to ultimately include all the population, so each number f_i/N will tend to a limiting value. Then, by allowing n to become arbitrarily large, it is possible to associate a number $P(x)$ with every value of x in $a \leq x \leq b$. The type of smooth limiting curve that results is illustrated in Fig. 16·4 (b). Because x is indeterminate within the interval $a \leq x \leq b$ associated with $P(x)$ it will be said to be a *random variable* having the distribution $P(x)$. It is important at this stage to distinguish between the general random variable x occurring in a continuous probability distribution of this type, which is a theoretical distribution, and its realization in terms of an actual measurement carried out as a result of an experiment. Later we look briefly at the problems of inferring properties of a governing probability distribution from measurements performed on a sample of the population.

This intuitive approach to continuous probability distributions must suffice for our purposes, and we shall not look deeper into the important matters of interpretation that arise in connection with the sense in which the limit f_i/N is to be understood. It is, nevertheless, clear from these arguments that a continuous probability distribution is one with the following properties.

Continuous distribution

Any continuous probability distribution $P(x)$ defined for some interval $a \leq x \leq b$ has

(i) a continuous probability density $P(x) \geq 0$ for $a \leq x \leq b$; is such that

(ii) $\displaystyle\int_a^b P(x)\,dx = 1$; and has

(iii) a *cumulative distribution function* $U(x)$ defined by

$$U(x) = \int_a^x P(t)\,dt.$$

$U(\alpha)$ has the property that it is the probability that the random variable x will be such that $x \leq \alpha$.

When the integrals are convergent, a continuous distribution has associated with it a

$$\text{mean } \mu = \int_a^b xP(x)\,dx, \qquad\qquad (16\cdot15)$$

and a

$$\text{variance } \sigma^2 = \int_a^b (x - \mu)^2 P(x)\, dx. \tag{16·16}$$

The simplest example of a continuous distribution is the *uniform* or *rectangular distribution* defined by

$$P(x) = \begin{cases} 1/(b - a) & \text{for } a \le x \le b \\ 0 & \text{elsewhere.} \end{cases}$$

This characterizes an event associated with a value of x that is equally likely to occur for any x in the interval $a \le x \le b$. From Eqns (16·15) and (16·16) it follows that for the uniform distribution

$$\mu = \frac{a + b}{2} \quad \text{and} \quad \sigma^2 = \frac{a^2 + ab + b^2}{3}.$$

Example 16·6 A target moves along a straight path of length l joining points A and B. The probability density that a single shot fired at it will score a hit is described by the uniform distribution over AB. If a hit is scored at a point C distant x from A, what is the probability that the ratio AC/CB lies between the two positive numbers k_1 and k_2 ($k_1 < k_2$).

Solution As C is distant x from A we know that $AC/CB = x/(l - x)$, so that

$$k_1 \le \frac{x}{l - x} \le k_2 \quad \text{or, equivalently,} \quad \frac{k_1 l}{1 + k_1} \le x \le \frac{k_2 l}{1 + k_2}.$$

Now the uniform probability density $P(x)$ over AB is simply $P(x) = 1/l$ for $0 \le x \le l$. Thus the desired probability P must be

$$P = \int_{k_1 l/1 + k_1}^{k_2 l/1 + k_2} \left(\frac{1}{l}\right) dx = \frac{k_2 - k_1}{(1 + k_1)(1 + k_2)}.$$

Normal distribution

By using more advanced methods than those we have at our disposal in this book it can be shown that

$$1 = \frac{1}{(2\pi)^{\frac{1}{2}}} \int_{-\infty}^{\infty} \exp\left(-\frac{X^2}{2}\right) dX, \tag{16·17}$$

and

$$1 = \frac{1}{(2\pi)^{\frac{1}{2}}} \int_{-\infty}^{\infty} X^2 \exp\left(-\frac{X^2}{2}\right) dX. \tag{16·18}$$

The fact that the integrand of Eqn (16·17) is essentially positive shows that the function

$$P(X) = \frac{1}{(2\pi)^{\frac{1}{2}}} \exp\left(-\frac{X^2}{2}\right) \quad \text{for } -\infty < X < \infty \tag{16·19}$$

can be regarded as a probability density function. Now $XP(X)$ is an *odd* function and tends rapidly to zero as $|X| \to \infty$, and so it follows immediately that

$$0 = \frac{1}{(2\pi)^{\frac{1}{2}}} \int_{-\infty}^{\infty} X \exp\left(-\frac{X^2}{2}\right) dX. \tag{16.20}$$

We have thus established that $P(X)$ as defined in Eqn (16·19) has a mean $\mu = 0$, and hence from Eqn (16·18) that $P(X)$ has a variance $\sigma^2 = 1$. When the random variable X satisfies the probability density (16·19) it is said to have a *standardized normal distribution*. The normal distribution is the most important of all the continuous probability distributions and for historical reasons it is also known as the *Gaussian distribution*.

Making the change of variable $x = \sigma X + \mu$ it is a simple matter to transform integral (16·17) and to deduce that the corresponding unstandardized normal distribution obeyed by x has the probability density function

$$P(x) = \frac{1}{(2\pi\sigma^2)^{\frac{1}{2}}} \exp\left(-\frac{(x-\mu)^2}{2\sigma^2}\right) \quad \text{for } -\infty < x < \infty. \tag{16·21}$$

By appeal to Eqns (16·18) and (16·20) it is then easily established that (16·21) described a *normal distribution* with mean μ and variance σ^2. It is customary to represent the fact that the random variable x obeys a normal distribution with mean μ and variance σ^2 by saying that x obeys the distribution $N(\mu, \sigma^2)$. In this notation the random variable X of (16·19) obeys the normal distribution $N(0, 1)$.

The cumulative standardized normal distribution is usually denoted by a special symbol, and we choose to use $\Phi(x)$, where

$$\Phi(x) = \frac{1}{(2\pi)^{\frac{1}{2}}} \int_{-\infty}^{x} \exp\left(-\frac{t^2}{2}\right) dt. \tag{16·22}$$

As before, $\Phi(x)$ is the probability that the random variable X is such that $X < x$.

The $N(0, 1)$ normal probability density (16·19) and the corresponding cumulative normal distribution function (16·22) are readily available in tabulated form in any statistical tables. An abbreviated table is given below for $x \geq 0$, and the corresponding graphs are illustrated in Fig. 16·5. To deduce values of $P(x)$ and $\Phi(x)$ for negative x from this table it is only necessary to use the fact that $P(x)$ is an even function, so that $P(-x) = P(x)$, when it then follows that $\Phi(-x) = 1 - \Phi(x)$.

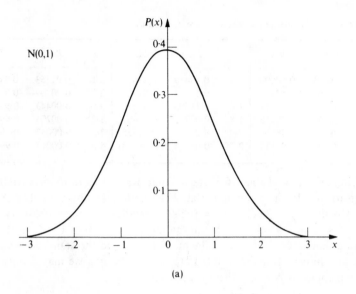

(a)

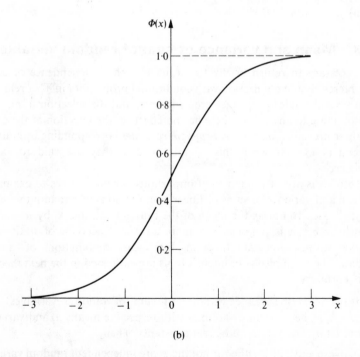

(b)

Fig. 16.5 (a) $N(0, 1)$ normal distribution. (b) Corresponding cumulative normal distribution.

Table 16·1 $N(0, 1)$ normal and cumulative distributions

x	$P(x)$	$\Phi(x)$	x	$P(x)$	$\Phi(x)$	x	$P(x)$	$\Phi(x)$
0	0·39894	0·50000	1·2	0·19419	0·88493	2·5	0·01753	0·99379
0·2	0·39104	0·57926	1·4	0·14973	0·91924	2·75	0·00909	0·99702
0·4	0·36827	0·65542	1·6	0·11092	0·94520	3·0	0·00443	0·99865
0·6	0·33322	0·72575	1·8	0·07895	0·96407	3·25	0·00203	0·99942
0·8	0·28969	0·78814	2·0	0·05399	0·97725	3·5	0·00087	0·99977
1·0	0·24197	0·84134	2·25	0·03174	0·98778	4·0	0·00013	0·99997

Examination of Table 16·1 shows that when the random variable X belongs to $N(0, 1)$, the probability that X lies within the interval $-1 \leq X \leq 1$ is $\Phi(1) - \Phi(-1) = 2\Phi(1) - 1 = 0·6827$. Similarly, the probability that $-2 \leq X \leq 2$ is $0·9545$ and the probability that $-3 \leq X \leq 3$ is $0·9973$. Recalling that when X is $N(0, 1)$ the mean $\mu = 0$ and the standard deviation $\sigma = 1$, and interpreting the results in terms of $N(\mu, \sigma^2)$, we may deduce that when x belongs to $N(\mu, \sigma^2)$ there is approximately

(i) a probability $\frac{2}{3}$ that $| x - \mu | \leq \sigma$,
(ii) a probability $0·95$ that $| x - \mu | \leq 2\sigma$,
(iii) a probability $0·997$ that $| x - \mu | \leq 3\sigma$.

16·3   Mean and variance of sum of random variables

It is necessary to return briefly to the idea of the independence of events encountered first in connection with conditional probability in Theorem 16·1. When several random variables are of interest and the selection of an event corresponding to any one of these has no effect on the selection of the others then these are called *independent random variables* corresponding to mutually independent events; when this is not the case they are said to exhibit *dependence*.

These ideas are of fundamental importance when it comes to examining the mean and variance of sums or linear combinations of n random variables $x_1, x_2, \ldots, x_n$. Denoting the mean of the random variable x_i by μ_i and the variance by σ_i^2 it is possible in a more advanced account of statistics to establish various important theorems for linear combinations of random variables. The main results of interest to us are expressed in the next theorem and its corollary.

THEOREM 16·3 (mean and variance of sum of random variables)   Let $x_1, x_2, \ldots, x_n$ be n random variables with respective means μ_i and variances σ_i^2, and let $a_1, a_2, \ldots, a_n, b$ be real constants. Then,

(i) irrespective of whether or not the x_i are independent random variables
mean of $(a_1x_1 + a_2x_2 + \ldots + a_nx_n + b) = a_1\mu_1 + a_2\mu_2$
$$+ \ldots + a_nx_n + b;$$

(ii) when the x_i are *independent* random variables

$$\text{variance of } (a_1 x_1 + a_2 x_2 + \ldots + a_n x_n + b) = a_1 \sigma_1{}^2$$
$$+ a_2 \sigma_2{}^2 + \ldots + a_n \sigma_n{}^2.$$

Corollary 16·3   If n independent random variables x_i are all drawn from distributions having the same mean μ and variance σ^2, then the random variable $(x_1 + x_2 + \ldots + x_n)/n$ has the same mean μ but the variance σ^2/n.

16·4   Statistics—Inference drawn from observations

Thus far we have considered the ideal situation involving random variables as characterized by probability theory. The situation confronting the user of statistics is somewhat different. He is usually required to draw some *inference* concerning the theoretical probability distribution governing some observations of interest from the imperfect data provided by only a sample taken from the possible population. Here, by *population*, we mean simply the totality of possible observational outcomes in the situation of interest.

For example, when attempting to assess the quality of a production line for 1000-ohm resistors it is customary to do so by a periodic check on a random sample of a given size. All resistors could be measured but the cost might be prohibitive. The situation is rather different with ammunition, for all shells could be tested by firing, but then there would be no useful output from the production line. These are, respectively, typical examples of non-destructive and destructive testing.

Let us first outline the simplest way in which data or *observations* may be categorized. In the first instance data may be grouped into categories, each with a value of a random variable associated with it, and a relative frequency polygon constructed similar to that in Fig. 16·4 (a), irrespective of whether or not the observations are governed by a continuous distribution. Thus we might take 5-ohm intervals about the desired value of a resistor and record the frequency of occurrence f_i with which the resistance of a resistor drawn from a given random sample of N falls in the ith such interval. The observed relative frequency of occurrence in that interval is then f_i/N and the results could be recorded in the form shown in Table 16·2.

Table 16·2

Interval i	1	2	3	. . .	i	. . .	n
Frequency	f_1	f_2	f_3	. . .	f_i	. . .	f_n
Relative frequency	f_1/N	f_2/N	f_3/N	. . .	f_i/N	. . .	f_n/N

By associating a nominal resistance x_i with the ith interval, equal, say, to the resistance at the mid-point of that interval, it is then possible to define the *sample mean* $\bar{x}$ in a natural manner by means of the expression

$$\bar{x} = \sum_{i=1}^{n} f_i x_i / N \quad \text{with} \quad N = \sum_{i=1}^{n} f_i. \tag{16.23}$$

It would then also seem reasonable to define the *sample variance* by the following expression

$$\sum_{i=1}^{n} (x_i - \bar{x})^2 f_i / N,$$

since this is in accord with (16·16). However, it can be shown that this definition does not take fully into account the fact that a finite sample size biases the estimate of the population variance. In advanced books on statistics it is established that this bias in the estimate is corrected if the *sample variance s^2* is computed from the modified expression

$$s^2 = \frac{1}{N-1} \sum_{i=1}^{n} (x_i - \bar{x})^2 f_i, \tag{16·24}$$

which is the one we shall use throughout this chapter. It is clear that this definition only differs significantly from the previous one when N is small, which is as would be expected, for then the bias is greatest. Computationally, it is usually best to combine (16·23) and (16·24), and to determine the sample variance from the result

$$s^2 = \frac{1}{N-1} \left[\sum_{i=1}^{n} f_i x_i^2 - N\bar{x}^2 \right], \tag{16·25}$$

which is equivalent to (16·24). It can be established that the estimate of the mean provided by (16·23) is free from bias due to sample size. As before, s is known as the sample *standard deviation*.

Other descriptive parameters associated with the relative frequency polygon that are sometimes used are:

(a) the *range* of the observations, which is the difference between the largest and smallest observations in the sample under consideration;

(b) the *mode*, which is equal to the random variable x_i corresponding to the most frequently occurring observation when this exists. It is not defined if more than one observation x_i has the maximum frequency of occurrence. This provides the simplest measure of the location of the bulk of the observations;

(c) the *median*, which is the value of the random variable x_i below which exactly half the observations lie. When there is no grouping and the number N of observations involved in the sample is even, the median is defined to be equal to the average of the two adjacent central observations. In the event that observations are grouped it usually becomes necessary to introduce a fictitious median by linear interpolation. This can also be found by inspection of a graph of the cumulative polygon in which adjacent points are joined by straight lines. The median provides a better measure of location for the observations than does the mode, and is usually preferable to the mean when the relative frequency polygon is badly asymmetric and the distribution is not known;

(d) the first, second and third *quartiles* Q_1, Q_2 and Q_3 which are an immediate extension of the idea underlying the median. They represent the values of the random variable x_i below which one quarter, one half and three quarters of the observations lie. The second quartile Q_2 is identical with the median;

(e) the *mean deviation* which, like the standard deviation, provides a measure of the spread or dispersion of the observations about their mean $\bar{x}$. It is defined as the number

$$\sum_{i=1}^{n} | x_i - \bar{x} | f_i/N. \tag{16·26}$$

Example 16·7   Find the range, mode, mean, median, variance, standard deviation, and mean deviation of the observations x_i occurring with frequencies f_i that are contained in the following table.

i	1	2	3	4	5	6	7	8
x_i	0·2	0·4	0·6	0·8	1·0	1·2	1·4	1·6
f_i	2	6	10	9	7	8	4	2
Σf_i	2	8	18	27	34	42	46	48

Solution   The range is $1·6 - 0·2 = 1·4$, the mode is $0·6$ and the mean

$$\bar{x} = (0·2 \times 2 + 0·4 \times 6 + 0·6 \times 10 + 0·8 \times 9 + 1·0 \times 7$$
$$+ 1·2 \times 8 + 1·4 \times 4 + 1·6 \times 2)/48$$
$$= 0·8625.$$

As there are 48 observations, inspection of the cumulative frequency in the bottom row of the table shows that a fictitious median is required corresponding to 24 observations. It must lie between 0·6 and 0·8, and the linearly interpolated value of the median is

$$0·6 + \left(\frac{24 - 18}{27 - 18}\right)(0·8 - 0·6) = 0·733.$$

The sample variance

$$s^2 = \frac{1}{N - 1} \left[\sum_{i=1}^{n} f_i x_i^2 - N\bar{x}^2 \right] = \tfrac{1}{47}[41·88 - 48 \times 0·8625^2] = 0·1313,$$

and the sample standard deviation $s = \sqrt{0·1313} = 0·362.$

The mean deviation is computed as follows:

$$\{2 \mid 0 \cdot 2 - 0 \cdot 863 \mid + 6 \mid 0 \cdot 4 - 0 \cdot 863 \mid + 10 \mid 0 \cdot 6 - 0 \cdot 863 \mid$$
$$+ 9 \mid 0 \cdot 8 - 0 \cdot 863 \mid + 7 \mid 1 - 0 \cdot 863 \mid + 8 \mid 1 \cdot 2 - 0 \cdot 863 \mid$$
$$+ 4 \mid 1 \cdot 4 - 0 \cdot 863 \mid + 2 \mid 1 \cdot 6 - 0 \cdot 863 \mid\}/48 = 0 \cdot 304.$$

The inferences drawn so far from the data of Table 16·2 are rather general and, for example, take no account of the fact that different samples drawn from the same population would vary, and so effect our conclusions. In what follows we examine a few basic statistical tests which take account of this variability, and look also at their possible application. We make no rigorous attempt to derive these tests since to do so would require aspects of probability theory it has been necessary to omit for reasons of mathematical simplicity.

In considering these tests we shall have in mind a *random sample* of given size N drawn from the possible population in such a way that each member of the population has an equal chance of inclusion in the sample. Usually some simple *statistical hypothesis* will be in mind which involves some property of the underlying probability distribution that is involved. We shall make some hypothesis or assumption about this distribution and then test to see if it is in agreement with the sample data. In statistical language such an hypothesis is called the *null hypothesis*, and it is usually abbreviated by the notation H_0. Thus H_0 might, for example, be the hypothesis that the population mean is 3·7. The test would then indicate with what probability this was in agreement with the calculated sample mean $\bar{x}$. Using statistical language, we would *accept* H_0 if the probability of agreement equalled or exceeded a predetermined probability; otherwise H_0 would be *rejected*. That is to say, when H_0 is rejected, we would conclude that the sample was not drawn from a population with mean 3·7.

Inherent in such tests is the notion of a confidence interval which has already been encountered at the end of Section 16·2, though this term was not then used. There we saw, for example, that with probability approximately 0·95 the random variable x drawn from the distribution $N(\mu, \sigma^2)$ will be contained in the interval $\mu - 2\sigma \leq x \leq \mu + 2\sigma$. For a given distribution the width of the confidence interval depends on the probability, or confidence, with which the location of the random variable x is required to be known. Thus for this same distribution we saw that to locate x with a probability of about 0·997 it is necessary to widen the interval to $\mu - 3\sigma \leq x \leq \mu + 3\sigma$.

Suppose that an arbitrary probability density function $P(x)$ for the random variable x is involved, that an estimate x_1 of x is obtained from such a sample and that a number r is given ($0 \leq r \leq 100$). Then, corresponding to a probability of occurrence $1 - r/100$ of x in the interval, the $100 - r$ per cent *confidence interval* for random variable x

$$x_1 - d \le x \le x_1 + d \qquad\qquad (16\cdot27)$$

is obtained by determining d from the equation

$$P(|\,x - x_1\,| \le d) = 1 - \frac{r}{100}$$

or, equivalently,

$$P(x_1 - d \le x \le x_1 + d) = 1 - \frac{r}{100}. \qquad\qquad (16\cdot28)$$

The number $1 - r/100$ is called the *confidence coefficient* for this $100 - r$ per cent confidence interval for x.

Let us suppose now that we have a random sample of n independent observations x_1. $x_2, \ldots, x_n$ drawn from a normal distribution $N(\mu, \sigma^2)$ in which σ^2 is known. Then their mean $\bar{x}$ is also a random variable, and the more advanced theory that gave rise to Theorem 16·3 also asserts that $\bar{x}$ has a normal distribution. The distirbution of $\bar{x}$ is known as the *sampling distribution of the mean*. Now from Corollary 16·3 we know that the population mean of $(x_1 + x_2 + \ldots + x_n)/n$ is μ, but that its variance is σ^2/n, and we are assuming σ^2 to be known. So, as the variance of the sample means is smaller by a factor $1/n$ than the variance of the x_i, we can take our sample mean $\bar{x}$ as a good estimate of the population mean μ if n is large ($n > 50$).

Thus $\bar{x}$ belongs to the distribution $N(\bar{x}, \sigma^2/n)$ and so we are immediately able to determine a confidence interval for μ. Following result (ii) from the end of Section 16·2 we know that with approximately 95 per cent confidence μ lies within the interval $\bar{x} - 2\sigma/\sqrt{n} \le \mu \le \bar{x} + 2\sigma/\sqrt{n}$. Alternatively, expressed in terms of Eqns (16·27) and (16·28), we have found that the sampling distribution of the mean $P(\bar{x})$ is $N(\bar{x}, \sigma^2/n)$ and that corresponding to $r = 5$, $x_1 = \bar{x}$, d is approximately $2\sigma/\sqrt{n}$. By interpolation in Table 16·1 the approximate results at the end of Section 16·2 may be made exact and turned into the following statistical rule.

Rule 1   Confidence interval for the population mean μ of a normal distribution with known variance

First select a $100 - r$ per cent confidence level and use this to determine α_r from the following table.

r	20	10	5	1	0·1
Confidence level	80%	90%	95%	99%	99·9%
α_r	1·282	1·645	1·960	2·576	3·291

Using this value of α_r, and denoting the computed sample mean by $\bar{x}$ and the known variance by σ^2, the $100 - r$ per cent confidence interval for the population mean μ is

$$\bar{x} - \alpha_r\sigma/\sqrt{n} \leq \mu \leq \bar{x} + \alpha_r\sigma/\sqrt{n}.$$

Example 16·8 A measuring microscope used for the determination of the length of precision components has a standard deviation $\sigma = 0·057$ cm. Find the 90 per cent and 99 per cent confidence intervals for the mean μ given that a sample mean $\bar{x} = 3·214$ cm was obtained as the result of 64 observations.

Solution Here $n = 64$, $\sigma = 0·057$ cm and for 90 per cent confidence interval $\alpha_{10} = 1·645$, whilst for the 99 per cent confidence interval $\alpha_1 = 2·576$. Thus $\alpha_{10}\sigma/\sqrt{n} = 0·0117$ cm and $\alpha_1\sigma/\sqrt{n} = 0·0184$ cm, and so from Test 1 the 90 per cent confidence interval for μ is $3·202$ cm $\leq \mu \leq 3·226$ cm and the 99 per cent confidence interval is $3·196$ cm $\leq \mu \leq 3·232$ cm.

The form of Rule 1 can be adapted to the problem of determining the confidence interval for the parameter p in the binomial distribution when n is large and $0·1 < p < 0·5$. It follows from this that as the mean and variance of the binomial distribution are simple functions of p, confidence intervals may also be deduced for them. The result we now discuss comes about because when n and p satisfy these criteria, the binomial distribution is closely approximated by the normal distribution with population mean $\mu = np$ and population variance $\sigma^2 = np(1 - p)$. That is, we may take the distribution $N(np, np(1 - p))$ as an approximation to the binomial distribution.

Because of this we can define a new variable $X = (x - np)/\sqrt{(np(1-p))}$ which obeys the normal distribution $N(0, 1)$. The $100 - r$ per cent confidence interval $-X_r \leq X \leq X_r$ for X is then given by solving

$$P(-X_r \leq X \leq X_r) = 1 - \frac{r}{100}.$$

As X belongs to the distribution $N(0, 1)$, the number X_r is precisely the number α_r defined in the table given in Rule 1. The $100 - r$ per cent confidence interval for X is thus given by

$$-\alpha_r \leq (x - np)/\sqrt{(np(1 - p))} \leq \alpha_r$$

or, equivalently, by

$$\frac{(x - np)^2}{np(1 - p)} \leq \alpha_r^2.$$

So, we arrive at an inequality determining p in terms of x of the form

$$(n + \alpha_r^2)p^2 - (2x + \alpha_r^2)p + \frac{x^2}{n} \leq 0. \tag{16.29}$$

Now if n is large and the frequency of occurrence of an event of interest is m, where both m and $n - m$ are large, we can reasonably approximate the random variable x by m. In terms of our criterion for p this is equivalent to requiring $n < 10m < 5n$. Inequality (16·29) then determines an interval $p_1 \le p \le p_2$ in which p must lie for (16·29) to be true. This is the $100 - r$ per cent confidence interval for p and the result gives us our next rule.

Rule 2   Confidence interval for p in the binomial distribution

This approximate rule applies to a series of n independent trials governed by the binomial distribution in which n is large and the frequency of occurrence m of an event of interest is such that $n < 10m < 5n$. Then the $100 - r$ per cent confidence interval for the parameter p is

$$p_1 \le p \le p_2,$$

where p_1 and p_2 are the roots of the equation

$$(n + \alpha_r{}^2)p^2 - (2m + \alpha_r{}^2)p + \frac{m^2}{n} = 0,$$

and α_r is determined from the table given in Rule 1.

Example 16·9 In a test, 240 microswitches were subjected to 25,000 switching cycles. Given that 28 failed to operate after this test, what is the 95 per cent confidence interval for the mean number of failures to be expected in similar batches of the same design of microswitch?

Solution The situation is governed by a binomial distribution, since a microswitch will either pass or fail the test. Here $n = 240$ is large, and m is such that $n < 10m < 5n$, so that for 95 per cent confidence Rule 2 applies with $\alpha_5 = 1·960$. Using these results, the 95 per cent confidence interval for p is found to be $0·082 \le p \le 0·163$. As the mean $\mu = np$, the corresponding 95 per cent confidence interval for the mean number of failures in further batches of 24 microswitches is $20 < \mu < 39$.

Often it is necessary to determine the confidence interval for the mean of a normal distribution when small samples ($n < 50$) are involved and the variance is not known. If in Rule 1 the population variance σ^2 were to be replaced by the sample variance s^2 given in Eqn (16·24) then a serious error could result. This is because of the dependence of that test on the fact that the population parameter σ^2 is assumed to be known. The difficulty can be overcome, however, by replacing the numbers contained in the table given in Rule 1 by a different set derived from another distribution known as the *t distribution* with $n - 1$ degrees of freedom. Because the originator of the *t*-distribution, W. S. Gosset, published his results under the pseudonym *Student* and denoted the random variable in his distribution by t this is also known as *Student's t-distribution*. This distribution does not depend on known population parameters and takes account of the fact that the number n of observations in the sample is not necessarily large.

18

The t-distribution is symmetric, but a little flatter and broader than the normal distribution, though it tends rapidly to the normal distribution as n becomes large. The number of *degrees of freedom* involved in this test can be thought of as the number of observations x_i whose values may be assigned arbitrarily when a value of $\bar{x}$ is specified. Clearly, in this case, as n observations are involved, the number of degrees of freedom $v = n - 1$. The modified rule is as follows.

Rule 3 Confidence interval for the population mean μ of a normal distribution with unknown variance

Given a sample of n observations, first select a $100 - r$ per cent confidence level and use this to determine α_r from the following table, looking under the nearest entry appropriate to the number of degrees of freedom $v = n - 1$ that is involved.

t-Distribution

	a_{10}	a_5	a_1		a_{10}	a_5	a_1
$100-r\%$	90%	95%	99%	$100-r\%$	90%	95%	99%
v				v			
2	2·920	4·303	9·925	20	1·725	2·086	2·845
4	2·132	2·776	4·604	30	1·697	2·042	2·750
6	1·943	2·447	3·707	40	1·684	2·021	2·704
8	1·860	2·306	3·355	50	1·678	2·011	2·682
10	1·812	2·228	3·169	∞	1·645	1·960	2·576

Then, using the computed sample mean $\bar{x}$ and variance s^2 determined by Eqn (16·25), the $100 - r$ per cent confidence interval for the population mean μ is

$$\bar{x} - \alpha_r s/\sqrt{n} \leq \mu \leq \bar{x} + \alpha_r s/\sqrt{n}.$$

Example 16·10 In a series of nine length measurements of components drawn at random from a production run the sample mean was found to be $\bar{x} = 11·972$ cm and the sample standard deviation $s = 0·036$ cm. If it is known that the errors are normally distributed, find the 99 per cent confidence limits for the population mean μ.

Solution As $n = 9$, the number of degrees of freedom involved is $v = 9 - 1 = 8$. At the 99 per cent confidence level, the appropriate α_r from the table in Rule 2 is $\alpha_1 = 3·355$. Now $\alpha_1 s/\sqrt{n} = 0·043$ cm, so that the 99 per cent confidence interval for the mean μ is $11·929$ cm $\leq \mu \leq 12·015$ cm.

We conclude this section by describing a rule and a statistical test which have frequent application. They employ the t-distribution and concern the difference between two sample means $\bar{x}_1, \bar{x}_2$ computed from two different independent samples of sizes n_1 and n_2, each of which is assumed to be drawn from a normal distribution with the same unknown variance. In the case of the test, the null hypothesis H_0 will be that the two unknown population means μ_1 and μ_2 are equal.

This situation can arise, for example, when two similar machines are used to produce the same components and their adjustment is in question. They will each be assumed to give rise to normally distributed errors, say in the length of the component, which have the same variance, though the length settings which affect the means may differ. The statistical problem is essentially one of deducing a $100 - r$ per cent confidence interval for the difference $\mu_1 - \mu_2$ of the two population means from $n_1, n_2, \bar{x}_1, \bar{x}_2, s_1$ and s_2. The test then follows by finding whether or not the confidence interval contains the point 0. If the point 0 is contained in the confidence interval, then the null hypothesis H_0 is possible and H_0 is *accepted*. That is, in our illustration, there is no reason to dispute the assumption that the means are equal, so that the machines need not be adjusted. However, if the point 0 lies outside the confidence interval then H_0 is *rejected*; that is we conclude that $\mu_1 \neq \mu_2$. In the case of our illustration we would conclude that the setting of the machines does require adjustment.

Rule 4   Confidence interval for the difference $\mu_1 - \mu_2$ between two means of normal distributions with the same variance

Given two samples comprising n_1 and n_2 observations, compute their sample means $\bar{x}_1, \bar{x}_2$ and their sample variances $s_1{}^2$ and $s_2{}^2$. Select a $100 - r$ per cent confidence level and use this to determine α_r from the table given in Rule 3, looking under the nearest entry appropriate to the number of degrees of freedom $v = n_1 + n_2 - 2$.

The $100 - r$ per cent confidence interval for the difference of the means $\mu_1 - \mu_2$ has for its two end points the numbers

$$\bar{x}_1 - \bar{x}_2 \pm \alpha_r d,$$

where

$$d = \left[\frac{(n_1 + n_2)\{(n_1 - 1)s_1{}^2 + (n_2 - 1)s_2{}^2\}}{n_1 n_2 (n_1 + n_2 - 2)} \right]^{\frac{1}{2}}.$$

The test follows directly from Rule 4 and provides an illustration of how the idea of a confidence interval may be turned into a statistical test of significance. The test takes the following form and we remark here that the number r used in Rule 4 is called the *significance level* of the test. The meaning of the significance level is that there is a probability $r/100$ that the null hypothesis H_0 will be accepted by mistake.

Test of the null hypothesis H_0 that $\mu_1 = \mu_2$ against the alternative $\mu_1 \neq \mu_2$ for the two means of normal distributions with the same variance

Given the information in Rule 4, the null hypothesis H_0 is to be

(i) *accepted* at the r per cent significance level

$$\text{if } |(\bar{x}_1 - \bar{x}_2)|/d < \alpha_r,$$

and

(ii) *rejected* at the r per cent significance level

$$\text{if } |(\bar{x}_1 - \bar{x}_2)|/d > \alpha_r.$$

Example 16·11 Two similar machines produce a certain component. A sample of 5 components from machine 1 has sample mean length $\bar{x}_1 = 11\cdot866$ cm and sample standard deviation $s_1 = 0\cdot071$ cm, while the corresponding quantities based on a sample of 7 components from machine 2 are $\bar{x}_2 = 11\cdot943$ cm and $s_2 = 0\cdot063$ cm. On the assumption that the samples are drawn from normal distributions with the same unknown variance, compute the 95 per cent confidence interval for the difference of the population means $\mu_1 - \mu_2$. Test at the 5 per cent significance level the hypothesis $\mu_1 = \mu_2$ against $\mu_1 \neq \mu_2$.

Solution Here $n_1 = 5$, $n_2 = 7$, $\bar{x}_1 = 11\cdot866$ cm, $\bar{x}_2 = 11\cdot943$ cm, $s_1 = 0\cdot071$ cm and $s_2 = 0\cdot063$ cm. The number of degrees of freedom $v = 5 + 7 - 2 = 10$ and the value of α_r from the table in Rule 3 corresponding to $r = 5$, the 95 per cent confidence level, is $\alpha_5 = 2\cdot228$. Using Rule 4 we find $d = 0\cdot03883$, so that the 95 per cent confidence interval for $\mu_1 - \mu_2$ has for its end points the lengths

$$(11\cdot866 - 11\cdot943) \pm 2\cdot228 \times 0\cdot03883 \text{ cm.}$$

Thus we know that with 95 per cent confidence

$$-0\cdot164 \text{ cm} \leq \mu_1 - \mu_2 \leq 0\cdot010 \text{ cm.}$$

The null hypothesis H_0 is to be accepted at the 5 per cent significance level, because

$$|(\bar{x}_1 - \bar{x}_2)/d| = |-0\cdot077/0\cdot03883| = 1\cdot983 < 2\cdot228.$$

Acceptance of H_0 at the 5 per cent significance level is, of course, implied by the 95 per cent confidence interval since this contains the point 0. Had the significance level been raised to 10 per cent, then the corresponding $\alpha_{10} = 1\cdot812$ and so that at this level H_0 must be rejected because $1\cdot983 > 1\cdot812$. This is reasonable, because the 95 per cent confidence interval is obviously wider than the 90 per cent confidence interval. Expressed another way, the larger the permitted probability of accepting H_0 in error, the closer together may the limits of the confidence interval be taken.

16·5   Linear regression

A very frequently occurring situation in engineering and science is one in which pairs of observations are recorded and then a straight line is fitted to the data. This is not always an easy task, because experimental errors tend to scatter the points so that a variety of different straight lines can often seem to provide an equally good visual fit to the data. *Linear regression* using least squares provides a way of fitting this data to a straight line, while with the aid of the *t*-distribution a confidence interval may be found for the gradient of the line. The gradient is usually a quantity of considerable physical significance in an experiment.

A typical physical example is provided by the experimental determination of the adiabatic exponent γ occurring in the law $pv^\gamma = $ constant, relating the pressure p and volume v of a fixed mass of gas. Setting $Y = \log p$ and $X = \log v$ the law takes the form $Y = -\gamma X + $ constant, showing that $-\gamma$ is the gradient of the straight line involved. By obtaining n pairs of experimental observations $(p_1, v_1), (p_2, v_2), \ldots, (p_n, v_n)$, and plotting their logarithms as indicated, we at once arrive at a linear regression problem. Knowledge of the confidence interval for the gradient then provides information about the accuracy with which γ can be determined.

Let us suppose that n pairs of observations $(x_1, y_1), (x_2, y_2), \ldots, (x_n, y_n)$ are given and that we wish to fit a linear relationship of the form

$$Y = aX + b. \tag{16·30}$$

Our approach using the method of *least squares* will be to fit this straight line to the n points (x_i, y_i) in such a manner that we minimize the sum of the squares of the n quantities $(Y(x_i) - y_i)^2$. That is, we minimize the sum of the squares of the deviations of these points from the straight line; the deviations being measured in the Y-direction. To achieve this we first define S to be the sum of the squares of the errors

$$S = \sum_{i=1}^{n} (ax_i + b - y_i)^2, \tag{16.31}$$

and observe that S is a function of the two unknown parameters a, b of Eqn (16·30). From the work of Section 12·5 we then know that the conditions for S as a function of the variables a, b to attain an extremum are $\partial S/\partial a = 0$ and $\partial S/\partial b = 0$. In terms of Eqn (16·31) these equations take the form

$$\sum_{i=1}^{n} x_i(ax_i + b - y_i) = 0 \tag{16·32}$$

and

$$\sum_{i=1}^{n} (ax_i + b - y_i) = 0. \tag{16·33}$$

Solving these inhomogeneous linear equations for the unknown parameters a, b then leads to the results

$$a = \frac{\sum_{i=1}^{n} x_i y_i - n\bar{x}\bar{y}}{\sum_{i=1}^{n} x_i^2 - n\bar{x}^2} \quad \text{and} \quad b = \bar{y} - a\bar{x}, \tag{16.34}$$

where $\bar{x}, \bar{y}$ denote the means of the x_i and y_i, respectively. The quantity S will be minimized by this choice of a, b because it is essentially positive, so that its smallest possible value will be zero which it will only attain when all the n points (x_i, y_i) lie on a straight line. It follows directly from (16·34) that Eqn (16·30) can be written in the form

$$Y - \bar{y} = a(X - \bar{x}), \tag{16·35}$$

which shows that the straight line that has been fitted passes through the point $(\bar{x}, \bar{y})$. In statistics, the line with equation (16·35), in which X is regarded as the independent variable, is called the *regression line* of Y on X.

If the errors in the y_i are all normally distributed with the same variance, and the n pairs of observations are independent, the t-distribution may be used to determine a $100 - r$ per cent confidence interval for the true gradient $\tilde{a}$ of the regression line of Y on X, of which a is the sample estimate. The rule by which this may be achieved is as follows.

Rule 5 Confidence interval for the gradient $\tilde{a}$ of the regression line of Y on X

Given n pairs of observations $(x_1, y_1), (x_2, y_2), \ldots, (x_n, y_n)$ select a $100 - r$ per cent confidence level and determine α_r from the table given in Rule 3 appropriate to $v = n - 2$ degrees of freedom. Then the $100 - r$ per cent confidence interval for the true gradient $\tilde{a}$ of the regression line of Y on X, of which a in Eqn (16·34) is the sample estimate, is

$$a - \delta \leq \tilde{a} \leq a + \delta,$$

where

$$\delta = \alpha_r \left\{ \frac{\sum_{i=1}^{n} [a(x_i - \bar{x}) - (y_i - \bar{y})]^2}{(n - 2)\left[\sum_{i=1}^{n} x_i^2 - n\bar{x}^2\right]} \right\}^{\frac{1}{2}}.$$

Example 16·12 Find the regression line of Y on X for the following six pairs of observations.

i	1	2	3	4	5	6
x_i	1	3	4	5	7	8
y_i	2	8	9	10	14	19

Find the 95 per cent confidence interval for the gradient of this line.

Solution It follows from the data that $n = 6$, $\bar{x} = 4\cdot667$, $\bar{y} = 10\cdot333$, $\Sigma x_i^2 = 164$, $\Sigma x_i y_i = 362$. Then from (16·34) we obtain $a = 2\cdot18$ and $b = 0\cdot16$ and $\Sigma[a(x_i - \bar{x}) - (y_i - \bar{y})]^2 = 8\cdot317$. As $v = 6 - 2 = 4$, we find from the table in Rule 3 corresponding to the 95 per cent confidence level that $\alpha_5 = 2\cdot776$. Thus $\delta = 0\cdot69$, so that the 95 per cent confidence interval for the true gradient $\tilde{a}$ of this regression line is $2\cdot18 - 0\cdot69 \leq \tilde{a} \leq 2\cdot18 + 0\cdot69$, or $1\cdot49 \leq \tilde{a} \leq 2\cdot87$.

PROBLEMS

Section 16·1

16·1 Toss a coin 50 times and plot the relative frequency of 'heads'.

16·2 Suggest a graphical representation for the sample space characterizing the score recorded in a trial involving the tossing of a die together with a coin which has faces numbered 1 and 2. Give examples of:
(a) two disjoint subsets of the sample space;
(b) two intersecting subsets of the sample space, indicating the points in their intersection.

16·3 A bag contains 30 balls of which 5 are red and the remainder are black. A trial comprises drawing a ball from the bag at random, recording the result and then replacing the ball and shaking the bag. This process is called sampling with replacement. If this process is repeated twice, what is the probability of selecting
(a) 2 red balls;
(b) 2 black balls;
(c) 1 red and 1 black ball?

16·4 By considering arrangements of the five letters A, B, C, D, E verify that $^5P_2 = 20$ and $\binom{5}{2} = 10$.

16·5 On the assumption that a participant in a raffle will buy either 2 or 4 numbered tickets, how many different sets of tickets may he choose from a book of 20 tickets.

16·6 A coin is biased so that the probability of 'heads' is $0\cdot52$. What is the probability that:
(a) 3 heads will occur in 6 throws;
(b) 3 or more heads will occur in 6 throws?

16·7 Shells fired from a gun have a probability $\frac{1}{3}$ of hitting the target. What is the probability of missing the target if 4 shells are fired?

16·8 Draw the probability density function for the binomial distribution in which $p = \frac{1}{2}$ and $n = 6$. Use your result to draw the corresponding cumulative distribution function.

16·9 By considering Fig. 16·1 (a) deduce and draw the probability density function describing the sum of the scores on the two dice.

16·10 A light bulb manufacturer knows that 0·3% of every batch of 2500 bulbs will fail before attaining their stated lifetime. What is the probability that out of a batch of 500 (a) no more than 6 will fail (b) exactly 5 will fail?

16·11 Using the information provided in Problem 16·10, find the probability that at least 2 but no more than 5 bulbs will fail out of a batch of 500 bulbs.

16·12 Prove that $P(x) = 3(b^2 - x^2)/4b^3$ defines a continuous distribution over the interval $-b \le x \le b$. Find its mean μ and variance σ^2, and determine the cumulative distribution $U(x)$.

16·13 Prove that $P(x) = ae^{-az}$ is a continuous distribution over the interval $0 \le x \le \infty$. Find the probability that $x \ge \mu$, where μ is the mean of this distribution.

16·14 Using the probability distribution $P(x)$ in Problem 16·13, find the variance σ^2 and determine the cumulative distribution $U(x)$. What is the probability that $x \ge 2\sigma$?

16·15 A target moves along a straight path of length l joining points A and B. The probability that a single shot fired at it will score a hit at a point distant x from A is $P(x) = 2x/l^2$. If a hit is scored at a point C what is the probability P that the ratio CB/AC exceeds the positive number k?

16·16 Find the range, mode, mean, median, and standard deviation of the observations x_i occurring with frequency f_i that are contained in the following table.

x_i	0·3	0·5	0·7	0·9	1·1
f_i	3	5	11	10	9

16·17 A sample containing 112 observations has a variance $\sigma = 0·079$ cm and a sample mean $\bar{x} = 4·612$ cm. Find the 99·9% confidence interval for the population mean μ.

16·18 In a mechanical shock test on 320 insulators 39 were found to have failed. What is the 99% confidence interval for the mean number of failures to be expected in similar sized batches of insulators?

16·19 In a series of eleven diameter measurements taken from a random sample of disk blanks the sample mean diameter was found to be $\bar{x} = 2·536$ cm. Given that the sample standard deviation $s = 0·027$ cm find the 95% confidence limits for the population mean μ.

16·20 Repeat the calculation in Problem 16·19 using the same values of $\bar{x}$ and s, but this time assuming that they had been obtained from a sample of four measurements. Use linear interpolation to determine α_r.

16·21 Repeat the calculations in text Example 16·11 using the same data, with the exception that this time it is assumed that a sample of 7 components was taken from machine 1 and a sample of 5 components from machine 2.

16·22 Find the regression line of Y on X for the following data.

x_i	0	2	5	7	9
y_i	−2	3	8	14	16

Find the 90% confidence interval for the gradient $\tilde{a}$ of this line and sketch the two possible extreme regression lines of Y on X that pass through $(\bar{x}, \bar{y})$.

16·23 Find the regression line of Y on X for the following data.

x_i	1	2	5	6	8	9
y_i	3	5	9	8	10	12

Find the 99% confidence interval for the gradient $\tilde{a}$ of this line and sketch the two possible extreme regression lines of Y on X that pass through $(\bar{x}, \bar{y})$.

Answers

Chapter 1

1·1 (a) $-3, 3, -4, 4, -5, 5, -6, 6$;

(b) $64, 125, 216$;

(c) $(0, 4), (0, -4), (4, 0), (-4, 0)$;

(d) $(0, 7, 7), (7, 0, 7), (0, 8, 8), (8, 0, 8), (1, 7, 8), (7, 1, 8)$;

(e) 1.

1·3 (a) $B \subset A$: B is proper subset of A; (b) $A \cap B = \phi$: disjoint; (c) $A \cap B = \phi$ disjoint.

Chapter 2

2·3 (a) $[1, 2]$: many-one; (b) $(5, 17)$: one-one; (c) $[1, 17]$: many-one; (d) $[1, 10]$: many-one.

2·5 $f(n) = 3$ for $n = 7, 8, 9, 10, 11, 12, 13$.

2·11 (a) neither even nor odd; (b) odd; (c) neither even nor odd; (d) odd; (e) even; (f) odd; (g) neither even nor odd. In (h) to (j) the first group of terms is even and the second is odd: (h) $f(x) = (1 + x \sin x) + x^3$ with the interval $[-2\pi, 2\pi]$; (i) $f(\dot{x}) = 1 + (x + | x | \sin x)$ with the interval $[-3\pi, 3\pi]$; (j) $f(x) = (1 + 2x^2) - (x - 4x^3)$ with the interval $[-3, 3]$.

2·13 Part of ellipse centred on origin with semi-major axis 2 drawn along x-axis and semi-minor axis 1 drawn along y-axis. Curve lies in the region $x \geq 0$.

2·17 Cross-sections by planes $x = $ constant are straight lines and cross-sections by planes $y = $ constant are parabolas.

Chapter 3

3·3 Limit points are:

0, corresponding to a sub-sequence with n odd;

1, corresponding to a sub-sequence with n a member of the sequence $\{4, 8, 12, \ldots\}$;

-1, corresponding to a sub-sequence with n a member of the sequence $\{2, 6, 10, \ldots\}$.

3·5 Limit points are 1, 0 and they are not members of the sequence.

3·7 nth term $u_n = 2^{((2^n - 1)/2^n)}$. $\lim u_n = 2$.

3·11 (a) 1; (b) $\frac{1}{2}$; (c) $\dfrac{\pi^2}{4} + \sin\dfrac{\pi^2}{4}$; (d) 0; (e) infinite.

3·13 (a) all x; (b) $(-\infty, -1), (-1, 1), (1, \infty)$; (c) all x; (d) $(-\infty, -4), (-4, \infty)$;
(e) everywhere apart from the infinity of points $x = n\pi/2$.

3·15 (a) No neighbourhood exists because of infinities along $x = -1$ and $y = 2$;
(b) Interior of circle $x^2 + y^2 = 1$. Defined at P but undefined everywhere on the boundary of this circular neighbourhood;
(c) The entire (x, y)-plane. Defined everywhere in neighbourhood, including the point P.

3·17 (a) $\sqrt{2}$; (b) 0; (c) $\cos x$; (d) $1/32$; (e) $\displaystyle\lim_{x\to0-} \frac{|\sin x|}{x} = -1$; $\displaystyle\lim_{x\to0+} \frac{|\sin x|}{x} = 1$.

Chapter 4

4·1 (a) $1, -1, i\sqrt{2}, -i\sqrt{2}$; (b) $i\sqrt{2}, -i\sqrt{2}, i\sqrt{3}, -i\sqrt{3}$; (c) $\sqrt{2}, -\sqrt{2}, \sqrt{3}, -\sqrt{3}$.

4·3 (a) $z_1 + z_2 = 7 + 6i$; (b) $z_1 + z_2 = -i$; (c) $z_1 + z_2 = 0$.

4·5 (a) $z_1z_2 = -1 + 5i$; (b) $z_1z_2 = 34$; (c) $z_1z_2 = 3 + 4i$.

4·7 (a) $z_1/z_2 = (1 + 5i)/2$; (b) $z_1/z_2 = 3$; (c) $z_1/z_2 = 2i$.

4·9 (a) $z_1 + z_2 = 1 + 5i, z_1 - z_2 = 3 + i$; (b) $z_1 + z_2 = 7 - i, z_1 - z_2 = -1 + i$;
(c) $z_1 + z_2 = 3, z_1 - z_2 = -3 + 8i$; (d) $z_1 + z_2 = -2, z_1 - z_2 = -4i$.

4·13 (a) $|z| = 5$, $\arg z = 2\cdot217$ radians; (b) $|z| = 5$, $\arg z = 4\cdot067$ radians; (c) $|z| = 3\sqrt{2}$, $\arg z = 3\pi/4$; (d) $|z| = 4$, $\arg z = 5\cdot761$ radians.

4·15 (a) $|z_1z_2| = 3/2$, $\arg z_1z_2 = \pi/2$: $|z_1/z_2| = 6$, $\arg(z_1/z_2) = -\pi/6$;
(b) $|z_1z_2| = 8$, $\arg z_1z_2 = -\pi/12$: $|z_1/z_2| = 2$, $\arg(z_1/z_2) = -7\pi/12$;

4·17 $z^{20} = -2^{19}(\sqrt{3} + i)$.

4·19 $w_k = 2^{1/4}\left[\cos\left(\dfrac{1 + 6k}{12}\right)\pi + i\sin\left(\dfrac{1 + 6k}{12}\right)\pi\right]$; $k = 0, 1, 2, 3$.

4·21 (a) $|\,\text{OP}\,| = \sqrt{6}$, $l = 2/\sqrt{6}$, $m = -1/\sqrt{6}$, $n = -1/\sqrt{6}$: $\theta_1 = 0\cdot625$ radians, $\theta_2 = 1\cdot992$ radians, $\theta_3 = 1\cdot992$ radians;
(b) $|\,\text{OP}\,| = 2\sqrt{5}$, $l = 2/\sqrt{5}$, $m = 0$, $n = 1/\sqrt{5}$: $\theta_1 = 0\cdot465$ radians, $\theta_2 = \frac{1}{2}\pi$, $\theta_3 = 1\cdot108$ radians;
(c) $|\,\text{OP}\,| = \sqrt{6}$, $l = -1/\sqrt{6}$, $m = 2/\sqrt{6}$, $n = 1/\sqrt{6}$: $\theta_1 = 1\cdot992$ radians, $\theta_2 = 0\cdot625$ radians, $\theta_3 = 1\cdot150$ radians.

4·23 (a) $\theta_1 = \frac{1}{6}\pi, \theta_2 = \frac{1}{2}\pi, \theta_3 = \frac{1}{3}\pi$; (b) $\theta_1 = \theta_2 = \theta_3 = 0\cdot956$ radians; (c) $\theta_1 = 1\cdot231$ radians, $\theta_2 = 1\cdot911$ radians, $\theta_3 = 0\cdot490$ radians.

4·25 (a) $\underline{\text{OP}} = \mathbf{i} + \mathbf{j} + \mathbf{k}$; (b) $\underline{\text{OP}} = -2\mathbf{i} + 3\mathbf{j} + 7\mathbf{k}$; (c) $\underline{\text{OP}} = 3\mathbf{i} - \mathbf{j} + 11\mathbf{k}$;
(d) $\underline{\text{OP}} = \mathbf{j}$.

4·27 (a) $\mathbf{a} + \mathbf{b} = 2\mathbf{i} - 4\mathbf{j} + 4\mathbf{k}, \mathbf{a} - \mathbf{b} = 4\mathbf{i} - 2\mathbf{k}$;
(b) $\mathbf{a} + \mathbf{b} = \mathbf{i} - 2\mathbf{j} + \mathbf{k}, \mathbf{a} - \mathbf{b} = -3\mathbf{i} + 6\mathbf{j} - 3\mathbf{k}$.

4·29 (a) $\mathbf{a} = \sqrt{14}\left(\dfrac{2}{\sqrt{14}}\mathbf{i} - \dfrac{1}{\sqrt{14}}\mathbf{j} + \dfrac{3}{\sqrt{14}}\mathbf{k}\right)$;

(b) $\mathbf{a} = \sqrt{19}\left(\dfrac{3}{\sqrt{19}}\mathbf{i} - \dfrac{3}{\sqrt{19}}\mathbf{j} + \dfrac{1}{\sqrt{19}}\mathbf{k}\right)$;

(c) $\mathbf{a} = -\dfrac{\sqrt{71}}{9}\mathbf{i} + \dfrac{1}{3}\mathbf{j} - \dfrac{1}{9}\mathbf{k}$.

4·31 (a) $\mathbf{a}\,.\,\mathbf{b} = -11,\ \theta = \arccos\left(\dfrac{-11}{3\sqrt{59}}\right)$;

(b) $\mathbf{a}\,.\,\mathbf{b} = 4,\ \theta = \arccos\left(\dfrac{4}{3\sqrt{34}}\right)$;

(c) $\mathbf{a}\,.\,\mathbf{b} = -28,\ \theta = \pi$.

4·35 (a) 8;   (b) 56;   (c) 32;   (d) 0.

4·37 (a) 1;   (b) -10;   (c) 0.

4·39 (a) $\left(\dfrac{7\mathbf{i} - 5\mathbf{j} + 2\mathbf{k}}{\sqrt{78}}\right)$;   (b) $\left(\dfrac{-\mathbf{i} - 6\mathbf{j} - 2\mathbf{k}}{\sqrt{41}}\right)$.

Results are unique apart from sign since if $\mathbf{n}$ is a unit normal, then so also is $-\mathbf{n}$.

4·43 $\mathbf{r} = (2\mathbf{i} + \mathbf{j} - \mathbf{k}) + \lambda(-3\mathbf{i} + 3\mathbf{k});\quad l = -3/\sqrt{18},\ m = 0,\ n = 3/\sqrt{18}$.

4·45 $p = \sqrt{24/11}$.

4·47 $\arccos\left(\dfrac{19}{7\sqrt{14}}\right)$.

4·49 $(\sqrt{6}x - \sqrt{6} - 4)^2 + (\sqrt{6}y - \sqrt{6} + 2)^2 + (\sqrt{6}z - 2\sqrt{6} - 2)^2 = 24$.

4·51 Resultant $= \dfrac{1}{\sqrt{2}}(5\mathbf{i} + 4\mathbf{j} + 3\mathbf{k})$; 5 units; $l = \dfrac{1}{\sqrt{2}},\ m = \dfrac{4}{5\sqrt{2}},\ n = \dfrac{3}{5\sqrt{2}}$.

4·55 $305/\sqrt{26}$ ft lbs.

4·57 $\mathbf{M} = -11\mathbf{i} + 7\mathbf{j} - 15\mathbf{k}$.

Chapter 5

5·3 (a) left-hand derivative is -3, right-hand derivative is 4;
(b) left-hand derivative is 2, right-hand derivative is 2.

5·5 (a) $f'(x) = \tfrac{1}{3}x^{-2/3} - 3\sin 3x$ for $-\tfrac{1}{2}\pi \le x < 0$ and $0 < x \le \pi$; $f(x)$ is non-differentiable at $x = 0$.
(b) $f'(x) = \sin 2x + 2x\cos 2x + \tfrac{5}{3}x^{2/3}$ for all x in $[-1, 3]$.
(c) $f'(x) = -\sin x$ for $0 \le x < \tfrac{1}{2}\pi$; $f'(x) = \sin x$ for $\tfrac{1}{2}\pi < x \le \pi$; $f(x)$ is non-differentiable at $x = \tfrac{1}{2}\pi$.

5·7 (a) $y' = 3(x + 1)(x^2 + 2x + 1)^{1/2}$;

(b) $y' = bx^2(a + bx^3)^{-2/3}$;

(c) $y' = 30 \cos 2x(2 + 3 \sin 2x)^4$;

(d) $y' = 6x^2 \cos (1 + 2x^3)$;

5·9 (a) $y' = \dfrac{-6 \sin x}{(1 - 3 \cos x)^3}$;

(b) $y' = \dfrac{1}{a^2(b^2 + x^2)^{1/2}} - \dfrac{x^2}{a^2(b^2 + x^2)^{3/2}}$;

(c) $y' = \dfrac{2x(1 + 2x^2) \sec^2 (1 + x^2 + x^4)}{\sin (1 + x^2)} - \dfrac{2x \cos (1 + x^2) \tan (1 + x^2 + x^4)}{\sin^2 (1 + x^2)}$;

5·11 $f(x) = \sin \dfrac{x}{2} \cos \dfrac{x}{2} = \frac{1}{2} \sin x$ so, setting $x_0 = \frac{1}{2}\pi$, we have $f(x_0) \geq f(x)$ for all x. Hence $x = \frac{1}{2}\pi$ is an absolute maximum of $f(x)$, though there is of course an infinity of other points x at which $f(x)$ also attains the value 0·5.

5·13 Critical points at $\xi_1 = 0$, $\xi_2 = 3$, and $\xi_3 = 3/2$.

5·17 (a) $\pi^2/2$; (b) 5; (c) 3; (d) -1.

5·19 (a) 0; (b) 1/5; (c) 0; (d) 3; (e) $2/\pi$; (f) -1.

5·21 $dV = 3\pi[\alpha R_0^2 H_0(1 + \alpha t)^2 - \beta r_0 h_0(1 + \beta t)^2]dt$.

5·23 $\lim\limits_{x \to 1-} f'(x) = \lim\limits_{x \to 1+} f'(x) = 5$; $\lim\limits_{x \to 1-} f''(x) = 14$, $\lim\limits_{x \to 1+} f''(x) = 10$.

5·25 (a) $x = 1$ is minimum; $x = -2$ is maximum; $x = -\frac{1}{2}$ corresponds to a point of inflection with gradient $-27/2$.

(b) neither maxima nor minima; point of inflection with zero gradient at $x = 0$.

(c) minimum at $x = 0$; maximum at $x = 6$; minimum at $x = 12$; points of inflection at $x_1 = 6 + \sqrt{12}$, $x_2 = 6 - \sqrt{12}$ with corresponding gradients $2x_i(x_i - 12)(2x_i - 12)$ for $i = 1, 2$.

5·27 (a) $f_x = 2x/y, f_y = -x^2/y^2$;

(b) $f_x = 6xy + (x + y)^2 + 2x(x + y), f_y = 3x^2 + 2x(x + y)$;

(c) $f_x = 2x \cos (x^2 + y^2), f_y = 2y \cos (x^2 + y^2)$.

5·29 (a) $f_x = 2xyz - \dfrac{1}{x^2yz^2}, f_y = x^2z - \dfrac{1}{xy^2z^2}, f_z = x^2y - \dfrac{2}{xyz^3}$;

(b) $f_x = \cos yz - yz \sin xz - yz \sin xy$,

$f_y = -xz \sin yz + \cos xz - xz \sin xy$,

$f_z = -xy \sin yz - xy \cos xz + \cos xy$;

(c) $f_x = -(2x + y) \sin (x^2 + xy + yz)$,

$f_y = -(x + z) \sin (x^2 + xy + yz)$,

$f_z = -y \sin (x^2 + xy + yz)$.

5·31 (a) $du = \left(\dfrac{-2}{x^3yz} + yz\right) dx + \left(\dfrac{-1}{x^2y^2z} + xz\right) dy + \left(\dfrac{-1}{x^2yz^2} + xy\right) dz$;

(b) $du = \sin(y^2 + z^2)\, dx + 2xy \cos(y^2 + z^2)\, dy + 2xz \cos(y^2 + z^2)\, dz$;

(c) $du = -3(1 - x^2 - y^2 - z^2)^{1/2}(x\, dx + y\, dy + z\, dz)$.

5·33 (a) $\dfrac{du}{dt} = 2[(1 + t^2)^{1/2} + 4t \cos(5t^2 + 1)] + 2t\left[\dfrac{t + (1 + t^2)^{1/2} \cos(5t^2 + 1)}{(1 + t^2)^{1/2}}\right]$;

(b) $\dfrac{du}{dt} = 3[1 + t^2(1 + t)^2 + t^6]^{1/2}[t(1 + t)(1 + 2t) + 3t^5]$;

(c) $\dfrac{du}{dt} = \dfrac{2t}{3}$.

5·35 (a) $\dfrac{du}{dx} = 3(x^2 + y) + \dfrac{3(x + y^2)(y \sin x - \cos y)}{(\cos x - x \sin y)}$;

(b) $\dfrac{du}{dx} = (2xy^2 + y \cos xy) + (2x^2y + x \cos xy)\dfrac{x}{2y}$.

5·37 $\dfrac{\partial u}{\partial x} = \dfrac{-(u + v)}{u^2 + v^2}, \qquad \dfrac{\partial u}{\partial y} = \dfrac{2v - 3u}{2(u^2 + v^2)}$,

$\dfrac{\partial v}{\partial x} = \dfrac{v - u}{u^2 + v^2}, \qquad \dfrac{\partial v}{\partial y} = \dfrac{2u + 3v}{2(u^2 + v^2)}$.

5·41 (a) $f'(x) = 2x \operatorname{arcsec}\left(\dfrac{x}{a}\right) + \dfrac{ax^2}{|x| \sqrt{(x^2 - a^2)}}$;

(b) $f'(x) = \dfrac{2x + 1}{\arcsin(x^2 - 2)} - \dfrac{2x(x^2 + x + 1)}{[\arcsin(x^2 - 2)]^2 \sqrt{[1 - (x^2 - 2)^2]}}$;

(c) $f'(x) = \frac{3}{2}(1 + x + \arccos 2x)^{1/2}\left(\dfrac{\sqrt{(1 - 4x^2)} - 2}{\sqrt{(1 - 4x^2)}}\right)$.

5·43 At $t = \frac{1}{2}\pi$, $\dfrac{dy}{dx} = 2$, $\dfrac{d^2y}{dx^2} = -2$.

5·45 $f_{xx}(1, 1) = 384$, $f_{xy}(1, 1) = 192$, $f_{yy}(1, 1) = 384$. $f_{xy} = f_{yx}$ everywhere because of Theorem 5·23.

Chapter 6

6·3 (a) $\frac{2}{3}$; (b) $\frac{1}{4}$; (c) $\frac{1}{8}$.

6·5 (a) $2e^{2x}/\sqrt{(1 - e^{4x})}$;   (b) $(e^x + xe^x + 1)/2\sqrt{(xe^x + x)}$;

(c) $(1 + x)e^x \cos(xe^x + 2)$; (d) $2e^x/(e^x + 1)^2$.

6·7 $\dfrac{\partial f}{\partial x} = -\dfrac{y}{x^2} e^{\sin y/x} \cos y/x$, $\dfrac{\partial f}{\partial y} = \dfrac{1}{x} e^{\sin y/x} \cos y/x$.

6·9 (a) 0; (b) $\frac{1}{4}\log 2$; (c) $\log 3$; (d) $\log 3/2$; (e) 1.

6·11 (a) $(1 + \log x)x^x$;

(b) $(\log \sin 2x + 2x \cot 2x)(\sin 2x)^x$;

(c) $\left(\cos x \log x + \dfrac{\sin x}{x}\right)x^{\sin x}$;

(d) $(\cot x \log 10)10^{\log \sin x}$.

6·13 $\dfrac{\partial u}{\partial x} = yz(xy)^{z-1}, \quad \dfrac{\partial u}{\partial y} = xz(xy)^{z-1}, \quad \dfrac{\partial u}{\partial z} = (xy)^z \log xy$.

6·17 (a) $2 \cosh x(\cosh 2x \cosh x + \sinh 2x \sinh x)$;

(b) $3 \sinh 3x \exp(1 + \cosh 3x)$;

(c) $\operatorname{sech} x \operatorname{cosech} x$;

(d) $-2x/\{(x^2 + \tfrac{1}{2})\sqrt{[1 - (x^2 + \tfrac{1}{2})^2]}\}, \qquad < x^2 < \tfrac{1}{2}$;

(e) $2 \cos 2x \sinh (\sin 2x)$.

6·21 (a) $\sqrt{2}e^{i\frac{1}{4}\pi}$;   (b) $\sqrt{2}e^{-i\pi/4}$;   (c) $16e^{-\pi i/3}$;   (d) $8e^{\pi i/2}$;   (e) $\sqrt{13}e^{i\alpha}$ with $\alpha = \arctan 3/2$.

6·23 $\cos^2 \theta \sin^3 \theta = \tfrac{1}{8} \sin \theta + \tfrac{1}{16} \sin 3\theta - \tfrac{1}{16} \sin 5\theta$.

6·25 $z = (2n + 1)\tfrac{1}{2}\pi$.

Chapter 7

7·1 $\underline{S}_{P_n} = \lambda(b - a)\left[a + \dfrac{(n - 1)(b - a)}{2n}\right], \quad \bar{S}_{P_n} = \lambda(b - a)\left[a + \dfrac{(n - 1)(b - a)}{2n}\right]$;

$\lim \underline{S}_{P_n} = \lim \bar{S}_{P_n} = \dfrac{\lambda}{2}(b^2 - a^2)$

7·3 $S_{P_n} = (e^{\lambda a} - e^{\lambda b})\left(\dfrac{\Delta}{1 - e^{\lambda \Delta}}\right)$, where $\Delta = (b - a)/n$.

Since, by L'Hospital's rule, $\lim_{\Delta \to 0} \{\Delta/(1 - e^{\lambda \Delta})\} = -1/\lambda$,

we have $\displaystyle\int_a^b e^{\lambda x}\, dx = \lim_{n \to \infty} S_{P_n} = \dfrac{1}{\lambda}(e^{\lambda b} - e^{\lambda a})$.

7·7 $\tfrac{1}{3}(e^{-12} - e^{-6})$.

7·9 $I = 7$.

7·11 Required area is a triangle $+ \displaystyle\int_1^\infty \dfrac{dx}{x^2} = \tfrac{1}{2} + 1 = 3/2$.

7·13 $\xi = \sqrt{\tfrac{7}{3}}$ is unique in first case. In second case ξ is not unique, for $\xi = \pm 2/\sqrt{3}$.

7·15 $4 + \pi + \dfrac{\pi^4}{4}$.

Chapter 8

8·1 (a) $-\frac{2}{3} \operatorname{arccoth} \frac{x}{2} + C$; (b) $-\frac{1}{3} \cos 3x + C$; (c) $\frac{1}{3} \operatorname{arctanh} \frac{x}{3} + C$ for $x^2 < 9$:

$\frac{1}{3} \operatorname{arccoth} \frac{x}{3} + C$ for $x^2 > 9$; (d) $\arctan \frac{x}{2} + C$; (e) $\frac{1}{12} \sin 4x + C$; (f) $3^x/\log 3$

$+ C$.

8·5 (a) $\frac{x^3}{3} - 3 \cos x + x + C$; (b) $4^x/\log 4 + \sin 2x + C$; (c) $4 \cosh x - \cos x$

$+ C$; (d) $\frac{1}{a} e^{ax} + 3x + C$.

8·7 $\frac{1}{2} \arccos \left(\frac{2}{x}\right)$ for $x > 2$.

8·9 $\arctan \sqrt{(\cosh x - 1)} + C$.

8·11 $\frac{1}{6}(3x^2 + 1)^6 + C$.

8·13 $\sqrt{(x^2 - 1)} - \arccos (1/x) + C$.

8·15 $8/105$.

8·17 $\log 121/25$.

8·19 $e^{ax} \left(\frac{x}{a} - \frac{1}{a^2}\right) + C$.

8·21 $x \log^2 x - 2x \log x + 2x + C$.

8·25 $\frac{13}{15} - \frac{\pi}{4}$.

8·27 $x + \log \left|\frac{x-3}{x-2}\right|^3 + C$.

8·29 $\frac{8}{49(x-5)} - \frac{27}{49(x+2)} + \frac{30}{343} \log \left|\frac{x-5}{x+2}\right| + C$.

8·31 $\log \left|\frac{\tan x/2 - 5}{\tan x/2 - 3}\right| + C$.

8·33 $\frac{1}{\sqrt{2}} \arcsin \left(\frac{4x-3}{5}\right) + C$.

8·35 $\log \left|\frac{x}{1 + \sqrt{(1 - x^2)}}\right| + C$.

8·37 $\frac{\sin x}{2} + \frac{\sin 5x}{20} + \frac{\sin 7x}{28} + C$.

8·39 $1/(1 - \lambda)$ if $\lambda < 1$; divergent if $\lambda \geq 1$.

8·41 π.

Chapter 9

9·3 (a) 9;   (b) 0;   (c) 15.

9·5 (a) Yes (7×9);   (b) No;   (c) Yes (1×1);   (d) Yes (3×4).

9·7 (a) Equal if $a = 1$, $b = 2$, $c = 4$;   (b) Cannot be made equal because no solution to the two equations $a = 1$, $a^2 = 4$;   (c) Equal if $a = 1$, $b = 3$, $c = -1$.

9·9 $\mathbf{AB} = \begin{bmatrix} 4 & 3 & 3 & 0 \\ 7 & 5 & 5 & 0 \\ 3 & 4 & 2 & 2 \end{bmatrix}$, $\mathbf{CD} = \begin{bmatrix} 20 \\ 29 \end{bmatrix}$.

9·11 $\mathbf{BAX} = \mathbf{BK} \Rightarrow \mathbf{IX} = \mathbf{BK} \Rightarrow \mathbf{X} = \mathbf{BK}$ so $x_1 = 15/8$, $x_2 = -3/8$, $x_3 = 1/4$.

9·13 $\mathbf{A}^2 = \begin{bmatrix} \cosh 2x & \sinh 2x \\ \sinh 2x & \cosh 2x \end{bmatrix}$, $\mathbf{A}^3 = \begin{bmatrix} \cosh 3x & \sinh 3x \\ \sinh 3x & \cosh 3x \end{bmatrix}$

and $\mathbf{A}^n = \begin{bmatrix} \cosh nx & \sinh nx \\ \sinh nx & \cosh nx \end{bmatrix}$.

9·17 (a) -1;   (b) 3;   (c) 0.

9·19 (a) Row $1 - 10$ Row 3, $|\mathbf{A}| = -31$;   (b) Remove factor 3 from Row 1 and a factor 2 from Row 2, $|\mathbf{A}| = -18$;   (c) Column 3 = Column 1 + 3 Column 2, $|\mathbf{A}| = 0$.

9·21 $M_{11} = -\frac{1}{3}$, $M_{12} = \frac{2}{3}$, $M_{13} = -\frac{2}{3}$, $M_{21} = -\frac{2}{3}$, $M_{22} = \frac{1}{3}$, $M_{23} = \frac{2}{3}$, $M_{31} = \frac{2}{3}$, $M_{32} = \frac{2}{3}$, $M_{33} = \frac{1}{3}$; $A_{11} = -\frac{1}{3}$, $A_{12} = -\frac{2}{3}$, $A_{13} = -\frac{2}{3}$, $A_{21} = \frac{2}{3}$, $A_{22} = \frac{1}{3}$, $A_{23} = -\frac{2}{3}$, $A_{31} = \frac{2}{3}$, $A_{32} = -\frac{2}{3}$, $A_{33} = \frac{1}{3}$.

9·23 (a) linearly independent;   (b) linearly independent;   (c) linearly dependent because Row 4 = Row 1 + Row 2 + 2 Row 3.

9·25 (a) $\begin{bmatrix} 6 & 1 & -5 \\ -2 & -5 & 4 \\ -3 & 3 & -1 \end{bmatrix}$ (b) $\begin{bmatrix} -7 & 6 & -1 \\ 1 & 0 & -1 \\ 1 & -2 & 1 \end{bmatrix}$ (c) $\begin{bmatrix} d & -b \\ -c & a \end{bmatrix}$.

9·27 $\mathbf{A}^{-1} = \begin{bmatrix} 1 & 2 & 3 \\ 2 & 5 & 7 \\ -2 & -4 & -5 \end{bmatrix}$.

9·29 $x_1 = 3$, $x_2 = 2$, $x_3 = 2$.

9·31 $x_1 = 1$, $x_2 = 2$, $x_3 = 1$, $x_4 = -1$.

9·33 Rank 2: $x_1 = (22 + x_3)/19$; $x_2 = (5 + 8x_3)/19$.

9·35 (a) unique solution;   (b) inconsistent;   (c) consistent;   (d) infinity of solutions.

9·39 $\lambda_1 = 2$, $\lambda_2 = -1 : \mathbf{X}_1 = \mu \begin{bmatrix} 1 \\ -1 \end{bmatrix}$, $\mathbf{X}_2 = \mu \begin{bmatrix} 1 \\ 2 \end{bmatrix}$ for arbitrary scalar μ.

Chapter 10

10·1 (a) semi-circle of radius 1 centred on origin, and in upper half plane.

(b) ellipse centred on origin with semi-axes (a, b).

(c) rectangular hyperbola to right of y-axis and symmetric about x-axis with apex at $z = 1$.

(d) circle radius 3 centred on $z = -2 + i$.

(e) circle radius 4 centred on origin.

10·3 Region in annulus in first quadrant between circles of radii 2 and 3 centred on origin. Points on radial lines to be included and points on annular boundary to be excluded.

10·5 The curve is that part of a circle of radius $1/\sqrt{2}$ drawn about $(-5/2, \frac{1}{2})$ as centre that lies above the x-axis. (i.e., drawn with $(-2, 0)$, $(-3, 0)$ as a chord which subtends an angle of $\frac{1}{4}\pi$ at circumference.) The region is semi-circular and is interior to this curve and to the line $y = \frac{1}{2}$ which is a diameter. Boundary points are to be included.

10·7 (a) all z; (b) for $z \neq 2$; (c) all z; (d) all z because $\sinh z = \sinh (x + iy) = \sinh x \cos y + i \cosh x \sin y$.

10·9 (a) $7 + 4i$; (b) $(2 + 4i)/5$; (c) $5(1 - i)$.

10·11 (a) Yes; (b) Yes; (c) No; (d) Yes; (e) No; (f) No.

10·13 (a) $a = 2, b = 1$; (b) $a = 3, b = 1$.

Chapter 11

11·1 (a) constant pitch helix with elliptic cross-section; (b) variable pitch helix with elliptic cross-section lying entirely above (x, y)-plane; (c) curve with parabolic projection on (x, y)-plane and cubic projection on (y, z)-plane.

11·3 (a) $\dfrac{d\mathbf{u}}{dt} = -2a\pi \sin 2\pi t\mathbf{i} + 2b\pi \cos 2\pi t\mathbf{j} + \mathbf{k}$,

$\dfrac{d^2\mathbf{u}}{dt^2} = -4a\pi^2 \cos 2\pi t\mathbf{i} - 4b\pi^2 \sin 2\pi t\mathbf{j}$;

(b) $\dfrac{d\mathbf{u}}{dt} = \mathbf{i} + 2t\mathbf{j} + 3t^2\mathbf{k}$,

$\dfrac{d^2\mathbf{u}}{dt^2} = 2\mathbf{j} + 6t\mathbf{k}$.

$\dfrac{d\mathbf{u}}{dt} = \dfrac{1}{r^2}\dfrac{d\mathbf{r}}{dt} - \dfrac{2}{r^3}\left(\dfrac{d\mathbf{r}}{dt}\right)\mathbf{r} + \left(\mathbf{a} \cdot \dfrac{d\mathbf{r}}{dt}\right)\mathbf{b} + \mathbf{a} \times \dfrac{d^3\mathbf{r}}{dt^3}$.

11·5 $\mathbf{T}(0) = \mathbf{i}$, $\mathbf{T}(1) = \dfrac{1}{\sqrt{14}} (\mathbf{i} + 2\mathbf{j} + 3\mathbf{k})$.

11·7 (a) $\frac{1}{2} \sinh 2t\mathbf{i} + \log t\mathbf{j} + \dfrac{t^4}{4}\mathbf{k} + \mathbf{C}$;

(b) $[(2 - t^2) \cos t + 2t \sin t]\mathbf{i} + e^t\mathbf{j} + t(\log t - 1)\mathbf{k} + \mathbf{C}$.

11·11 $\left(\dfrac{d\mathbf{r}}{dt}\right)^2 = -\Omega^2 r^2 + \text{const.}$

11·13 transverse velocity $= \dfrac{u \cos \theta}{\cos \theta - \sin \theta}$;

radial acceleration $= \dfrac{-u^2 \cos \theta}{ae^\theta (\cos \theta - \sin \theta)^2}$;

transverse acceleration $= \dfrac{2u^2(\cos \theta - \sin \theta) + 2u^2 \sin \theta \cos \theta}{ae^\theta(\cos \theta - \sin \theta)^2}$

11·15 (a) $\frac{1}{3}(6x + y^2) + \frac{2}{3}(2xy + z) - \frac{2}{3}y$;

(b) $\frac{2}{3}xyz + \frac{2}{3}(x^2 z - \sin y) - \frac{2}{3}x^2 y$;

(c) $-\dfrac{1}{3x^2 yz} - \dfrac{2}{3xy^2 z} + \dfrac{2}{3xyz^2}$.

11·17 $n = \dfrac{1}{3\sqrt{2}} (\mathbf{i} + 4\mathbf{j} - \mathbf{k})$; $x + 4y - z = 8$.

Chapter 12

12·1 (a) $\dfrac{2n + 1}{4^n}$; (b) $\dfrac{2n}{3n + 2}$; (c) $\dfrac{1 \cdot 3 \cdot 5 \ldots (2n - 1)}{1 \cdot 4 \cdot 7 \ldots (3n - 2)}$.

12·3 (a) 47/12, remainder $\dfrac{8}{3} \left(\dfrac{1}{4}\right)^n + \dfrac{5}{4} \left(\dfrac{1}{5}\right)^n$;

(b) 17/12, remainder $\dfrac{8}{3} \left(\dfrac{1}{4}\right)^n - \dfrac{5}{4} \left(\dfrac{1}{5}\right)^n$.

12·7 (a) divergent; (b) convergent $R_6 < 1/\log 6$.

12·9 (a) convergent, $R_4 < \dfrac{4}{3} \left(\dfrac{2 \cdot 5 \cdot 8 \cdot 11 \cdot 14}{1 \cdot 5 \cdot 9 \cdot 13 \cdot 17}\right)$; (b) convergent, $R_4 < \dfrac{9}{5\sqrt{3}} \left(\dfrac{1}{1 - \sqrt{3}}\right)$

(c) divergent; (d) convergent, $R_4 < 5^{30}/5!$; (e) divergent.

12·11 (a) $-e < x < e$; (b) $-4 < x < 4$; (c) $-\infty < x < \infty$; (d) $-\infty < x < \infty$;

(e) $-16 < x < 2$.

12·13 $\arcsin x = x + \dfrac{1}{2} \cdot \dfrac{x^3}{3} + \dfrac{1 \cdot 3}{2 \cdot 4} \dfrac{x^5}{5} + \dfrac{1 \cdot 3 \cdot 5}{2 \cdot 4 \cdot 6} \dfrac{x^7}{7} + \cdots$, radius of convergence

$r = 1$, interval of convergence $-1 < x < 1$, divergent at $x = \pm 1$.

12·15 (a) $x + \displaystyle\sum_{n=2}^{\infty} \dfrac{(-1)^{n-1} 2^{n-1} x^n}{(n - 1)!}$, $-\infty < x < \infty$;

(c) $8 + 3 \displaystyle\sum_{n=1}^{\infty} \left(\dfrac{1 + 2^n + 3^{n-1}}{n!}\right) x^n$, $-\infty < x < \infty$;

(e) $x - \dfrac{1}{2} \dfrac{x^3}{3} + \dfrac{1 \cdot 3}{2 \cdot 4} \dfrac{x^5}{5} - \dfrac{1 \cdot 3 \cdot 5}{2 \cdot 4 \cdot 6} \dfrac{x^7}{7} + \cdots$, $-1 \le x \le 1$.

12·17 (a) $\dfrac{x^3}{3!} \cosh (a + \xi)$, $a < \xi < a + x$; (b) $-\dfrac{x^3}{3!} \cos (a + \xi)$, $a < \xi < a + x$.

12·19 $|e^x - P_5(x)| \le (\frac{1}{2})^5 \dfrac{e^{1/2}}{5!}$ for $0 \le x \le \frac{1}{2}$.

12·21 $f(x, y) = y + xy + (3x^2y - y^3)/3!.$

12·23 $4\pi^2/e.$

12·25 $3(\log 2)^2/2.$

12·27 $4·5826.$

12·29 $1·895.$

12·31 min at $(0, 0)$, saddle points at $(5, 6)$, $(-5, 6)$, $(5, -6)$, $(-5, -6)$.

12·33 $(0, 0)$ saddle point; $(1, 1)$ minimum.

12·35 $f(x) = 2 \sum_{n=1}^{\infty} (-1)^{n+1} \dfrac{\sin nx}{n}.$

12·37 $f(x) = \dfrac{2}{\pi} - \dfrac{4}{\pi} \sum_{n=1}^{\infty} \dfrac{\cos 2nx}{4n^2 - 1}.$

12·39 $f(x) = \dfrac{1}{\pi} \sum_{n=1}^{\infty} (-1)^n \dfrac{\cos (2n-1)x}{2n-1} + \dfrac{1}{2\pi} \sum_{n=1}^{\infty} \dfrac{(1 - (-1)^n)}{n} \sin 2nx$
$$+ \dfrac{1}{\pi} \sum_{n=1}^{\infty} \dfrac{\sin (2n-1)x}{2n-1}.$$

12·41 $f(x) = \dfrac{\pi^2}{3} + 4 \sum_{n=1}^{\infty} (-1)^n \dfrac{\cos nx}{n^2}.$

12·43 $f(x) = \dfrac{3}{2} - \dfrac{12}{\pi^2} \sum_{n=1}^{\infty} \dfrac{1}{(2n-1)^2} \cos \left[\dfrac{(2n-1)\pi x}{3} \right].$

12·45 $f(x) = \dfrac{8}{\pi} \sum_{n=1}^{\infty} \dfrac{n \sin 2nx}{4n^2 - 1}.$

Chapter 13

13·1 (a) order 3, degree 1;
    (c) order 2, degree 2;
    (e) order 3, degree 1.

13·3 $m \, dv/dt = -mg - \lambda v^2.$

13·5 (a) $x + 2y \dfrac{dy}{dx} = 0;$ (b) $y'(x + y'^2) - y = 0;$

    (c) $2xy' - 3y - x = 0;$ (d) $(1 + x)y' \left[1 + \log \left(\dfrac{1 + x}{y} \right) \right] - y = 0;$

    (e) $y'' - 3y' + 2y = 0;$ (f) $y'' - 4y' + 4y = 0.$

13·6 (b) No; (d) Yes; (f) Yes.

13·7 (b) Yes; (d) No; (f) Yes.

13·9 $2·4354$, -3 per cent.

13·11 $1·1404$, -3 per cent.

13·13 Node.

Chapter 14

14·1 (a) $\csc^2 y = \sec^2 x + C$;

    (b) $x^2 + y^2 = \log | Cx^2 |$;

    (d) $\tan y = C(1 - e^x)^3$; $x = 0$.

14·2 (a) $y = x/\{\log | x | + \frac{1}{2}\}$;

    (c) $(x - C)^2 - y^2 = C^2$;

    (e) $y = x \operatorname{arccoth} \left\{ \log \left| \dfrac{C}{x} \right| \right\}$.

14·3 (b) $xy + \sin xy + x^2 = C$.

14·4 (a) $\mu = x^2$; $3x^3y^2 + x^4y^2 + x^3y = C$;   (c) No;   (e) No.

14·5 (a) $y = Cx + 2x^2$;

    (c) $y = \dfrac{3(x + 1)^4}{2} + C(x + 1)^2$;

    (e) $y = \dfrac{1 + x}{\cos x}$;

    (g) $y = \dfrac{1}{1 + Ce^{x^2/2}}$.

14·7 $y = \dfrac{1}{x} [1 + e^x(x - 1)]$.

Chapter 15

15·1 (a) $P(\lambda) = \lambda^2 + 5\lambda - 14$; $y = C_1e^{2x} + C_2e^{-7x}$;

    (c) $P(\lambda) = \lambda^2 + 4\lambda + 3$; $y = C_1e^{-3x} + C_2e^{-x}$;

    (e) $P(\lambda) = \lambda^3 + 7\lambda^2 + 12\lambda$; $y = C_1 + C_2e^{-3x} + C_3e^{-4x}$;

    $y = 2 + e^{-3x} + 3e^{-4x}$.

15·3 The general solution is $y = C_1e^x + C_2e^{2x} + C_3e^{3x}$.
The Wronskian $W = 2e^{6x}$ is non-vanishing for finite x proving that the functions e^x, e^{2x}, and e^{3x} are linearly independent for finite x.

15·5 (a) $y = C_1e^x + C_2 \cos x + C_3 \sin x$;

    (c) $y = (C_1 + C_2x + C_3x^2)e^{ax}$;

    (e) $y = (C_1 \cos \sqrt{(2x)} + C_2 \sin \sqrt{(2x)})e^{-x} + (C_3 \cos \sqrt{(2x)} + C_4 \sin \sqrt{(2x)})e^x$.

15·6 (b) $y = (C_1 + C_2x + C_3x^2)e^x - 6$;

    (d) $y = e^{-2x}(C_1 \cos x + C_2 \sin x) + e^x \sin 2x$;

    (f) $y = C_1 \cos 2x + C_2 \sin 2x + \frac{1}{4}x \sin 2x$;

    (h) $y = e^{-x}(C_1 \cos 2x + C_2 \sin 2x) + \dfrac{1}{8} e^x + \dfrac{4}{13} e^{2x}$;

    (j) $y = C_1e^{-2x} + C_2e^{3x} - \dfrac{1}{6} e^x + \dfrac{1}{50} (\cos x - 7 \sin x)$.

15·7 (a) $x = C_1e^t + C_2e^{-t}$; $y \doteq C_1e^t - C_2e^{-t}$.

(c) $x = (C_1 + C_2t)e^{-t} + 5t - 9$; $y = (C_2 - 2C_1 - 2C_2t)e^{-t} - 6t + 14$.

15·8 General solution: $x = C_1e^{-t} + C_2e^{2t}$; $y = -(C_1 + C_3)e^{-t} + C_2e^{2t}$;
$z = C_3e^{-t} + C_2e^{2t}$.

Particular solution: $x = \frac{1}{3}(e^{-t} + 2e^{2t})$; $y = \frac{1}{3}(e^{-t} + 2e^{2t})$;
$z = \frac{2}{3}(e^{2t} - e^{-t})$.

Chapter 16

16·3 (a) $1/36$; (b) $25/36$; (c) $5/18$.

16·5 $\binom{20}{2} + \binom{20}{4}$.

16·7 $16/81$.

16·11 0.4349.

16·13 $\int_0^\infty P(x)\, \mathrm{d}x = 1$; $\mu = 1/a$, $P(x \le \mu) = 1/e$.

16·15 $P = (1 + k)^{-2}$.

16·17 $4.587 \text{ cm} \le \mu \le 4.637 \text{ cm}$.

16·19 $2.518 \text{ cm} \le \mu \le 2.554 \text{ cm}$.

16·21 $-0.166 \text{ cm} \le \mu_1 - \mu_2 \le 0.012 \text{ cm}$: Accept H_0.

16·23 $a = 1.00$, $b = 2.63$, $Y = X + 2.63$; $0.41 \le \tilde{a} \le 1.59$.

Index